Springer Series in Optical Sciences

Volume 168

For further volumes:
http://www.springer.com/series/624

Springer Series in Optical Sciences

The Springer Series in Optical Sciences, under the leadership of Editor-in-Chief William T. Rhodes, Georgia Institute of Technology, USA, provides an expanding selection of research monographs in all major areas of optics: lasers and quantum optics, ultrafast phenomena, optical spectroscopy techniques, optoelectronics, quantum information, information optics, applied laser technology, industrial applications, and other topics of contemporary interest.
With this broad coverage of topics, the series is of use to all research scientists and engineers who need up-to-date reference books.
The editors encourage prospective authors to correspond with them in advance of submitting a manuscript. Submission of manuscripts should be made to the Editor-in-Chief or one of the Editors. See also www.springer.com/series/624

Arnaud Zoubir
Editor

Raman Imaging

Techniques and Applications

Springer

Editor
Arnaud Zoubir
HORIBA Jobin Yvon S.A.S.
Villeneuve d'Ascq
France

ISSN 0342-4111 ISSN 1556-1534 (electronic)
ISBN 978-3-642-28251-5 ISBN 978-3-642-28252-2 (eBook)
DOI 10.1007/978-3-642-28252-2
Springer Heidelberg New York Dordrecht London

Library of Congress Control Number: 2012941425

Printed on acid-free paper

Springer is part of Springer Science+Business Media (www.springer.com)

Preface

Raman spectroscopy has now become one of the key analytical tools for an extremely diverse range of applications. Although initially limited to research applications, it is now increasingly used for routine testing and quality control as well as process control applications. As the technology reaches such a level of maturity, it is an interesting exercise to look back and retrace the history that led to what is now one of the fastest growing molecular spectroscopy techniques.

In 1928, Sir C.V. Raman was awarded the Nobel Prize for discovering what will then be known as the Raman effect which relies on the fact that when light interacts with matter, the incoming wavelength is shifted as vibrational transitions are excited.

For the following 20 years, Raman scattering was used, in conjunction with infrared spectroscopy (IR), as a means to confirm the molecular configuration of a sample. For instance, the controversy about the structure of benzene was solved by Raman and IR spectroscopies. If the bonds between the carbon atoms in the ring alternated between single and double bonds, then the molecule would have D_{3h} symmetry, but not a center of symmetry. Without a center of symmetry, no vibration could be both Raman and IR active. If the single and double bonds resonate, then the molecule has D_{6h} which does have a center of symmetry and the "mutual exclusion selection rule" holds. Looking at the IR and Raman spectra overlaid, one will find that there are no overlapping bands. Therefore, benzene has a center of symmetry, meaning that it has the D_{6h} configuration. The complementary nature of Raman to infrared spectroscopy continues to lead to strong demand from many of the same end-users in industries such as pharmaceuticals, polymers and chemicals.

During this early period, it was easier to do Raman spectroscopy than IR because the IR sources and detectors could not provide high sensitivity levels. That all changed in the 1960's when the commercial Fourier Transform IR systems were developed with better detectors. The FTIR systems were quite sensitive and easy to use. The laser, also introduced during this period, was believed to provide a better source for Raman which would improve its sensitivity to study "real world" samples. Unfortunately, that did not happen because of the fluorescence interference from the samples. Curiously, during the early years, when spectroscopy was used to determine structure, samples had been extensively purified which meant that fluorescence from impurities was not seen as an interference. When studying real world samples, one does not usually have the option to purify the samples. The laser that was used during this period was an argon laser, with lines between 514 and 457 nm. As it happens, these lines are also ideal for exciting impurity fluorescence. As a consequence, Raman was a disappointment for many industrial applications.

Since then, tremendous progress was made in the instruments design to circumvent the two main limitations of this technique, namely: low sensitivity and fluorescence. The Raman effect tends to be a rather inefficient one, producing signals typically million times weaker than the exciting beam. Concurrently, fluorescence is a much more efficient effect, which can interfere with the measurement, often to the point where it would mask the Raman spectrum when it occurs. Successive implementation of more powerful and more diversified laser sources, together with more sensitive detectors have made Raman spectrometers commercially available and ready for industrial applications. This technique would then allow molecular analysis on a great variety of samples, non-destructively, with no sample preparation, and at ambient pressure and temperature. Raman spectra represent chemical fingerprints of samples under study, allowing the operator to easily identify unknown components, distinguish between several polymorphs and crystalline structures of a given compound, trace the production and consumption of chemical species in chemical reactions, etc.

Another significant step in the history of Raman spectroscopy is the coupling of a Raman spectrograph to an optical microscope. This seminal work was accomplished in 1974 by Professor Michel Delhaye at the Technical University of Lille in northern France[1] with the development of the MOLE™. The method involved bathing the sample in laser light over a large area and then transferring an image of the sample onto a camera after a Raman wavelength was selected with the spectrograph. It was recognized that the Raman microscope would complement the elemental information from the electron microprobe by providing information on molecular bonding between atoms. For example, if there was an organic impurity on an integrated circuit, the electron microprobe would tell you that there was carbon in the impurity. Whether that carbon was in the form of carbon black, graphite, polymer, photoresist, or any other of the possible organic molecules that might be in the circuit's environment, the electron microscope could not tell you. But a Raman spectrum could provide this critical information. Further developments made during the following decades led to better integrated instruments with higher spatial resolution thanks to the confocal design, greater versatility with more lasers wavelengths available and faster measurement speeds thanks to higher power laser sources, more efficient designs and more sensitive detectors. Its ability to localize and identify impurities at the micron scale has finally made it a workhorse analytical tool for both Research and routines QC/QA laboratories alike.

[1] M Delhaye; P Dhamelincourt; *J Raman Spectrosc.* **1975** 3, 33–43.

It is then no wonder that Raman imaging has found an incredibly wide array of applications, from semiconductors to the pharmaceuticals, from art conservation to security, from biology to geology. Raman imaging provides a way to represent a sample in two or even three dimensions with information-rich content: chemical nature, molecular orientation, crystallinity ratios and polymorph content, strain and stress, all being accessible from the Raman spectra available for each pixel of the Raman image.

The first challenge in editing this book was to set its scope by drawing a perimeter in a wide field of applications while reflecting such diversity. The second challenge was to give a good picture of the state of the art for benchtop laboratory Raman instruments and common methods of so-called classical Raman imaging, while allowing the reader to get a grip on the most innovative developments being made at the research level. Choice was then made to also include work from some of the most prominent researchers from the technological frontier where stand near-field Raman imaging, coherent anti-Stokes Raman spectroscopy (CARS) and stimulated Raman spectroscopy (SRS).

The intended result is a book that is addressed to students, researcher and end-users alike, whether they are focused on one specific subject or working in a central laboratory brushing on very diverse applications. The goal is to give an overview of this powerful technique with concrete 'real world' examples directly from experts in the field and with a strong inclination towards application. We hope that this book will convey part of their experience and dispense some methodology and analytical recipes that are not always available from textbooks.

Acknowledgments

I would like to thank the following people for their kind assistance in helping me during the editing of this book: Fran Adar, Andreas Thoss, Ruth Geiger, Yasushi Nakata, Renata Lewandowska, Eloïse Lancelot, Alexandra Rapaport. Needless to say that I am also grateful to the authors for their patience and the diligence they put in their contributions, and for supplying the flesh of this book. Above all, I would like to acknowledge Hans-Juergen Reich for his inspiring role in the developments of Raman imaging instrumentation.

Villeneuve d'Ascq Arnaud Zoubir

Contents

Chapter 1
Imaging Modes

Eunah Lee

Abstract In this chapter, we cover fundamentals of Raman imaging. The different imaging methods, such as point mapping, line scanning and global imaging are described in terms of their advantages and limitations, as well as new imaging modes allowing faster image acquisition. Important definitions of the experiment design, such as spot size, spatial and axial resolution are reviewed. Important concepts enabling to push the limits in terms of spatial resolution are addressed. Finally, basic and more advanced data-processing methods are discussed, as means to separate and eliminate undesirable spectral features. Analysis methods and chemometric tools are introduced, including univariate and multivariate analysis, image and morphological analysis, as well as statistical representation methods. Illustrations are shown with "real-life" samples like pharmaceutical tablets or carbon nanotubes. It concentrates on the empirical 'good' practices that can be used in experiment designs.

1.1 Introduction

The term "hyperspectral imaging" is originally defined in the remote sensing field as "to operate remote sensors that can have high spectral resolution" [1]. The term multispectral imaging [2] was already in use to refer images that were recorded as intensity images at discrete wavelength bands. The spectral resolution was determined by the bandwidth, and $0.01\,\mu\text{m}(= 10\,\text{nm})$ was considered "high" spectral resolution. To provide a perspective, 10 nm spectral resolution in a Raman spectrum recorded with a green laser ($\lambda = 532\,\text{nm}$) corresponds to $346.8\,\text{cm}^{-1}$. A modern commercial confocal Raman microscope can reach the spectral resolution of $\sim 1\,\text{cm}^{-1}$ when the same green laser is used.

E. Lee (✉)
HORIBA Instruments Incorporated, 3880 Park Avenue,
Edison, NJ 08820-3012, USA
e-mail: eunah.lee@horiba.com

A. Zoubir (ed.), *Raman Imaging*, Springer Series in Optical Sciences 168,
DOI: 10.1007/978-3-642-28252-2_1,

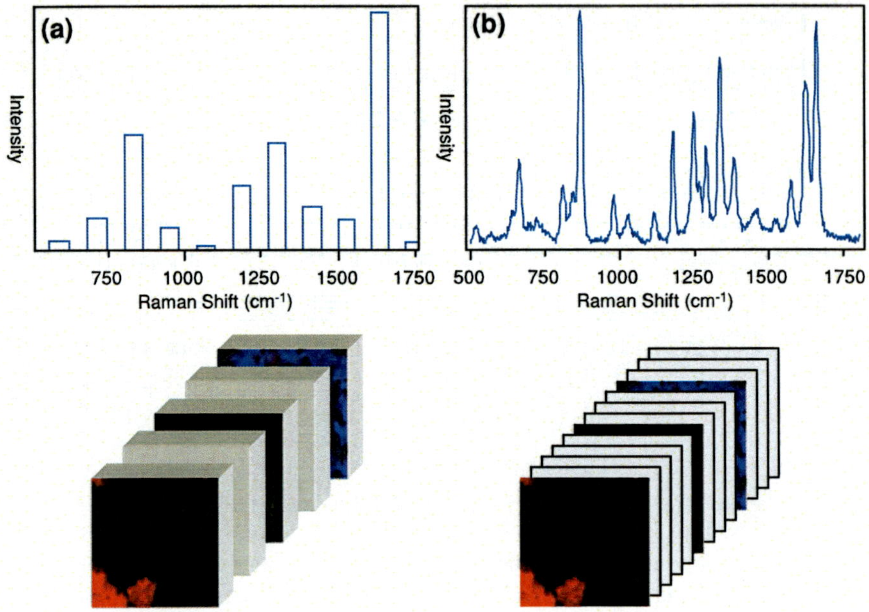

Fig. 1.1 Simulated examples of (**a**) a multispectral data and (**b**) a hyperspectral data

The term "hyperspectral imaging" was introduced when the modern technology made it possible to record images with higher spectral resolution. Figure 1.1 shows the difference between multispectral imaging and hyperspectral imaging. Hyperspectral imaging became widespread to refer any spectral imaging methods whose spectral quality (e.g. spectral resolution, spectral range, the number of spectral data points, etc.) is comparable to that of traditionally recorded spectra.

Raman hyperspectral imaging can be defined, therefore, as a method to record a Raman hypercube whose spectral quality is comparable to that of traditionally obtained Raman spectra. A Raman hypercube is an intensity data set made of $I(x, y, v)$, where x and y are the image pixel coordinates and v is the wavenumber. Typically, Raman hyperspectral imaging is performed with a Raman microscope at a high spatial resolution, and collects numerous Raman spectra from different locations of the sample. It can be considered as a collection of spatially resolved Raman spectra from a spectroscopist point of view, or a collection of spectrally resolved Raman images from a microscopist point of view. Figure 1.2 shows the conventional experiment schematics from bright field imaging to Raman imaging.

Raman hyperspectral imaging often requires multivariate analysis.

The main reason to perform Raman hyperspectral imaging is to characterize chemically and/or spatially complex samples. Raman spectra measured from a complex sample are more likely to be complex themselves. First, spectral features would be mixed at varying degree with contributions from multiple chemical/spectral species. The contribution will depend on the local concentration and intrinsic Raman

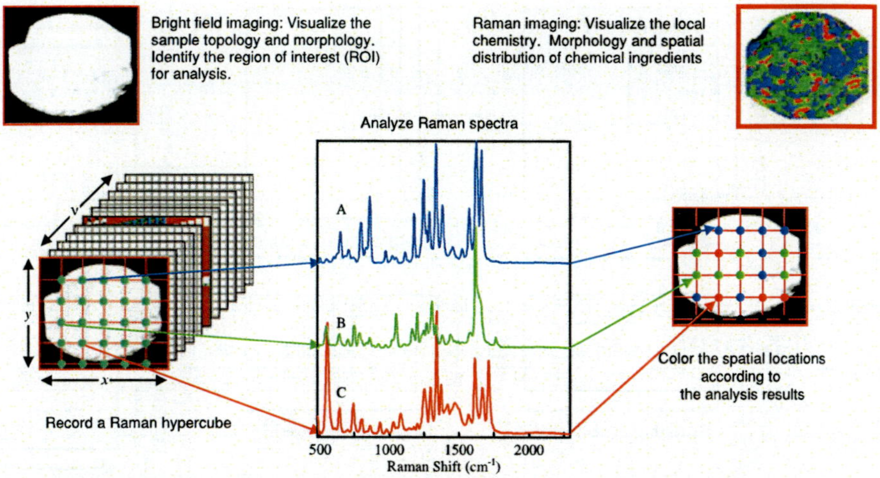

Fig. 1.2 Schematic diagrams from bright field imaging to Raman imaging. The hypercube is a collection of Raman intensities as a function of x, y and v. At a fixed spatial position (x_i, y_i), it is a localized Raman spectrum. At a fixed spectral position v_k, it is a Raman intensity map

scattering cross section of each ingredient. Second, overall and relatively intensities can be affected by crystalinity, orientation, or particle domain size of each ingredient. Third, there would be added interference from fluorescence background, scattering and noise. Multivariate analysis can help separate Raman signals from varying sources and ingredients for qualitative and quantitative analysis.

The volume of data and information in a hypercube is another reason to employ multivariate analysis. A hypercube often consists of thousands or tens of thousand spectra. It is practically impossible to examine all spectra in a Raman hypercube one by one, let alone comparing multiple Raman hypercubes. Multivariate analysis applies statistical method to Raman hypercubes to extract scientifically meaningful information.

It must be noted that, because a multivariate analysis is a numerical analysis, there is no general solution(s) that applies to all data sets or samples. Rather, there are good practices (details of a good practice will be discussed later in this chapter with examples) to develop an optimum method for each project with respect to the experiment objective (e.g. classification or concentration prediction) and data behavior (e.g. signal-to-noise ratio or the number of spectra). One key point of a good practice is verification. A numerical analysis always presents the chance for misleading, incorrect or artificial results. Only way to safe-guard against it is to exercise spectroscopy. The results must be verified with actual spectra. Another key point of a good practice is validation. Even if a method "worked" for the data from of one sample, there is no guarantee that it will work for the data from all other samples. The method must be tested by applying to the data from other samples of known characteristics. If the results derived from the method are in a good agreement

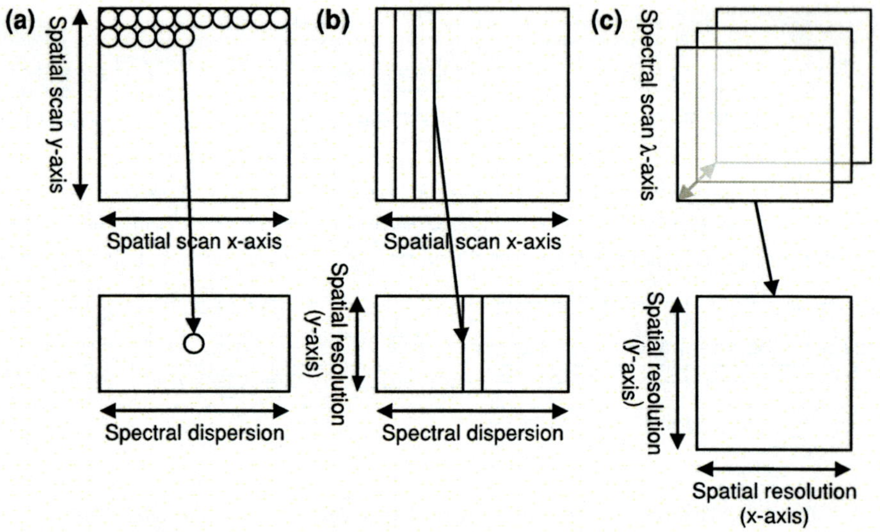

Fig. 1.3 Schematic diagrams of (**a**) point mapping, (**b**) line scanning and (**c**) global imaging. The top row shows the illumination pattern on the sample and the bottom row shows the image onto the detector

with the actual sample characteristics, the method is "validated" and (relatively) safe to use to analyze data from unknown samples. If they are not, the method must be modified or a new method must be developed.

1.2 Data Measurement Methodologies

Figure 1.3 shows the schematic diagrams of point mapping, line scanning and global imaging (Fig. 1.4).

1.2.1 Point Mapping

This is the simplest (and often the best) way to record Raman hypercubes of high spectral and spatial resolutions. The mapping area and measurement locations are determined by the operator, most often in a rectangular shape and grid pattern. A Raman spectrum is recorded at each point, and the sample is moved to the next point using a mapping stage. It requires scanning along two spatial axes (x- and y-axes). Because it is a point-by-point method, any experiment condition that is controlled by

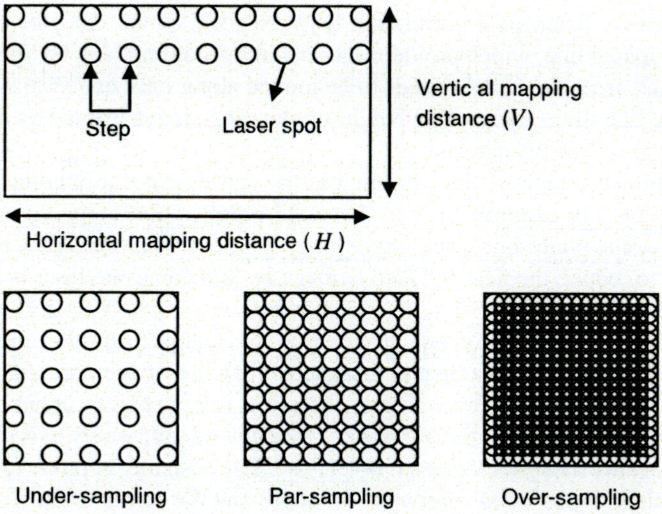

Fig. 1.4 Schematics for terminology conventions

the computer can be programmed, if desired. This is the most flexible method and, so far, the most popular Raman hyperspectral imaging method.

Even though point mapping typically scans the sample, it is possible to scan the laser. Similarly to the laser scanning confocal microscopy, mirrors can be placed in the optical path so both incident and scattering beams pass. Mirrors are moved either at a high speed or at a high precision. When the mirrors are moved at a high speed, they can illuminate a predetermined area (bigger than the laser spot size) of the sample virtually uniformly. Even if the mirrors are moving at a high speed, because Raman is scattering phenomenon, the signal returns instantaneously, and the scattering beam path coincides with the incident beam path. The result is an averaged Raman spectrum of the illuminated area. The advantage of this method is to measure large areas of the sample with a few number of measurements. It is favorable for a sample with large particles or agglomerates, or finding a small sample on a large substrate. The mirrors can be moved in a stop-and-go fashion to excite individual points within the field of view. This method allows measurements at a very fine grid, which is useful to map a small sample (see Sect. 1.5.1 for details).

1.2.2 Line Scanning

There is more to line scanning than the optical design, but the principle is the same. The sample is illuminated in a line, which is imaged onto the entrance slit of the spectrograph, and then onto the detector along the short axis (column). Raman signal along the line is dispersed along the long axis (row) of the detector for spectral resolution. There is a correlation between rows of the detector and points in the illuminated line, which provides the spatial resolution along the illuminated line. To map an area, the sample needs to be moved along only one spatial axis, orthogonal to the line illumination. The height of mapping area is limited by the objective lens and the size of the detector.

The advantage of line scanning is the speed, especially compared to the point mapping. For example, to map 25×25 points, point mapping requires 625 measurements, while line scanning requires 25 measurements. Also, line scanning can help to reduce the heat-induced damage by spreading the laser beam over the illuminated line.

The disadvantage of line scanning is the spatial resolution. Because the signal propagates from the sample to the detector in a line shape, a confocal hole cannot be used to reject out-of-focus signal. Therefore, line scanning cannot be truly confocal. Even though the entrance slit offers a measure of confocality, it is inferior to the true confocality using the confocal hole. Also, line scanning is prone to the non-uniform illumination, especially between the edges and the center of the illuminated line.

1.2.3 Global Imaging

A finite area of the sample is illuminated, which is then imaged onto the detector. A dispersive spectrograph cannot be used for wavelength selection since both x- and y-dimensions of the detector are dedicated for spatial resolution. Spectral resolution is achieved with a liquid crystal tunable filter (LCTF) or a combination of LCTFs.

Global imaging performs Raman multispectral imaging by recording Raman intensity images at a few predetermined wavelengths (e.g. peak positions of characteristic Raman bands). This takes a much shorter time than recording a full Raman hypercube. However, without a full range spectrum, the source of intensity (e.g. Raman, fluorescence, scattering) cannot be determined, and the authenticity of "Raman" image cannot be verified.

To record a full spectrum, filter(s) must be tuned at a series of wavelengths. The number of points along the spectral axis and the speed of tuning from one wavelength to the next determine the speed of the measurement.

The spectral resolution is limited by the filters. Unlike a dispersive spectrograph used in a point mapping or a line scanning method, the spectral resolution cannot be adjusted with LCTFs. Global imaging is prone to the non-uniform illumination, especially between the circumference and the center of the illuminated area.

1.3 Instrumentation and Terminology Conventions

A confocal Raman microscope with an automated stage can perform various operations such as depth profiling, kinetic reaction monitoring and high throughput multiwell screening. In this chapter, however, we will concentrate on XY Raman imaging (two dimensional lateral Raman imaging) using point mapping. The sampling area per point measurement is often approximated with the laser spot size. Theoretical approximation of the laser spot size is explained in detail in Sect. 1.5.

The terminologies used in this chapter are defined below:

Small sample: sample size is smaller than the laser spot size

Large sample: sample size is larger than the laser spot size

Exposure time (t_{expoure}): the duration during which CCD accumulates Raman signal from the sample before read out

Number of accumulations ($n_{\text{accumulations}}$): the number of repeat readouts of CCD

Acquisition time ($t_{\text{expoure}} \times n_{\text{accumulations}}$): the total duration during which Raman signal is acquired from a point of the sample

Instrument time: the time required to perform operations other than signal accumulation such as the detector readout (t_{readout}), data transfer (t_{transfer}), communication between hardware and software, and hardware operation

Raman map: a collection of spatially resolved Raman spectra recorded from predetermined points of two dimensional area of the sample; a conventional name for a Raman hypercube;

Mapping distances: the horizontal (H) and vertical (V) dimensions of the map

Mapping area ($H \times V$): the area the map covers

Row: a horizontal line of measurements at a fixed vertical position

Column: a vertical line of measurements at a fixed horizontal position

Number of points (n_{points}): the number of measurement points, the same as the number of spectra

Steps: the distance between two neighboring measurement points

Laser spot: the diameter (d_{laser}) is approximated as (airy disk approximation); an approximation to the sampling area of a single measurement

Measured area ($n_{\text{points}} \times \pi r_{\text{laser}}^2$): portion of the mapping area from which Raman spectra were measured; is a radius of the laser spot, $d_{\text{laser}}/2$

Sampling ratio ($(n_{\text{points}} \times \pi r_{\text{laser}}^2) / (H \times V)$): ratio of the measured area of the map with respect to the mapping area

Under-sampling: the step size is larger than that of the laser spot

Par-sampling: the step size is the same as the laser spot

Over-sampling: the step size is smaller than the laser spot

1.4 Data Measurement Speed

The speed of the system must be evaluated with respect to the performance of the system (e.g. sensitivity, precision, stability and reproducibility). For example, a conventional CCD [3, 4] can be read at a different speed. A high speed read out mode tends to suffer a high noise, which may require a long exposure time, negating the advantage of the high speed read out. Similarly, a mechanical mapping stage [5] can be configured to move at a high speed. However, it may compromise the precision of the measurement locations, generating scientifically meaningless data. In fact, the maximum travel speed (without compromising the precision) is one of specifications for a mechanical mapping stage. A fast system, therefore, requires carefully selected components.

The speed of measurement must be determined with respect to the signal-to-noise ratio (S/N). Unless the adequate level of S/N is achieved, the data has no scientific meaning, and therefore useless. The "adequate" level of S/N can vary depending on the goal and form of the analysis. In a hyperspectral imaging where each data set contains numerous spectra, the requirement of S/N can be somewhat lower than that of single point measurements. Even then, the absolute minimum S/N is 2:1, preferably better than 5:1 (see Sect. 8 for details).

It is not uncommon for the Raman cross section of a compound to be 1,000 times greater or smaller than another compound. So do the acquisition time necessary to achieve the same S/N. For example, if compound A requires 1 ms acquisition time to achieve 5-to-1 S/N while compound B takes 1 s to achieve the same S/N, to record a map made of 10,000 points, it will take at least 10 s for A, and 2.8 h for B, assuming all other conditions are the same. A system with a good optical design, therefore, is a fast system because a good optical design ensures the high sensitivity to shorten the necessary exposure for the adequate S/N given a sample.

1.4.1 Standard Scanning

In the standard scanning mode, the stage accelerates, reaches a steady speed, decelerates, and then stops for a measurement. During the spectrum acquisition, preprogrammed operations can be performed. The flowchart of the standard scanning mode is shown in Fig. 1.5a, and the total experiment time is calculated according to the Eq. 1.1. Because stage stops before each measurement, the measurement position is well defined and well controlled, offering the best spatial resolution and precision. The area measured at each point is the same for every point. The spectral quality of a single measurement is not affected by the mapping distance or the number of points. On the other hand, it requires a longer instrument time ($t_{\text{stage}}, t_{\text{readout}}, t_{\text{transfer}}$) compared to the fast scanning (see Sect. 1.4) per point adding to the total experiment time.

$$t_{\text{experiment}} = n_{\text{rows}} \times \left\{ n_{\text{points per row}} \times \left[t_{\text{stage, next point}} + n_{\text{accumulation}} \times \left(t_{\text{exposure}} + t_{\text{readout}} + t_{\text{transfer}} \right) \right] + t_{\text{stage, next row}} \right\}. \tag{1.1}$$

1.4.2 Fast Scanning

The fast scanning mode is designed to minimize the instrument time (schematics shown in Fig. 1.5b). For each row of the measurement, the stage moves continuously

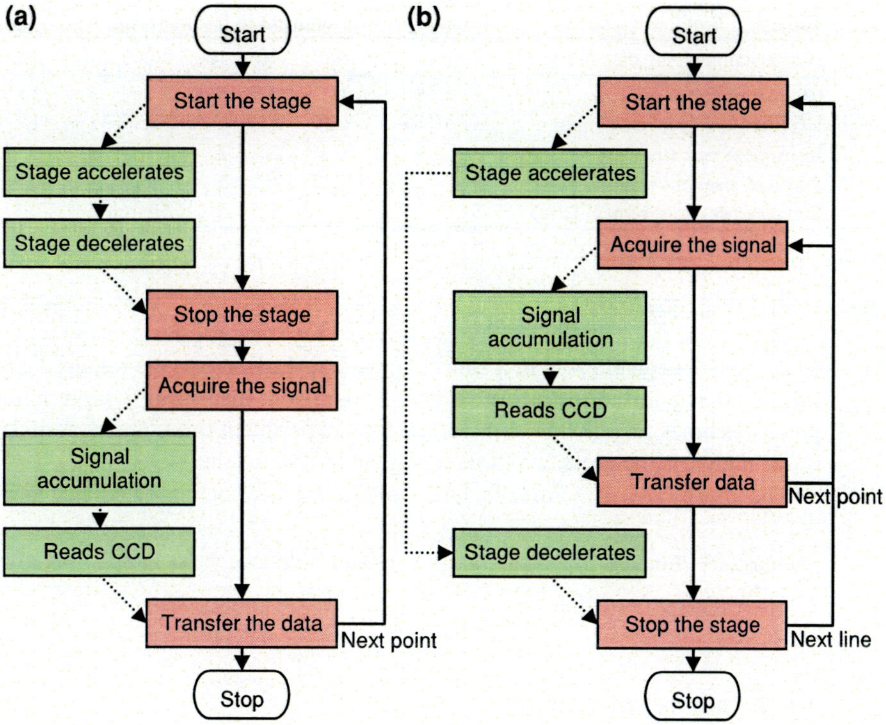

Fig. 1.5 Flowcharts for (**a**) standard scanning mode and (**b**) fast scanning mode

at the predetermined speed while CCD is read out at a predetermined interval. The speed of the stage, v, is determined by the horizontal mapping distance, H, the number of points per row, the exposure time and the instrument time (Eq. 1.2).

$$v_{\text{stage}} = H/\left\{n_{\text{points per row}} \times \left(t_{\text{exposure}} + t_{\text{readout}} + t_{\text{transfer}}\right)\right\}. \tag{1.2}$$

The signal accumulates while the stage is moving, so the measurement is performed over a line segment instead of a stationary point. The measurement area is proportional to the length of the line segment, which is a function of the horizontal mapping distance, exposure time and the number of points per row (Eq. 1.3).

$$l_{\text{measured distance per point}} = v_{\text{stage}} \times t_{\text{exposure}}. \tag{1.3}$$

A longer line segment translates to the worse spatial resolution compared to the standard scanning. A few simulated scenarios are summarized in Table 1.1. It demonstrates that short horizontal mapping distance, high number of points per row and the short exposure time per point measurement are preferable for maintaining a good spatial resolution.

Table 1.1 Simulated calculation of distance per measurement in fast scanning

Horizontal mapping distance (μm)	100	100	100	10	100	1,000	100	100	100
Number of points per row	5	5	5	5	5	5	5	50	500
Exposure time per point (ms)	1	10	100	1	1	1	1	1	1
Instrument time (ms)	3	3	3	3	3	3	3	3	3
Speed of stage movement (ms/μm)	5	1.54	0.19	0.5	5	50	5	0.5	0.05
Distance per measurement (μm)	5	15.38	19.42	0.5	5	50	5	0.5	0.05

In general, fast scanning mode is preferred for strong Raman scatterers that can achieve an acceptable S/N in a short exposure time. Because the required exposure time is short and the instrument time is minimized, measuring a large number of points becomes practical in terms of the total experiment time. The combination, as stated above, is the most favorable to a good spatial resolution.

The total experiment time of a fast scanning measurement is calculated as Eq. 1.4.

$$t_{\text{experiment}} = n_{\text{rows}} \times \left\{ n_{\text{points per row}} \times \left(t_{\text{exposure}} + t_{\text{readout}} + t_{\text{transfer}} \right) + t_{\text{stage, next row}} \right\} \tag{1.4}$$

1.5 Experiment Design

A good experiment design takes the full advantage of the Raman hyperspectral imaging at the minimum time penalty. The first step of a good experiment design is to define the objective of the experiment as clearly as possible. For example, a small laser spot translates to a high spatial resolution and a high power density (watts for a unit area). The spectrum is recorded from a small area without the undesired contribution from the surrounding materials. On the other hand, a small laser spot with the same number of measurements translate to the small measured area, skipping the large area. This could cause the data to be less than healthy statistical representation of the sample. Therefore, the objective must be clear. To continue the same example, experiment design can be quite different when a high spatial resolution is required vs. when a statistical representation of the sample is required.

1.5.1 Laser Spot Size

The sampling area of a single measurement (point mapping and standard scanning) is often approximated by the laser spot size. The laser spot can be approximated with an airy disk, whose diameter is given as $d = 1.22\lambda/\text{N.A.}$, where λ is the wavelength of the light and N.A is the numerical aperture of the objective lens. Table 1.2 summarizes the theoretical laser spot size for common lasers and objectives.

Table 1.2 The laser spot size approximated the diameter to the first null of an airy disk $d(\mu m) = 1.22\lambda/\text{N.A.}$

λ (nm)	N.A.						
	0.1	0.25	0.5	0.75	0.95	1.2	1.4
244	2.98	1.19	0.60	0.40	0.31	0.25	0.21
325	3.97	1.59	0.79	0.53	0.42	0.33	0.28
473	5.77	2.31	1.15	0.77	0.61	0.48	0.41
488	5.95	2.38	1.19	0.79	0.63	0.50	0.43
514	6.27	2.51	1.25	0.84	0.66	0.52	0.45
532	6.49	2.60	1.30	0.87	0.68	0.54	0.46
633	7.72	3.09	1.54	1.03	0.81	0.64	0.55
638	7.78	3.11	1.56	1.04	0.82	0.65	0.56
785	9.58	3.83	1.92	1.28	1.01	0.80	0.68
830	10.13	4.05	2.03	1.35	1.07	0.84	0.72
1,024	12.49	5.00	2.50	1.67	1.32	1.04	0.89

The actual laser spot size degrades from the theoretical laser spot size due to the imperfection of optical components, the finite bandwidth of the laser, and the finite penetration depth into the sample. The imperfection in the measurement of the spot size, due to the less than perfect alignment and focus, also manifests as degradation from the theoretical spot size. An empirical way of measuring the actual spot size is to measure a series of Raman spectra across a sharp interface between two materials. It is preferable to have the flat surface made of opaque materials without a difference in heights. Intensity of a characteristic band will increase or decrease as the measurement crosses the interface. The laser spot size is approximated by the distance between the points with the 10 and 90% of the intensity range. This method has the benefit of requiring no additional data processing. Alternatively, the (1st) derivative of the intensity profile can be fit with a Gaussian curve, and the laser spot size is approximated by the full width at half maximum of the fitted curve. (Figure 1.6) A previous study [6] reported laser spot size measurements using various objectives (laser wavelength = 457.9 nm) in comparison to the theoretical value. The best agreement of the study showed the measured laser spot size be ~107 % of the theoretical size. The worst agreement of the study showed the measured value be about twice the theoretical one.

1.5.2 *Spatial Resolution*

Strictly speaking, spatial resolution must be defined as a function of both the instrument performance and the sample characteristics. The spatial resolution for an instrument is conventionally defined by assuming that sample sizes are infini-

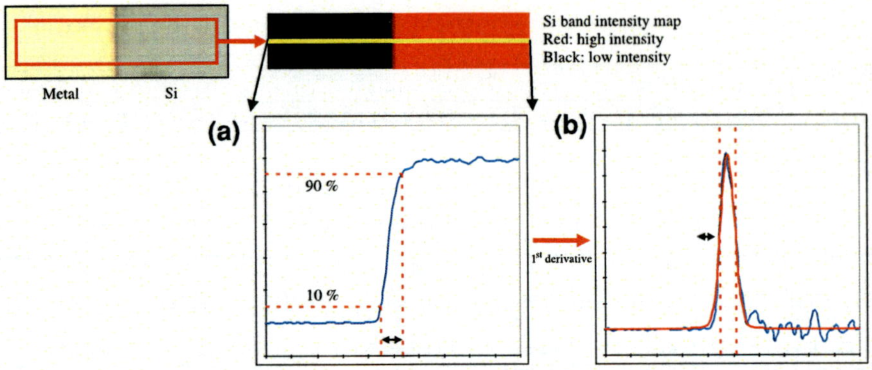

Fig. 1.6 Empirical laser spot measurements (**a**) 10/90 intensity criterion (**b**) curve fitting 1st derivative of intensity profile

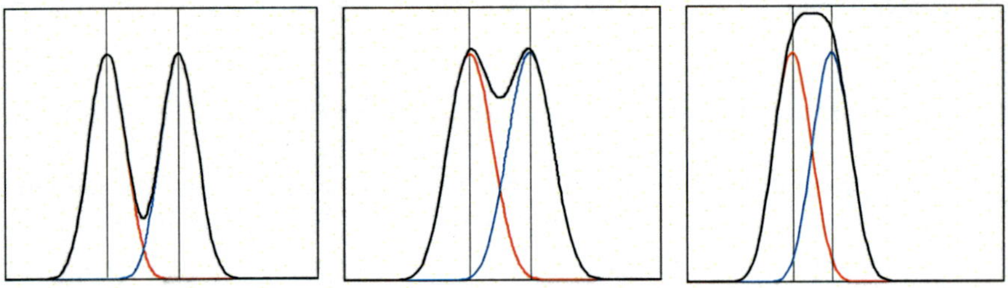

Fig. 1.7 Schematic diagram demonstrating (**a**) Rayleigh (**b**) 80 separation and (**c**) Sparrow criteria of resolution. Distances between two objects are 0.61 λ/N.A., 0.42 λ/N.A. and 0.33 λ/N.A., respectively

tesimal. Under this convention, the spatial resolution can be defined as the shortest distance between two infinitesimal objects that can resolved.

There is more than one criterion to qualify two objects as resolved. One of the most often used criteria is Ralyleigh criterion where two objects are deemed as resolved when the second one is placed at the first null of the first one's intensity profile (Fig. 1.7a). According to the Rayleigh criterion, the spatial resolution is one half of the laser spot size (0.61 λ/N.A.). Sparrow criterion deems two objects resolved when the first and the second derivatives of the sum of two intensity profiles reach zero (Fig. 1.7c) between two objects. The spatial resolution according to the Sparrow criterion is 0.33 λ/N.A. In this chapter, two objects are deemed resolved if the minimum between two objects is 80% or less of the maximum of the sum of two intensity profile (Fig. 1.7b). The spatial resolution according to this criterion is 0.42 λ/N.A., for a confocal microscope. Table 1.3 summarizes the spatial resolution of a confocal microscope according to the 80% separation criterion for common lasers and objectives.

Table 1.3 Spatial resolution according to the 80% separation; $R_r(\mu m) = 0.42\,\lambda/(\text{N.A.})$

λ (nm)	N.A.						
	0.1	0.25	0.5	0.75	0.95	1.2	1.4
244	1.02	0.41	0.20	0.14	0.11	0.09	0.07
325	1.37	0.55	0.27	0.18	0.14	0.11	0.10
473	1.99	0.79	0.40	0.26	0.21	0.17	0.14
488	2.05	0.82	0.41	0.27	0.22	0.17	0.15
514	2.16	0.86	0.43	0.29	0.23	0.18	0.15
532	2.23	0.89	0.45	0.30	0.24	0.19	0.16
633	2.66	1.06	0.53	0.35	0.28	0.22	0.19
638	2.68	1.07	0.54	0.36	0.28	0.22	0.19
785	3.30	1.32	0.66	0.44	0.35	0.27	0.24
830	3.49	1.39	0.70	0.46	0.37	0.29	0.25
1,046	4.39	1.76	0.88	0.59	0.46	0.37	0.31

Table 1.4 The sampling depth approximated as the distance between the first minima of an airy disk $\Delta = 4n\,\lambda/(\text{N.A.})$ [2]

λ (nm)	N.A.						
	0.1	0.25	0.5	0.75	0.95	1.2	1.4
244	97.68	15.63	3.91	1.74	1.08	0.90	0.75
325	130.10	20.82	5.20	2.31	1.44	1.20	1.00
473	189.35	30.30	7.57	3.37	2.10	1.75	1.46
488	195.36	31.26	7.81	3.47	2.16	1.81	1.51
514	205.76	32.92	8.23	3.66	2.28	1.90	1.59
532	212.97	34.08	8.52	3.79	2.36	1.97	1.64
633	253.40	40.54	10.14	4.50	2.81	2.34	1.96
638	255.40	40.86	10.22	4.54	2.83	2.36	1.97
785	314.25	50.28	12.57	5.59	3.48	2.91	2.43
830	332.27	53.16	13.29	5.91	3.68	3.07	2.57
1,024	409.93	65.59	16.40	7.29	4.54	3.79	3.17
n			Air:1.00			Water:1.33	Oil: 1.52

1.5.3 Sampling Depth and Axial Resolution

Even though the focus of this chapter is XY Raman imaging, the sampling depth and axial resolution are highly relevant and important. Using the airy disk approximation, the sampling depth, defined as the distance between the first minima, is $\Delta = 4n\,\lambda/(\text{N.A.})$ [2]. According to the 80% separation criterion, the axial resolution is $1.46n\lambda/\text{N.A.}$ [2]. for a confocal microscope.

Table 1.4 summarizes the sampling depths, and Table 1.5 the axial resolutions of a confocal microscope using common laser wavelengths and objective N.A.s

Table 1.5 Axial spatial resolution according to the 80 % separation $R_z(\mu m) = 1.46n\ \lambda/(\text{N.A.})$ [2]

λ (nm)	N.A.						
	0.1	0.25	0.5	0.75	0.95	1.2	1.4
	0.1	0.25	0.5	0.75	0.95	1.2	1.4
244	35.62	5.70	1.42	0.63	0.39	0.33	0.28
325	47.45	7.59	1.90	0.84	0.53	0.44	0.37
473	69.06	11.05	2.76	1.23	0.77	0.64	0.54
488	71.25	11.40	2.85	1.27	0.79	0.66	0.55
514	75.04	12.01	3.00	1.33	0.83	0.69	0.58
532	77.67	12.43	3.11	1.38	0.86	0.72	0.60
633	92.42	14.79	3.70	1.64	1.02	0.85	0.72
638	93.15	14.90	3.73	1.66	1.03	0.86	0.72
785	114.61	18.34	4.58	2.04	1.27	1.06	0.89
830	121.18	19.39	4.85	2.15	1.34	1.12	0.94
1,046	152.72	24.43	6.11	2.71	1.69	1.41	1.18
n			Air: 1.00		Water: 1.33	Oil: 1.52	

1.5.4 Small Target (Target Size < Laser Spot Size) Analysis

The smallest target that can be measured is not only a function of the laser spot size (or the sampling area) but also of the Raman cross section of the sample. Consider a small sample that is "captured" in the laser spot. The entire sample is being excited. If the sample is a strong Raman scatterer, it may generate enough signals for detection. Now consider two samples of the same material, both smaller than the laser spot, but one bigger than the other. The bigger sample contains more molecules than the smaller sample. Because both samples are smaller than the laser spot, when "captured" in the laser spot, the entire sample is being excited. The bigger sample with more molecules will generate more Raman signal than the smaller sample. This stays true till the sample size reaches the laser spot size. Once the sample size gets bigger than the laser spot size, the laser spot size becomes the limiting factor in how much sample is being excited, generating Raman signals and thus being analyzed. Figure 1.8 illustrates the relationship of spectral intensity and the sample size when the sample is smaller than the laser spot size and made of the same material.

The best example is a carbon nanotube (CNT). Compare to the laser spot size that is in the order of micrometers, the diameter of a CNT is in the order of nanometers. Because a CNT is a very strong Raman scatterer, as long as the laser spot "captures" the CNT, enough Raman signals are generated for detection. When a map of CNTs is recorded (Fig. 1.9), the locations and lengths (this could be in the order of microns) of CNTs can be obtained by analyzing the Raman image. However, the diameters of CNTs cannot be. The diameter of single-walled carbon nanotubes (SWCNTs) are obtained from the frequencies of radial breathing modes (RBMs) of Raman spectra.

The sampling frequency (the number of measurement points per unit distance) is a parameter for the sampling ratio and the mapping speed. For a large sample

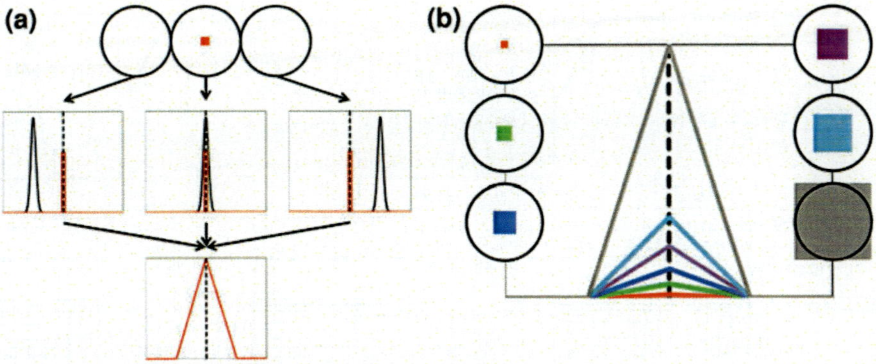

Fig. 1.8 Schematic diagram illustrating the spectral intensity for samples smaller than the laser spot. **a** simulates the convolution of laser intensity (*black*) with the sample (*red*) at three points (*left*, on *top* and *right* of the sample) **b** displays the simulation results as a function of sample sizes

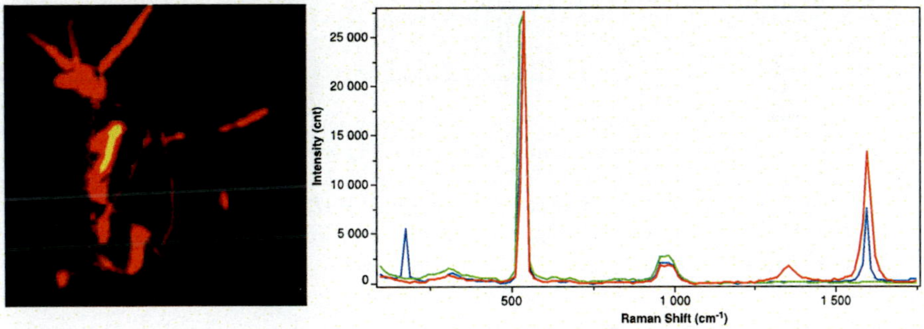

Fig. 1.9 A map of CNTs grown at the end of a metal electrode. *Red* spectrum is of MWCNTs (no RBM), *blue* of SWCNTs. and *green* of the Si substrate

(sample size > laser spot size), par sampling or (carefully planned) under sampling is recommended to reduce the experiment time. For a small sample, on the other hand, over sampling is recommended.

Let's consider a sample whose size is similar to the laser spot size. Figure 1.10a illustrates two sets of par-sampling grids that are overlapping but not overlaying. The sample size is often approximated from the FWHM of the intensity profile across the sample. FWHMs estimated from two grids (dotted lines) are different, resulting in the error when estimating the sample size from the image. Figure 1.10b illustrates affect and benefit of over-sampling. Compared to par-sampling, 5× over-sampling shows a significant improvement (better definition of the sample shape and size), and 10× over-sampling shows even more improvement. Figure 1.10c illustrates two sets of 10× over-sampling grids that are overlapping but not overlaying. FWHMs estimated from two grids are nearly identical, and the error negligible.

The benefit of the over-sampling must be balanced against the time penalty because the large number of measurements increases the time required to record

(a)

(b)

(c)

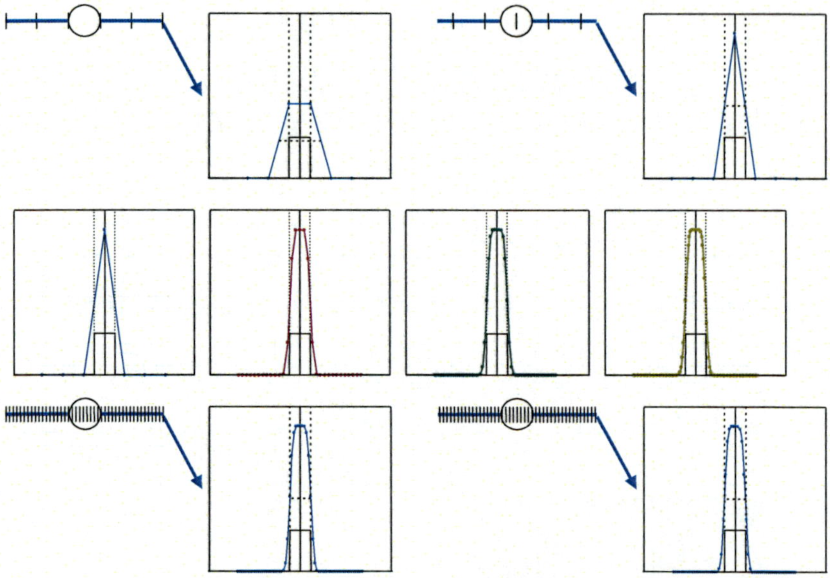

Fig. 1.10 Illustrates the effect of different grids and sampling frequency. The target size (*tophat*) is similar to the laser spot size (*circle*). **a** Par sampling (step size = laser spot size), not overlapping (*left*) and overlaying (*right*) grids. **b** From left to right: par sampling; 5× over-sampling (step size = 1/5 of laser spot size); 10× over-sampling (step size = 1/10 of laser spot size); 20× over-sampling (step size = 1/20 of laser spot size). **c** 10× over-sampling, not overlapping (*left*) and overlaying (*right*) grids

the map significantly. Figure 1.9b illustrates that the improvement in 20× over-sampling is marginal compared to 10× over-sampling, while it takes four times (in an x-y image) longer to record the map. Unless there is an extenuating circumstance, more than 10× over-sampling lacks the merit to justify the time penalty. Note that this is for a sample whose size is similar to the laser spot size.

The threshold for the optimum merit cost ratio for the sampling frequency depends on the sample size. Figure 1.11 illustrates the effect of the sampling frequency as a function of the sample size. These graphs represent par-sampling, purple 5× over-sampling, teal 10× over-sampling and dark yellow 20× over-sampling. Figure 1.11a simulates the sample whose size is one fifth of the laser spot size, Fig. 1.11b one half, Fig. 1.11c the same, Fig. 1.11d twice and Fig. 1.11e five times. Par sampling becomes reasonably agreeable to 5× over-sampling for the sample five times larger than the laser spot size. 5× over-sampling shows dramatic improvement over par sampling for other samples. 10× over-sampling shows noticeable improvement over 5× sampling for samples whose sizes are the same of or smaller than the laser spot size. 20× sampling is nearly identical to 10× sampling for all samples. This indicates that the sample whose size is the same order as laser spot size require 5× – 10× sampling for sufficiently precise measurements.

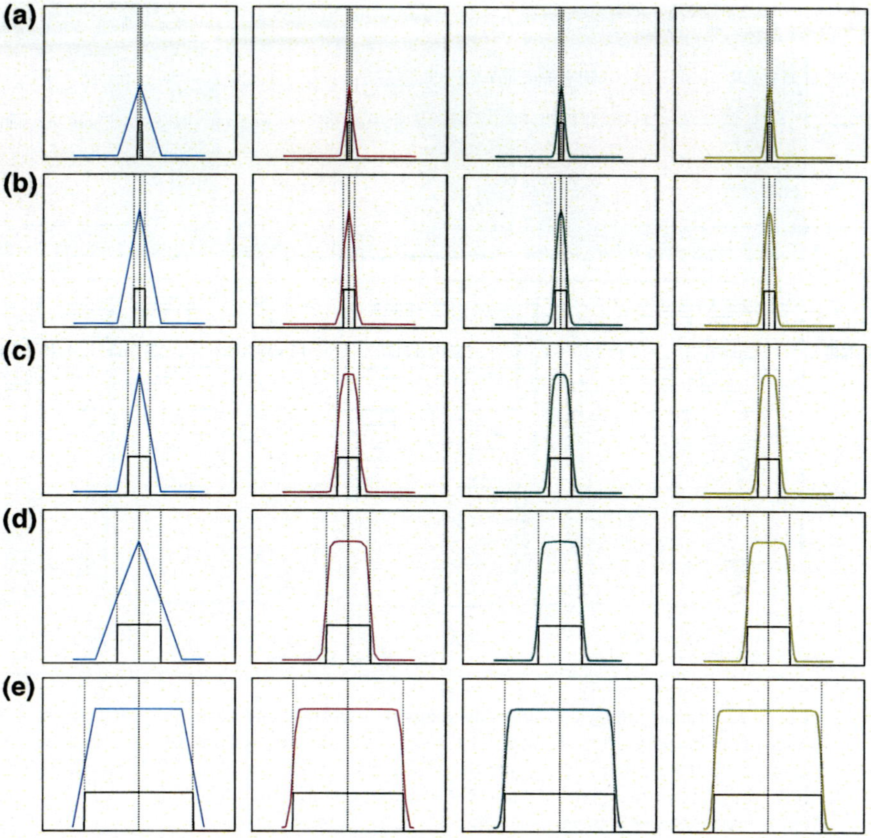

Fig. 1.11 Illustrating the effect of sampling frequency as a function of the target size. is similar to the laser spot size. The sample sizes are (**a**) one fifth of, (**b**) one half of, (**c**) the same as, (**d**) twice of and (**e**) five times of the laser spot size. From left to right: par sampling (step size = laser spot size); 5× sampling (step size = 1/5 of laser spot size); 10 × sampling (step size = 1/10 of laser spot size); 20 × sampling (step size = 1/20 of laser spot size)

Figure 1.12 shows a real life example. The sample is a polystyrene sphere placed on a Si wafer. The sample diameter is 6 μm, which is ~8× larger than the theoretical laser spot diameter, 0.72 μm (λ = 532 nm; N.A. = 0.9). Three Raman maps were recorded and Raman images were produced as intensity maps of polystyrene characteristic bands (~1,000 cm^{-1}, red) and a Si band (520 cm^{-1}, green). Step sizes were 1 μm (Fig. 1.12a), 500 nm (Fig. 1.12b) and 100 nm (Fig. 1.12c), respectively. The images recorded with 100 nm and 500 nm steps (over-sampling) show dramatic improvements in the image quality over the one with 1 μm step (under-sampling). The image recorded with the 100 nm step shows noticeable improvements over the one with 500 nm step.

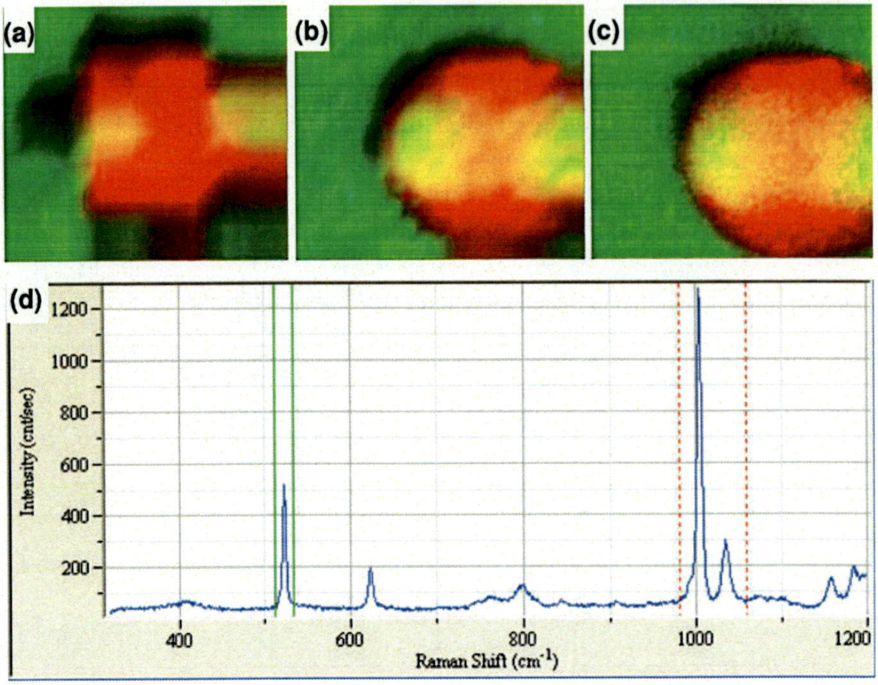

Fig. 1.12 Raman images as intensity maps of a polystyrene bead (dia ~6 μm) on a Si wafer. Intensity of the polystyrene band at ~1,000 cm^{-1} is colored in red, and that of Si band at ~520 cm^{-1} in green. The (theoretical) laser spot is d ~0.72 μm (λ = 532 nm, N.A. = 0.9). Step sizes are (**a**) 1 μm (0.72× under-sampling) (**b**) 500 nm (1.4× over-sampling) and (**c**) 100 nm (~7.2× over-sampling). (**d**) A representative Raman spectrum

It is imperative to understand that the over-sampling does not improve spatial resolution. It does not improve the accuracy of the measurement. It does, however, help the interpretation and potentially improves the precision of the measurement.

1.5.5 Entire Sample Analysis

Imagine a theoretical sample where all ingredients are completely mixed. Any spectrum measured from anywhere in the sample is mixed (spectral features from all ingredients are present), and identical. Only one spectrum is required to characterize the entire sample. The sample is completely homogeneous. The benefit of performing Raman hyprespectral imaging is minimal.

Imagine another theoretical sample where all ingredients are completely separated. Any spectrum measured from a given location of the sample is pure showing spectral features from only one ingredient. Spectra from different locations are either identical (of the same ingredient) or completely different (of the different ingredients). A collection of Raman spectra from the entire sample is required to characterize the sample. The sample is completely heterogeneous. Raman hyperspectral imaging is not only beneficial, it is necessary.

A real sample falls between these two extremes. The number of measurements required (e.g. the number of measurements, the size of mapping area, the sampling ratio) to characterize the sample depends on the degree of the sample heterogeneity. Conversely, the sample heterogeneity can be determined by the amount of the measurement required.

There is more than one aspect to the sample heterogeneity. The degree of mixing described above is the most often discussed [7]. The underlying assumption is that domains (discrete areas of a pure ingredient) are very small compared to the laser spot size. Let's call these samples as Type I

Samples with large domains behave differently, and the heterogeneity (or homogeneity) needs to be accessed differently. Consider a sample where domains of one ingredient are embedded in a matrix of other ingredients. Raman spectra measured from a domain will be definitely different from those from the matrix. The sample is heterogeneous. Characterizing the entire sample requires multiple Raman spectra from different locations. Now let's consider the distribution of domains. If all domains are evenly distributed throughout the sample, characterizing a fraction of the sample will be equivalent to characterizing the entire sample. If all domains are agglomerated to one location, characterizing a fraction of the sample will not be sufficient to characterizing the entire sample. The former can be categorized as homogenous, and the latter heterogeneous, not locally but grossly. The portion of the sample whose Raman spectrum is equivalent to that of the entire sample is an indicator of the degree of the sample heterogeneity. Conversely, the more heterogeneous the sample is, a larger portion of the sample must be analyzed to characterize the entire sample. Let's call these samples as Type II.

Now consider a sample whose domain size is large but the concentration is low. Even if domains are evenly distributed throughout the sample, the number of domains will be small. When a given number of spectra are measured, the chance of measuring one of domains will be low. Similarly when a given area is analyzed, the chance of the area containing one of domains will be low. To "capture" the domain, a large portion of the sample must be analyzed and/or a large number of spectra must be measured. In other words, even though the sample is grossly homogeneous, it behaves like a heterogeneous sample regarding the low concentration component. Let's call these samples Type III.

The degree of sample heterogeneity, in and of itself, is an interesting characteristic. For example, if a pharmaceutical tablet is highly heterogeneous, the dissolution rate of active pharmaceutical ingredient (API) would deviate from the desired rate. Since the dissolution rate is a major factor in determining the bioavaility rate, this results in a poor performance of the drug. In a simple example, an analgesic that was designed a six hour pain relief without drowsiness may last only four hours with a spike during the first half an hour or so, causing drowsiness.

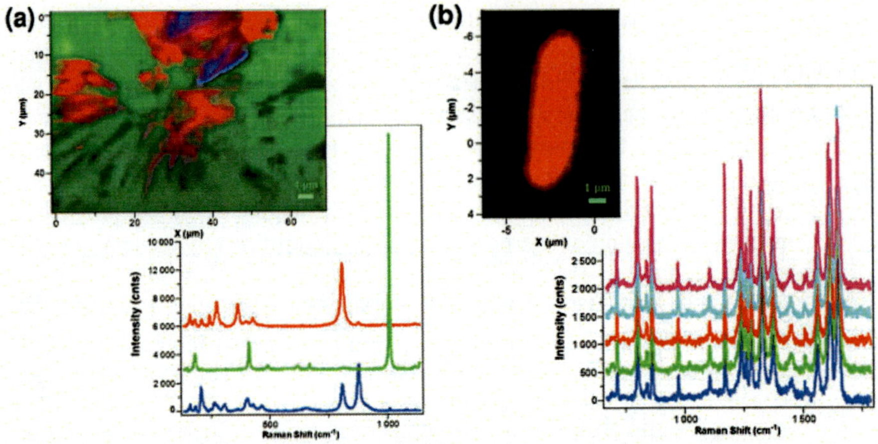

Fig. 1.13 Examples of (**a**) a highly heterogeneous sample and (**b**) a pure (thus highly homogeneous) sample. In the highly heterogeneous sample, a spectrum (*red*) measured from red areas, a spectrum (*green*) from green areas, and a spectrum (*blue*) from blue areas are all different. The mean spectrum (*pink*) shows the spectral features from all spectra. In the pure sample, spectra (*blue* to *cyan*) are the same regardless where they are measured, and the same as the mean spectrum (*pink*)

The degree of sample heterogeneity, in an experiment design of Raman hyper-spectral imaging, is, on the other hand, an interesting characteristic because it is a major factor in determining the minimum requirements for a Raman map be a healthy representation of the entire sample. The maximum requirement is par-sampling of the entire sample (or at least the entire surface of the sample, for Raman microscopy is mainly a surface analysis method). With the laser spot size on the order of micrometers, and the sample often on the order of millimeter or centimeters, this may mean recording 10^6 to 10^8 spectra per sample, which can take a prohibitingly long time. The minimum requirement minimizes the time-cost to analyze a sample without compromising the integrity of the analysis.

Figure 1.13 illustrates Type I samples. For a heterogeneous sample, a sulfate similar to gypsum and thenardite with a high arsenic (As) content is selected. For a homogeneous sample, it is easy to obtain a completely homogeneous mixture in solid state. Pure ingredients is completely homogenous. A single aspirin crystal is selected as a substitute for a homogeneous sample.

Theoretically, the minimum requirement to characterize the entire aspirin crystal is one spectrum. (A good practices in the lab recommends to record a few spectra from different locations to confirm that they indeed are the same, especially if it is an unknown material.) To demonstrate the minimum measurement requirements for a healthy statistical representation of the sulfate sample, the same data set is truncated to successively smaller maps. Mean spectra from each map is compared to the mean spectrum of the entire map. For each map, the mean spectrum is subtracted from the mean spectrum of the entire map, and normalized with respect to the maximum intensity of the mean spectrum of the entire map. (Figure 1.14). The maximum

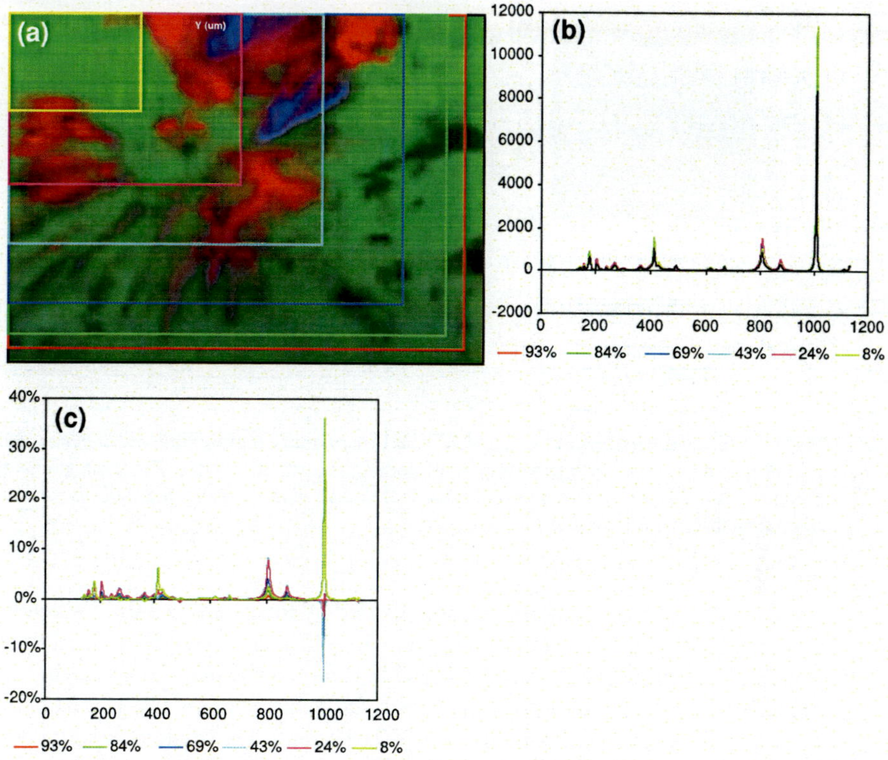

Fig. 1.14 Determining the minimum requirement for a good statistical representation of the sample This is a simulation based on a single data. Instead of recording multiple maps, the large map (100% of the sample) is truncated to simulate small maps. **a** Rectangles indicate the mapping areas in descending order. **b** Mean spectra of each map. **c** Difference between the mean spectrum of a small map (<100% of the sample) and the large map (100% of the sample), normalized by the maximum intensity of the mean spectrum of the large map. *Black graphs* are for 100% map, *red* for the 93% map, *green* for the 84% map, *blue* for the 69% map, *cyan* for the 43% map, *pink* for the 24% map and *dark yellow* for the 8% map

normalized differences are 1% for 93% map, 2% for 84% map, 4% for 69% map, 8% for 43% map, 8% for 24% map and 36% for 8% map. The criterion for a "healthy" statistical representation must be determined by the accuracy/precision requirement of the analysis. Assuming the accuracy threshold is 95%, at least 69% or more of the sample must be mapped to meet the criterion.

An over-the-counter (OTC) analgesic tablet was selected for an example of Type 2 samples. It contains three major ingredients, two of which forms discrete domains. Spatial distribution of one ingredient is assigned to red, the other to blue (Fig. 1.15a). Binary images classifying domains are created. "Open" filter was applied before domains were analyzed [8]. For red domains (Fig. 1.15b), the mean domain size is 0.088 mm^2 and the standard deviation of domain sizes is 0.177 mm^2. The relative

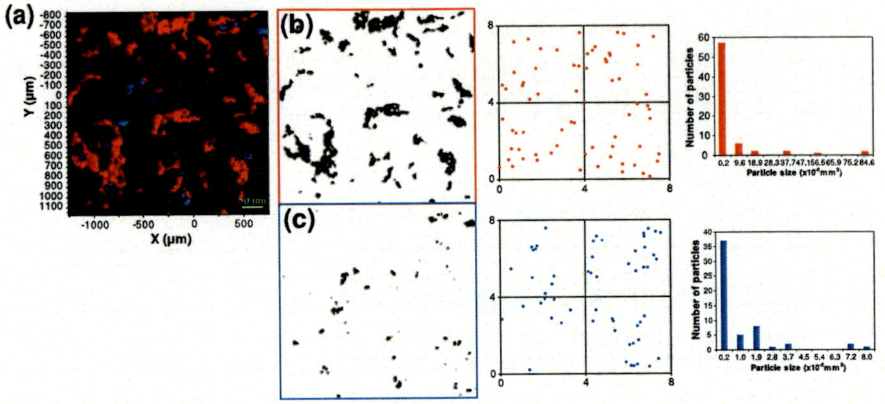

Fig. 1.15 **a** Raman images of two ingredients **b** Binary image highlighting *red domains*, the scatter plot of their locations, and the histogram of domain size distribution. **c** The same graphs for *blue domains*. *Red domains* are more evenly distributed than *blue domains*. Sizes of *blue domain* vary more than those of *red domains*

standard deviation (standard deviation divided by the mean) is 2.021. When the map is divided into four quadrants, there were 16 (left top), 16 (right top), 19 (right bottom) and 18 (left bottom) particles in each quadrant (one particle was at the edge and excluded). For blue domains (Fig. 1.15c), the mean domain size is 0.014 mm^2 and the standard deviation of domain sizes is 0.019 mm^2. The relative standard deviation is 1.363. The numbers of particles in each quadrant are 10 (left top), 20 (right top), 17 (right bottom) and 9 (left bottom). Red domains show more heterogeneity in the domain sizes than blue domains based on the relative standard deviation. The distribution of red domains is more even than that of blue domains based on the numbers of particles in each quadrant. There are other statistical indicators for heterogeneity of domain sizes and distribution such as the d_{10}, d_{50} and d_{90} or the average distance between nearest neighbors.

Powder compacts are selected as examples for Type 3 samples. Theophylline and lactose are mixed at 50:50, and 99:1 by weight/weight, and then compacted. The entire surface on one side area is measured and used as reference. Figure 1.16a shows API (red) images [9] of 1% and 50% compact recorded at par-sampling. As expected, API (red) shows much higher abundance in the 50% sample than the 1% sample. Note that the population of "hot" spots (spots where API concentration is higher than the surrounding area, bright red) is much higher in the 50% sample than the 1% sample. Given the same number of spectra, the chance to "capture" API is much lower in the 1% (low concentration) sample than the 50% (high concentration) sample.

The mean scores of these images are taken as reference values. Sub-sampling causes the mean API score deviate from the reference value. As the sub sampling becomes severe, so does the deviation from the reference value. First, sub-sampling is achieved by reducing the mapping area, without changing the mea-

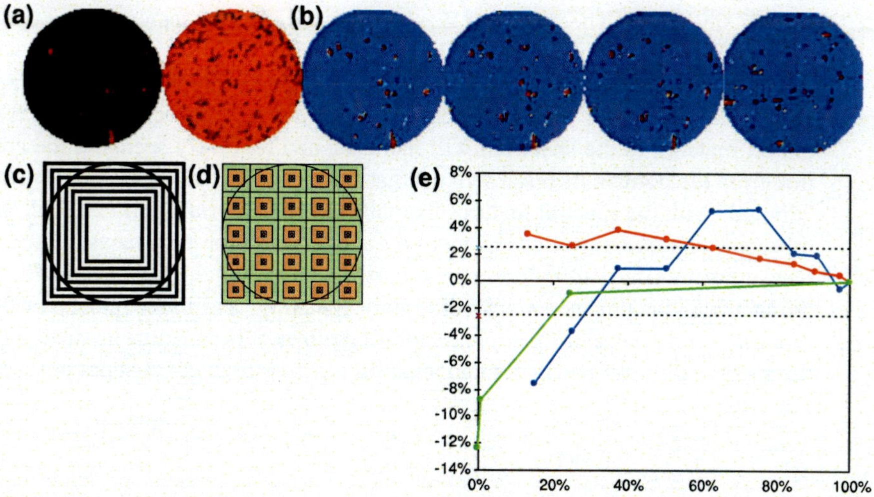

Fig. 1.16 **a** API distribution of 1% and 50% samples. **b** API distribution of 1% sample. The image on the left top is par-sampling. The rest of images are under-sampling; 25%, 10%, and less than 1%, respectively. Schematic diagrams of sub-sampling **c** by reducing the mapping area, and **d** by reducing the measured area per point. **e** The percentage deviation from the 100% sampling plotted with respect to the sub-sampling ratio. *Blue*: 1% sample, successive sub-sampling by reducing the mapping area; *Red*: 50% sample, successive sub-sampling by reducing the mapping area.; *Green*: 1% sample, successive sub-sampling by reducing the measured area

surement area per point or step size between two neighboring measurements. The number of spectra is also reduced. In the 50% sample, mapping areas were gradually reduced. The deviation of the mean API scores from the reference API score is plotted with respect to the sub-sampling ratio (Fig. 1.16e, red graph). Assuming ±2.5% tolerance, characterizing 63% of the sample yields the equivalent results to characterizing 100% of the sample. On the other hand, 1% sample requires 85% or higher sub-sampling.

Second, sub-sampling is achieved by reducing the measurement area per point. The number of spectra and the step size between two neighboring measurements are the same. (Figure 1.16e, green graph) Analyzing 25% of the sample yields the equivalent results to characterizing is the entire sample. While this approach allows the smaller portion of the sample to represent the whole, the number of spectra recorded is kept the same, and thus the experiment time. Therefore, the advantage of this approach is speed. So why would one do this at all? The advantage of reducing the measurement area point at risk of the less than a good statistical representation is reduced spectral dilution.

Let's consider a domain (diameter $\sim 1\,\mu$m) of one material embedded in a matrix of another material without any other neighboring domains. Image the measurement centered at the domain. If the measurement size is $\sim 1\,\mu$m, the spectrum would be (virtually) purely from the domain material. Now imagine the measurement area is, let's say, 5 μm. The spectrum would show the spectral features from the matrix material as well as those from the domain material.

The spectrum is not mixed. Now imagine the measurement area increasing gradually to 100 μm. As the measurement area increases, the contribution from the matrix material to the spectrum will increase, and relatively speaking the contribution from the domain material will decrease. In other words, the domain spectrum is "diluted" with the spectral features from the matrix material. This is called "spectral dilution," and one of reasons to measure a small area for each measurement. If the measurement size is too large, or the dilution is too much, the spectral features of domain may become too small, relatively speaking, and unrecognizable. Spectral dilution is relative. For example, if the domain size is 100 μm in stead of 1 μm, the spectral dilution would not happen until the measurement area per point exceeds 100 μm.

1.5.6 Small Targets on a Large Substrate

This is one of the most difficult samples for which to design the experiment. If the location is known, a small target can be analyzed with a map using a small measurement spot and a small step (over-sampling) can applied. Unless the entire substrate is analyzed, the measurement may miss a small target, and the location may be never found. However, analyzing the entire substrate with a small spot and a small step can take an extremely long time. This is a place to exercise a common sense.

Microscopy in Raman microscopy can provide very useful information. Given enough contrast, micrographs can provide the location, size and morphology of small targets. Once the location and size is determined, Raman measurements can be optimized for each target. Bright field microscopy is usually the standard method. Dark field microscopy, polarized light microscopy (PLM), differential interference contrast (DIC) microscopy and phase contrast (PH) microscopy can also be added to a Raman microscope [10] to enhance the contrast and thus help identify small targets.

Successive "zooming" measurements can save time while providing "concentrated" information. Imagine 10 targets of $1 \times 1 \mu m^2$ spread on a 1×1 cm^2 substrate. Let's say a Raman map measuring $300 \times 300 \mu m^2$ per point is recorded. The entire substrate can be analyzed with 33×33 (1,089 in total) spectra. Statistically, 10 (or less) points ($300 \times 300 \mu$ m^2/point) would be identified for being "different." Let's say a Raman map measuring $10 \times 10 \mu m^2$ per point for each point. Each $300 \times 300 \mu m^2$ point can be analyzed with 30×30 (900 in total) spectra. Again, statistically, 10 points ($10 \times 10 \mu m^2$/point) would be identified for being "different." Now, the experiment is $1 \times 1 \mu m^2$ target in a $10 \times 10 \mu m^2$ substrate times 10, which can be performed using a small measurement area and a small step. The total number of spectral recorded to identify 10 of $10 \times 10 \mu m^2$ areas containing 10 targets is 10,089. The number of

Fig. 1.17 Schematics of typical Raman imaging data processing

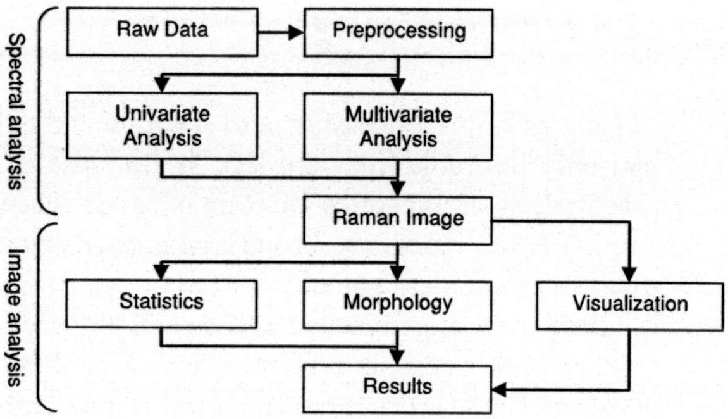

spectra to map to map $1 \times 1\,\mathrm{cm}^2$ using $10 \times 10\mu\mathrm{m}^2$ spot is 1,000,000, which is more than 99 times of 10,089, while providing the same information.

1.6 Data Processing

As was in previous sections, this section concentrates on empirical aspects of data processing only. Theoretical backgrounds are discussed elsewhere in this book. Note that chemometrics and multivariate analyses methods mentioned in this chapter enjoy an extremely wide range of applications [11]. Any statement regarding an algorithm or a method is pertaining to the particular example. Also note that any general statements are for typical, predominant or popular cases. Exceptions are stipulated but not discussed.

Once the data is recorded, the first and foremost thing, is to look at the data; examine spectra from various locations and examine intensity maps of various Raman bands. The importance of this step cannot be emphasized enough. A digital datum is independent of its neighboring numbers. Any cohesion (e.g. the 1,000 numbers forming a spectrum) is awarded by the scientist. A numerical analysis, similarly, offers no constraint. Any boundary conditions (e.g. the 1,000 numbers being spectral intensities and therefore always positive) is applied by the scientist. Therefore, the scientist must monitor the progress of digital data throughout the numerical analysis to ensure their integrity.

Figure 1.17 shows the data processing procedure a typical Raman imaging data analysis takes. There are a few important things to remember:

(1) The data are digital and the process is numerical.
(2) Preprocessing must be taken very seriously.
(3) There are two “axes” in data analysis of a Raman hypercube—spectral axis and image axis.

1.6.1 Preprocessing

One goal of preprocessing is to separate and eliminate undesirable spectral features so only desirable ones are left. In the case of Raman hyperspectral imaging, the desired spectral features are Raman bands. The undesirable spectral features include spikes from cosmic ray, broad backgrounds from fluorescence or stray light, and the baseline from dark signal.

Preprocessing must be taken and undertaken as seriously, if not more, as the analysis itself. It is easy to introduce, and difficult to detect an artifacts in preprocessing.

Dark signal is the photon counts that come from the detector itself, and is a function of exposure time and temperature. It is a detector (type and unit) specific constant (plus random noise). Subtracting a constant (e.g. the minimum intensity) from the spectrum can remove the dark signal. This is often called "offset correction."

Cosmic ray is generated from high energy particles from the outer space. A cosmic ray can be identified by appearing at random. Even if two spectra are recorded consecutively under the identical conditions from the same sample, a cosmic ray would appear in only one spectrum, not in both. Live correction compares two repeat spectra point-to-point, and removes the point with a sudden and significant intensity increase. Alternatively, a cosmic ray can be identified by delta-function like increase/decrease of intensity instead of the gradual increase/decrease of a Raman band (Lorentzian/Gaussian curve). An example of an after-the-fact correction compares a window of spectral points, identifies a point with sudden and significant change intensity and replaces it with the mean or median of neighboring points.

Fluorescence bands are much stronger and nearly always much broader than Raman bands. It is best to select the excitation wavelength that avoids or minimizes the fluorescence. When that is not entirely applicable, and the level of fluorescence is tolerable, it can be corrected by fitting and subtracting the baseline (e.g. polynomial functions). However, baseline fitting is an example where an artifact can be introduced during data processing. A Raman hypercube is a collection of numerous, and often heterogeneous spectra. Baseline may vary in each and every spectrum. The quality of fitting can vary a great deal. Over- or under-correction, therefore, can introduce an artificial feature.

An alternative way to correction baseline is to perform 1st or 2nd derivatives followed by subtracting the mean spectrum. This process removes the offset (1st derivative, Eq. 1.6), linear baseline (2nd derivative, Eq. 1.7) or 2nd order polynomial curve baseline (2nd derivative and then subtract the mean spectrum, Eq. 1.8).

$$\text{Rawspectrum} : I = R\left(s\right) + as^2 + bs + c \tag{1.5}$$

$$1^{\text{st}}\text{derivativespectrum} : I' = R'\left(s\right) + 2as + b \tag{1.6}$$

$$2^{\text{nd}}\text{derivativespectrum} : I'' = R''\left(s\right) + 2a \tag{1.7}$$

$$\text{Subtractthemeanspectrum} : I'' - \overline{I''} = R''\left(s\right) - \overline{R''}\left(s\right). \tag{1.8}$$

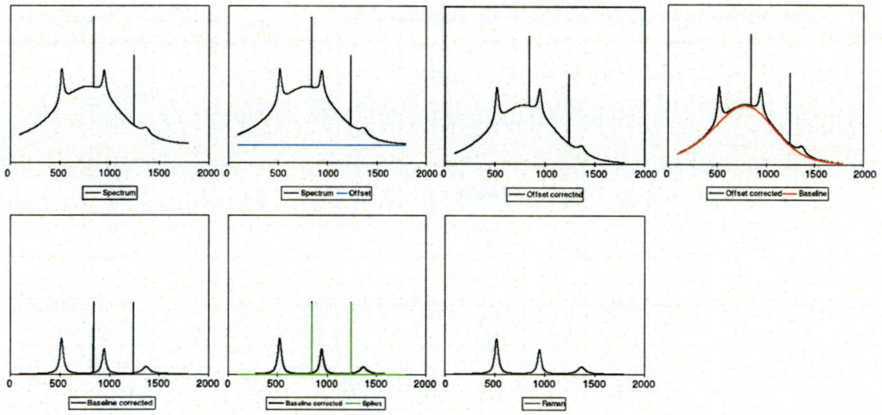

Fig. 1.18 Illustration of a typical preprocessing

The derivative is usually performed using Savitzky–Golay algorithm. It requires two input parameters: the order of polynomial and the number of points. The number of points specifies the spectral window to be fitted. The window is fitted with the specified order of polynomial curve. The fitted polynomial curve is then used to calculate the derivative. Because of this, performing nth order derivative requires $(n + 1)$th or higher order polynomial. For example, taking 1st derivative requires the order of polynomial be two and higher. For 2nd derivative, the minimum polynomial order is three and higher. Increasing the order of polynomial will result in sharper bands in the result. The minimum number of points is three. Increasing the number of points results in broader bands in the results. Edges of a spectrum are compromised when the number of points becomes less than the half of the specified.

Figure 1.18 illustrates the impact of these preprocessing steps on the raw data using a simulated spectrum.

Another goal of preprocessing is to separate and eliminate variation that is not chemically originated. The preprocessing methods that can be used for this objective includes smoothing and normalization.

Smoothing must be performed only when necessary, because it always changes the spectral features, significantly or insignificantly. It is best to reduce S/N by increasing the acquisition time for more Raman signals, whenever possible. There are many ways to smooth including median or mean filters, Savitzky–Golay smoothing, fast fourier transform (FFT) and SMISDA.

There are many ways to normalize spectra. The most widely used ones are:

(1) Normalize with respect to the sum of intensity: $I_i^{\text{norm}} = I_i / \sum I_i$.
(2) Normalize with respect to the total spectrum area: $I_i^{\text{norm}} = I_i / \sum I_i \Delta S_i$ (If the spectral axis is evenly spaced ($\Delta S_i = \Delta S$ for all i's), this is the same as normalizing with respect to the sum of intensity).

(3) Normalize with respect to the norm of the spectrum as a vector: $I_i^{\text{norm}} = I_i / \sqrt{\sum I_i^2}$.
(4) Normalize with respect to the maximum intensity: $I_i^{\text{norm}} = (I_i - I_{\text{min}}) / (I_{\text{max}} - I_{\text{min}})$.
(5) Normalize with respect to the variance of the spectrum intensity (standard normal variate, SNV): $I_i^{\text{norm}} = (I_i - \mu) / \sigma$ where $\mu = \sum I_i / n$ and $\sigma = \sqrt{\sum (I_i - \mu)^2 / n}$.

I_i is the ith intensity at the spectral point of the spectrum. ΔS_i is the spectral pixel width at the ith spectral point of the spectrum (spectral axis of a Raman spectrum is unevenly spaced).

Normalization for a qualitative analysis (e.g. classification, identification) scales spectral intensity to be the same. Unit area normalization (# 2) or height normalization (# 4) suffices.

Normalization for a quantitative analysis aims to separate and remove intensity variation from other than chemical composition. Let's consider a clear solution. Imagine a collimated laser beam passing through the sample generating Raman signals. Raman signals are proportional to the number of molecules excited. A high number of molecules exist in a unit volume in a high concentration solution. Given a fixed path length, an increase in the concentration will increase Raman signals. A long path length exposes a high number of molecules to the laser beam. Given a fixed concentration, an increase in the path length will increase Raman signals. To predict the concentration from Raman signal, the intensity must be normalized with respect to the path length.

Now consider an ideal solid sample under a Raman microscope and a perfectly focused laser beam. In an ideal scenario, Raman signals are generated by molecules within the focal volume only. Increasing the number of molecules within the focal volume (e.g. increasing the local concentration or density) will increase Raman signals. In the same ideal scenario, all photons from the laser are focused into the focal volume without loss to excite molecules. Increasing the number of photons (e.g. increasing the laser power) or decreasing the focal volume (e.g. using a high N.A. objective) will increase Raman signals.

Now imagine a real sample with penetration, reflection, scattering, and/or diffusion with uneven surface with physical defects. At a given point, the focal volume would not be well defined, and vary. Raman signals are generated from out-of-focus portion of the sample, albeit with less intensity, as well as within the focal volume. The volume from which Raman signals are generated (this is a scattering volume) is different from the focal volume. The volume from which Raman signals reach the detector (this is information volume) is different from the focal volume and scattering volume. The spectral intensity is affected by the information volume as well as the chemical composition, and it is not trivial to estimate the information volume with high accuracy. Without the information volume, normalization corrects for spectral intensity variation from all sources, including concentration variation. Therefore, for quantitative analysis, performing normalization is optional. If it is decided to

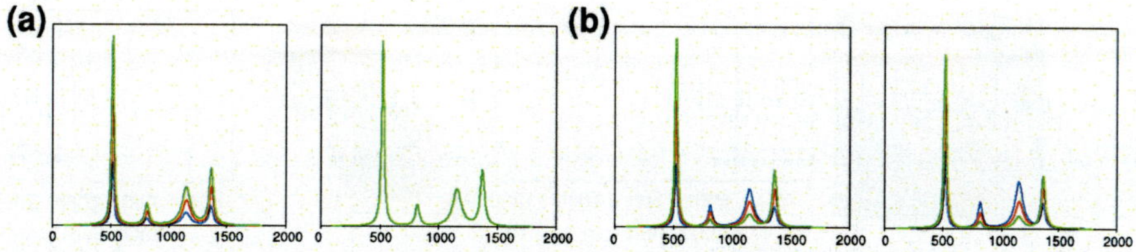

Fig. 1.19 Effect of normalization on spectra of (**a**) the same composition and (**b**) different compositions. Spectra on the left are before normalization and show the overall intensity variation. Spectra on the right are after normalization, and show different behavior. Those simulate the same chemical composition become identical while those simulate different chemical compositions still are different

include normalization in preprocessing, it must be decided carefully or empirically which type of normalization will be performed. Once normalization is performed, the analysis steps must be chosen accordingly. For example, a relative intensity of Raman bands from different ingredients instead of an absolute intensity of a Raman band is a better indicator for a change in the chemical composition. Similarly, multivariate analysis (using all bands in the spectrum) is less sensitive to the normalization than a univariate analysis, and uses normalization routinely. Figure 1.19 illustrates the effect of normalization on a simulated data. Figure 1.19a simulates spectra of the same chemical composition. Figure 1.19b simulates spectra of different chemical compositions. After normalization, spectra in Fig. 1.19a become identical (the same chemical composition) while spectra in Fig. 1.19b still show the difference (different chemical compositions).

Another goal of preprocessing is to separate a part of data for analysis to increase accuracy or sensitivity. This is different from separating signals from desired sources versus undesired sources (described above). The processing steps in this category aim to identify and isolate the part of data that is the most relevant to the objective of the analysis. The amount of information contained in a Raman hypercube is tremendous. Not only there are thousands of spectra representing local chemistry, but also the variation between and statistics of these spectra provide another level of information for the mapping area (e.g. chemically specific morphology). Some of information is more closely related to a part of a Raman hypercube than the rest. Identifying, isolating and separating the pertinent part of the data can increase accuracy and sensitivity of the analysis.

Spatial truncation and masking classify a subset of spectra as a group to be processed or ignored together. As an example, let's consider a polystyrene (PS) bead on a silicon (Si) wafer. Let's say the objective is to analyze the PS bead, and the Si wafer was a substrate of convenience. Spatial masking classifies PS bead spectra to be analyzed, and Si spectra to be ignored. Ignoring Si spectra may render a large portion of the map irrelevant to the analysis. Spatial truncation removes irrelevant parts and creates a new map, which is a subset of the original map, for the analysis. The resulting data maintains all data pertinent to the PS bead, and minimizes data pertinent to the Si wafer.

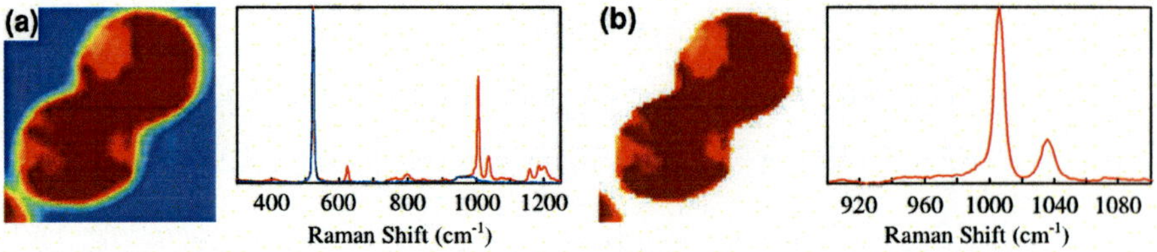

Fig. 1.20 An example of spatial masking and spectral truncation. **a** The original data with full range spectrum and the entire map area. Red spectrum is from a red area and shows both the Si band (~520 cm^{-1}) and PS bands (the strongest at ~1,005 cm^{-1}). The image is the pseudo color intensity map of a PS band at ~1,005 cm^{-1}. High intensity (*PS*) is colored in *red*, and the low intensity (Si) in *blue*. **b** Spatial masking removed SI area (white, not spectra). **c** Spectral truncation removed the Si band at ~520 cm^{-1}

Spectral truncation and masking, similarly, classifies a section of the spectral range to be processed or ignored together. As an example, let's consider the PS bead on the Si wafer. If the size of the PS bead is small enough (e.g. diameter ~5μm), spectra recorded from the PS bead are mixed, showing spectral features from both PS and Si. Truncating the spectral axis to the small region of the PS band excludes the Si band. The resulting data maintains all data pertinent to the PS bead, and minimizes data pertinent to the Si wafer.

Figure 1.20 shows an example of the spatial masking followed by the spectral truncation. Figure 1.20a shows the original data with full range spectrum and the entire map area. The strongest Si band centers at ~520 cm^{-1}. The strongest PS band centers at ~1,005 cm^{-1}. The image is the pseudo color intensity map at ~1,005 cm^{-1}. High intensity (PS) is colored in red. The low intensity (Si) is colored in blue. Spatial masking removed Si area (white) leaving only the PS beads. The Si area no longer contains spectra. Spectral truncation isolated the strongest PS band at ~1,005 cm^{-1} removing the Si band at ~520 cm^{-1}.

Subtracting mean spectrum can increase the subtle contrast due to small change in chemical composition. Consider a mixture sample where one ingredient is predominant. The spectra will also be predominated by the spectral features of the ingredient. Let's say the objective of analysis is to investigate the subtle variation in the local chemistry. Because spectra are predominated by the spectral features of the predominant ingredient, the spectral variation due to the change in local chemistry under these circumstances will also be subtle. In other words, individual spectra would be very similar to each other, as would be the mean spectrum. If the mean spectrum is subtracted from individual spectra, the common features, which are mostly from the predominant ingredient, would be largely eliminated, leaving spectral features that represent spectral variation be highlighted. For the subtraction to be the most effective, the Raman map may require normalization before it is performed.

Figure 1.21 shows the example described. The sample is PS beads on Si wafer. The left most are spectra without any preprocessing. The red spectrum is from a point of a PS bead. Note that both Si and PS bands are present. The Si band at ~520 cm^{-1} is stronger than the PS band at ~1,005 cm^{-1}. The spectra at the center are from

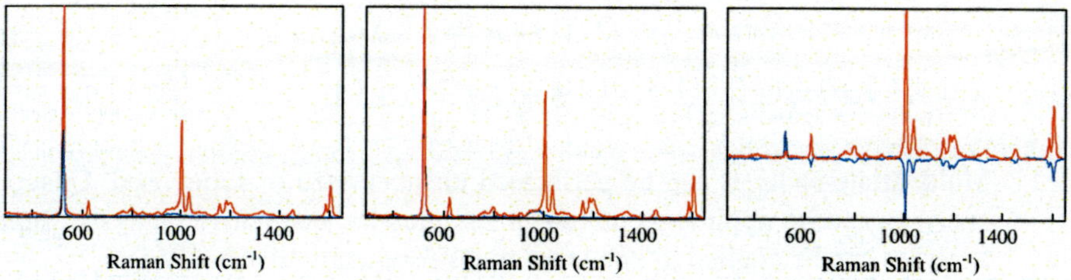

Fig. 1.21 An example of subtracting the mean spectrum proceeded by normalizing to the maximum intensity. The sample is PS beads on Si wafer. *Left*: spectra without any preprocessing. At the center are spectra from normalizing to the maximum intensity. *Right*: spectra from subtracting the mean spectrum. Note that the ratio of the PS band intensity to the Si band intensity increased because the Si band is minimized. Red spectra are from a PS bead, and blue from Si

normalizing to the maximum intensity. The spectra on the right are after the mean spectrum is subtracted. Note that the PS band at $\sim$1,005 cm^{-1} is now stronger than the Si band at $\sim$520 cm^{-1}.

1.6.2 Analysis

1.6.2.1 Spectral Analysis

Univariate analysis uses one variable at a time, including peak position (absolute or shift), peak height (absolute, relative or normalized), band width (absolute or broadening) or area (absolute, relative or normalized).

Multivariate analysis treats a spectrum as a vector, and a Raman map as a matrix. One type of multivariate analyses is performed by comparing each element of a spectrum with the corresponding element of another spectrum. An example is the Euclidean distance (Eq. 1.9). I and J are spectra. I_i and J_i are ith element of spectra I and J, respectively. The operation is performed element-to-element.

$$\|I - J\| = \sqrt{\sum_{i=1}^{k} (I_i - J_i)^2}. \tag{1.9}$$

Another type of multivariate analyses is based on the covariance. Principal component analysis (PCA) is the best known example (Eq. 1.10). It performs eigenvector decomposition on the covariance matrix created from the Raman map. D is the Raman map. X is the covariance matrix. L is the matrix of eigenvectors. Δ is the matrix of eigenvalues. Eigenvectors are called principal components. Scores, S, are the projection of principal components onto the Raman map.

$$D^T D = X = L^T \Delta L, \tag{1.10}$$

$$S = DL^T \left(LL^T\right)^{-1}. \tag{1.11}$$

Multivariate analysis can be performed unsupervised or supervised. Unsupervised analyses can be applied to unknown samples for qualitative analysis. Supervised analyses can be applied to known samples for quantitative analysis.

1.6.2.2 Image Analysis

A Raman image is an image whose contrast is based on Raman spectra. The value at each point must be based on the scientifically reasonable analysis of the Raman spectrum from that point. Otherwise, the image is not scientifically meaningful and cannot be called a Raman image.

Any conventional image analysis can be applied to a Raman image. For example, domains in a Raman image can be identified based on the spectral contrast, and then highlighted in a binary image or images. The binary image or images can be analyzed, using standard image analysis tools [8], for the location, size and shape of each domain.

1.6.3 Combined and Complimentary Analysis

1.6.3.1 Statistical Analysis

Statistics offers a way to perform quantitative analysis under a certain circumstance. Let's consider a sample with big domains that are evenly distributed throughout the sample. Imagine a Raman map of the entire sample surface. Each spectrum would be relatively pure, and be exclusively classified to one ingredient. After classification, the number of spectra belonging to each ingredient (the population) can be counted. The ratio of the population of one ingredient to the total number of spectra would be proportional to the concentration of the ingredient. Therefore, the population ratio provides relative abundance, if not absolute concentration. In other words, if two similar samples are analyzed and compared, the population ratio will show which sample contains higher concentration than the other. However, it is not sufficient to predict their concentration with accuracy, because the population ratio is proportional to the area ratio, not weight, volume or molar ratio.

1.6.3.2 Morphological Analysis

Particle analysis of discrete targets can be performed using bright field or dark field imaging. The targets are identified based on the intensity contrast, and then high-

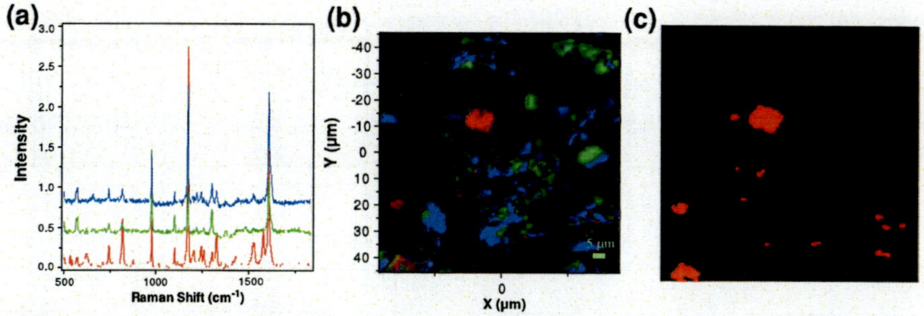

(d)

	Area	Diameter	X	Y	Circularity	Aspect Ratio
1	888	33.62	122.66	107.63	0.8	1.3
2	62	8.88	86.24	105.52	0.9	1.6
3	18	4.79	89.50	165.00	1.0	1.4
4	54	8.29	117.02	170.54	0.8	2.0
5	21	5.17	97.50	200.50	0.9	2.3
6	186	15.39	22.99	211.39	0.7	1.4

Fig. 1.22 **a** Reference spectra. **b** Scores image with respect to reference spectra. **c** Binary image highlighting red domains. **d** Particle sizes, locations and shape descriptors

lighted in a binary image or images. The binary image is then analyzed for location, size and shape of each target [12].

The same procedure can be applied to Raman images to analyze chemical domains. Sizes, locations and shape descriptors of domains can be obtained. The statistical distributions of sizes and locations of domains are indicative of sample heterogeneity.

Performing image analysis on a Raman image compliments the spectral information with the morphological information. Conversely, performing Raman analysis on discrete targets compliments the morphological information with the spectral information.

Figure 1.22 shows an example performing morphological analysis. Figure 1.22a shows manually selected reference spectra. Figure 1.22b shows scores image with respect to reference spectra. The scores image with respect to the red reference spectrum is selected. A binary image (Fig. 1.22c) is created highlighting high scores areas. Morphological information is obtained including area (number of pixels scaled to physical unit, and diameter of the circle with the same area), locations (centers of mass), circularity and aspect ratio (the ratio of shortest line possible to the longest line possible).

1.6.3.3 Chemical Contrast

Similar to having color contrast, having chemical contrast clarifies the interface between neighboring domains objects better than not having the contrast. Figure 1.23 illustrates the affect of chemical contrast in Raman images. Let's consider two small neighboring domains. Imagine a Raman map including both domains. If domains

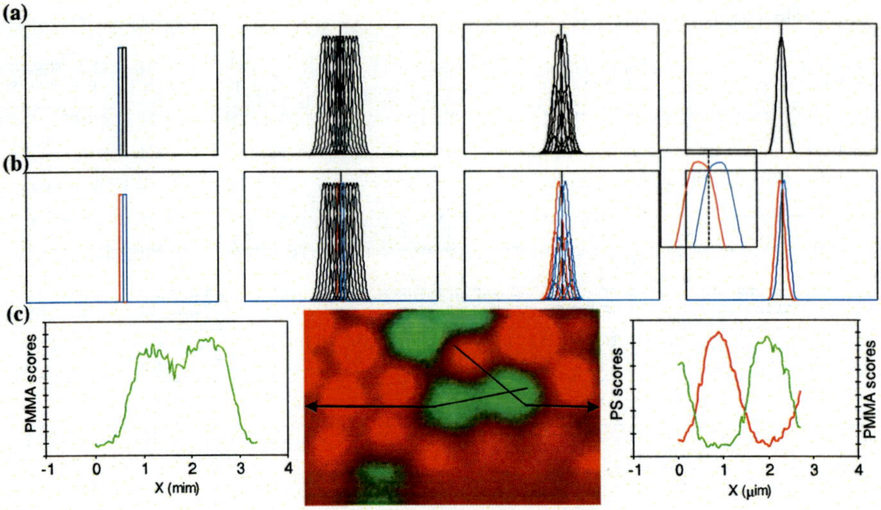

Fig. 1.23 Illustrates Raman intensity profiles overt two small domains (**a**) of the same chemistry, and (**b**) of different chemistry. **c** An example demonstrating the impact of chemical contrast. Raman map of mixed microspheres made of polystyrene (*PS, red*) and polymethyl methacrylate (*PMMA, green*). Intensity profile across two beads of different chemistry (on the *right*) shows a much clearer than that across two beads of the same chemistry (on the *left*)

are of the same chemistry (Fig. 1.23a), Raman spectra measured from both domains are the same. Imagine an intensity map of a Raman band. There will be chemical contrast, and the interface between two domains will be invisible. If domains are of different chemistry (Fig. 1.23b), Raman spectra measured from each domain are different. Imagine an intensity map of a Raman band of one domain. The intensity will vary at the interface, providing the contrast. Imagine two intensity maps of Raman bands, one each from each domain, assign to different colors, overlaid. The interface is clearly visualized in color contrast. It is important to note that having chemical contrast does not improve the spatial resolution of the Raman microscope. However, it increases the precision of a measurement by clearly defining the edges of objects.

Figure 1.23 illustrates Raman intensity profiles over two small domains of the same vs. different chemistry. If two domains are of the same chemistry, no discernible interface between two domains is present. If two domains are of different chemistry, overlaying two intensity profiles (one for each spectral species) presents the interface between two domains. For a real life sample, mixed microspheres made of polystyrene (PS, red) and polymethyl methacrylate (PMMA, green) were mapped and analyzed. An intensity profile across two beads of different chemistry shows a much clearer interface than that across two beads of the same chemistry does.

1.6.3.4 Signal-to-Noise Ratio

Loadings are not statistical representation of a spectral species. Multiple copies of a spectral species would be present in a Raman map, and thus a loading may appear a higher S/N than spectra. Conversely, a Raman map can be recorded at a (relatively) low S/N, and still produce recognizable loadings.

An example (Fig. 1.24) demonstrates the S/N of loadings versus that of spectra. Sb_2O_3 crystals were dispersed on a Si wafer and a series of Raman maps were recorded using different exposure times (1 ms, 5 ms, 10 ms and 50 ms, respectively) to vary the S/N (20-to-1, 40-to-1, 70-to-1 and 120-to-1, respectively). S/N is obtained dividing the maximum intensity (~267 cm^{-1}) of the strongest spectrum (Fig. 1.24a) by the range of intensity variation of the spectral region without Raman bands (1,000–1,800 cm−1). PCA is applied. Two principal components (PCs) representing Si and Sb_2O_3 were retained (Fig. 1.24b). Even if S/Ns are significantly different in original spectra, PCs are virtually identical.

The merit of a long acquisition time (thus achieving a high S/N) is in the score images. The green particles are less sharp and the red areas show noise-like variation in values in thee score image produced from the Raman map using 1 ms/point exposure time (Fig. 1.24d). In the score image produced from the Raman map using 10 ms/point exposure time (Fig. 1.24f) shows a noticeable improvement in the image quality; particles are better defined and areas smoother. On the other hand, the score image produced from the Raman map using 50 ms/point exposure time (Fig. 1.24g) does not show noticeable improvement that can justify five times longer experiment time.

Because loadings are statistical representation of spectral species, the presence of more copies of the same spectral species produces a high S/N looking loading, even if the S/N of original maps are the same. Figure 1.25 shows two Raman maps, both recorded using 5 ms/point exposure time, and thus achieving equivalent S/Ns. One Raman map is composed of 6,000 spectra. The other Raman map is composed of 29,800 spectra, approximately five times more spectra than the former. At the first glance, principal components (Fig. 1.25a, b) appear equivalent. When the spectral region with weak or no bands are zoomed in (Fig. 1.25c), the principal components show a noticeable difference in S/N, even though the only difference between two Raman maps are the numbers of spectra.

1.7 Conclusions

It cannot be stressed more strongly that, no matter how advanced a technology is, how sophisticated an instrument is, or how elegant a software is, it is ultimately the scientist's intelligence and integrity that ensures a scientifically meaningful investigation. It is not necessarily proportional to the amount of information or knowledge, even though they help. It is about the humility to accept the ignorance, the sense of adventure to explore the new, the patience to suspend the conclusion till all data

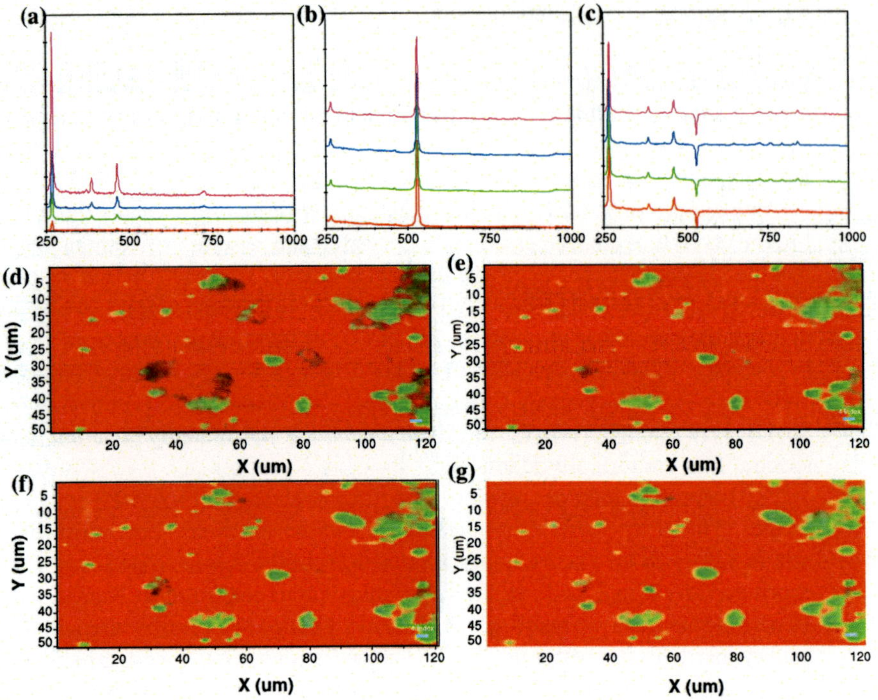

Fig. 1.24 **a** The strongest spectrum of four Raman maps. *Red spectrum*: 1 ms exposure time; *green spectrum*: 5 ms; *blue spectrum*: 10 ms; *pink spectrum*: 50 ms. **b** 2nd principal components. Score images of 1st (*red*) and 2nd (*green*) principal components from the Raman maps recorded using **d** 1 ms/point, **e** 5 ms/point, **f** 10 ms/point and **g** 50 ms/point exposure time

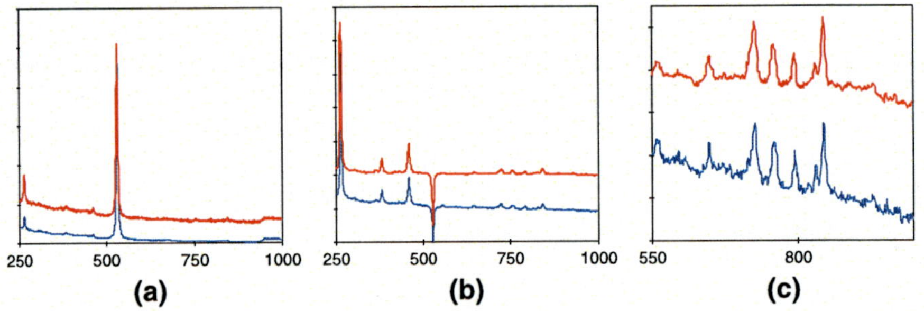

Fig. 1.25 **a** 1st principal components. PC1 (*top*) is produced from a Raman map containing 29,800 spectra. PC2 (*bottom*) is produced from a Raman map containing 6,000 spectra. **b** 2nd principal components. **c** Weak/no band region of 2nd principal components to examine noise level

are in, the tenacity to explore all possible interpretations, the maturity to accept the results as they are instead of trying to fit them into the comfortable convention, and the passion to enjoy every step of it.

References

1. http://rst.gsfc.nasa.gov/Sect13/Sect13_5.html
2. http://rst.gsfc.nasa.gov/Intro/Part2_11.html
3. http://www.andor.com/pdfs/specifications/Andor_iDus_420_Specifications.pdf
4. http://www.horiba.com/fileadmin/uploads/Scientific/Documents/OSD/OSD0058256FIVS.pdf
5. http://www.marzhauser.com/en/products/measuring-stages/motorized-measuring-stages.html
6. I. De Wolf et al., J. Appl. Phys. **79**(9), 7148 (1996)
7. R. Lyon et al., AAPS. PharmSciTech **3**(3), 1–15 (2002)
8. ImageJ, http://rsbweb.nih.gov/ij/docs/menus/process.html#binary
9. ISys 5, http://www.malvern.com/LabEng/products/sdi/software/isys_software.htm
10. http://www.horiba.com/in/scientific/products/raman-spectroscopy/raman-systems/analytical-raman/xplora/xplora-tm-124/
11. http://eigenvector.com/applications.php
12. http://www.horiba.com/us/en/scientific/products/particle-characterization/particle-size-analysis/details/psa300-604/

Chapter 2
Raman Imaging in Semiconductor Physics: Applications to Microelectronic Materials and Devices

Antoine Tiberj and Jean Camassel

Abstract The unique versatility of micro-Raman spectroscopy (μRS) in semiconductor physics remains in Raman imaging. Numerous applications cover the whole development of modern electronic and optoelectronic devices: from semiconductor growth to advanced device inspection tools. In this chapter, a wide variety of semiconductors (SiC, graphene, GaN, GaAs, SiGe, strained Si, sSOI, SGOI) and devices (FETs, lasers, MEMS) are addressed. First, it will be shown how Raman mapping enables to check the crystalline quality, the composition, the doping, and the uniformity of as-grown semiconductors. Then, we will focus on the most popular application in microelectronics: strain measurements either at the device or at the full wafer scale. Finally, we will show how μRS imaging can be used for final device inspection through the temperature mapping of operating devices (FETs, lasers, actuators).

Since the first report by Raman in 1928 [1, 2], Raman spectroscopy has become increasingly popular in materials science and, especially, in semiconductor physics and microelectronics. Basically Raman scattering probes the inelastic scattering of a monochromatic light (incoming-photons) by the atomic vibrations in a medium (solid, liquid, or gas). In crystalline solids, the atomic vibrations are quantized (phonons) and they are very sensitive to internal and external perturbations, such as doping and stress. The frequency of the scattered light (out-coming photons) is then a local probe of the perturbation experienced (or not) by the medium.

Today, the large number of results collected on semiconductors, combined with a good theoretical understanding of the scattering mechanisms, allows to predict reli-

A. Tiberj (✉) · J. Camassel
Laboratoire Charles Coulomb UMR 5221, Université Montpellier 2,
34095 Montpellier, France
e-mail: Antoine.Tiberj@univ-montp2.fr

A. Tiberj · J. Camassel
Laboratoire Charles Coulomb UMR 5221, CNRS,
34095 Montpellier, France
e-mail: Jean.Camassel@univ-montp2.fr

A. Zoubir (ed.), *Raman Imaging*, Springer Series in Optical Sciences 168,
DOI: 10.1007/978-3-642-28252-2_2, © Springer-Verlag Berlin Heidelberg 2012

ably the effect of an external perturbation on the electronic and vibrational properties of the investigated medium (either ordered or disordered). Raman spectroscopy can then be used to identify the constituting species, study the compositional uniformity, crystallinity, doping level, and to probe locally the temperature and stress.

At the industrial level, in modern clean room facilities micro-Raman spectroscopy (μRS) has become (like many other optical techniques) a very attractive characterization tool. This is because of its contactless and nondestructive nature. Thanks to recent turnkey Raman systems, one can perform repetitively large area mapping with spatial resolution down to 300 nm. The use of several laser wavelengths enables then to probe the in-depth profile of a given semiconductor or device. All these features confirm the unique versatility and potentialities of μRS imaging in microelectronics.

In this chapter, we illustrate how this unique characterization tool covers the whole process flow of fabrication of modern electronic and optoelectronic devices. The first section will give some background considerations on Raman imaging in semiconductor physics. The second section will be devoted to the study of advanced wafers manufacturing and epitaxial layers development, such as graphene, $Si_{1-x}Ge_x$ alloys, and strained Si wafers. The third section will focus on the most popular application of Raman imaging in microelectronics: stress monitoring. Finally the fourth one will address how Raman imaging can contribute to the final device inspection.

2.1 Background Considerations

In this section, we remind briefly the basic principles of Raman scattering in a semiconductor. We limit ourselves to first-order Raman scattering because higher order processes are seldom used for Raman imaging purpose. For a more detailed introduction and description of μRS in materials science, and especially in semiconductors, please refer to [3–5]. For a general review of μRS in nanomaterials, including SiC optical fibers for telecommunications, see [6].

First-order Raman scattering is nothing but a two-photon event that involves the simultaneous annihilation of an incident photon and the creation of a scattered photon. Consider, for instance, a monochromatic light beam of frequency ν_i reaching a crystal. Most of the incident photons will be either reflected or transmitted or absorbed following the standard laws of reflection and transmission. Among the last part, some will be elastically (Rayleigh) scattered. The remaining ones will be inelastically scattered by the emission (or absorption) of a phonon having a frequency ν_p. The emission of a phonon corresponds to the Stokes process. The absorption corresponds to the anti-Stokes process. In both cases, this inelastic scattering must satisfy the conservation of energy and momentum. This means that the scattered photons have a frequency $\nu_s = \nu_i \pm \nu_p$. Wave vector conservation means also that the first-order spectra display only a discrete set of peaks, corresponding to phonon energy values at the center of the Brillouin zone. Usually, the linewidth is very sharp and any (weak) perturbation induced by a structural change, by fluctuations of the composition, the doping level or a local strain can be easily detected.

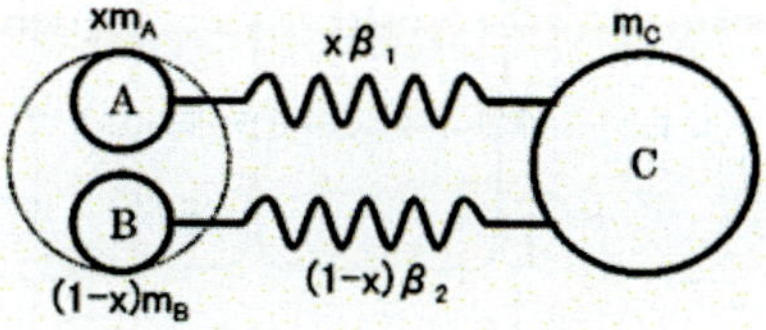

Fig. 2.1 Pseudo-unit-cell of a ternary alloy of the form $A_xB_{1-x}C$. The effective masses are respectively xm_A, $(1-x)m_B$ and m_C. The corresponding force constants are $x\beta_1$ and $(1-x)\beta_2$ [7]

2.1.1 Compositional Changes

Compositional changes are important parameters in the case of opto-electronic devices (like IR detectors for instance) in which the active part is made of a ternary or quaternary alloy in order to tune the optical response. In such systems, the simplest approach is to define a pseudo-unit cell (PUC) such that for a ternary alloy of the form $A_xB_{1-x}C$, for instance, all atoms C are surrounded by A and B atoms with effective mass xm_A and $(1-x)m_B$, respectively [7]. This is shown in Fig. 2.1. The corresponding force constants are $x\beta_1$ and $(1-x)\beta_2$ and the composition dependence of phonon frequencies can be obtained from a simple 2×2 matrix like:

$$\begin{vmatrix} b_{11} + m_{11}\omega^2 & b_{12} + m_{12}\omega^2 \\ b_{21} + m_{21}\omega^2 & b_{22} + m_{22}\omega^2 \end{vmatrix} = 0 \tag{2.1}$$

Since Eq. 2.1 is a quadratic equation, there are always two solutions for a given composition x. The usual classification in terms of one-mode, two-modes or mixed modes behavior comes from the consideration of the different coupling coefficients with the incoming light.

In group IV binary alloys, like $Si_{1-x}Ge_x$ for instance, the situation appears more complicated. Any substitutional atom can sit on any of the two sublattices and, in both cases, there is a strong local distortion that accompanies the substitution. In such a case, both strain and alloying must be simultaneously taken into account. To this end, Rücker and Methfessel have developed an anharmonic Keating model [8] that predicts for the Si-Si Raman mode on the {001} face of $Si_{1-x}Ge_x$ [9]:

$$\Delta\omega_{\text{Si-Si}} = -69x - 780\varepsilon \tag{2.2}$$

which is in excellent agreement with the experimental results summarized in Ref. [10]:

$$\Delta\omega_{\text{Si-Si}} = -68x - 830\varepsilon \tag{2.3}$$

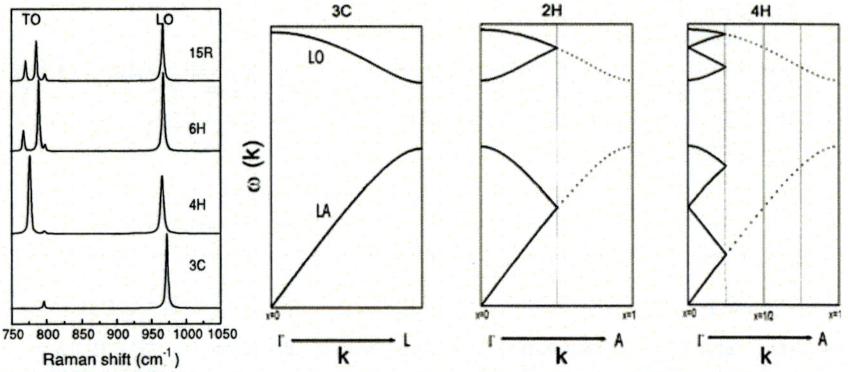

Fig. 2.2 Raman spectra of the most common SiC polytypes (3C, 4H, 6H, 15R). Their increased complex Raman fingerprint allows an easy polytype identification and can be used to image polytype inclusion in SiC substrates. Dispersion curves for the 3C, 2H, 4H polytypes. The increased complexity comes from the folding of the (3C) unit cell as the length of the unit cell increases along the *c* axis

2.1.2 Change in Crystal Structure

Structural changes are seldom observed in III–V alloys but may appear in some cases, as the composition reaches $x = 0.5$. This is, for instance, the triclinic (Cu-Pt like) form of $GaInP_2$. Structural changes also appear in binary compounds, like GaN or SiC, as the total energy difference between two crystallographic structures is small. This is the case of the cubic and wurtzite forms of GaN. This is also the case of the numerous polytypes of SiC [5, 11].

As a matter of fact, SiC exhibits a very large number of polytypes (250) with different structural and physical properties. All polytypes have the same chemical composition but presents different crystallographic structures (cubic, hexagonal, rhombohedric) that differ only by the stacking sequence of N elemental Si-C bilayers in the unit cell. For a review see [12]. The most investigated (crystalline) forms are 4H, 6H, and 3C-SiC. Among them, 4H-SiC has long been recognized for a large variety of high-power and high-temperature electronic applications. 6H-SiC is more considered as substrate for the growth of graphene or nitride semiconductors. 3C-SiC could replace both of them but suffers from the lack of substrate. In polycrystalline form, SiC nanowires have demonstrated interest for field emission devices [13] while SiC fibres have been used to strengthen nano-composite materials [6].

Whatever the application, one of the main concern for SiC wafers manufacturers is to check the absence of polytype inclusion, the crystalline quality [14] and the doping uniformity [15]. Figure 2.2 shows an illustration of the first application. On the left panel, we display four different Raman spectra which appear for the 3C, 4H, 6H and 15R polytypes, with increasingly complex Raman fingerprints. This comes from the folding of the cubic (3C) unit cell as the length of the unit cell increases along the *c* axis (right panel).

2.1.3 Doping and Doping Uniformity

Raman imaging can be used to measure the doping level and, even more important, the doping uniformity in polar semiconductors. This spatial homogeneity is a crucial point since inhomogeneous doping often impacts the thermal process steps used to manufacture the devices and degrades their performances.

The carrier concentration measurements are based on the strong influence of the phonon–plasmon coupling on the Raman shift and lineshape of the longitudinal optical phonons in polar media. As the carrier's density increases, the LO phonon mode shifts to higher frequency and broadens asymmetrically. This is because the carrier's collective modes of oscillations (called plasmons) modify the dielectric constant. The macroscopic electric field associated with the plasmons interacts with the longitudinal optical phonons. This interaction modifies the Raman cross-sections and this leads to coupled modes of mixed phonon–plasmon character (LPP mode) as the plasmon frequency approaches the phonon ones. The Raman shape of this mode is more and more distorted. In SiC [16] or GaN [17], it occurs for concentrations of technological interests in the 10^{17}–$10^{19}\,\mathrm{cm}^{-3}$ range.

The lineshape of a LPP mode can be simply modeled using [16]:

$$I(\omega) = SA(\omega)\, Im\left(\frac{-1}{\epsilon(\omega)}\right) \tag{2.4}$$

where

$$A(\omega) = 1 + \frac{2C\omega_{\mathrm{TO}}^2\left[\omega_p^2\gamma\left(\omega_{\mathrm{TO}}^2-\omega^2\right)-\omega^2\Gamma\left(\omega^2-\gamma^2-\omega_p^2\right)\right]}{\Delta} + \frac{C^2\omega_{\mathrm{TO}}^4\left\{\omega_p^2\left[\gamma\left(\omega_{\mathrm{LO}}^2-\omega_{\mathrm{TO}}^2\right)+\Gamma\left(\omega_p^2-2\omega^2\right)\right]+\omega^2\Gamma\left(\omega^2+\gamma^2\right)\right\}}{\Delta\left(\omega_{\mathrm{LO}}^2-\omega_{\mathrm{TO}}^2\right)} \tag{2.5}$$

and

$$\Delta = \omega_p^2\gamma\left[\left(\omega_{\mathrm{TO}}^2-\omega^2\right)^2+(\omega\Gamma)^2\right]+\omega^2\Gamma\left(\omega_{\mathrm{LO}}^2-\omega_{\mathrm{TO}}^2\right)\left(\omega^2+\gamma^2\right) \tag{2.6}$$

ω_{LO} and ω_{TO} are the longitudinal and transverse optical phonon frequencies, γ is the plasmon dampening constant which is related to the carrier's mobility, Γ is the phonon dampening constant, and C is the Faust-Henry coefficient. S is a proportionality factor. For 4H-SiC, the following values are commonly used: $\omega_{\mathrm{TO}} = 783\,\mathrm{cm}^{-1}$, $\omega_{\mathrm{LO}} = 965\,\mathrm{cm}^{-1}$, and $C = 0.43$.

The dielectric function $\epsilon(\omega)$in Eq. 2.4 can be written as the sum of a phonon and a plasmon contribution:

$$\epsilon(\omega) = \epsilon_\infty \left(1 + \frac{\omega_{\mathrm{LO}}^2 - \omega_{\mathrm{TO}}^2}{\omega_{\mathrm{TO}}^2 - \omega^2 - i\omega\Gamma} - \frac{\omega_p^2}{\omega(\omega + i\gamma)} \right) \tag{2.7}$$

where ω_p is the plasmon frequency. For carriers in a single, isotropic and parabolic band the plasmon frequency is given by:

$$\omega_p^2 = \frac{ne^2}{m^* \epsilon_\infty} \tag{2.8}$$

where n is the carrier concentration, m^* their effective mass, and ϵ_∞ the high-frequency dielectric function.

2.1.4 Strain

All semiconductor devices are subjected to strain or stress. The stress can be internal (residual stress) or external (applied stress) but, in any case, the associated strain changes the equilibrium position of the atoms in the lattice. This alters the phonon properties due to the anharmonic components of the interatomic potential and the phonon frequencies corresponding to the deformed crystal are shifted. The shifts are of the order of 0.1 % of the mode frequency and can be calculated as a perturbative effect [18].

For a cubic crystal like Si, the new phonon frequencies and polarization vectors will be given respectively by the eigenvalues and eigenvectors of the secular equation [19]:

$$\begin{vmatrix} p\varepsilon_{xx} + q(\varepsilon_{yy} + \varepsilon_{zz}) - \lambda & 2r\varepsilon_{xy} & 2r\varepsilon_{xz} \\ 2r\varepsilon_{xy} & p\varepsilon_{yy} + q(\varepsilon_{xx} + \varepsilon_{zz}) - \lambda & 2r\varepsilon_{yz} \\ 2r\varepsilon_{xz} & 2r\varepsilon_{yz} & p\varepsilon_{zz} + q(\varepsilon_{xx} + \varepsilon_{yy}) - \lambda \end{vmatrix} = 0 \tag{2.9}$$

with $\lambda = \omega^2 - \omega_0^2$. For small values of strain, this can be simplified and gives:

$$\Delta\omega = \omega - \omega_0 \sim \frac{\lambda}{2\omega_0} \tag{2.10}$$

p, q, r are the so-called phonon deformation potentials which are related to the changes in the spring constant due to the strain, ω_0 is the frequency of the triply degenerated phonon and ω is the new frequency of the phonon for the strained crystal. The reference axes x, y and z correspond to [100], [010], and [001], respectively. To deduce from the selection rules, which phonon mode can be observed, the experimental scattering configuration must be taken into account. The intensity of a given Raman mode is then given by:

$$I \propto \sum_j \left| \boldsymbol{e}_i \cdot \boldsymbol{R}_j \cdot \boldsymbol{e}_s \right|^2 \tag{2.11}$$

where $\boldsymbol{e}_i$ and $\boldsymbol{e}_s$ are the polarization vectors of the incident and scattered light while $\boldsymbol{R}_j$ are Raman tensors. For a cubic crystal like Si, they are given (in the absence of stress) by:

$$\boldsymbol{R}_x = \begin{pmatrix} 0 & 0 & 0 \\ 0 & 0 & d \\ 0 & d & 0 \end{pmatrix} \boldsymbol{R}_y = \begin{pmatrix} 0 & 0 & d \\ 0 & 0 & 0 \\ d & 0 & 0 \end{pmatrix} \boldsymbol{R}_z = \begin{pmatrix} 0 & d & 0 \\ d & 0 & 0 \\ 0 & 0 & 0 \end{pmatrix} \tag{2.12}$$

As an example, let us consider strained silicon grown on top a relaxed SiGe layer. It is subjected to a biaxial tensile stress in the *x–y* plane, with stress components $\sigma_{xx} = \sigma_{yy} = \sigma$. The corresponding strain components are given by the Hooke's law:

$$\begin{aligned} \varepsilon_{xx} &= \varepsilon_{yy} = (S_{11} + S_{12})\sigma \\ \varepsilon_{zz} &= 2S_{12}\sigma \\ \varepsilon_{xy} &= \varepsilon_{xz} = \varepsilon_{yz} = 0 \end{aligned} \tag{2.13}$$

Solving the (diagonal) secular equation 2.9 then gives:

$$\begin{aligned} \Delta\omega_x &= \Delta\omega_y = \frac{(p+q)S_{11} + (p+3q)S_{12}}{2\omega_0}\sigma \\ \Delta\omega_z &= \frac{qS_{11} + (p+q)S_{12}}{\omega_0}\sigma \end{aligned} \tag{2.14}$$

Clearly, in this simple case, the phonon polarizations and the Raman tensors $\boldsymbol{R}_j$ remain unchanged. Moreover, from backscattering from a {001} surface, only the phonon frequency ω_z can be probed. Using the numerical values given in Table 2.1 yields then a numerical relationship between the stress, the strain and the Raman shift:

$$\Delta\omega_z(\text{cm}^{-1}) = -4.6\sigma(\text{GPa}) = -830\varepsilon_{xx} \tag{2.15}$$

In a more general case, given a stress (or strain) tensor and an experimental configuration, one can (easily) deduce the relationship between the Raman frequency shifts and the strain level. Unfortunately, performing the reciprocal is quite impossible. Six independent Raman data are needed to determine exactly the strain tensor without any assumptions. These six data are the frequency shifts and intensities of the three Raman modes which can seldom be simultaneously measured and mapped at the μRS scale. To obtain a quantitative estimation of the strain, one has to presuppose a strain model or to compare the experimental with the strain calculated from finite-element modeling [20–23].

Table 2.1 Phonon deformation potentials and elastic constants of silicon [24] and diamond [25]

	p/ω_0^2	q/ω_0^2	r/ω_0^2	S_{11} (GPa^{-1})	S_{12} (GPa^{-1})	S_{44} (GPa^{-1})
Silicon	−1.85	−2.31	−0.71	7.69×10^{-3}	-2.14×10^{-3}	12.6×10^{-3}
Diamond	−2.81	−1.77	−1.9	0.952×10^{-3}	-0.099×10^{-3}	1.737×10^{-3}

2.2 Application to Advanced Wafers Manufacturing and Epitaxial Layers Development

In the literature, Raman imaging has often been used to check the uniformity of wafers or epitaxial layers such as InN [26] or graphene [27]. Investigations of the strain present in conformal GaAs grown on Si wafers [28] or GaN/AlN multilayers grown on a mesh-patterned Si(111) substrate [29, 30] have also been reported. In this section, we will focus on three different examples: the development of SiC wafers for power device applications, the growth of graphene on SiC (one of the most exciting material system today) and, finally, the development of strained-silicon layers to increase the channel mobility in C-MOS technology.

2.2.1 Development of 4H-SiC Wafers

In the recent years, much work has been done in this field. Focusing on a few typical examples: in Japan, Nakashima et al. studied modulation-doped 6H-SiC crystals by Raman imaging [31] and extracted from the Raman spectra the carrier's density and mobility profiles of the as-grown crystals; in Europe, more emphasize was made on the parallel development of *n*-type and *p*-type SiC wafers. This is shown in Fig. 2.3 which displays two different series of spectra collected on 2 in 4H-SiC wafers [14, 15]. From top to bottom are:

1. A series of experimental spectra fitted with the model of Sect. 2.1.3. Standard fitting procedures enable to determine the plasmon frequency ω_p and its dampening constant γ. The carrier's density and mobility can then be extracted [31]. This extraction can be simplified by calibrating the shift and/or broadening of the LO phonon with the doping [14] or by considering only the high frequency coupled plasmon–phonon mode in the limit of no dampening. However, the latter technique tends to slightly underestimate the carrier's density. This can be seen in Fig. 6 of Ref. [15].
2. Application of this procedure to determine the doping profile across two 4H-SiC wafers is shown in Fig. 2.3b. Measuring along the diameter of 2 in wafers, the shift of the LO phonon gives an estimate of the nitrogen concentration which is plotted in Fig. 2.3b. The spatial dependence varies from one wafer to another and is not symmetric with respect to rotations of the wafer. Moreover, the doping level at the edges is much lower than the nominal nitrogen concentration noted

in the legend. These results were compared to resistivity maps performed on the same wafers. The maps confirmed the high nonuniformity of doping and were in good agreement with the Raman data.

3. For comparison, in Fig. 2.3c, d two examples of data collected on *p*-type wafers are shown [14]. The strain field surrounding two screw dislocations is imaged in Fig. 2.3c by measuring the strain-induced frequency shift of the E_2(TO) mode. Optical microscopy evidenced that these screw dislocations are associated with two micropipes, which are device killers for SiC electronics. The Raman map of the LO phonon broadening (Fig. 2.3d) performed in the same area exhibits a sixfold symmetry along the crystallographic ⟨1-210⟩ directions in the hexagonal SiC lattice. This LO phonon broadening was attributed to a variation of the carrier's density due to a dopant depletion/accumulation in the vicinity of the dislocation network induced by the micropipes.

2.2.2 *Graphene Growth on SiC*

To date, among the emerging research materials, graphene has been identified as a promising candidate to complement silicon in forthcoming generations of high frequency circuits [32]. As a matter of fact, graphene has outstanding electronic, optical, thermal, and mechanical properties [33–35] and, recently, Geim and Novoselov shared the 2010 Nobel Prize in Physics for groundbreaking experiments [36].

Graphene is nothing but a single sheet of carbon atoms packed in a honeycomb lattice in which strong sp^2 covalent bonds are formed. Excepting diamond, it constitutes the 2D building block of all carbon allotropes. It can be wrapped into 0D fullerenes, rolled into 1D carbon nanotubes, or stacked into 3D graphite [37]. Most of the electronic properties of graphene (including mass-less Dirac fermions, high carriers mobility, and ballistic transport properties at room temperature) come from its linear band structure with 6 double cones at the inequivalent K and K′ points of the Brillouin zone. This linear dispersion was predicted a long time ago by Wallace in 1947 [38] and gives rise to an unusual quantum Hall effect [39, 40] that remains at room temperature [41]. Graphene can then be used to develop new resistance standard for metrology applications [42]. It is also transparent, stronger than steel, and also has a very high thermal conductivity. Therefore, a wide range of potential applications has been identified like: flexible electronics, gas sensors, transparent conductor for touch screens, liquid crystal displays, photovoltaic cells, and organic light-emitting diodes.

However, and because of the so-called Kohn anomaly (which is nothing but the failure of the usual adiabatic Born–Oppenheimer approximation in zero-gap semiconductors [43]) it is also a perfect example to illustrate all the applications of μRS that have been mentioned before. Raman spectroscopy on few layers graphene (FLG) can not only evaluate the crystalline quality but also the thickness, the stacking order of graphene sheets, the doping level and, finally, the residual strain.

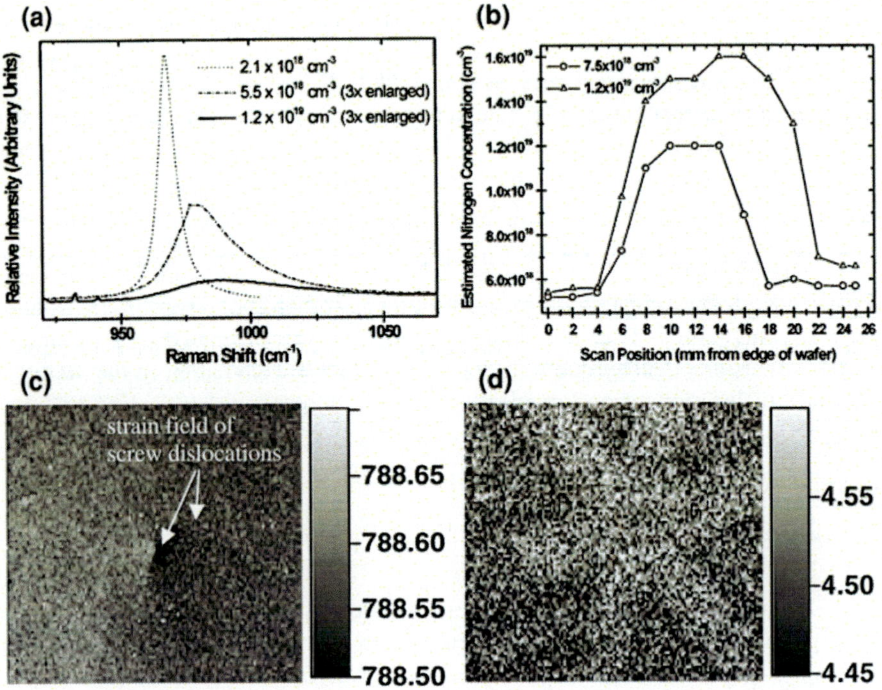

Fig. 2.3 **a** Influence of carrier's density on the Raman spectrum of the SiC A_1 LO phonon mode. As the carrier's density is increased from 2.1×10^{18} to $1.2 \times 10^{19}\,cm^{-3}$, the A_1 LO phonon shifts to higher frequency and asymmetrically broadens; **b** Raman imaging can then be applied to evaluate the profile of the nitrogen concentration across the diameter of 2 4H-SiC wafers. The nominal doping levels are indicated in the legend. Reprinted with permission from [15]. Copyright 1998, American Institute of Physics; **c** micro-Raman map of the strain-induced frequency shift of the E_2 TO mode around two screw dislocations. These dislocations are associated with the presence of two micropipes which are surrounded by a dislocation network having a sixfold symmetry [14]; **d** micro-Raman map of the LO phonon broadening which can be explained by an acceptor accumulation/depletion around this dislocation network. These doping fluctuations shows clearly the sixfold symmetry of this dislocation network. Reprinted from [14]. Copyright 2005, with permission from Wiley-VCH

To date, graphene has been produced using many different ways:

1. Mechanical exfoliation of graphite using scotch tape [37].
2. Chemical exfoliation using either redox reaction [44] or solvent [45].
3. Chemical vapor deposition (CVD) on metal surfaces such as Ru [46], Ir [47, 48], Ni [49–51], and Cu [52, 53].
4. High-temperature annealing (>1,100 °C) of a SiC surface [54, 55].

The last technique is one of the most promising method to get large-scale graphene wafers of high quality and uniformity. It consists in a controlled sublimation of few Si atomic layers from a single crystalline SiC surface [54–60]. During the sublimation, the remaining C atoms rearrange themselves and form FLG which are called "epitaxial graphene" (EG) layers. Such EG samples can be grown either on the Si face (0001) of a SiC substrate or on the C face (000$\bar{1}$). The main differences between EG grown on the C and the Si face come from the SiC surface reconstruction and its interaction with the first graphene sheet.

On the Si face, the graphene planes are Bernal (AB) stacked on top of a $6\sqrt{3} \times \sqrt{3}R30$ SiC surface with a so-called "buffer layer" in between. This "buffer layer" is composed of an intermediate C-rich layer with remaining covalent bonds with Si atoms of the SiC surface [59–61]. As a consequence, EG on the Si-face is (usually) highly n-type doped (around 10^{13} cm^{-2}) with a low carrier mobility (few thousands cm^2 V^{-1} s^{-1}).

On the C-face the situation is completely different. There is no need for a "buffer layer" anymore but two different pristine surface reconstructions have been evidenced below the graphene layers: $(2 \times 2)_C$ and (3×3) SiC surface reconstructions. The graphene layers may have several orientations on top of these surface reconstructions [62–64] and the stacking order can be either Bernal or turbostratic-like. Because of these different surface reconstructions, the interaction between the graphene layers and the C-face of SiC substrate is reduced compared to the one existing on the Si-face. Moreover, graphene grown on a (3×3) SiC surface experiences the weakest interaction with the underlying substrate. This weaker interaction may be one of the reasons for the better carrier mobility measured on EG on the C-face (27,000 cm^2 V^{-1}s^{-1}) [54].

In this section, we review some recent Raman imaging results collected on EG grown on the C and Si-faces of 6H-SiC substrates [65, 66]. First, we briefly describe the growth techniques and the experimental set-up used for micro-Raman and micro-transmission imaging. Then, we discuss results collected on self-organized graphene ribbons grown on the C-face of 6H-SiC substrates and we show how the thicknesses, stacking order, and a rough estimate of the doping level can be obtained. Finally, FLG grown on the Si-face are investigated and we show how the compressive stress experienced by such FLGs can be estimated.

Growth Technology

In this section, we will only consider samples grown using the processes described in [27, 65, 66], with a graphite cap to increase the C and Si partial pressures over the SiC surface. Such a graphite coverage lowers the Si out-diffusion rate during the growth and enables to perform FLG growth at higher temperature. This promotes better SiC surface reconstruction and on the C face of 6H-SiC substrates after 15 min annealing at 1,700 °C in a secondary vacuum, the growth of long (self-organized) graphene ribbons can be done [65]. These ribbons are 5 μm wide and 150 μm long and fully occupy a single terrace of the heavily reconstructed (step-bunched) SiC surface [27]. On the Si face, the temperature was slightly higher (1,750 °C for 20 min under argon). A full graphene coverage was obtained with a similar (pronounced) step-like morphology of the SiC substrate [66]. The average terrace width was again 5 μm with a remarkable homogeneity of width and orientation over a scale of 1 cm^2.

Micro-Raman and Micro-Transmission Spectroscopy

Raman spectra were collected at room temperature, using a Jobin-Yvon Horiba T64000 spectrometer operated in the confocal mode. The 514 nm line of an Ar-Kr ion laser was used for excitation. With a x100 microscope objective, the spot diameter was about ~1 μm with, typically, 1 mW power focussed on the sample. To combine micro-Raman spectroscopy with micro-transmission experiments, a low noise photodiode was inserted between the SiC substrate and the XYZ piezoelectric stage. For details, see [27]. In this way, it was possible to measure (at the same time and using the same laser beam as probe) the power transmitted through the sample and the associated micro-Raman spectrum. The true FLG's spectra were obtained by subtracting the SiC reference signal from the experimental results.

The graphene extinction was deduced from the following equation:

$$\eta = \frac{T_0 - T}{T_0} \tag{2.16}$$

in which T_0 is the bare SiC transmittance value and T the modified one which epitaxial FLGs on top. From the work of Ref. [67], this can be expressed as:

$$T = \frac{4n}{\left|1 + n + \frac{N\sigma}{\epsilon_0 c}\right|^2} \text{ in which } \sigma = \frac{\pi e^2}{2h} \tag{2.17}$$

In this equation, $n = 2.68$ is the SiC refractive index, N the number of graphene layers and σ the optical conductivity of a single (isolated) graphene sheet that was confirmed experimentally [34]. The relative (theoretical) extinction of a monolayer and a bilayer graphene on top of a SiC substrate is then 1.23 and 2.44 %, respectively.

Raman Imaging of Isolated Graphene Ribbons Grown on the C-Face

In Fig. 2.4, we show the results of the analysis of a large ($20 \times 100\,\mu m^2$) map collected on two neighboring graphene ribbons. The step size was 0.5 μm for the X direction and 2 μm for the Y direction. Six individual maps are shown. The first one corresponds to the extinction values, the second one to the integrated intensity of the G band normalized to the HOPG peak. The third one gives the normalized integrated intensity of the 2D band, while the fourth and fifth ones give the shift of the G and 2D bands, respectively. Finally, the last one corresponds to the absolute value of the Fermi level computed from the previous results. Of course, because of the limited range of the XY piezostage ($100 \times 100\,\mu m^2$) the two ribbons could not be completely probed. However, a first point to be noticed is that, on both ribbons, no D band map could be given. This is shown in greater detail in Fig. 2.5a, b and demonstrates the excellent crystalline quality of these graphene samples.

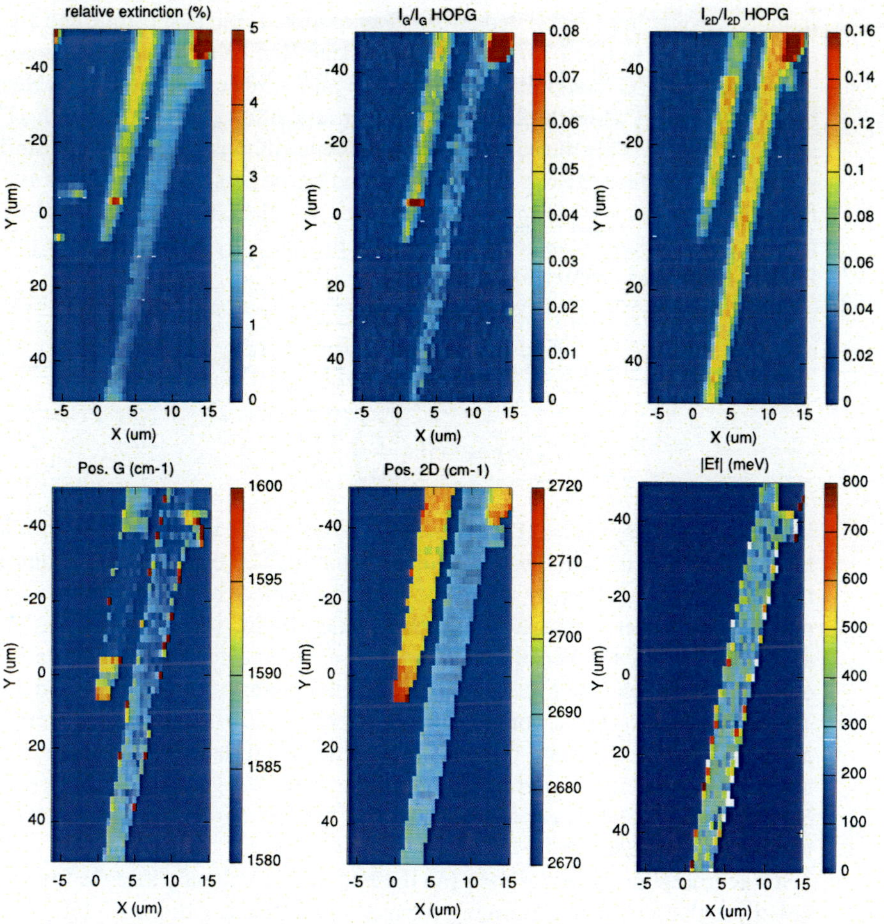

Fig. 2.4 $20 \times 100\,\mu m^2$ maps of two graphene ribbons grown on the C-face of 6H-SiC. The step sizes are 0.5 and $2\,\mu m$ for the *X* and *Y* axes, respectively. The relative extinction, the normalized intensities, and Raman shifts of the G and 2D band are shown. The right left ribbon corresponds to a monolayer (bilayer) graphene. The absolute value of the Fermi level is evaluated from the ratio between the intensities of the 2D and G bands only for the monolayer. It corresponds to a doping level between 3×10^{12} and $9 \times 10^{12}\,cm^{-2}$ with an average of $6 \times 10^{12}\,cm^{-2}$

Let us now consider everything more in details. The first (extinction) map shows that both ribbons have an excellent thickness uniformity. However, and because the relative extinction is different, it shows also that the left one is thicker than the right one. Concerning the absolute values, from these extinction maps complemented by additional point by point measurements, we find that the thinner (right) ribbon corresponds with relative extinction values η ranging from 1.2 to 1.4 %. This shows that we deal with a true epitaxial monolayer graphene (MLG) ribbon. To ascertain this result, a Raman spectrum collected at the center of ribbon is shown in Fig. 2.5a (lower spectrum). It is clearly similar to the one reported in the literature [68] for monolayers graphene exfoliated on top of an oxidized silicon substrate and all spectra collected on the same ribbon exhibited the same Raman fingerprint.

Typically, the G band falls between 1,583 and 1,587 cm^{-1}, with a FWHM of the order of 13 cm^{-1}, while the 2D band ranges from 2,682 to 2,688 cm^{-1} with a FWHM around 25 cm^{-1}. This means that these graphene ribbons are strain-free (unlike EG grown on the Si-face of SiC). This is confirmed by the presence of wrinkles evidenced by AFM [65]. For such graphene monolayers, the absolute value of the Fermi level (and the doping level) can be extracted from the ratio I_G/I_{2D} between the integrated intensities of the G band and the 2D band [69][1]:

$$|E_f| = \frac{\gamma_{ep}}{0.06}\left(\sqrt{\frac{I_G}{I_{2D}}\frac{1}{0.26}} - 1\right) \tag{2.18}$$

$$|n| = \frac{1}{\pi}\left(\frac{E_f}{\hbar v_0}\right)^2 \tag{2.19}$$

In these equations, $\gamma_{ep} = 21$ meV is proportional to the electron–phonon scattering rate [70], $v_0 = 10^6\,m\,s^{-1}$ is the electron velocity. The 0.06 factor is deduced from the function $f(r_S)$ [69] by using the effective dielectric constant $\varepsilon_{eff} = 5.33$ for our graphene layer comprised between air and 6H SiC ($\varepsilon_{6H-SiC} = 9.66$ [71]). The absolute value of the Fermi level was then evaluated for all data points collected on the right ribbon (see Fig. 2.4), giving absolute values between 200 and 350 meV. This corresponds with a doping level between 3×10^{12} and $9 \times 10^{12}\,cm^{-2}$ with an average of $6 \times 10^{12}\,cm^{-2}$. We have not checked directly on the same ribbon but transport measurements performed on similar ones [27] gave a hole concentration of $5 \times 10^{12}\,cm^{-2}$. Such a concentration is in excellent agreement with the G and 2D band positions [72].

Concerning the second ribbon (on the left side), as already said we found a twice larger relative extinction. Ranging from 2.6 to 2.8 %, it indicates a bilayer system. The relative extinction and the G band intensity both indicate that this ribbon is a bilayer with an excellent thickness uniformity. On the contrary, the 2D band intensity map in Fig. 2.4 reveals sharp variations and, basically, the ribbon can be divided into three different domains: the top and bottom part having a less intense 2D band intensity than the middle one. Since these variations are correlated with shifts of the 2D and G bands positions, we assume that there are some doping level fluctuations. Indeed, it has been recently demonstrated that the G Raman band depends strongly on its electrostatic environment [73, 74]. If the top graphene sheet has a different doping level than the bottom one, the doping difference changes the Raman shift and intensity of the G band. It also breaks the inversion symmetry and activates antisymmetric modes (that are usually Raman inactive). This results in a splitting of the G band which was observed experimentally [75–77]. Our interpretation is

[1] The ratio between the integrated intensities of the 2D and G band depends on the experimental setup. In our case $I_{2D}/I_G = 1.02$ for our HOPG reference sample. For the lack of a better knowledge, we assume that our setup is similar to the ones of Refs. [69, 70].

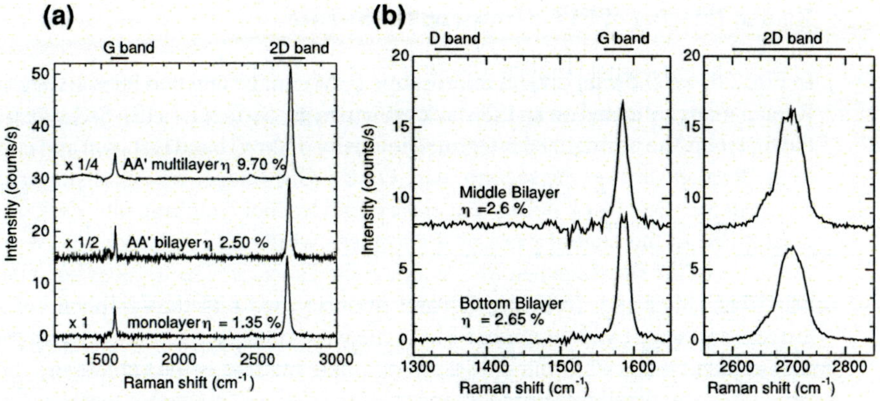

Fig. 2.5 **a** Raman spectra of different ribbons: a monolayer, a misoriented bilayer and a turbostratic multilayer with their corresponding relative extinction. A misoriented bilayer has a similar Raman spectrum than a monolayer with a twice intensity. The multilayer corresponds to eight graphene sheets that are all disoriented with respect to each other. Therefore the single Lorentzian shape of the 2D band cannot be used as a proof to assert the monolayer character of a FLG sample; **b** Raman spectra of different area of the AB bilayer ribbon. One is extracted from the bottom, the second from the middle of the ribbon. The bottom part of the ribbon exhibits a double G peak that can correspond to a different doping level between the top and bottom graphene sheet

strengthened by the two Raman spectra shown in Fig. 2.5b. In both cases, the 2D band exhibits the characteristic shape of AB (Bernal) stacking [68] but the G band is different. On the upper spectrum a single G peak is observed while on the lower one (collected in the bottom part of the ribbon) a clear G band splitting shows that both layers are not evenly doped.

Bernal stacking is not that usual for EG grown on the C-face of SiC substrates. Most of the time the graphene planes are slightly misoriented, corresponding to turbostratic stacking. In this work, we also found misoriented ribbons (not shown). The presence of rotational stacking faults between the two (or more) successive graphene planes results in Raman spectra similar to the monolayer one as shown in Fig. 2.5a. The line shape is not modified. Simply the intensity increases as the number of graphene sheets increases. In Fig. 2.5a we mentioned the relative extinction measured on these ribbons. We found 2.5 and 9.7 % that corresponds respectively to a bilayer and a 7 or 8 misoriented layers stack. These spectra correspond to FLG where all graphene sheets are disoriented with respect to each other. The spectra of folded Bernal stacked multilayers can be even more complex [78]. Therefore, the fact that the 2D band has a single Lorentzian shape can definitively not be used as a proof to assert the monolayer character of FLG flakes. The combination of μRS with micro-transmission measurements appears then as a most necessary tool to discriminate (without any ambiguity) between true MLG and misoriented multilayers. Of course, to perform such reproducible intensity measurements, any laser power fluctuation has to be carefully corrected. In this work, this was done using an additional low noise photodiode that measured continuously the laser power during the Raman map acquisitions.

Raman Imaging of FLG Grown on the Si-Face

In Fig. 2.6, we show an optical microscope (OM) image and two $30 \times 40\,\mu m^2$ micro-Raman maps collected on an EG stack grown on the Si-face of a 6H-SiC substrate. On the first map, the normalized integrated intensity of the G band is shown and compared to the Raman shift on the second map. On the OM image, we can distinguish three different features: black points that correspond to thick graphitic pits, dark areas that correspond to the central parts of the terraces while the bright areas correspond to the edges of the step bunched SiC reconstructed surface. The terraces are 5 μm wide and 10 nm high and, from the G band intensity, we find that graphene covers all the SiC surface. It is then impossible to measure directly the relative extinction and (consequently) the FLG thickness. An alternate method is to evaluate the thickness from the normalized integrated intensity of the G band, assuming that the average G band normalized intensity of a monolayer is between 0.025 and 0.03. Beware that these values depend on the experimental configuration and must be calibrated. For thin FLG (less than five layers) the error can be of 1 layer and, for thicker FLGs, the estimated thickness may have a factor two error. Unlike the G band, the 2D band intensity cannot be used to evaluate the thickness for the reasons discussed before. The 2D band intensity strongly depends on the stacking for multilayers and on the Fermi level for monolayers [69]. In this way, one finds about 5 layers in the center of terraces (green–blue areas in Fig. 2.6) and about 11 layers on the stripes close to the edge of terraces. On the black points of the optical image, the G band intensity is much intense. This means that they correspond to thick graphitic pits. The Raman spectra of these pits exhibit a strong D band, characteristic of a bad crystalline quality. Such pits are probably induced by an increased growth rate coming from the presence of structural defects, like threading dislocations, in the SiC wafer.

Raman spectra collected in the middle of the terraces and on the stripes are shown in Fig. 2.7a. Unlike the graphitic pits, no D band can be observed. This shows that most of the grown FLG have an excellent crystalline quality. The 2D band is broad with a lower frequency shoulder. This shoulder is more pronounced for the thickest FLG which have a 2D band shape similar to the HOPG one. This asymmetric 2D band is a clear indication of Bernal stacking even for the thinner FLG, where the low-frequency shoulder is known to become less visible [79].

These Raman spectra also reveal that both G and 2D bands are shifted to high frequencies. For the thinnest FLG the G band falls between 1,590 and 1,600 cm^{-1} and the 2D band between 2,750 and 2,760 cm^{-1}. Such high blueshift cannot be explained by doping but rather by a compressive stress experienced by the FLGs. This stress comes from the differential dilatation coefficient between the FLG and the SiC substrate when cooling down the sample after the growth. Thanks to the Grüneisen parameters that have been recently measured [80, 81], this biaxial stress/strain can be estimated from the coefficients listed in Table 2.2. We find strain values between −0.3 and −0.4 %, corresponding to local stress values ranging from −3 and − 5 GPa. However, for thicker FLGs, the G and 2D band are less shifted. This reveals a partial strain relaxation as the graphene thickness increases, which can be easily seen in Figs. 2.6 and 2.7a. The thicker the FLG, the less blueshifted the spectrum.

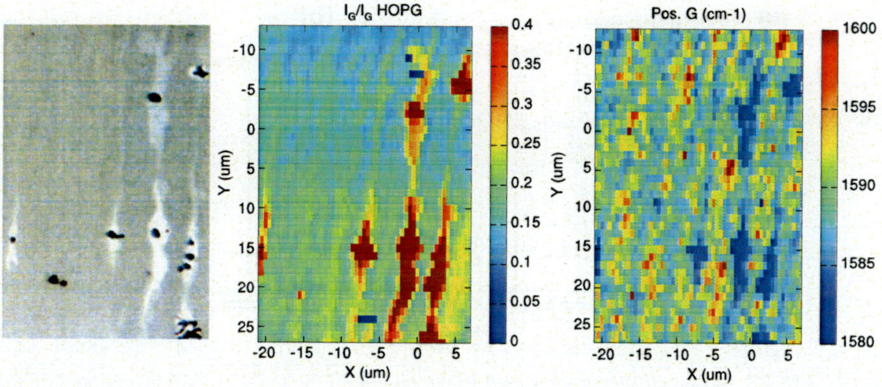

Fig. 2.6 $30 \times 40\,\mu m^2$ optical image of the graphene surface and the corresponding Raman maps of the G band intensity and Raman shift. The intensity of the G band is integrated and normalized by the G band of an HOPG reference sample. A full graphene coverage of the surface is observed with thickness inhomogeneities. FLG are thicker at the step edges (about 11 layers) than in the middle of the terraces (about 5 layers). On the edges we can clearly observe stripes: *bright areas* on the OM image and *red areas* on the G band intensity map. On the OM image, we can also see *black points* that correspond to C-rich graphite pits induced by an increased growth rate due to the presence of crystalline defects. On the G band intensity map, *blue points* mark the presence of Si clusters where the Raman fingerprint of silicon was observed. Finally, the G band is shifted to higher frequencies indicating that FLG are compressively stressed. This stress is progressively relaxed as FLG are thicker

Table 2.2 Frequency shifts of the G and 2D bands for a biaxial strain of 1 % or a biaxial stress of 1 GPa [80, 81]

$\varepsilon_{biax} = 1\,\%$	$\sigma_{biax} = 1\,GPa$
$\Delta\omega_G^{biax} = -60\,cm^{-1}$	$\Delta\omega_G^{biax} = -4.8\,cm^{-1}$
$\Delta\omega_{2D}^{biax} = -153\,cm^{-1}$	$\Delta\omega_{2D}^{biax} = -12.3\,cm^{-1}$

From these results, we confirm that there is a strong difference between graphene grown on the C-face and graphene grown on the Si-face of SiC substrates. On the Si-face graphene strongly interacts with the underlying SiC lattice. This interaction leads to the formation of the so-called buffer layer, which is covalently bound to the SiC lattice [60, 82]. This buffer layer interacts also with the graphene layers that are grown subsequently. It induces a downward shift ($E_D = -0.4\,eV$) of the K point corresponding to a n-type doping. This interaction causes also deviations from the linear band dispersion leading to a parabolic dispersion with an apparent gap of $\sim$0.25 eV. This strong interaction is also responsible for the thermal stress experienced by these graphene layers. This is no longer true for the C-face, on which it has been shown that graphene interacts very weakly with the underlying substrate [63, 64]. This is the weakness of this interaction that explains why rotational stacking faults can easily occur and why the graphene sheets can relax the thermal stress by forming wrinkles or pleats.

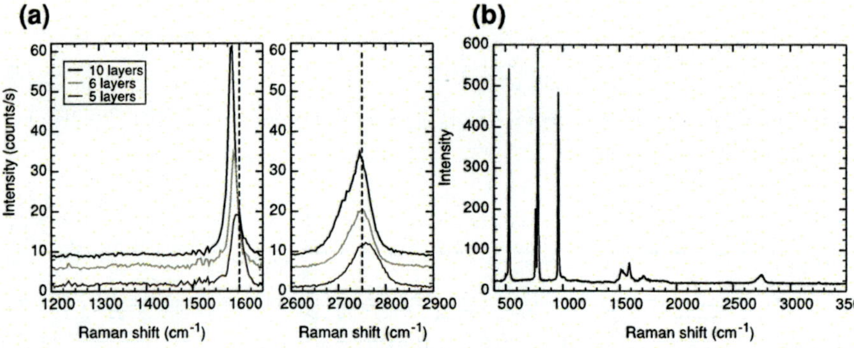

Fig. 2.7 **a** Raman spectra collected in the middle of the terraces (5–6 layers) and on the stripes close to the step edges (11 layers). No D band can be observed confirming the excellent crystalline quality of these FLG. The asymmetric shape of the 2D band (that is more pronounced for the thicker FLG) reveals a Bernal stacking of the graphene planes. Finally, both bands are *blueshifted*. Such shift can only be explained by a compressive strain of the graphene lattice coming from the differential dilatation during the cooling down of the sample. A partial strain relaxation occurs for thicker FLG since the thicker the less shifted; **b** uncorrected Raman spectrum extracted from the Raman mapping that corresponds to one of the *blue point* in the G band intensity map. The first-order Raman scattering of SiC correspond to the bands at 764, 786, and 964 cm^{-1}. Its second overtone falls between 1,400 and 2,000 cm^{-1} with the sharp G band around 1,590 cm^{-1}. The 2D band is around 2,780 cm^{-1}. No D band can be seen on this point. The sharp and intense band around 532 cm^{-1} correspond to a crystalline Si cluster that is highly compressively stressed by the SiC substrate

Finally, on the G band intensity map we can see several points marked in blue. These blue points correspond to area where crystalline silicon clusters were found. One of the corresponding uncorrected Raman spectrum is shown in Fig. 2.7b. The presence of these crystalline silicon (c-Si) clusters is evidenced by the sharp and intense band around 532 cm^{-1} blueshifted compared to bulk silicon. This high blueshift is again due to a strong compressive stress induced by the SiC substrate (-2 GPa or a strain of -1.3 %). First-order Raman scattering of the SiC substrate corresponds to the two TO modes of E_2 symmetry at 764 and 786 cm^{-1} and the A_1(LO) phonon at 964 cm^{-1}. Its second overtone with its characteristic fingerprint [83] falls between 1,400 and 2,000 cm^{-1} under the sharp G band of FLG around 1,590 cm^{-1}. The 2D band is around 2,780 cm^{-1}. No D band can be seen on these points. Moreover, we can see that no significant variations of the G band intensity and Raman shift can be observed close to these Si clusters. This means that these Si clusters do not alter the graphene growth. These clusters are located close to the step edges, like the graphitic pits. There might be a link between the presence of different defects at the step edges like these clusters, the higher growth rate, and the clear electrical anisotropy that has been evidenced by magnetoresistance experiments performed on several Hall bars with different orientations [66].

Summarizing: from recent Raman imaging experiments performed on EG grown on the C and Si face of 6H SiC substrates, we have shown the benefits of combining

Raman spectroscopy with micro-transmissions measurements. Provided the relative extinction of FLGs can be obtained, this enables to determine (without any ambiguity) the thickness, homogeneity, and stacking order (Bernal or turbostratic). On the C-face of SiC substrates we have shown that long, self-ordered, graphene ribbons can be grown. These ribbons have excellent crystalline quality and are strain relaxed. They are up to 600 μm long and 5 μm wide. They are mainly monolayers and Bernal stacked bilayers but turbostratic bi and multilayer areas have also been found. Finally and for monolayers graphene, we also illustrated how the absolute value of the Fermi level can be found, in good agreement with electrical results. On the opposite, on the Si face, a full graphene coverage of the SiC surface has been found. The surface is still heavily step bunched but a high compressive thermal strain and n-type doping was observed. It confirms that on the Si face a strong interaction exists between the graphene layers and the underlying SiC substrate. FLG on the Si face exhibit Bernal stacking with thickness inhomogeneity. Thin (5 layers) FLGs were grown in the middle of terraces, while thicker graphene stripes grew close to the step edges. In the vicinity of these steps disordered graphite pits and crystalline Si clusters were found. There might be a link between the presence of these defects, the thickness inhomogeneity, and the clear electrical anisotropy that has been recently evidenced by magnetoresistance experiments.

2.2.3 Strained Silicon Wafers Manufacturing

Many advanced substrates have been developed recently, based on the idea of integrating any semiconductor with silicon (Anything on Si). This is a very attractive idea, since the technologies developed on these semiconductors would become directly compatible with the mainstream Si technology. Among these new "engineered" or "smart" materials, epitaxial $Si_{1-x}Ge_x/Si$ heterostructures have attracted considerable interest, because of the possibility to enhance the performances of complementary metal oxide semiconductor (CMOS) devices. This is a very important point since, for a long time, the performance improvements have been done by miniaturizing individual devices. Unfortunately, this so-called scaling down technology is facing significant physical and economical limitations. As a consequence, since the 90 nm technology node, an alternative approach called "strain engineering" has been incorporated to maintain the Moore's law predictions.

As a matter of fact, introducing strain in the active region of a device results in large mobility enhancement of, both, the electrons and holes. Such strained film can be either strained Si (sSi), strained Ge (sGe), or even strained SiGe Alloys (sSiGe). This depends only on the final application: MOSFETs, SiGe-based heterojunction bipolar transistors (HBT), optoelectronic devices like optical waveguide, quantum well infrared detectors, and quantum cascade emitters [85–92].

Basically, two types of techniques can be used for strain introduction. First, the "global strain" approach uses a wafer with a sSi film on top. This can be either sSi on a SiGe virtual substrate or sSi on SiGe on Insulator or sSi on Insulator. These three

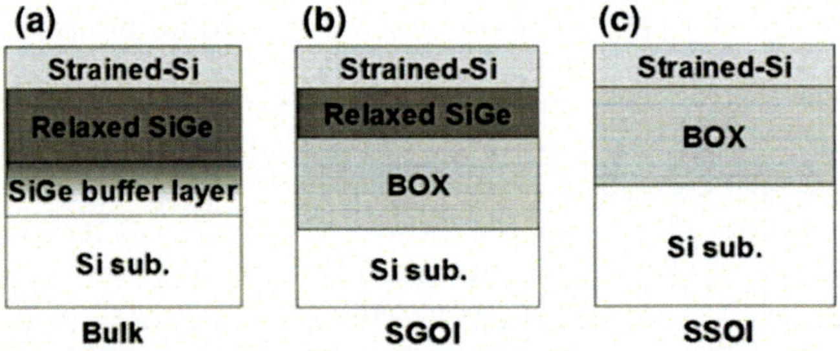

Fig. 2.8 Schematic drawing of three types of wafers used for introducing strain in the channel of field effect transistors. **a** A strained silicon film is grown on top of a relaxed SiGe layer. The relaxation of the SiGe layer is made by growing a graded buffer with an increasing Ge content; **b** SGOI wafer with a sSi film on a relaxed SiGe layer on an insulator (silicon oxide); **c** sSOI wafers are made of a strained silicon film on top of a buried oxide layer. Reprinted from [84]. Copyright 2007, with permission from Elsevier

wafer stacks are sketched in Fig. 2.8 in which, every time, the sSi film is produced by growing a thin Si layer on top of a strain-relaxed SiGe buffer layer (virtual substrate). The second technique, called "local strain" approach, consists in introducing strain in a selected region during LSI fabrication. It will be discussed in the next section.

Focusing on the first "global strain" approach, and whatever is the final option, the fabrication of this virtual substrate is a key issue. Indeed, the properties of the overgrown Si film are strongly influenced by the quality of the buffer layer and, since the strain relaxation of the SiGe buffer layer (grown on a Si handle wafer) generates dislocations, several techniques have been developed to prevent dislocation extension to the surface. The most established ones are using a compositional graded buffer [94] or a low-temperature buffer layer [95]. In both cases, they lead to threading dislocations density below 10^{-5} cm^{-2}.

Bulk sSi/$Si_{1-x}Ge_x$/Si Heterostructures

The ability to control (and monitor) the composition and relaxation of a SiGe alloy is crucial to improve the growth process and engineer the electrical and optical properties of the final heterostructures. In this case, since the frequencies of the alloy Raman peaks depend on the composition, μRS is of great interest and has been used as a straightforward tool to determine the depth profile of the alloy composition.

Basically, depth profiling in Raman imaging can be done in several ways. The first one uses different laser wavelengths to change the penetration depth of the incident beam and the Raman cross-section. In this way, depth profiles of stress [96, 97] and lattice damage in annealed GaAs [98] have been measured. The second one is based on confocal microscopy and consists in moving the sample vertically [99]. The third one consists in measuring the Raman profile along a cross-section of the sample, on a cleaved or bevelled surface. This method is destructive but gives a very

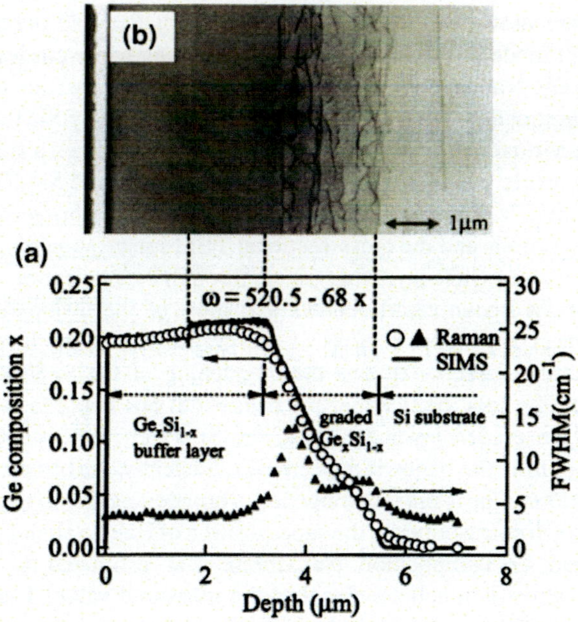

Fig. 2.9 Characterization of the depth profile of a strained silicon film grown on a relaxed $Si_{0.8}Ge_{0.2}$ virtual substrate. Raman image has been acquired on the cleaved surface of the graded buffer. **a** The Ge composition determined from Raman and SIMS analyses are plotted against the depth position and are in perfect agreement. The Raman linewidth of the Si-Si Raman mode is also represented against the depth position. The broadening has maxima at the lower and upper parts of the graded layer. It comes from the high density of dislocations evidenced by the **b** cross-sectional TEM image of the same structure. Reprinted with permission from [93]. Copyright 2004, American Institute of Physics

detailed image of the in-depth stress profile in SOI substrates [100–105]. It has been also used to study heteroepitaxial InAs grown on GaAs or InP substrates [106], the carrier density at the ZnSe/GaAs interface [107], and even to image quantum wells [108–110].

Concerning sSi/SiGe/Si, Nakashima et al. [93] applied the cross-section technique to study the growth of a strained Si film on top of a relaxed SiGe virtual substrate. The main concern was to determine the depth profile of the Ge composition in the graded buffer and in the uniform SiGe buffer layers. They combined μRS with cross-sectional transmission electron microscopy (TEM) and secondary ion mass spectroscopy (SIMS) to determine the Ge content of the SiGe alloys and to study its crystalline quality. The results are shown in Fig. 2.9. Assuming that the strain is completely relaxed in the SiGe layer, the shift of the Raman frequency $\omega_{\text{Si-Si}}$ of the Si-Si Raman peak was used to determine the Ge content x through the relationship (see Sect. 2.1.1):

$$\omega_{Si\text{-}Si} = 520.5 - 68x \tag{2.20}$$

Results are shown in Fig. 2.9a. The Ge composition and the broadening of the Si-Si Raman mode are plotted against the depth as white circles and black triangles, respectively. Notice that the composition determined by Raman spectroscopy (open dots) is in excellent agreement with the one obtained by SIMS (full line). On top of the SiGe virtual substrate, the Raman Si-Si peak reaches a frequency of 507 cm^{-1} which is nearly equal to the frequency of an unstrained $Si_{0.8}Ge_{0.2}$ layer. They also checked by UV Raman spectroscopy that the top Si film was fully strained. These two observations and the cross-sectional TEM image shown in Fig.2.9b confirm the relaxation of the SiGe layer that occurs through the nucleation of misfit dislocations.

The TEM image clearly shows that most of the dislocations are concentrated in the lower and upper regions of the graded $Si_{1-x}Ge_x$ layer. This high density of crystalline defects explains the broadening of the Si-Si Raman mode. This is shown in Fig. 2.9a. Notice that the bandwidth has maxima at the lower and upper parts of the graded layer, while, in the Si wafer, the $Si_{1-x}Ge_x$ buffer layer and the top sSi film, the bandwidth decreases. Basically, the broadening comes from two contributions. One is the inhomogeneous broadening due to the laser spot size (about 1 μm) which corresponds to the superposition of Raman signals coming from regions of different Ge composition. Nakashima et al. estimated this contribution around 3 cm^{-1}, which is much smaller than the measured value. The second contribution comes from the broadening induced by the defects. As the lattice looses translation symmetry, the wave vector conservation breaks down and the photons can be scattered by phonons that have wavevectors far away from the zone center. This happens in mildly imperfect crystals with a small concentration of defects or in solid solutions, alloys, and amorphous solids [6]. For lightly disordered lattices, this manifests as an asymmetrical broadening of the Raman peak and, most of the time, the peak broadens at lower frequency [101]. Of course, as the material becomes amorphous, the Raman spectrum consists of a few broad bands which are correlated to the density of states. S. Nakashima et al. used this broadening to estimate qualitatively the crystalline quality of the $Si_{1-x}Ge_x$ alloy. It can also be used to characterize the damage induced by ion implantation in silicon on insulator (SOI) technology [111] or the crystallite size in micro and polycrystalline Si [112].

Up to now, we have seen that the strain relaxation of $Si_{1-x}Ge_x$ alloys involves the generation of misfit dislocations at the SiGe/Si substrate interface [113] and yields surface roughening. The surface has a well-known crosshatch pattern running along orthogonal (110) crystalline directions, which is characteristic of the low misfit regime (lattice mismatch lower than 2 %). It is, for instance, easily evidenced by atomic force microscopy observations [114–116] and, in SiGe layers, develops through the combination of two mechanisms which depend on the film thickness: (i) dislocations-induced surface steps due to shear displacements caused by dislocations glide and (ii) the growth of self-organized periodic ridges [117].

To study this crosshatch pattern, Chen et al. [118] performed Raman mapping on a 500 nm thick $Si_{0.75}Ge_{0.25}$ epitaxy grown on top of a 50 nm low temperature Si (LT-Si) buffer layer. The micro-Raman spectra were measured in the backscattering configuration, using the 514.5 nm line of an Ar laser. Such wavelength enables to probe simultaneously the SiGe epitaxial film, the LT-Si layer, and the Si substrate underneath. As a consequence, the spectra exhibited 4 peaks corresponding to the Si substrate, the Ge-Ge, Si-Ge, and Si-Si bond vibrations in the epitaxial SiGe layer. The frequency maps of these peaks evidenced clearly the crosshatch pattern along the ⟨110⟩ crystalline directions and they demonstrated that the frequency shifts were due to strain fluctuations coming directly from misfit dislocations nucleated at the SiGe/Si interface. They also demonstrated that the Ge composition of the SiGe layer was uniform and independent of these strain undulations.

Following this first series of experiments, Sawano et al. [119–122] showed that this crosshatch morphology is almost the same for the surface roughness evidenced by AFM and for the in-plane strain distribution observed by Raman imaging. This inhomogeneous strain field comes from the underlying misfit dislocation and exists all over the SiGe film even after surface planarization made by chemical mechanical polishing (CMP). It exists also in sSi layers grown on a thick SiGe virtual substrate. Strain fluctuation and surface roughness, particularly its wavelength, depend strongly on the buffer thickness: the thicker the buffer, the longer the crosshatch wavelength. Since TEM experiments indicate that the misfit dislocation network is confined in the SiGe graded layer, this means that the crosshatch wavelength depends only on the distance between the surface and the dislocation network. The crosshatch surface roughness appears again after homoepitaxial regrowth of SiGe on planarized SiGe buffer layers because the growth kinetics and especially the surface diffusion of adatoms is perturbed by the strain field fluctuation existing on the SiGe surface. Therefore, increasing the buffer thickness does definitely not solve this issue. Although, the wavelength increases and the uniformity is improved, long range strain fluctuation still exists on buffer layers as thick as 7 μm.

Strained Silicon Films on Insulator

Strained silicon films on SiGe epilayers on bulk Si substrates suffer from technological limitations coming (most of the time) from the graded buffer layer. In this case, Ge diffusion, strain relaxation, nucleation, and migration of misfit and threading dislocations are some of the issues that limit the flexibility of CMOS integration. To overcome these limitations, merging strained silicon with SOI technologies has been suggested [123]. This has many advantages. For partially depleted SOI device architectures, relaxed SiGe on insulator with a tensile strained silicon layer on top (SGOI wafers in Fig. 2.8) can boost circuit performance up to 30 % while tensile-strained silicon directly on insulator (sSOI wafers in Fig. 2.8) combines the performance and scalability of fully depleted ultra-thin SOI with the mobility enhancement of strained silicon.

Several technologies allow producing such hybrid substrates. The preferred one is the so-called Smart Cut™ technology patented by SOITEC for mass production of SOI wafers. It uses hydrogen implantation to form a buried (weakened) zone in a so-called donor wafer A. Then wafer A is bonded to an handle wafer B by wafer bonding. Prior to the bonding step, an insulating layer such as SiO_2 is formed at least on one of the two wafers. The splitting step that follows occurs inside the weakened

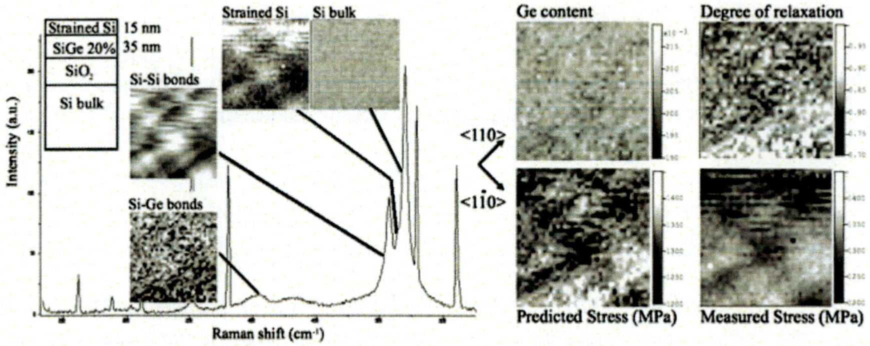

Fig. 2.10 Raman spectrum collected on a SGOI wafer. The wafer composition is sketched in the *upper left corner*. Starting from the high frequencies, we clearly observe the Raman modes of the bulk Si substrate, the strained Si cap, the Si-Si and Si-Ge bond vibrations in the relaxed $Si_{0.8}Ge_{0.2}$. The *sharp lines* correspond to Rayleigh scattered plasma lines of the Argon laser. They are used for frequency calibration. From a $20 \times 20\,\mu m^2$ Raman mapping, one determines from the frequency of each mode, the Ge content, the degree of relaxation of the SiGe layer, and the stress in the strained Si film. All the maps except the Ge content exhibit clearly the crosshatch pattern. This pattern comes from the network of the misfit dislocations that nucleate during the SiGe growth. It manifests itself by a roughening of the SiGe surface, and by a strain fluctuation in the SiGe lattice. This strain fluctuation is kept after the layer transfer process in the SGOI wafers. From the SiGe layer properties, one calculates the SiGe lattice constant and predicts the stress in the strained Si cap. In the *bottom maps*, we compare the predicted and measured stress. The stress difference is lower than 100 MPa for all points of the map. It confirms the perfect lattice matching between the SiGe and the sSi films without misfit dislocation nucleation [124]

zone and gives rise to the transfer of a thin layer from the donor substrate A onto the handle wafer B. A final treatment removes the rough surface left after splitting. This concept can be applied to almost any semiconductor such as: Si, SiC, GaN, GaAs, InP, SiGe, Ge, ... to obtain combinations of interest like: SOI, SOQ, GaNOI, SiCOI, GaAs/Si, InP/Si, etc....

To produce SGOI or sSOI wafers, the donor substrates are, respectively, a relaxed SiGe virtual substrates or a strained Si film on top of a relaxed SiGe virtual substrate. The final stacks are sketched in Fig. 2.8. To produce a SGOI wafer, only a thin part of the relaxed SiGe buffer is transferred, followed by a strained silicon film epitaxy. For sSOI wafers, a bilayer containing the strained Si and the relaxed SiGe films is transferred. The remaining SiGe film is removed by selective etching (for instance).

One can wonder how this layer transfer technology affects the strain field coming from the crosshatch. Indeed, since the misfit dislocation network is not transferred in the SGOI and sSOI wafers, this is not obvious. To answer this question, Raman mapping on donor wafers with a strained Si cap and on SGOI wafers have been done [124]. Typical results are shown in Fig. 2.10. The measurements were done using the 488 nm line of an Ar laser which enabled to probe the strained Si cap, the $Si_{0.8}Ge_{0.2}$ film, and the handle Si substrate simultaneously. The Raman maps were $20 \times 20\,\mu m^2$ large. Raman shifts variations of both film (SiGe and strained Si) exhibited the same crosshatch pattern as the one observed on the donor wafer. From these Raman shifts, we determined the stress in the strained Si film, the Ge content x, and the degree of relaxation of the SiGe layer through the following relationships [124]:

$$\omega_{\text{Si-Si}} = 521 - 68x - 815\epsilon_{\|} \tag{2.21}$$

$$\omega_{\text{Si-Ge}} = 399.5 + 14.2x - 575\epsilon_{\|} \tag{2.22}$$

while the degree of relaxation is given by:

$$\text{degree of relaxation} = 1 - \frac{\epsilon_{\|\text{exp}}}{\epsilon_{\|\text{th}}}. \tag{2.23}$$

In these expressions, $\epsilon_{\|\text{exp}}$ is the residual strain measured in the SiGe layer and $\epsilon_{\|\text{th}}$ is the theoretical strain of a fully strained $\text{Si}_{1-x}\text{Ge}_x$ layer grown on a Si substrate.

It should be noticed that a controversy still exists concerning the exact values of the numerical coefficients [125–127] and that the Eq. 2.22 is only valid for Ge content lower than 30 %. Moreover, the Si-Ge frequency does not vary linearly versus Ge content [128]. The Ge content x and the degree of relaxation of the SiGe maps are shown in Fig. 2.10. These results confirm that the Raman shift variations come only from strain modulation in the SiGe layer and not from a composition fluctuation. This strain network remains in the SGOI wafer and also in the sSOI wafers (see Fig. 2.11c) even in the absence of the misfit dislocation network. They also confirm the influence of the strain modulation induced by the dislocations on the growth kinetics of the SiGe and strained Si films.

From the SiGe composition and its degree of relaxation, the SiGe in-plane lattice parameter was deduced. Assuming that the sSi film is grown with perfect lattice matching, its in-plane lattice parameter should be the same as the SiGe one. Therefore, one can deduce the induced strain/stress of the strained Si cap. This predicted stress is compared to the experimental one in Fig. 2.10. Notice the very good agreement with a stress difference lower than 100 MPa. This confirms the perfect lattice matching between the Si cap and the strain relaxed SiGe buffer. It was also observed that the bulk substrate was subjected to a small in-plane tensile stress (around 100 MPa) due to the high-temperature annealing steps and the difference between the thermal expansion coefficients of the buried oxide and silicon [102–105].

UV Raman Approach

Until now, we have seen that the strain field induced by the network misfit dislocations affects the local growth rate. It causes also the crosshatch surface roughening which remains in SGOI and sSOI wafers after layer transfer. As this strain field might also impact the process parameters used to manufacture the devices (and therefore their performances [129, 130]), it is crucial to develop characterization tools that can monitor the strain field continuously before and during the device fabrication.

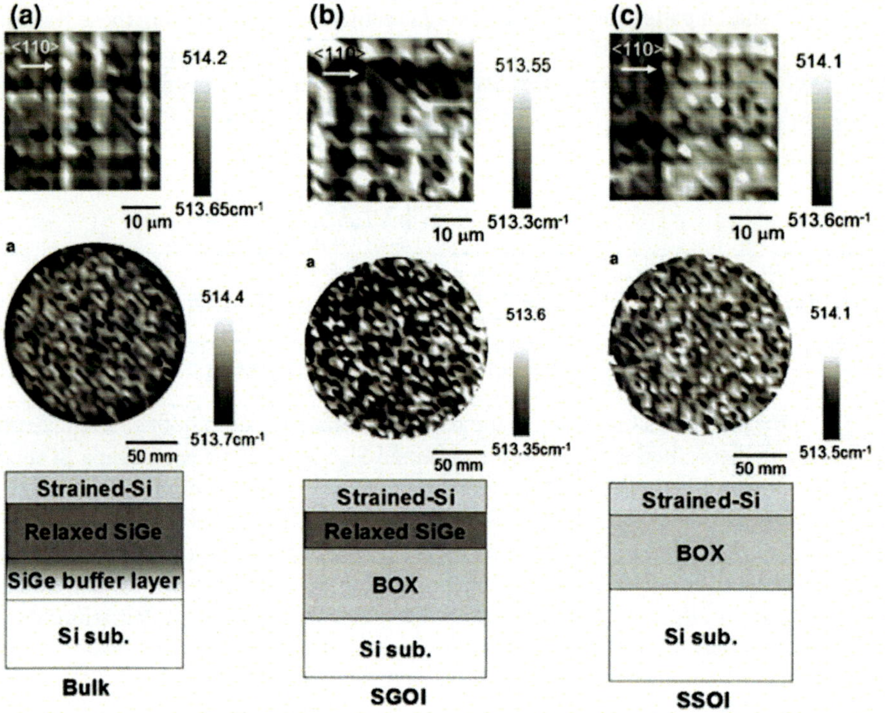

Fig. 2.11 $40 \times 40\,\mu m^2$ and full wafer map of the Raman frequency of the strained Si layer measured on 200 mm wafers of **a** strained Si on a relaxed SiGe virtual substrate; **b** SGOI; **c** sSOI. We clearly observe the crosshatch pattern on the $40 \times 40\,\mu m^2$ maps. It shows that the strain network induced by the misfit dislocations network is kept after the layer transfer process step. The Raman mode of the strained silicon film exhibits a $0.6\,cm^{-1}$ frequency fluctuations. The same fluctuations are observed for the full wafer map of the 200 mm wafers. We can deduce from these measurements that the strain is almost uniform across the surface of the wafer. It is indeed impossible to discriminate on the full wafer map if the strain fluctuations come only from the crosshatch strain modulation or from long range wafer inhomogeneities. Reprinted from [84]. Copyright 2007, with permission from Elsevier

To answer this need Ogura et al. [131] built a specific UV-Raman setup. Basically, the 363.8 nm line of an Ar laser is focused by a 90x objective lens giving a spatial resolution of $0.5\,\mu m$. A $300 \times 400\,mm^2$ translation stage combined to an autofocus system enabled to map 12 in wafers. A high resolution spectrograph (2 m focal length) is coupled to a charge couple device (CCD) detector and a dual focusing mirror was used to perform real-time calibration of the laser frequency. The dual concave mirror enabled to measure simultaneously (on the same CCD detector) the Raman signal coming from the wafer and the Rayleigh scattering of the incident laser which was used as an internal frequency reference. Finally, to ensure a good reproducibility and repeatability, no optic (including mechanical components in the spectrograph) was moved during the measurement. The spectral resolution obtained was $0.1\,cm^{-1}$.

An important point to notice is that the 363.8 nm line of the Ar laser is (by far) the best one to perform Raman imaging of a thin silicon film. The penetration depth is less than 10 nm and the wavelength corresponds to a very efficient resonant Raman scattering process [132]. The Raman cross-sections is almost 100 times higher than for the visible lines of the Ar laser and 10 times higher than for the 325 nm line of an HeCd laser. This intense signal enables to reduce the acquisition time to few seconds and therefore to perform fast Raman mapping of $40 \times 40\,\mu m^2$ and of the whole wafer surface. Figure 2.11 displays several examples of mappings collected on a strained Si on a relaxed SiGe virtual substrate, a SGOI, and a sSOI wafer. We clearly observe the crosshatch pattern on the $40 \times 40\,\mu m^2$ maps. The Raman mode of the strained silicon film exhibits a $0.6\,cm^{-1}$ frequency fluctuations. The same fluctuations are observed for the full wafer map of the 200 mm wafers. This image was obtained by collecting a Raman spectrum every 5 mm. More than 1,000 spectra were acquired. We can deduce from these measurements that the strain is almost uniform across the surface of the wafer. It is indeed impossible to discriminate on the full wafer map if the strain fluctuations come only from the crosshatch strain modulation or from long range wafer inhomogeneities. To get a better image of the strain homogeneity across the wafer, the crosshatch strain fluctuations must be averaged either by performing $40 \times 40\,\mu m^2$ maps each 5 mm, or by illuminating the same $40 \times 40\,\mu m^2$ surface with galvano mirrors vibrating at high frequency. They also studied the influence of rapid thermal annealing on these wafers with the same system [84]. It has been shown that the sSOI wafers properties remains almost unchanged after high temperature annealing [84, 123]. This is not the case for SGOI and SiGe virtual substrate since Ge diffusion takes place. Finally, recent results [133] report that relaxed SiGe layers can be realized by heavy ion implantation without the formation of crosshatch strain fluctuations.

2.3 Stress Monitoring in Production Lines

Mechanical stress has always been a major concern for integrated circuits (IC) technology because of the detrimental effect on devices performance and integrity. Indeed, many problems of defective ICs can be ascribed to stress-induced defects generated during device manufacturing: nucleation and propagation of dislocations, voids, and cracks around local oxidation structures (LOCOS) or shallow trench isolation (STI) [29, 30, 134–136], open or short-circuits failures in interconnection metal lines due to crack, voids and hillock formation. These reliability issues became even more challenging with the constant trend in device downscaling. As Raman imaging is a nondestructive and a relatively fast technique, it has been used for routine analyses and to help designers at the early stage of device development. As already mentioned in the preceding section, the step further was to incorporate strain engineering in the ITRS roadmap [32]. In the "local strain" approach, it consists in introducing and controlling the strain in the channel of a field effect transistor (FET). This can be done by almost all elements constituting and surrounding the FETs: SiGe source drain contacts for a uniaxially compressed channel, trench isolations, silicides, silicon nitride spacers which can induce both tensile or compressive stress in the channel [85, 137]. Since strain and stress are now considered as key parameters to improve the device performance, the most popular application of Raman imaging in Microelectronics has become stress measurement. In this section, we focus on two examples: stress measurements around LOCOS and STI structures.

2.3.1 Stress Mapping of Local Oxidation Structures

Local oxidation of silicon (LOCOS) is the most popular technique to achieve lateral isolation in very large-scale integration (VLSI) devices. It consists in defining thick isolation regions between small active device parts by growing silicon dioxide in such isolation regions, also called field regions. LOCOS in field regions is achieved by masking the regions where no oxide must be grown with a silicon nitride (Si_3N_4) layer. Since nitride oxidizes at a much lower rate than silicon, a thick oxide insulating layer is grown in the field regions, while the areas where the active devices are to be fabricated, are not oxidized. The main reasons for the success of LOCOS in semiconductor technology are its simplicity, effectiveness, and compatibility with all other processing steps.

The process sequence involved in LOCOS isolation is as follows. After growing a thin SiO_2 layer called pad oxide, a Si_3N_4 film is deposited and patterned. This defines the oxidation mask. The wafers are next put in an oxidation ambient to grow a rather thick field oxide (FOX) in the regions where the nitride layer was removed. Since oxidation of silicon occurs also laterally, the growing oxide penetrates under the edges of the nitride mask, lifts them up and gives rise to the formation of a specific geometrical feature called "bird's peak". At the end of the process, the nitride film is removed and the wafers planarized.

The apparent simplicity of this process hides several technological challenges, in which mechanical stress plays a significant role. For instance, stress influences the diffusion of oxidant species underneath the nitride edges. It can also induce the nucleation of dislocations at the bird's peak tip [136]. Altogether, this explains why LOCOS structures have been extensively studied by Raman spectroscopy [134, 135, 138–141]. An illustration is given in Fig. 2.12 from [138, 139]. It focuses on the stress field induced by a poly-buffered LOCOS (PBLOCOS) process. PBLOCOS is an alternative LOCOS structure, in which a polycrystalline silicon (poly-Si) film is added between the pad oxide and the silicon nitride film. This Si_3N_4/poly-Si/SiO_2 sandwich reduces the field oxide encroachment in the active region. The poly-Si film limits the bird's peak formation, and absorbs the stress from the nitride.

To monitor the stress-induced Raman shift of the Si phonon during the manufacturing of the PBLOCOS, de Wolf et al. measured 1D profile across the Si_3N_4 stripe before the field oxidation [138, 139]. They also performed a Raman mapping of $4 \times 4\,\mu m^2$ active area [134]. A spectrum was recorded each $0.2\,\mu m$ using the 457.9 nm line of an Ar laser and a DILOR XY-800 spectrometer coupled to a micro-

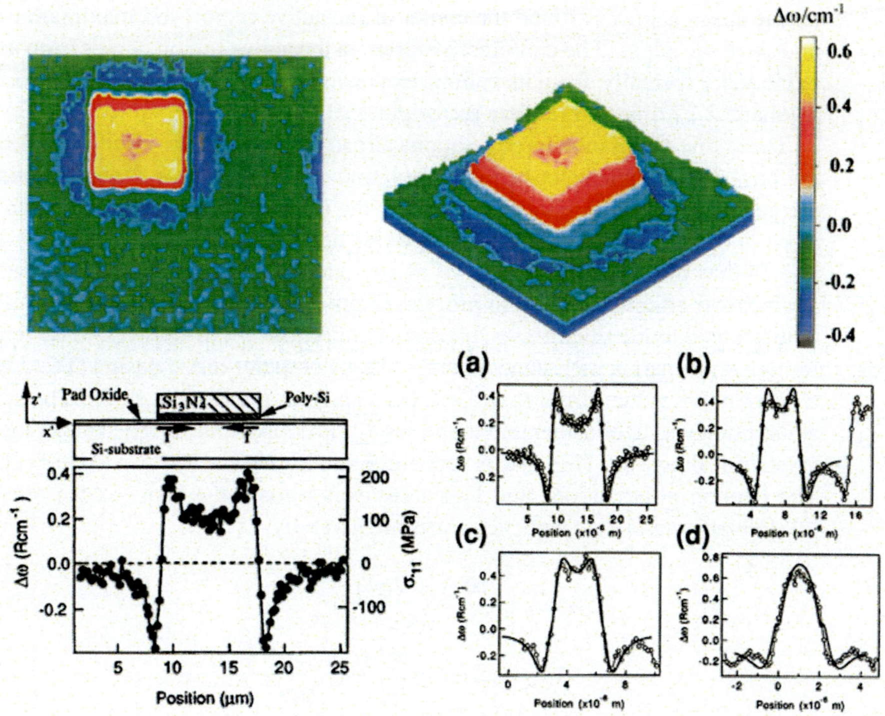

Fig. 2.12 2D map of the stress-induced Raman shift of the Si mode measured around a $4 \times 4\,\mu m^2$ active area surrounded by a PBLOCOS isolation structure. The 2D Raman map was measured across a $11.5 \times 13\,\mu m^2$ area with a spectrum recorded each $0.2\,\mu m$. The Si experiences a tensile stress at the sidewall of the oxide/silicon interface. The edges of the active regions are subjected to a maximum compressive stress that relaxes in the middle of the active region. Reprinted from [134]. Copyright 1999, with permission from Wiley-VCH. 1D profile of the stress-induced Raman shift across the Si_3N_4 stripe before the field oxidation for different widths. In the *left hand side* profile, the shift is converted in stress assuming an uniaxial stress along x' (110). In the *right hand side* profiles (marked *a, b, c, d*) the *solid line* corresponds to the theoretical shift of the Raman Si mode derived from an analytical model. It gives an excellent agreement with the experimental data plotted with *open circles*. Reprinted with permission from [138, 139]. Copyright 1996, American Institute of Physics

scope and a XY translation stage. To perform such stress measurements, special care must be taken to avoid artifacts. In Fig. 2.12, plasma lines were used to calibrate the spectrometer and the position of the Si Raman peak was evaluated by fitting it with a single Lorentzian function. The 2D Raman map reveals clearly that the shift starts from zero far from the active region and becomes negative ($\sim -0.3\,cm^{-1}$) in the vicinity of the active region. This tensile stress comes from the sidewall of the oxide/silicon interface and from the bird's peak. In the active region, the stress becomes compressive. It reaches a maximum value at the edge (especially in the corner) and relaxes slightly in the middle of the active region. On the contrary, the tensile stress is lower around the corner of the active region and maximum near the middle of the edges. The compressive stress in the active region comes from the field oxide and, especially, from its volume expansion during Si oxidation. SiO_2 occupies a volume 2.27 times larger than the original silicon one.

These first qualitative interpretations already help to get an intuitive understanding how the stress develops in such isolation structure. Stress depends directly on the geometry of the device. For instance, the (compressive) stress increases as the slope of the sidewall between the field oxide and the silicon is steeper [134]. Extensive studies has been carried out to investigate how stress is affected by all these geometrical effects, and all the process parameters (temperature, nitride thickness, field and pad oxide thicknesses, oxidation conditions ...). Moreover, to correlate the Raman experiments with standard technology computer-aided design (TCAD) tools, a quantitative value of the stress was derived from the Raman shifts. The standard assumption was to consider that the stress σ_{xx} is uniaxial and aligned along the $\langle 100\rangle$ crystalline direction. This case is very similar to the biaxial one treated in Sect. 2.1.4. The phonon polarizations and Raman tensors remain the same as the ones for a relaxed lattice and the frequency shifts are given by:

$$\begin{aligned}\Delta\omega_x &= \frac{pS_{11} + 2qS_{12}}{2\omega_0}\sigma \\ \Delta\omega_y = \Delta\omega_z &= \frac{qS_{11} + (p+q)S_{12}}{2\omega_0}\sigma \end{aligned} \tag{2.24}$$

For pure backscattering from a 001 surface, only the ω_z mode is probed. The relationship between the Raman shift and the stress is then:

$$\Delta\omega_z(\mathrm{cm}^{-1}) = -2.3\sigma_{xx}(\mathrm{GPa}) = -300\varepsilon_{xx} \tag{2.25}$$

This assumption cannot be used to evaluate the stress in the 1D profile shown in the lower left corner of Fig. 2.12. This profile was measured across a Si_3N_4 stripe before the field oxidation. The structure and composition of the stripe is sketched above the Raman profile. Since this stripe is aligned along the $\langle -110\rangle$ direction, the preceding assumption is not valid. However, if we assume an uniaxial stress σ aligned along the $\langle 110\rangle$ crystalline direction, the strain tensor becomes:

$$\begin{aligned}\varepsilon_{xx} = \varepsilon_{yy} &= \frac{S_{11} + S_{12}}{2}\sigma \\ \varepsilon_{zz} &= S_{12}\sigma \\ \varepsilon_{xy} &= \frac{S_{44}}{4}\sigma \\ \varepsilon_{xz} = \varepsilon_{yz} &= 0 \end{aligned} \tag{2.26}$$

Solving the secular equation gives then three different Raman shifts in which the difference with the $\langle 100\rangle$ uniaxial stress comes only from the shear component ε_{xy}.

$$\Delta\omega_1 = \frac{(p+q)S_{11} + (p+3q)S_{12} + rS_{44}}{4\omega_0}\sigma$$
$$\Delta\omega_2 = \frac{(p+q)S_{11} + (p+3q)S_{12} - rS_{44}}{4\omega_0}\sigma$$
$$\Delta\omega_3 = \frac{qS_{11} + (p+q)S_{12}}{2\omega_0}\sigma \quad (2.27)$$

associated with polarization vectors:

$$v_1 = \frac{1}{\sqrt{2}}(1\ 1\ 0) \quad v_2 = \frac{1}{\sqrt{2}}(-1\ 1\ 0) \quad v_3 = (0\ 0\ 1) \quad (2.28)$$

The Raman tensors become:

$$\boldsymbol{R}_1 = \frac{1}{\sqrt{2}}\begin{pmatrix} 0 & 0 & d \\ 0 & 0 & d \\ d & d & 0 \end{pmatrix} \quad \boldsymbol{R}_2 = \frac{1}{\sqrt{2}}\begin{pmatrix} 0 & 0 & d \\ 0 & 0 & -d \\ d & -d & 0 \end{pmatrix} \quad \boldsymbol{R}_3 = \begin{pmatrix} 0 & d & 0 \\ d & 0 & 0 \\ 0 & 0 & 0 \end{pmatrix} \quad (2.29)$$

With standard backscattering from the 001 surface, only the third mode can be observed which still corresponds to the phonon polarized along the z axis. Moreover the strain induced shift $\Delta\omega_z$ remains the same. This latter remark confirms the difficulty in determining the strain tensor from Raman experiments. It is indeed impossible to discriminate if the stress is uniaxial, biaxial and to determine its orientation with respect to the crystalline directions. Using Eq. 2.25 gives only a semiquantitative estimation of the stress in the sample. To get a more accurate determination of the nature and the value of the stress, it is necessary to use a more tedious method. The strain and stress tensors must be determined in all points of the sample probed by Raman imaging. They can be computed by analytical models [136] or by finite element simulations [142]. The next step is then to determine how this stress distribution influences the Raman scattering (frequency shifts, polarization vectors and Raman tensors). The Raman tensors indicate which Raman modes are allowed and give their intensity. The final step is to average the Raman scattering of the probed volume that is given by the penetration depth and the spot size of the laser. "Theoretical" Raman maps can then be determined and compared to the experimental ones.

This procedure was applied to fit the 1D Raman profile recorded across Si_3N_4 stripes of different widths [138, 139]. To analyze the data, the edge force model of Hu [143] was used. This model describes the effect of the Si_3N_4 film on the substrate by a force $\boldsymbol{f}$ parallel to the surface and starting from the edges as shown in Fig. 2.12. It confirms that the stress is uniaxial only in the center of the stripe and that shear stress and stress in the $\langle 001 \rangle$ direction can not be neglected close to the edges. The Raman shifts calculated with this analytical model are plotted by a straight line and compared to the experimental ones represented in open circles. Despite the simplicity of the model, the results are remarkably good confirming that the model gives a stress distribution close to the real one. There are still small discrepancies. The model predicts larger Raman shifts than observed at the edges. It might come from a small defocusing of the laser spot or to an overestimation of the stress given by the model. As the stripe width decreases, the relaxation of the compressive stress in the center of the stripe decreases and vanishes for the 2.1 μm stripe. This is only due to the limitation coming from the spatial resolution. Since the 457.9 nm line of the Ar laser is focused to a spot of around 1 μm and penetrates up to 300 nm deep into silicon, the stress measured is a weighted average over the penetration depth and the laser spot size. Moreover, the analytical model fails for this narrow stripe and gives an unexpected shoulder at the edges. This small discrepancy emphasizes the pitfall to univocally correlate the stress tensor to the experimental strain-induced Raman shifts. Some answers will be given in the next section to fix partially this issue.

2.3.2 Stress Mapping of Shallow Trench Isolation

Scaling down LOCOS to smaller dimensions is mainly limited by the lateral encroachment of the FOX under the nitride mask that reduces the active area dimensions as compared to the dimensions of the nitride mask. Moreover, if the bird's peak length is larger than half the nitride length, the bird's peak originating from both sides of the nitride will overlap and a thick oxide will be grown underneath the nitride. This phenomenon is known as bird's peak punch through or nitride lifting. If the pad oxide thickness increases significantly owing to this phenomenon, it will be impossible to remove it during the removal of the oxidation mask, and the active region will remain covered with an oxide layer, thus preventing the fabrication of devices. This is why LOCOS isolation techniques are no longer adequate for sub-0.25 μm integrated circuit processing [32] and has to be replaced by STI. The process sequence is very similar. The active areas are defined in the same way with the combination of a pad oxide and a Si_3N_4 film on top. The isolation trenches are etched by reactive ion etching (RIE). Then a thermal oxide is grown at high temperature on the trench walls. The process conditions of this step must be optimized to getter the defects induced by RIE and to smoothen the edges of the trenches. Finally, several oxidation and oxide deposition processes are combined to fill the trenches and minimize the stress induced by such high temperature steps.

The most important benefit of STI is that it allows an independent control of the isolation depth with respect to the isolation width. The width is primarily defined by lithography, while the isolation depth is solely controlled by the trench etch. STI reduces also the final topography induced by the LOCOS process. Because of the nature of silicon oxidation, about 50 % or greater of the oxide thickness is above the silicon surface. This topography is especially problematic for optical lithography with limited depth of focus. Despite these advantages, STI is one of the most critical technology modules from the point of view of defect generation during the process. The defect generation is directly related to the mechanical stress surrounding the trench. This is why the effect of processing conditions and trench geometry on the stress magnitude has been extensively studied in the literature [134, 135, 144–151].

Since the switch from LOCOS to STI is driven by the shrinking of the pattern size, the weighted average over the penetration depth and the laser spot size becomes even more detrimental to study the stress induced by such small structures. The first answer is then to perform Raman imaging with a smaller wavelength to decrease both the penetration depth and the laser spot size. In this way, Dombrowski et al. [144, 146] studied the evolution of the stress in STI stripes of different widths during their fabrication. They evidenced the benefits of using UV laser (363.8 nm) compared to visible laser (457.9 nm). The main advantage was to decrease the penetration depth from 300 nm down to 10 nm. This smaller penetration depth enabled to detect a very large stress that was confined in the vicinity of the surface. The spatial resolution was also improved, the spot size decreasing from 1 down to 0.7 μm. With such experimental improvements, Dombrowski et al. identified a critical process step which increases strongly the compressive stress at the edges. After the sidewall oxidation, if a wet oxidation process is used instead of a dry oxidation, it results in an increased sidewall oxide thickness and in the formation of a large bird's peak under the nitride. This observation was confirmed by scanning electron microscopy (SEM) experiments. Although the stress was qualitatively estimated from the usual uniaxial assumption, this example is an excellent demonstration that Raman imaging is a powerful tool for process monitoring. It can identify easily and nondestructively a potentially dangerous process step. Moreover, it avoided to perform tedious TEM experiments to identify this issue.

According to Poborchii et al. [150, 152, 153], using oil-immersion objective lenses with the 363.8 nm wavelength improves the spatial resolution down to 150 nm. Thanks to this improvement, they performed Raman mapping of the STI structures sketched in Fig. 2.13. These structures were composed of Si stripes surrounded by SiO_2 trenches. The Si stripes were oriented along the ⟨1-10⟩ axis and were 250 nm wide. The oxide trenches were 300 nm deep and 1 μm wide. For this study, a specific system of axes was defined. It is represented in Fig. 2.13, where a, b and c are, respectively, parallel to the ⟨110⟩, ⟨001⟩ and ⟨1-10⟩ crystalline directions. These structures were measured in the backscattering geometry with two different laser incidences. The first was realized in the standard configuration, the laser being aligned along the *b* axis [153]. The second was performed on the cross-section of the STI (on the cleaved edge of the sample) [150]. The laser was in this case aligned along the *c* axis. Combining different polarizations of the incident and scattered light enables to measure five different Raman modes: aa, ca, cc from the standard backscattering geometry; aa, and ab from the backscattering on the cross-section. The first letter stands for the direction of the incident polarization and the second for the direction of the scattered one. Notice that the ca Raman mode is forbidden by the selection rules for the three stress examples described before. This means that the shear strains ε_{xz} and ε_{yz} must be different from zero to explain the observation of this mode. Moreover, it is very intense at the edges of the stripes and vanishes outside. This was confirmed by finite element stress simulations [149, 154] which show that, in addition to the ⟨110⟩ stress, contributions of the ⟨001⟩ stress and shear stress become important in the edge area.

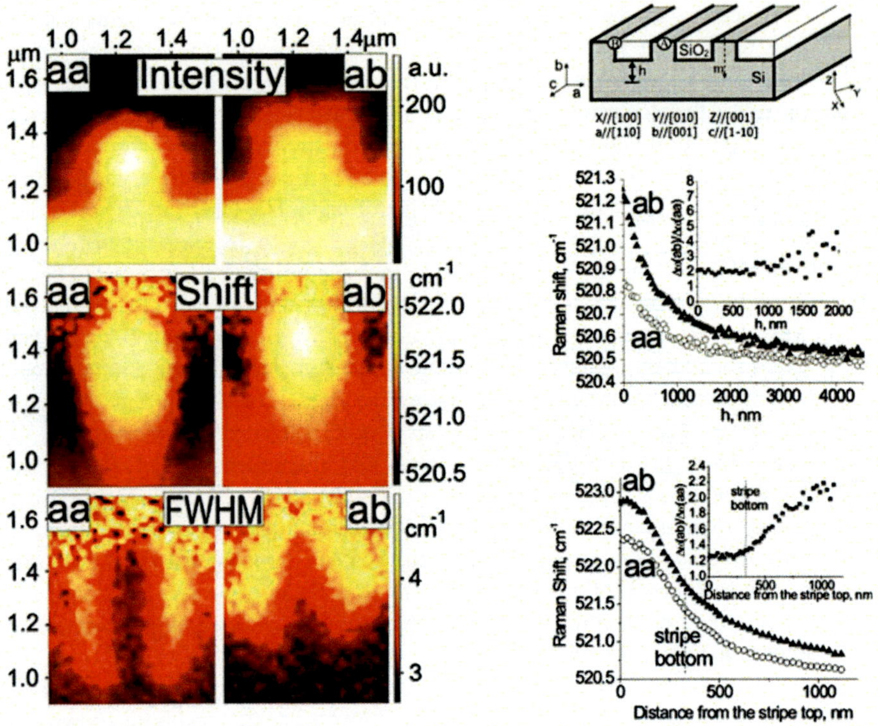

Fig. 2.13 Raman maps of the cross-section of STI structures sketched in the *top right corner*. Raman spectra were recorded with an oil-immersion objective each 20 nm and fitted by a simple Lorentz curve. Intensities, Raman shifts, and broadening were evaluated for two polarization configurations aa and ab. The first letter stands for the direction of the incident polarization and the second for the direction of the scattered one. The 1D profiles shown in the *right-hand side* were extracted from these maps along *h* and *m*. Reprinted with permission from [150]. Copyright 2007, American Institute of Physics

Raman maps in Fig. 2.13 were collected on the cross-section of the STI structures with aa and ab polarization configurations. Spectra were recorded each 20 nm and fitted with a single Lorentz curve. The shape of the stripe given by the intensity maps depends on the polarization configuration. It comes from a stronger Raman scattering at the stripe edges and corners in the ab configuration than in the aa configuration. For both polarizations, a compressive stress is observed in the stripe area, while the bottom corners and parts of the sidewalls experience a tensile stress. Finally, the broadening is the highest at the stripe edges, and on the sidewalls since the stress varies rapidly in these area as it is usually shown by FEM simulations. Poborchii et al. interested themselves to the ratio between the Raman shifts of the aa and ab modes because it gives information about the stress tensor shape and can discriminate between uniaxial stresses aligned along the ⟨001⟩ or the ⟨110⟩ axes. The stress-induced frequency shifts are indeed given by the following relationships depending on the uniaxial stress σ_{110} or σ_{001}:

$$\begin{aligned}
\Delta\omega_{ab} = \Delta\omega_1 &= \frac{(p+q)S_{11} + (p+3q)S_{12} + rS_{44}}{4\omega_0}\sigma_{110} = -2.88\sigma_{110} \\
\Delta\omega_{aa} = \Delta\omega_3 &= \frac{qS_{11} + (p+q)S_{12}}{2\omega_0}\sigma_{110} = -2.3\sigma_{110} \\
\Delta\omega_{ab} = \Delta\omega_{x,y} &= \frac{qS_{11} + (p+q)S_{12}}{2\omega_0}\sigma_{001} = -2.3\sigma_{001} \\
\Delta\omega_{aa} = \Delta\omega_z &= \frac{pS_{11} + 2qS_{12}}{2\omega_0}\sigma_{001} = -1.13\sigma_{001}
\end{aligned} \tag{2.30}$$

It means that the ratio $\Delta\omega_{ab}/\Delta\omega_{aa}$ is equal to 1.25 for $\langle 110 \rangle$ uniaxial stress and to 2 for $\langle 001 \rangle$ uniaxial stress. From the FEM simulations, the stress should be $\langle 110 \rangle$ uniaxial in the middle of the Si stripe just below the surface, and it should be $\langle 001 \rangle$ uniaxial in the Si substrate beneath the stripes and the trenches. Two profiles were extracted from these maps. The first is along h. It starts from the middle of the trench oxide and goes down in the Si substrate. The ratio $\Delta\omega_{ab}/\Delta\omega_{aa}$ is constant and equal to 2. This confirms experimentally that the stress is $\langle 001 \rangle$ uniaxial. It is compressive and equal to −310 MPa near the trench bottom and relaxes completely 3 μm away in the Si substrate. At 1 μm it is only equal to −90 MPa. The second profile was extracted along m. It starts from the surface and the middle of the Si stripe. The ratio $\Delta\omega_{ab}/\Delta\omega_{aa}$ is increasing from 1.25 up to 2. It confirms also that the stress is $\langle 110 \rangle$ uniaxial in the stripe and gradually transforms into the $\langle 001 \rangle$ uniaxial stress in the substrate under the stripe bottom. The compressive stress in the middle of the Si stripe was estimated to −830 MPa.

Combining polarized Raman maps from the surface and the cross-section of these STI structures, gives an excellent estimation of the stress fields in such structures that is in good agreement with the FEM simulations [149, 154]. A light $\langle 001 \rangle$ compression occurs in the substrate just under the oxide trenches. Strong $\langle 110 \rangle$ compressive stress in the middle of the Si stripe is coupled to weaker tensile stress on the sidewalls and strong shear stress around the top and bottom corners. Finally, a strong Raman intensity enhancement was observed if the incident polarization is aligned along the c axis (i.e. the stripe length) [151, 152]. This enhancement results from a high confinement of the electromagnetic field at the top corners of the Si stripes. Finite difference time domain (FDTD) simulation reveals that the light is localized in a very small area (<20 nm) around the edge [152]. This light confinement increases with the real part of the dielectric constant of silicon. This effect is therefore quite strong for near UV experiments, while it is very weak for visible light. Enhancement factor of 100 times has been experimentally demonstrated by measuring Si stripes 20 nm wide and 10 nm high [152]. These recent and very promising results push away the main limitation of Raman imaging in microelectronics: the spatial resolution that prevents measuring the stress in the actual FET's channel. This edge-enhanced Raman scattering (EERS) is very similar to the elegant solution to improve spatial resolution detailed in the following chapters: tip enhanced Raman spectroscopy (TERS) which

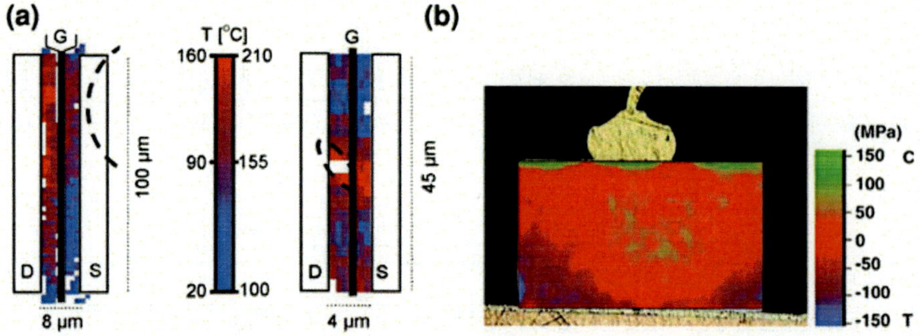

Fig. 2.14 Temperature maps of two operating AlGaN/GaN field effect transistors. The local heating comes from the presence of a micropipe underneath the source contact of device (**a**) and of a microcrack in the channel of device (**b**). The defects location is marked by the *dashed line*. Reprinted with permission from [160]. Copyright 2007, American Institute of Physics. Stress map of a silicon chip bonded on a Cu substrate. Reprinted from [159]. Copyright 2001, with permission from Elsevier

takes advantage of the huge electric field confinement induced by a sharp metallic tip. Another alternative technique combining Raman spectroscopy with near field techniques: scanning near-field optical Raman microscopy (SNORM) still suffers from a low level of signal and does not significantly improve the spatial resolution compared to oil-immersion objective [155]. To complete this review, stress imaging can also be performed by several interesting techniques such as X-ray micro-diffraction [156], convergent beam electron diffraction [145, 147], and electron holography [157, 158].

2.4 Final Device Inspection

Raman imaging contributes to the final device inspection since, both, the performances and reliability are influenced by many unknown parameters during the whole cycle of fabrication. Among them, the presence of temperature or stress fields may cause soft or hard failures during operation. For instance, thermal stress is still a critical reliability issue in ICs packaging and in microelectromechanical systems (MEMS). From μRS the nice point is that it can be easily monitored because the required spatial resolution is on the micron scale or larger.

For instance, Raman imaging has been used to study the stress induced by the anodic bonding of a silicon membrane on a glass substrate, by the thinning, dicing, and packaging of IC [159]. Figure 2.14 shows an illustration of the stress induced by bonding a silicon chip on a Cu substrate. This image was collected on the cross-section and reveals, at the same time, the stress introduced by the solder bump on top of the silicon chip.

Raman imaging can also be used to monitor the temperature in operating devices [26, 160–162]. In this case, several methods are available to calculate the lattice temperature from Raman spectra. The first one consists in measuring the intensity ratio between the Stokes and anti-Stokes processes as it depends directly on the Bose–Einstein phonon occupation number but, despite its apparent simplicity, this method is time-consuming and prone to systematic errors. For details, see Compaan and Trodhal [163] who reviewed the corrections needed to extract properly the temperature from this ratio in silicon. The other methods rely on the influence of the temperature on the phonons frequencies and lifetimes. Usually, a Raman mode shifts to lower frequencies and broadens with increasing temperature. The broadening arises from the decrease of the phonon lifetime, due to its decay into phonon with lower energies.The downshift of the Raman frequency is mainly due to the effects of thermal expansion and anharmonic coupling to other phonons. The temperature can then be estimated either from the peak shift or from the broadening. However, the peak shift depends both on stress and temperature variations, while the broadening depends only on temperature. This renders possible to measure simultaneously the temperature and stress [162].

In Fig. 2.14, we display two temperature maps collected on AlGaN/GaN field effect transistors. They evidence local self-heating that occurs in the active region of FETs [160]. It is due to a micropipe located underneath the source contact for device (a) and a microcrack in the channel for device (b). These defects are indicated as dashed line in the figure. The presence of such defects reduces locally the thermal conductivity and, therefore, the heat dissipation. Although defects might also influence the current flow in the defect vicinity, these devices had normal DC electrical characteristics and Raman imaging was, in this particular case, of great interest since it could detect such reliability issue.

2.5 Conclusion

In this chapter, we have shown that Raman imaging applications cover the whole cycle of fabrication of modern electronic and optoelectronic devices. It enables to check the crystalline quality and the presence of defects during the semiconductor growth, to measure the alloy composition and the doping level. It can also contribute to the final device inspection through temperature mapping of the operating devices. However, its main application is to monitor and control the stress introduced by the different steps in the process flow chart.

References

1. C.V. Raman, A new class of spectra due to secondary radiation. Indian J. Phys. **2**, 387–400 (1928)
2. C.V. Raman, A change of wave-length in light scattering. Nature **121**, 619 (1928)
3. P.Y. Yu, M. Cardona, *Fundamentals of Semiconductors: Physics and Materials Properties*, 3rd edn. (Springer, Berlin, 2005)

4. W.H. Weber, R. Merlin (eds.), *Raman Scattering in Materials Science*, Springer Series in Materials Science, vol. 42 (Springer, Berlin, 2000)
5. S. Nakashima, Raman imaging of semiconductor materials: characterization of static and dynamic properties. J. Phys. Condens. Matter **16**(2), S25–S37 (2004)
6. G. Gouadec, P. Colomban, Raman spectroscopy of nanomaterials: how spectra relate to disorder, particle size and mechanical properties. Prog. Cryst. Growth Charact. Mater. **53**(1), 1–56 (2007)
7. R. Zheng, M. Matsuura, Electron-phonon interaction in mixed crystals. Phys. Rev. B **59**(23), 15422–15429 (1999)
8. P.N. Keating, Effect of invariance requirements on the elastic strain energy of crystals with application to the diamond structure. Phys. Rev. **145**, 637–645 (1966)
9. H. Rücker, M. Methfessel, Anharmonic Keating model for group-IV semiconductors with application to the lattice dynamics in alloys of Si, Ge, and C. Phys. Rev. B **52**, 11059–11072 (1995)
10. J. Menendez, P. Gopalan, G.S. Spencer, N. Cave, J.W. Strane, Raman-spectroscopy study of microscopic strain in epitaxial $Si_{1-x-y}Ge_xC_y$ alloys. Appl. Phys. Lett. **66**(10), 1160–1162 (1995)
11. H. Okumura, E. Sakuma, J.H. Lee, H. Mukaida, S. Misawa, K. Endo, S. Yoshida, Raman scattering of SiC: Application to the identification of heteroepitaxy of SiC polytypes. J. Appl. Phys. **61**(3), 1134–1136 (1987)
12. J. Camassel, S. Contreras, J.L. Robert, SiC materials: a semiconductor family for the next century. Compte. Rendu. de l'Acad. des Sci. Paris **1**(Série IV), p5–p21 (2000)
13. K. Zekentes, K. Rogdakis, SiC nanowires: material and devices. J. Phys. D Appl. Phys. **44**(13), 133001 (2011)
14. P.J. Wellmann, T. Straubinger, U. Kunecke, R. Muller, S.A. Sakwe, M. Pons, A. Thuaire, A. Crisci, M. Mermoux, L. Auvray, J. Camassel, Optical mapping of aluminum doped p-type SiC wafers. Phys. Status Solidi A **202**(4), 598–601 (2005)
15. J.C. Burton, L. Sun, M. Pophristic, S.J. Lukacs, F.H. Long, Z.C. Feng, I.T. Ferguson, Spatial characterization of doped SiC wafers by Raman spectroscopy. J. Appl. Phys. **84**(11), 6268–6273 (1998)
16. H. Harima, S.-I. Nakashima, T. Uemura, Raman scattering from anisotropic LO-phonon-plasmon-coupled mode in n-type 4H- and 6H-SiC. J. Appl. Phys. **78**(3), 1996–2005 (1995)
17. P. Perlin, J. Camassel, W. Knap, T. Taliercio, J.C. Chervin, T. Suski, I. Grzegory, S. Porowski, Investigation of longitudinal-optical phonon-plasmon coupled modes in highly conducting bulk GaN. Appl. Phys. Lett. **67**(17), 2524–2526 (1995)
18. S. Ganesan, A.A. Maradudin, J. Oitmaa, A lattice theory of morphic effects in crystals of the diamond structure. Ann. Phys. **56**(2), 556–594 (1970)
19. M.H. Grimsditch, E. Anastassakis, M. Cardona, Effect of uniaxial stress on the zone-center optical phonon of diamond. Phys. Rev. B **18**(2), 901–904 (1978)
20. B. Dietrich, V. Bukalo, A. Fischer, K.F. Dombrowski, E. Bugiel, B. Kuck, H.H. Richter, Raman-spectroscopic determination of inhomogeneous stress in submicron silicon devices. Appl. Phys. Lett. **82**(8), 1176–1178 (2003)
21. A. Steegen, I. De Wolf, K. Maex, Characterization of the local mechanical stress induced during the Ti and Co/Ti salicidation in sub-0.25 μm technologies. J. Appl. Phys. **86**(8), 4290–4297 (1999)
22. K. Pinardi, S.C. Jain, M. Willander, A. Atkinson, H.E. Maes, R. Van Overstraeten, A method to interpret micro-Raman experiments made to measure nonuniform stresses: application to local oxidation of silicon structures. J. Appl. Phys. **84**(5), 2507–2512 (1998)
23. M. Bowden, D.J. Gardiner, D. Wood, J. Burdess, A. Harris, J. Hedley, Raman and finite-element analysis of a mechanically strained silicon microstructure. J. Micromech. Microeng. **11**(1), 7–12 (2001)
24. E. Anastassakis, A. Cantarero, M. Cardona, Piezo-Raman measurements and anharmonic parameters in silicon and diamond. Phys. Rev. B **41**(11), 7529–7535 (1990)

25. M. Mermoux, B. Marcus, A. Crisci, A. Tajani, E. Gheeraert, E. Bustarret, Micro-Raman scattering from undoped and phosphorous-doped (111) homoepitaxial diamond films: stress imaging of cracks. J. Appl. Phys. **97**(4), 043530 (2005)
26. X.D. Pu, J. Chen, W.Z. Shen, H. Ogawa, Q.X. Guo, Temperature dependence of Raman scattering in hexagonal indium nitride films. J. Appl. Phys. **98**(3), 033527 (2005)
27. N. Camara, A. Tiberj, B. Jouault, A. Caboni, B. Jabakhanji, N. Mestres, P. Godignon, J. Camassel, Current status of self-organized epitaxial graphene ribbons on the C face of 6H-SiC substrates. J. Phys. D Appl. Phys. **43**(37), 374011 (2010)
28. O. Martinez, L.F. Sanz, J. Jimenez, A. Martin-Martin, B. Gerard, E. Gil-Lafon, Stress distribution mapping of GaAs on Si conformal layers. J. Appl. Phys. **101**(5), 054901 (2007)
29. C.-H. Chen, C.-M. Yeh, J. Hwang, T.-L. Tsai, C.-H. Chiang, C.-S. Chang, T.-P. Chen, Band gap shift in the GaN/AlN multilayers on the mesh-patterned Si(111). Appl. Phys. Lett. **88**(16), 161912 (2006)
30. C.-H. Chen, C.-M. Yeh, J. Hwang, T.-L. Tsai, C.-H. Chiang, C.-S. Chang, T.-P. Chen, Stress relaxation in the GaN/AlN multilayers grown on a mesh-patterned Si(111) substrate. J. Appl. Phys. **98**(9), 093509 (2005)
31. S. Nakashima, H. Harima, N. Ohtani, M. Katsuno, Raman characterization of local electrical properties and growth process in modulation-doped 6H-SiC crystals. J. Appl. Phys. **95**(7), 3547–3552 (2004)
32. The Latest International Technology Roadmap for Semi-conductors (ITRS), http://www.public.itrs.net/
33. A.H. Castro Neto, F. Guinea, N.M.R. Peres, K.S. Novoselov, A.K. Geim, The electronic properties of graphene. Rev. Mod. Phys. **81**(1), 109–162 (Jan 2009)
34. R.R. Nair, P. Blake, A.N. Grigorenko, K.S. Novoselov, T.J. Booth, T. Stauber, N.M.R. Peres, A.K. Geim, Fine structure constant defines visual transparency of graphene. Science **320**(5881), 1308 (2008)
35. N. Ferralis, Probing mechanical properties of graphene with Raman spectroscopy. J. Mater. Sci. **45**(19), 5135–5149 (2010)
36. http://www.nobelprize.org/nobel_prizes/physics/laureates/2010/sci.html
37. A.K. Geim, K.S. Novoselov, The rise of graphene. Nat. Mater. **6**(3), 183–191 (2007)
38. P.R. Wallace, Phys. Rev. **71**, 476 (1947)
39. K.S. Novoselov, A.K. Geim, S.V. Morozov, D. Jiang, M.I. Katsnelson, I.V. Grigorieva, S.V. Dubonos, A.A. Firsov, Two-dimensional gas of massless dirac fermions in graphene. Nature **438**, 197 (2005)
40. Y. Zhang, Y.-W. Tan, H.L. Stormer, P. Kim, Experimental observation of the quantum hall effect and berry's phase in graphene. Nature **438**, 201 (2005)
41. K.S. Novoselov, Z. Jiang, Y. Zhang, S.V. Morozov, H.L. Stormer, U. Zeitler, J.C. Maan, G.S. Boebinger, P. Kim, A.K. Geim, Room-temperature quantum Hall effect in graphene. Science **315**(5817), 1379 (2007)
42. A. Tzalenchuk, S. Lara-Avila, A. Kalaboukhov, S. Paolillo, M. Syväjärvi, R. Yakimova, O. Kazakova, T.J.B.M. Janssen, V. Fal'ko, S. Kubatkin, Towards a quantum resistance standard based on epitaxial graphene. Nat. Nanotechnol. **5**, 186 (2010)
43. L.M. Malard, M.A. Pimenta, G. Dresselhaus, M.S. Dresselhaus, Raman spectroscopy in graphene. Phys. Rep. **473**(5–6), 51–87 (2009)
44. G. Eda, G. Fanchini, M. Chhowalla, Large-area ultrathin films of reduced graphene oxide as a transparent and flexible electronic material. Nat. Nanotechnol. **3**, 270 (2008)
45. Y. Hernandez, V. Nicolosi, M. Lotya, F.M. Blighe, Z. Sun, S. De I.T. McGovern, B. Holland, M. Byrne, Y.K. Gun'Ko, J.J. Boland, P. Niraj, G. Duesberg, S. Krishnamurthy, R. Goodhue, J. Hutchison, V. Scardaci, A.C. Ferrari, J.N. Coleman, High-yield production of graphene by liquid-phase exfoliation of graphite. Nat. Nanotechnol. **3**, 563 (2008)
46. P.W. Sutter, J.-I. Flege, E.A. Sutter, Epitaxial graphene on ruthenium. Nat. Mater. **7**(5), 406–411 (2008)
47. I. Pletikosić, M. Kralj, P. Pervan, R. Brako, J. Coraux, A.T. N'diaye, C. Busse, T. Michely, Dirac cones and minigaps for graphene on Ir(111). Phys. Rev. Lett. **102**(5), 056808 (2009)

48. J. Coraux, A.T. N'Diaye, C. Busse, T. Michely, Structural coherency of graphene on Ir(111). Nano Lett. **8**(2), 565–570 (2008)
49. Y. Lee, S. Bae, H. Jang, S. Jang, S.-E. Zhu, S.H. Sim, Y.I. Song, B.H. Hong, J.-H. Ahn, Wafer-scale synthesis and transfer of graphene films. Nano Lett. **10**(2), 490–493 (2010)
50. A. Reina, X. Jia, J. Ho, D. Nezich, H. Son, V. Bulovic, M.S. Dresselhaus, J. Kong, Large area, few-layer graphene films on arbitrary substrates by chemical vapor deposition. Nano Lett. **9**(1), 30–35 (2009)
51. K.S. Kim, Y. Zhao, H. Jang, S.Y. Lee, J.M. Kim, K.S. Kim, J.-H. Ahn, P. Kim, J.-Y. Choi, B.H. Hong, Large-scale pattern growth of graphene films for stretchable transparent electrodes. Nature **457**(7230), 706–710 (2009)
52. S. Bae, H. Kim, Y. Lee, X. Xu, J.-S. Park, Y. Zheng, J. Balakrishnan, T. Lei, H.R. Kim, Y.I. Song, Y.-J. Kim, K.S. Kim, B. Ozyilmaz, J.-H. Ahn, B.H. Hong, S. Iijima, Roll-to-roll production of 30-inch graphene films for transparent electrodes. Nat. Nanotechnol. **5**(8), 574–578 (2010)
53. X. Li, W. Cai, J. An, S. Kim, J. Nah, D. Yang, R. Piner, A. Velamakanni, I. Jung, E. Tutuc, S.K. Banerjee, L. Colombo, R.S. Ruoff, Large-area synthesis of high-quality and uniform graphene films on copper foils. Science **324**(5932), 1312–1314 (2009)
54. C. Berger, Z. Song, X. Li, X. Wu, N. Brown, C. Naud, D. Mayou, T. Li, J. Hass, A.N. Marchenkov, E.H. Conrad, P.N. First, W.A. de Heer, Electronic confinement and coherence in patterned epitaxial graphene. Science **312**(5777), 1191–1196 (2006)
55. C. Berger, Z.M. Song, T.B. Li, X.B. Li, A.Y. Ogbazghi, R. Feng, Z.T. Dai, A.N. Marchenkov, E.H. Conrad, P.N. First, W.A. de Heer, Ultrathin epitaxial graphite: 2D electron gas properties and a route toward graphene-based nanoelectronics. J. Phys. Chem. B **108**(52), 19912–19916 (2004)
56. P. Sutter, Epitaxial graphene: how silicon leaves the scene. Nat. Mater. **8**(3), 171–172 (2009)
57. C. Virojanadara, M. Syväjarvi, R. Yakimova, L.I. Johansson, A.A. Zakharov, T. Balasubramanian, Homogeneous large-area graphene layer growth on 6H-SiC(0001). Phys. Rev. B **78**(24), 245403 (2008)
58. K.V. Emtsev, A. Bostwick, K. Horn, J. Jobst, G.L. Kellogg, L. Ley, J.L. McChesney, T. Ohta, S.A. Reshanov, J. Roehrl, E. Rotenberg, A.K. Schmid, D. Waldmann, H.B. Weber, T. Seyller, Towards wafer-size graphene layers by atmospheric pressure graphitization of silicon carbide. Nat. Mater. **8**(3), 203–207 (2009)
59. S. Kim, J. Ihm, H.J. Choi, Y.-W. Son, Origin of anomalous electronic structures of epitaxial graphene on silicon carbide. Phys. Rev. Lett. **100**(17), 176802 (2008)
60. Y. Qi, S.H. Rhim, G.F. Sun, M. Weinert, L. Li, Epitaxial Graphene on SiC(0001): more than just honeycombs. Phys. Rev. Lett. **105**(8), 085502 (Aug 2010)
61. F. Varchon, P. Mallet, J.-Y. Veuillen, L. Magaud, Ripples in epitaxial graphene on the Si-terminated SiC(0001) surface. Phys. Rev. B **77**(23), 235412 (Jun 2008)
62. H.E. Hoster, M.A. Kulakov, B. Bullemer, Morphology and atomic structure of the SiC(000(1)over-bar)3 $\times$ 3 surface reconstruction. Surf. Sci. **382**(1–3), L658–L665 (June 1997)
63. F. Hiebel, P. Mallet, F. Varchon, L. Magaud, J.-Y. Veuillen, Graphene-substrate interaction on 6H-SiC(000$\bar{1}$): a scanning tunneling microscopy study. Phys. Rev. B **78**(15), 153412 (2008)
64. F. Hiebel, P. Mallet, L. Magaud, J.-Y. Veuillen, Atomic and electronic structure of monolayer graphene on 6H-SiC(000$\bar{1}$)(3 $\times$ 3): a scanning tunneling microscopy study. Phys. Rev. B **80**(23), 235429 (2009)
65. N. Camara, J.-R. Huntzinger, G. Rius, A. Tiberj, N. Mestres, F. Pérez-Murano, P. Godignon, J. Camassel, Anisotropic growth of long isolated graphene ribbons on the C face of graphite-capped 6H-SiC. Phys. Rev. B **80**(12), 125410 (2009)
66. B. Jouault, B. Jabakhanji, N. Camara, W. Desrat, A. Tiberj, J.-R. Huntzinger, C. Consejo, A. Caboni, P. Godignon, Y. Kopelevich, J. Camassel, Probing the electrical anisotropy of multilayer graphene on the Si face of 6H-SiC. Phys. Rev. B **82**(8), 085438 (2010)
67. T. Stauber, N.M.R. Peres, A.K. Geim, Optical conductivity of graphene in the visible region of the spectrum. Phys. Rev. B **78**(8), 085432 (2008)

68. A.C. Ferrari, J.C. Meyer, V. Scardaci, C. Casiraghi, M. Lazzeri, F. Mauri, S. Piscanec, D. Jiang, K.S. Novoselov, S. Roth, A.K. Geim, Raman spectrum of graphene and graphene layers. Phys. Rev. Lett. **97**(18), 187401 (2006)
69. D.M. Basko, S. Piscanec, A.C. Ferrari, Electron-electron interactions and doping dependence of the two-phonon raman intensity in graphene. Phys. Rev. B **80**(16), 165413 (2009)
70. C. Casiraghi, Doping dependence of the raman peaks intensity of graphene close to the dirac point. Phys. Rev. B **80**(23), 233407 (2009)
71. L. Patrick, W.J. Choyke, Static dielectric constant of SiC. Phys. Rev. B **2**(6), 2255–2256 (1970)
72. A. Das, B. Chakraborty, S. Piscanec, S. Pisana, A.K. Sood, A.C. Ferrari, Phonon renormalization in doped bilayer graphene. Phys. Rev. B **79**(15), 155417 (2009)
73. P. Gava, M. Lazzeri, A.M. Saitta, F. Mauri, Probing the electrostatic environment of bilayer graphene using raman spectra. Phys. Rev. B **80**(15), 155422 (2009)
74. T. Ando, M. Koshino, Field effects on optical phonons in bilayer graphene. J. Phys. Soc. Jpn. **78**(3), 034709 (2009)
75. M. Bruna, S. Borini, Observation of raman *g*-band splitting in top-doped few-layer graphene. Phys. Rev. B **81**(12), 125421 (2010)
76. J. Yan, T. Villarson, E.A. Henriksen, P. Kim, A. Pinczuk, Optical phonon mixing in bilayer graphene with a broken inversion symmetry. Phys. Rev. B **80**(24), 241417 (2009)
77. L.M. Malard, D.C. Elias, E.S. Alves, M.A. Pimenta, Observation of distinct electron-phonon couplings in gated bilayer graphene. Phys. Rev. Lett. **101**(25), 257401 (2008)
78. P. Poncharal, A. Ayari, T. Michel, J.-L. Sauvajol, Effect of rotational stacking faults on the raman spectra of folded graphene. Phys. Rev. B **79**(19), 195417 (2009)
79. D. Graf, F. Molitor, K. Ensslin, C. Stampfer, A. Jungen, C. Hierold, L. Wirtz, Spatially resolved raman spectroscopy of single- and few-layer graphene. Nano Lett. **7**(2), 238–242 (2007)
80. N. Ferralis, Probing mechanical properties of graphene with raman spectroscopy. J. Mater. Sci. **45**, 5135–5149 (2010). doi:10.1007/s10853-010-4673-3
81. T.M.G. Mohiuddin, A. Lombardo, R.R. Nair, A. Bonetti, G. Savini, R. Jalil, N. Bonini, D.M. Basko, C. Galiotis, N. Marzari, K.S. Novoselov, A.K. Geim, A.C. Ferrari, Uniaxial strain in graphene by raman spectroscopy: *g* peak splitting, grüneisen parameters, and sample orientation. Phys. Rev. B **79**(20), 205433 (2009)
82. F. Varchon, R. Feng, J. Hass, X. Li, B. Ngoc Nguyen, C. Naud, P. Mallet, J.-Y. Veuillen, C. Berger, E.H. Conrad, L. Magaud, Electronic structure of epitaxial graphene layers on SiC: effect of the substrate. Phys. Rev. Lett. **99**(12), 126805 (2007)
83. J.C. Burton, L. Sun, F.H. Long, Z.C. Feng, I.T. Ferguson, First- and second-order raman scattering from semi-insulating 4H-SiC. Phys. Rev. B **59**(11), 7282–7284 (1999)
84. A. Ogura, D. Kosemura, K. Yamasaki, S. Tanaka, Y. Kakemura, A. Kitano, I. Hirosawa, Measurement of in-plane and depth strain profiles in strained-Si substrates. Solid-State Electron. **51**(2), 219–225 (2007)
85. D.J. Paul, Si/SiGe heterostructures: from material and physics to devices and circuits. Semicond. Sci. Technol. **19**(10), R75–R108 (2004)
86. S.-I. Takagi, J.L. Hoyt, J.J. Welser, J.F. Gibbons, Comparative study of phonon-limited mobility of two-dimensional electrons in strained and unstrained Si metal-oxide-semiconductor field-effect transistors. J. Appl. Phys. **80**(3), 1567–1577 (1996)
87. R. Oberhuber, G. Zandler, P. Vogl, Subband structure and mobility of two-dimensional holes in strained Si/SiGe MOSFET's. Phys. Rev. B **58**(15), 9941–9948 (1998)
88. T. Irisawa, S. Tokumitsu, T. Hattori, K. Nakagawa, S. Koh, Y. Shiraki, Ultrahigh room-temperature hole Hall and effective mobility in $Si_{0.3}Ge_{0.7}/Ge/Si_{0.3}Ge_{0.7}$ heterostructures. Appl. Phys. Lett. **81**(5), 847–849 (2002)
89. B. Rössner, D. Chrastina, G. Isella, H. von Känel, Scattering mechanisms in high-mobility strained Ge channels. Appl. Phys. Lett. **84**(16), 3058–3060 (2004)
90. C.W. Leitz, M.T. Currie, M.L. Lee, Z.-Y. Cheng, D.A. Antoniadis, E.A. Fitzgerald, Hole mobility enhancements and alloy scattering-limited mobility in tensile strained Si/SiGe surface channel metal-oxide-semiconductor field-effect transistors. J. Appl. Phys. **92**(7), 3745–3751 (2002)

91. M.T. Currie, C.W. Leitz, T.A. Langdo, G. Taraschi, E.A. Fitzgerald, D.A. Antoniadis, Carrier mobilities and process stability of strained Si n- and p-MOSFETs on SiGe virtual substrates. J. Vac. Sci. Technol. B **19**(6), 2268–2279 (2001)
92. H. Rho, H.E. Jackson, B.L. Weiss, Raman imaging of stress in a SiGe/Si photoelastic optical channel waveguide structure. Appl. Phys. Lett. **75**(9), 1287–1289 (1999)
93. S. Nakashima, T. Yamamoto, A. Ogura, K. Uejima, T. Yamamoto, Characterization of Si/Ge_xSi_{1-x} structures by micro-Raman imaging. Appl. Phys. Lett. **84**(14), 2533–2535 (2004)
94. M.L. Lee, E.A. Fitzgerald, M.T. Bulsara, M.T. Currie, A. Lochtefeld, Strained Si, SiGe, and Ge channels for high-mobility metal-oxide-semiconductor field-effect transistors. J. Appl. Phys. **97**(1), 011101 (2005)
95. J.H. Li, C.S. Peng, Y. Wu, D.Y. Dai, J.M. Zhou, Z.H. Mai, Relaxed $Si_{0.7}Ge_{0.3}$ layers grown on low-temperature Si buffers with low threading dislocation density. Appl. Phys. Lett. **71**(21), 3132–3134 (1997)
96. V. Paillard, P. Puech, R. Sirvin, S. Hamma, P. Roca i Cabarrocas, Measurement of the in-depth stress profile in hydrogenated microcrystalline silicon thin films using Raman spectrometry. J. Appl. Phys. **90**(7), 3276–3279 (2001)
97. P. Puech, G. Landa, R. Carles, P.S. Pizani, E. Daran, C. Fontaine, Strain relaxation in [001]- and [111]-GaAs/CaF_2 analyzed by Raman spectroscopy. J. Appl. Phys. **77**(3), 1126–1132 (1995)
98. P.S. Pizani, C.E.M. Campos, Raman probing of thermal damage depth profile in annealed GaAs. J. Appl. Phys. **84**(12), 6588–6591 (1998)
99. M.J. Matthews, J.W.P. Hsu, S. Gu, T.F. Kuech, Carrier density imaging of lateral epitaxially overgrown GaN using scanning confocal Raman microscopy. Appl. Phys. Lett. **79**(19), 3086–3088 (2001)
100. J. Camassel, N. Planes, L. Falkovsky, H. Moller, M. Eickhoff, G. Krotz, SOL thickness dependence of residual strain in SOI material. Electron. Lett. **35**(15), 1284–1286 (1999)
101. J. Camassel, L.A. Falkovsky, N. Planes, Strain effect in silicon-on-insulator materials: investigation with optical phonons. Phys. Rev. B **63**(3), 035309 (2000)
102. A. Tiberj, J. Camassel, N. Planes, Y. Stoemenos, H. Moriceau, O. Rayssac, Effect of silicon nitride and silicon dioxide bonding on the residual stress in layer-transferred SOI, in *Proceedings: Silicon-On-Insulator Technology and Devices XI*, ISBN 1-56677-375-X (Electrochemical Society, Pennington, 2003), p. 117
103. A. Tiberj, B. Fraisse, C. Blanc, S. Contreras, J. Camassel, Evaluation of strain induced by implantation in soi materials. Phys. Status Solidi (c) **0**, 1060 (2003)
104. A. Tiberj, B. Fraisse, C. Blanc, S. Contreras, J. Camassel, Process-induced strain in silicon-on-insulator materials. J. Phys. Condens. Matter **14**(48), 13411 (2002)
105. J. Camassel, A. Tiberj, Strain effects in device processing of silicon-on-insulator materials. Appl. Surf. Sci. **212–213**, 742–748 (2003) (11th International Conference on Solid Films and Surfaces)
106. M. Ichimura, Y. Moriguchi, A. Usami, M. Tabuchi, A. Sasaki, Characterization of GaAs heterolayers by micro-Raman spectroscopy. J. Cryst. Growth **149**(3–4), 167–174 (1995)
107. O. Pagés, M.A. Renucci, O. Briot, R.L. Aulombard, Raman study of band bending at ZnSe/GaAs interfaces. J. Appl. Phys. **77**(3), 1241–1248 (1995)
108. T. Zhang, M.L. Hsieh, W.R. Branford, M.J. Steer, R.A. Stradling, L.F. Cohen, Observation of bevelled GaSb/InAs quantum wells by Raman mapping. J. Raman Spectrosc. **36**(10), 978–983 (2005)
109. R. Srnanek, J. Geurts, M. Lentze, G. Irmer, D. Donoval, P. Brdecka, P. Kordos, A. Förster, B. Sciana, D. Radziewicz, M. Tlaczala, Study of δ-doped GaAs layers by micro-Raman spectroscopy on bevelled samples. Appl. Surf. Sci. **230**(1–4), 379–385 (2004)
110. R.M.B. Agaiby, S.H. Olsen, P. Dobrosz, H. Coulson, S.J. Bull, A.G. O'Neill, Nanometer strain profiling through Si/SiGe quantum layers. J. Appl. Phys. **104**(1), 013507 (2008)
111. C. Villeneuve, K.K. Bourdelle, V. Paillard, X. Hebras, M. Kennard, Raman spectroscopy study of damage and strain in (001) and (011) Si induced by hydrogen or helium implantation. J. Appl. Phys. **102**(9), 094905 (2007)

112. H. Richter, Z.P. Wang, L. Ley, The one phonon Raman spectrum in microcrystalline silicon. Solid State Commun. **39**(5), 625–629 (1981)
113. W.-P. Huang, H.H. Cheng, G. Sun, R.-F. Lou, J.H. Yeh, T.-M. Shen, The characteristic of strain relaxation on SiGe virtual substrate with thermal annealing. Appl. Phys. Lett. **91**(14), 142102 (2007)
114. J.W.P. Hsu, E.A. Fitzgerald, Y.H. Xie, P.J. Silverman, M.J. Cardillo, Surface morphology of related Ge_xSi_{1-x} films. Appl. Phys. Lett. **61**(11), 1293–1295 (1992)
115. J.M. Hartmann, B. Gallas, J. Zhang, J.J. Harris, Gas-source molecular beam epitaxy of SiGe virtual substrates: II. Strain relaxation and surface morphology. Semicond. Sci. Technol. **15**(4), 370–377 (2000)
116. S.Y. Shiryaev, F. Jensen, J.W. Petersen, On the nature of cross-hatch patterns on compositionally graded $Si_{1-x}Ge_x$ alloy layers. Appl. Phys. Lett. **64**(24), 3305–3307 (1994)
117. T. Spila, P. Desjardins, J. D'Arcy-Gall, R.D. Twesten, J.E. Greene, Effect of steady-state hydrogen coverage on the evolution of crosshatch morphology during $Si_{1-x}Ge_x$/Si(001) growth from hydride precursors. J. Appl. Phys. **93**(4), 1918–1925 (2003)
118. H. Chen, Y.K. Li, C.S. Peng, H.F. Liu, Y.L. Liu, Q. Huang, J.M. Zhou, Q.-K. Xue, Cross-hatching on a SiGe film grown on a Si(001) substrate studied by Raman mapping and atomic force microscopy. Phys. Rev. B **65**(23), 233303 (2002)
119. K. Sawano, S. Koh, Y. Shiraki, N. Usami, K. Nakagawa, In-plane strain fluctuation in strained-Si/SiGe heterostructures. Appl. Phys. Lett. **83**(21), 4339–4341 (2003)
120. K. Sawano, N. Usami, K. Arimoto, S. Koh, K. Nakagawa, Y. Shiraki, Observation of strain field fluctuation in SiGe-relaxed buffer layers and its influence on overgrown structures. Mater. Sci. Semicond. Process. **8**(1–3), 177–180 (2005)
121. K. Sawano, N. Usami, K. Arimoto, K. Nakagawa, Y. Shiraki, Strain field and related roughness formation in SiGe relaxed buffer layers. Thin Solid Films **508**(1–2), 117–119 (2006)
122. K. Sawano, N. Usami, K. Arimoto, K. Nakagawa, Y. Shiraki, Thickness dependence of strain field distribution in SiGe relaxed buffer layers. Jpn. J. Appl. Phys. **44**(12), 8445–8447 (2005)
123. B. Ghyselen, J.-M. Hartmann, T. Ernst, C. Aulnette, B. Osternaud, Y. Bogumilowicz, A. Abbadie, P. Besson, O. Rayssac, A. Tiberj, N. Daval, I. Cayrefourq, F. Fournel, H. Moriceau, C. Di Nardo, F. Andrieu, V. Paillard, M. Cabié, L. Vincent, E. Snoeck, F. Cristiano, A. Rocher, A. Ponchet, A. Claverie, P. Boucaud, M.-N. Semeria, D. Bensahel, N. Kernevez, C. Mazure, Engineering strained silicon on insulator wafers with the Smart Cut^{TM} technology. Solid-State Electron. **48**(8), 1285–1296 (2004)
124. A. Tiberj, V. Paillard, C. Aulnette, N. Daval, K.K. Bourdelle, M. Moreau, M. Kennard, I. Cayrefourcq. Stress metrology: the challenge for the next generation of engineered wafers, in *High-Mobility Group-IV Materials and Devices, Proceedings of MRS Symposium*, vol. 809, San Francisco, pp. 97–102, 2004
125. J.C. Tsang, P.M. Mooney, F. Dacol, J.O. Chu, Measurements of alloy composition and strain in thin Ge_xSi_{1-x} layers. J. Appl. Phys. **75**(12), 8098–8108 (1994)
126. F. Pezzoli, E. Bonera, E. Grilli, M. Guzzi, S. Sanguinetti, D. Chrastina, G. Isella, H. von Känel, E. Wintersberger, J. Stangl, G. Bauer, Phonon strain shift coefficients in $Si_{1-x}Ge_x$ alloys. J. Appl. Phys. **103**(9), 093521 (2008)
127. J. Groenen, R. Carles, S. Christiansen, M. Albrecht, W. Dorsch, H.P. Strunk, H. Wawra, G. Wagner, Phonons as probes in self-organized SiGe islands. Appl. Phys. Lett. **71**(26), 3856–3858 (1997)
128. H.K. Shin, D.J. Lockwood, J.-M. Baribeau, Strain in coherent-wave SiGe/Si superlattices. Solid State Commun. **114**(10), 505–510 (2000)
129. S.H. Olsen, A.G. O'Neill, D.J. Norris, A.G. Cullis, N.J. Woods, J. Zhang, K. Fobelets, H.A. Kemhadjian, Strained Si/SiGe n-channel MOSFETs: impact of cross-hatching on device performance. Semicond. Sci. Technol. **17**(7), 655–661 (2002)
130. N. Sugii, Thermal stability of the strained-Si/$Si_{0.7}Ge_{0.3}$ heterostructure. J. Appl. Phys. **89**(11), 6459–6463 (2001)
131. A. Ogura, K. Yamasaki, D. Kosemura, S. Tanaka, I. Chiba, R. Shimidzu, UV-Raman spectroscopy system for local and global strain measurements in Si. Jpn. J. Appl. Phys. **45**(4B), 3007–3011 (2006)

132. J.B. Renucci, R.N. Tyte, M. Cardona, Resonant Raman scattering in silicon. Phys. Rev. B **11**(10), 3885–3895 (1975)
133. K. Sawano, Y. Ozawa, A. Fukumoto, N. Usami, J. Yamanaka, K. Suzuki, K. Arimoto, K. Nakagawa, Y. Shiraki, Strain-field evaluation of strain-relaxed thin SiGe layers fabricated by ion implantation method. Jpn. J. Appl. Phys. **44**(43), L1316–L1319 (2005)
134. I. De Wolf, Stress measurements in Si microelectronics devices using Raman spectroscopy. J. Raman Spectrosc. **30**(10), 877 (1999)
135. I. De Wolf, Micro-Raman spectroscopy to study local mechanical stress in silicon integrated circuits. Semicond. Sci. Technol. **11**(2), 139–154 (1996)
136. S.M. Hu, Stress-related problems in silicon technology. J. Appl. Phys. **70**(6), R53–R80 (1991)
137. K. Mistry, C. Allen, C. Auth, B. Beattie, D. Bergstrom, M. Bost, M. Brazier, M. Buehler, A. Cappellani, R. Chau, C.-H. Choi, G. Ding, K. Fischer, T. Ghani, R. Grover, W. Han, D. Hanken, M. Hatttendorf, J. He, J. Hicks, R. Huessner, D. Ingerly, P. Jain, R. James, L. Jong, S. Joshi, C. Kenyon, K. Kuhn, K. Lee, H. Liu, J. Maiz, B. McIntyre, P. Moon, J. Neirynck, S. Pei, C. Parker, D. Parsons, C. Prasad, L. Pipes, M. Prince, P. Ranade, T. Reynolds, J. Sandford, L. Schifren, J. Sebastian, J. Seiple, D. Simon, S. Sivakumar, P. Smith, C. Thomas, T. Troeger, P. Vandervoorn, S. Williams, K. Zawadzki, A 45 nm logic technology with high-k plus metal gate transistors, strained silicon, 9 Cu interconnect layers, 193 nm dry patterning, and 100% Pb-free packaging, in *2007 IEEE International Electron Devices Meeting*, vols. 1–2, pp. 247–250, 2007
138. I. De Wolf, E. Anastassakis, Addendum: "Stress measurements in silicon devices through Raman spectroscopy: bridging the gap between theory and experiment" [J. Appl. Phys. **79**, 7148 (1996)]. J. Appl. Phys. **85**(10), 7484–7485 (1999)
139. I. De Wolf, H.E. Maes, S.K. Jones, Stress measurements in silicon devices through Raman spectroscopy: bridging the gap between theory and experiment. J. Appl. Phys. **79**(9), 7148–7156 (1996)
140. I. De Wolf, J. Vanhellemont, A. Romano-Rodríguez, H. Norström, H.E. Maes, Micro-Raman study of stress distribution in local isolation structures and correlation with transmission electron microscopy. J. Appl. Phys. **71**(2), 898–906 (1992)
141. I. De Wolf, H. Norström, H.E. Maes, Process-induced mechanical stress in isolation structures studied by micro-Raman spectroscopy. J. Appl. Phys. **74**(7), 4490–4500 (1993)
142. S.C. Jain, A.H. Harker, A. Atkinson, K. Pinardi, Edge-induced stress and strain in stripe films and substrates: a two-dimensional finite element calculation. J. Appl. Phys. **78**(3), 1630–1637 (1995)
143. S.M. Hu, Film-edge-induced stress in substrates. J. Appl. Phys. **50**(7), 4661–4666 (1979)
144. K.F. Dombrowski, B. Dietrich, I. De Wolf, R. Rooyackers, G. Badenes, Investigation of stress in shallow trench isolation using UV micro-Raman spectroscopy. Microelectron. Reliab. **41**(4), 511–515 (2001)
145. V. Senez, A. Armigliato, I. De Wolf, G. Carnevale, R. Balboni, S. Frabboni, A. Benedetti, Strain determination in silicon microstructures by combined convergent beam electron diffraction, process simulation, and micro-Raman spectroscopy. J. Appl. Phys. **94**(9), 5574–5583 (2003)
146. K.F. Dombrowski, I. De Wolf, B. Dietrich, Stress measurements using ultraviolet micro-Raman spectroscopy. Appl. Phys. Lett. **75**(16), 2450–2451 (1999)
147. I. De Wolf, V. Senez, R. Balboni, A. Armigliato, S. Frabboni, A. Cedola, S. Lagomarsino, Techniques for mechanical strain analysis in sub-micrometer structures: TEM/CBED, micro-Raman spectroscopy. X-ray micro-diffraction and modeling. Microelectron. Eng. **70**(2–4), 425–435 (2003)
148. R. Liu, M. Canonico, Applications of UV-Raman spectroscopy and high-resolution X-ray diffraction to microelectronic materials and devices. Microelectron. Eng. **75**(3), 243–251 (2004)
149. E. Bonera, M. Fanciulli, G. Carnevale, Raman stress maps from finite-element models of silicon structures. J. Appl. Phys. **100**(3), 033516 (2006)

150. V. Poborchii, T. Tada, T. Kanayama, Study of stress in a shallow-trench-isolated Si structure using polarized confocal near-UV Raman microscopy of its cross section. Appl. Phys. Lett. **91**(24), 241902 (2007)
151. E. Bonera, M. Fanciulli, M. Mariani, Raman spectroscopy of strain in subwavelength microelectronic devices. Appl. Phys. Lett. **87**(11), 111913 (2005)
152. V. Poborchii, T. Tada, T. Kanayama, Edge-enhanced Raman scattering in Si nanostripes. Appl. Phys. Lett. **94**(13), 131907 (2009)
153. V. Poborchii, T. Tada, T. Kanayama, High-spatial-resolution Raman microscopy of stress in shallow-trench-isolated Si structures. Appl. Phys. Lett. **89**(23), 233505 (2006)
154. S.M. Hu, Stress from isolation trenches in silicon substrates. J. Appl. Phys. **67**(2), 1092–1101 (1990)
155. M. Yoshikawa, M. Murakami, K. Matsuda, R. Sugie, H. Ishida, R. Shimizu, Stress characterization of si by a scanning near-field optical Raman microscope with spatial resolution and with penetration depth at the nanometer level, using resonant Raman scattering. Jpn. J. Appl. Phys. **45**(17–19), L486–L489 (2006)
156. C.E. Murray, M. Sankarapandian, S.M. Polvino, I.C. Noyan, B. Lai, Z. Cai, Submicron mapping of strained silicon-on-insulator features induced by shallow-trench-isolation structures. Appl. Phys. Lett. **90**(17), 171919 (2007)
157. F. Hüe, M. Hÿtch, H. Bender, F. Houdellier, A. Claverie, Direct mapping of strain in a strained silicon transistor by high-resolution electron microscopy. Phys. Rev. Lett. **100**(15), 156602 (2008)
158. M. Hytch, F. Houdellier, F. Hue, E. Snoeck, Nanoscale holographic interferometry for strain measurements in electronic devices. Nature **453**(7198), 1086-U5 (2008)
159. I. De Wolf, C. Jian, W.M. van Spengen, The investigation of microsystems using Raman spectroscopy. Opt. Lasers Eng. **36**(2), 213–223 (2001)
160. J.W. Pomeroy, M. Kuball, D.J. Wallis, A.M. Keir, K.P. Hilton, R.S. Balmer, M.J. Uren, T. Martin, P.J. Heard, Thermal mapping of defects in AlGaN/GaN heterostructure field-effect transistors using micro-Raman spectroscopy. Appl. Phys. Lett. **87**(10), 103508 (2005)
161. T.J. Ochalski, D. Pierścińska, K. Pierściński, M. Bugajski, J.W. Tomm, T. Grunske, A. Kozlowska, Complementary thermoreflectance and micro-Raman analysis of facet temperatures of diode lasers. Appl. Phys. Lett. **89**(7), 071104 (2006)
162. T. Beechem, S. Graham, S.P. Kearney, L.M. Phinney, J.R. Serrano, Invited article: simultaneous mapping of temperature and stress in microdevices using micro-Raman spectroscopy. Rev. Sci. Instrum. **78**(6), 061301 (2007)
163. A. Compaan, H.J. Trodahl, Resonance Raman scattering in Si at elevated temperatures. Phys. Rev. B **29**(2), 793–801 (1984)

Chapter 3
Raman Mapping for the Investigation of Nano-phased Materials

G. Gouadec, L. Bellot-Gurlet, D. Baron and Ph. Colomban

Abstract Nanosized and nanophased materials exhibit special properties. First they offer a good compromise between the high density of chemical bonds by unit volume, needed for good mechanical properties and the homogeneity of amorphous materials that prevents crack initiation. Second, interfaces are in very high concentration and they have a strong influence on many electrical and redox properties. The analysis of nanophased, low crystallinity materials is not straigtforward. The recording of Raman spectra with a geometric resolution close to 0.5 μm^3 and the deep understanding of the Raman signature allow to locate the different nanophases and to predict the properties of the material. Case studies are discussed: advanced polymer fibres, ceramic fibres and composites, textured piezoelectric ceramics and corroded (ancient) steel.

3.1 Raman Spectroscopy and (Nano)materials Properties

Nanosized and nanophased materials possess a high concentration of interfaces, which results in unique optical and redox properties, as well as a good resistance to crack propagation. Not all conventional materials science research techniques are

G. Gouadec (✉) · L. Bellot-Gurlet · D. Baron · Ph. Colomban
Laboratoire de Dynamique, Interactions et Réactivité (LADIR), UMR 7075,
CNRS et Université Pierre et Marie Curie (UPMC - Paris 6),
4 place Jussieu, 75252, Paris, Cédex 05, France
e-mail: gwenael.gouadec@upmc.fr

L. Bellot-Gurlet
e-mail: ludovic.bellot-gurlet@upmc.fr

D. Baron
e-mail: denis.baron@upmc.fr

Ph. Colomban
e-mail: philippe.colomban@upmc.fr

A. Zoubir (ed.), *Raman Imaging*, Springer Series in Optical Sciences 168,
DOI: 10.1007/978-3-642-28252-2_3,

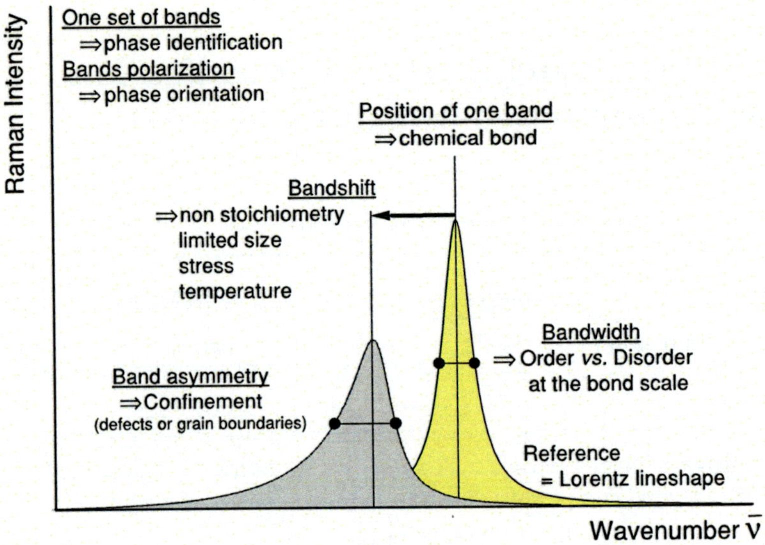

Fig. 3.1 The basics of Raman spectra interpretation. The vibrational wavenumbers are often called Raman shifts as they correspond to a Raman-induced wavenumber shift of the scattered photons with respect to the laser excitation

applicable to study the structure–properties relationship in such materials and Raman microspectroscopy (μRS) is one of them (Fig. 3.1) [1, 2]:

- The actual probes are vibrational modes, which strongly depend on atoms mass, bond lengths and the angles between atom-sharing bonds.
- In the most frequent case where a visible laser is used for the excitation (IR and UV-Raman are also possible), the excitation and collection of the signal is performed through a standard optical microscope, which makes a selection of the analysis point very easy.
- Even low crystallinity materials like those often found at interfaces (which seldom reach thermodynamic equilibrium during the synthesis process) are available for analysis.

More generally, Raman spectroscopy being a non contact, non destructive, and relatively fast technique, it offers the possibility to characterize a material under working conditions and assess how external perturbations (mechanical stress, thermal cycling, electrochemical cycling, corroding atmosphere, ...) may affect it.

Modern Raman microspectrometers are equipped with motorized tables and acquisition softwares allowing for the automated mapping of a surface. The user ultimately gets a "hyperspectral map", which is a collection of individual Raman spectra, each being associated with a given point on the sample surface (Fig. 3.2). Note that *Raman mapping* must not be confused with *Raman imaging*, where a large area is lit by a defocused laser beam and photons from only one selected wavenumber

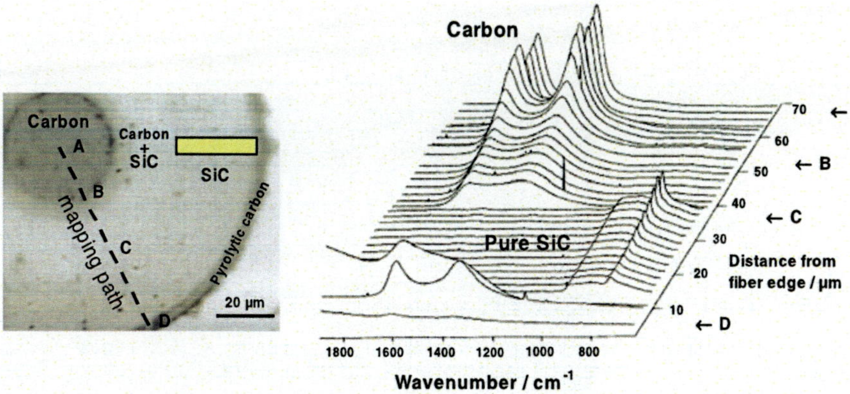

Fig. 3.2 In-line Raman map of a SCS-6 fiber radius (adapted from Ref. [6]). The *shaded rectangle* in the optical picture indicates the region mapped in Fig. 3.16

range form an image of that area on a mosaic detector (CCD) [3]. The possibility exists to use hyperspectral maps data to build a "map" of the investigated area where the color of each point indicates the integrated Raman intensity in a wavenumber Range Of Interest (ROI). Such a map may be useful to locate a phase with strong Raman scattering in the ROI (Figs. 3.3 and 3.4) but a much more sophisticated analysis is achieved with maps of spectral parameters (position, width, area or intensity of a given band) obtained through the mathematical fitting of the Raman spectra. If a physical property can be expressed as a function of one or several such parameters, it is eventually possible to draw a "smart map" [4, 5] of that property based on the treatment of Raman spectra (see for instance Figs. 3.14, 3.16–3.19, and 3.21c).

3.2 Recording a Raman Map

Raman mapping is most valuable to point out subtle spectral modifications taking place from one place of a sample to another (Figs. 3.2–3.4 and 3.17). A large number of good quality spectra are thus necessary and the overall recording time (t_{rec}) will possibly reach several hours. Some preliminary thinking is advisable to make the most of it.

3.2.1 The Lateral Resolution of μRS

First of all, the microscope objective should be chosen for the resulting probing to match the sample details of interest. From a practical point of view, microscope objectives are usually characterized by the numerical aperture NA [8]:

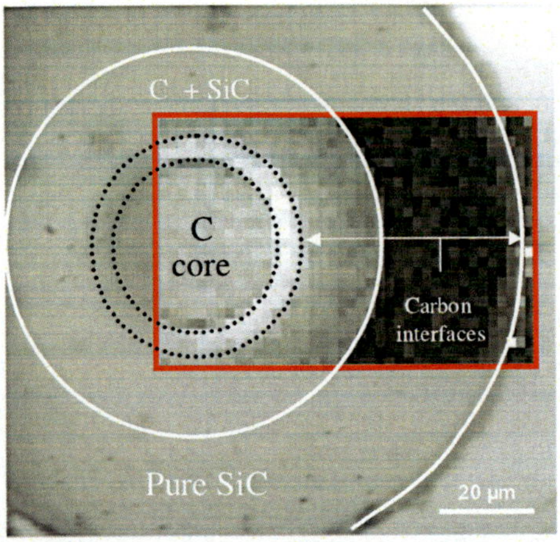

Fig. 3.3 The selection of graphite signal (ROI at 1300–1600 cm^{-1}) in a Raman map (*boxed rectangle*; 1 μm × 1 μm *square pixels*) reveals the different regions of an SCS-6 fiber (the darker a pixel, the lower the graphite signal)

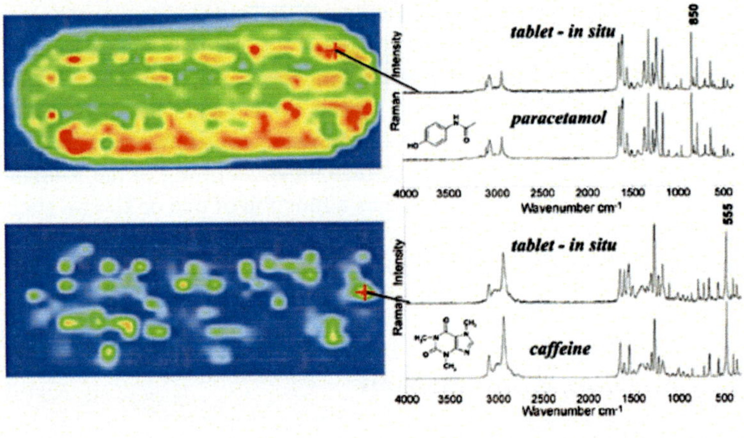

Fig. 3.4 The Raman mapping of a drug tablet (18.0 × 8.5 mm) makes it possible to discriminate paracetamol (ROI at 850 cm^{-1}) from caffeine (ROI at 555 cm^{-1}). Reprinted with permission from Elsevier [7]

$$NA = n \sin(\theta_{back}) \tag{3.1}$$

In Eq. (3.1), θ_{back} is the maximum collection angle for the backscattered light and n is the refractive index in the medium between the sample and the microscope lens (Fig. 3.5).

The numerical aperture is a key parameter because it sets the resolution R of the microscope, defined as the shortest spacing for two points on a sample surface to be resolved with λ wavelength observation [9]:

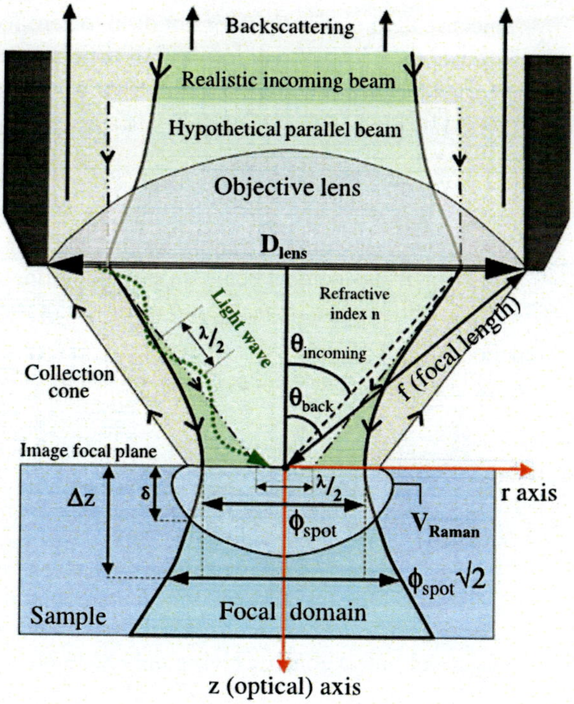

Fig. 3.5 Basic schematic of the Raman microscope in backscattering configuration. The labels signification is given throughout the text

$$R \sim \frac{0.61\lambda}{\text{NA}} \tag{3.2}$$

Equation (3.2) is usually referred to as Rayleigh criterion and follows from Abbe theory of the diffraction pattern created by distinct point sources observed through a microscope [10]. The best objectives (oil immersion objectives) having numerical apertures around 1.4, Rayleigh criterion is sometimes referred to as a "$\lambda/2$ *diffraction limit*". The focusing of incoming laser beams, even perfectly parallel ones, is diffraction limited by their finite diameter. Rayleigh criterion therefore forbids to obtain a laser spot smaller than about half the laser wavelength (Fig. 3.5), even though smaller objects will still give a Raman signal. Figure 4.2 in the next chapter illustrates this last point: 170 nm diameter GaN nanowires were indeed observed with a lateral resolution of, at best, 230 nm (confocal microscope—see further; 0.9 NA; $\lambda = 514.5$ nm) [11]. Besides, the polarization sensitivity of GaN nanowires Raman modes demonstrates the possibility to analyze growth directions and crystallographic orientations in nanophases.

Actual laser beams are not perfectly parallel (which has been exaggerated in Fig. 3.5) and their focusing through a microscope lens actually gives an elongated volume, which is called the focal domain or focal cylinder [12, 13]. The diameter $\phi(z)$ of the focal domain at z axial coordinate must be defined arbitrarily as the electric field obviously does not drop to zero for a definite r distance away from the optical axis. The radial decrease of the electric field in a laser beam actually obeys a Gaussian law and $\phi(z)$ is often taken where the maximum field $E_0(z, 0)$ is divided by e [13]:

$$E(z, r) = E_0(z, 0)e^{-4(r/\phi(z))^2}, \quad \text{with} \quad \phi_{\text{spot}} = \phi(z = 0) \tag{3.3}$$

We will assume for simplicity that the lens (focal length: f; Diameter: D_{lens}) is fully lit by the incident beam. In such conditions (the incoming beam diameter equals D_{lens} and $\theta_{\text{incoming}} = \theta_{\text{back}} = \theta$ in Fig. 3.5), it can be shown that [12, 13]:

$$\phi_{spot} \sim \frac{4\lambda f}{\pi D_{\text{lens}}} = \frac{2\lambda}{\pi \sin\theta} = \frac{2n\lambda}{\pi \text{NA}} \tag{3.4}$$

Combining Eqs. (3.2) and (3.4), one finds:

$$\frac{\phi_{\text{spot}}}{R} = \frac{2n}{0.61\pi} \sim n \tag{3.5}$$

ϕ_{spot} is thus very close to the diameter of the smallest circle in which all incoming energy can, in theory, be focused. Remember however that, according to Eq. (3.3), ϕ_{spot} delimits a zone receiving only 86 % of the total laser irradiance [13].

3.2.2 The Axial Resolution of μRS

Similar to the lateral resolution, the in-depth or axial resolution of Raman spectroscopy cannot be defined unambiguously. Indeed, the laser intensity does not drop to zero for a given z value and one has to choose an arbitrary threshold. It is common to use the *depth of field* Δz (Fig. 3.5), which is the distance along which the irradiance is cut by half (or, equivalent, ϕ(z) is multiplied by $\sqrt{2}$) [13]. Note that Raman spectroscopists often improperly refer to Δz as a *depth of focus* (which should refer to the focus tolerance in the image plane of the microscope). For the lens focusing of a Gaussian beam, one finds [12, 13] (there is a missing λ factor in Eq. (11) from Ref. [13]):

$$\Delta z \sim \frac{8\lambda}{\pi} \times \left(\frac{f}{D_{\text{lens}}}\right)^2 = \frac{2n^2\lambda}{\pi \text{NA}^2} \tag{3.6}$$

The calculation in Eq. (3.6) is based on the laser focal cylinder but the actual probing depth (δ in Fig. 3.5) can be significantly lower in absorbent materials (δ is inversely proportional to the linear absorption (or attenuation) coefficient [14]), making μRS a surface technique.

In the confocal microscopy setup, pinholes are placed in the microscope at intermediate image planes, resulting in a better in-depth discrimination power [15].

In that case, the integrated intensity coming from a given plane perpendicular to the optical axis is no longer a constant but, rather, decreases by 50 % between the focal plane ($z = 0$) and $\Delta z_{\text{confocal}}$ (the half width of the so-called point spread function), which is well approximated with [16, 17]:

$$\Delta z_{\text{confocal}} \sim \frac{4.4 n \lambda}{2\pi \text{NA}^2} \tag{3.7}$$

The use of confocal pinholes goes with a lateral resolution improvement of about one third [17] but confocal microscopy interest mainly consists in the possibility to select sample layers axially.

Note that Δz and $\Delta z_{\text{confocal}}$ should obviously not be compared as the former only considers intensity variations along the optical axis, whereas the latter is based on the total intensity distribution at a given z value. Moreover, it might seem surprising that Eq. (3.7) does not involve the confocal pinholes diameter but it is implicitly assumed to be at an optimum corresponding to R_{confocal} (the lateral resolution in confocal mode) times the total magnification (usually greater than the nominal magnification of the objective) at the pinhole image plane [18].

3.2.3 Practical Aspects of Raman Mapping

Because μRS is sensitive to chemical bonds (through their vibrations), the technique obviously provides useful information about the nano-organization of matter. However, even with the best dry ($n = n_{\text{air}} = 1$, NA $= 0.95$) or oil immersion ($n = 1.51$, NA $= 1.4$) objectives, and in the most favorable case of a blue excitation ($\lambda \sim 400$ nm), ϕ_{spot} will remain above 275 nm (Eq. 3.4). Unless nano-objects are extremely dispersed in the sample (and the signal is still strong enough for analysis, which can be obtained by Surface Enhanced Raman Scattering—SERS [19, 20]), each Raman spectrum will therefore give the average response of a very large number of nano-objects. Mathematical simulations of the spectra should then ideally include statistical distributions of size or size-governed properties [21–32].

As a matter of fact, there is a possibility to break the λ/2 diffraction limit of optical microscopes and make Raman lateral resolution nanometric by maintaining a sharp metallic tip extremely close to the sample, with micro-manipulation tools borrowed from Atomic Force and Scanning Tunneling Microscopes (AFM/STM) [33, 34]. This Tip-Enhanced Raman Scattering (TERS) became a very hot topic in the past decade but its implementation still requires too much experimental and theoretical expertise for a generalized use [35]. We shall leave TERS aside and focus on the nano-related information that can be retrieved using conventional μRS. For that matter, let us first stress that the V_{Raman} volume drawn in Fig. 3.5 to represent the region probed in the sample neither matches ϕ_{spot}, nor Δz. Equations (3.4) and (3.6) may indeed give good estimates for the lateral and axial resolution of μRS but a number of factors will affect the actual values:

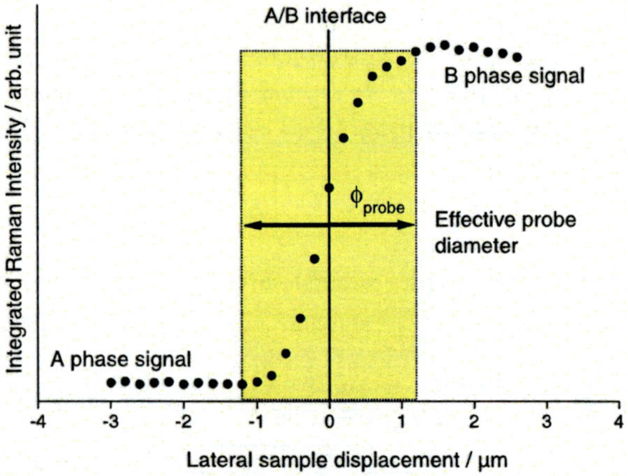

Fig. 3.6 Determination of the spatial resolution of a Raman microscope by crossing the interface between phases A and B at right angles. The wavenumber domain over which intensity is integrated must be chosen so as to maximize the difference in scattering efficiency between A and B phases

- Optical aberrations of the optical parts (lenses, mirrors, etc).
- Imperfect laser alignment along the optical axis of the microscope.
- Parasite refractions at the sample/light propagation medium interface [36, 37].
- A poor phase contrast on the sample surface.
- Light absorption in the sample (if $\delta << \Delta z$).
- Multiple scattering.

It might be advisable, especially in view of large-scale mapping campaigns, to get experimental spatial resolution values in the actual working conditions. A precise measurement of the spot diameter is easily achieved with a line-mapping across the interface of two phases (for instance a metal piece polished in a resin block) having different scattering cross-sections in a given wavenumber window [5, 38–40]. The plot of the intensity recorded in that window as a function of the displacement across the interface will resemble that shown in Fig. 3.6, thus providing the effective probe diameter ϕ_{probe}. As for the axial resolution, it can be measured from the intensity profile recorded when a mirror replacing the sample is taken away from the image focal plane of the microscope [41, 42].

3D-Raman mapping is possible in low absorbance samples or when small scatterers are dispersed in a transparent matrix [43], but 2D-raster mapping with ΔX and ΔY displacement steps of the sample-baring motorized table is much more common (Fig. 3.7a). However, if the microscope has the autofocus option, the axial degree of liberty may still serve to compensate the lack of planarity and/or horizontality of the sample surface. We shall further focus on Raman mapping with square pixels ($\Delta Y = \Delta X$, as shown in Fig. 3.7a), which obviously optimizes the sample surface coverage by a circular spot. The condition guaranteeing full probing of the mapped area (A_{map}) is:

$$\Delta X = \frac{\phi_{\text{probe}}}{C}; \quad \text{with } C \geq \sqrt{2} \tag{3.8}$$

In Eq. (3.8), $\sqrt{2}$ is the limit value for the laser spot to just reach the corners of each targeted "*pixel*" (the sample zone to which the recorded spectrum will be attached), but even in that case (mapping with ΔX_1 step in Fig. 3.7a), some part of the signal will actually come from neighboring pixels. Increasing C (decreasing ΔX) will raise the proportion of light backscattered from untargeted pixels (see the mapping with ΔX_2 step in Fig. 3.7a) but it will also improve the rendering of the sample details.

As a matter of fact, Raman intensity measured at pixel coordinate (x_p, y_p) is a convolution of the Raman scattering efficiency (σ function) by the (Gaussian) lineshape of the laser beam cross-section (half width at half maximum w):

$$I\left(\bar{\nu}, x_p, y_p\right) \propto \int\int \sigma\left(\bar{\nu}, x, y\right) \times e^{-\ln 2 \frac{(x-x_p)^2+(y-y_p)^2}{w^2}} \mathrm{d}x\mathrm{d}y \qquad (3.9)$$

According to Eq. (3.9), the exact position of any heterogeneity at the surface of the mapped area (for instance the L-wide inclusion in Fig. 3.7b) will necessarily be somewhere in the pixel showing the strongest heterogeneity-related Raman contribution. Details will thus be located with a precision of ΔX. Yet, even in the limit of $\Delta X \to 0$, they will appear about as large as ϕ_{probe}. This can be understood with the situation depicted in Fig. 3.7b: the L-wide inclusion will clearly contribute to the targeted pixel even though it is located outside the region associated with the said pixel.

Only a mathematical deconvolution of the hyperspectal map by the laser lineshape will "isolate" the signal truly coming from each pixel [4, 5, 39] (Fig. 3.8) but it will not be reliable if the phase repartition changes for depths lower than the laser penetration (the case illustrated for the irregular inclusion in Fig. 3.7b). Most 2D-imaging software programs offer the possibility to interpolate data points but it is only a smoothing of the images (Figs. 3.4, 3.14, 3.16, 3.20 and 3.21), which does not actually involve any deconvolution procedure [44].

If t_{spec} is the time to be devoted to each spectrum acquisition,[1] the total number of spectra N_{spec} in a Raman map will be:

$$N_{\text{spec}} = \frac{t_{\text{rec}}}{t_{\text{spec}}} = \frac{A_{\text{map}}}{\Delta X^2} \qquad (3.10)$$

And the combination of Eqs. (3.8) and (3.10) yields:

$$\phi_{\text{probe}} = C\sqrt{\frac{A_{\text{map}}\, t_{\text{spec}}}{t_{\text{rec}}}} \qquad (3.11)$$

Mapping a sample with a given objective, that is to say a given ϕ_{probe} value, will involve adjusting parameters C (hence ΔX step), A_{map}, t_{spec} and t_{rec} so as to satisfy

---

[1] t_{spec} must include the effective acquisition time but also the time-out for the displacements of the motorized table (usually negligible) for the (optional) auto-focus sequence, for the gratings rotation (in case of multi-window recording) and for data saving.

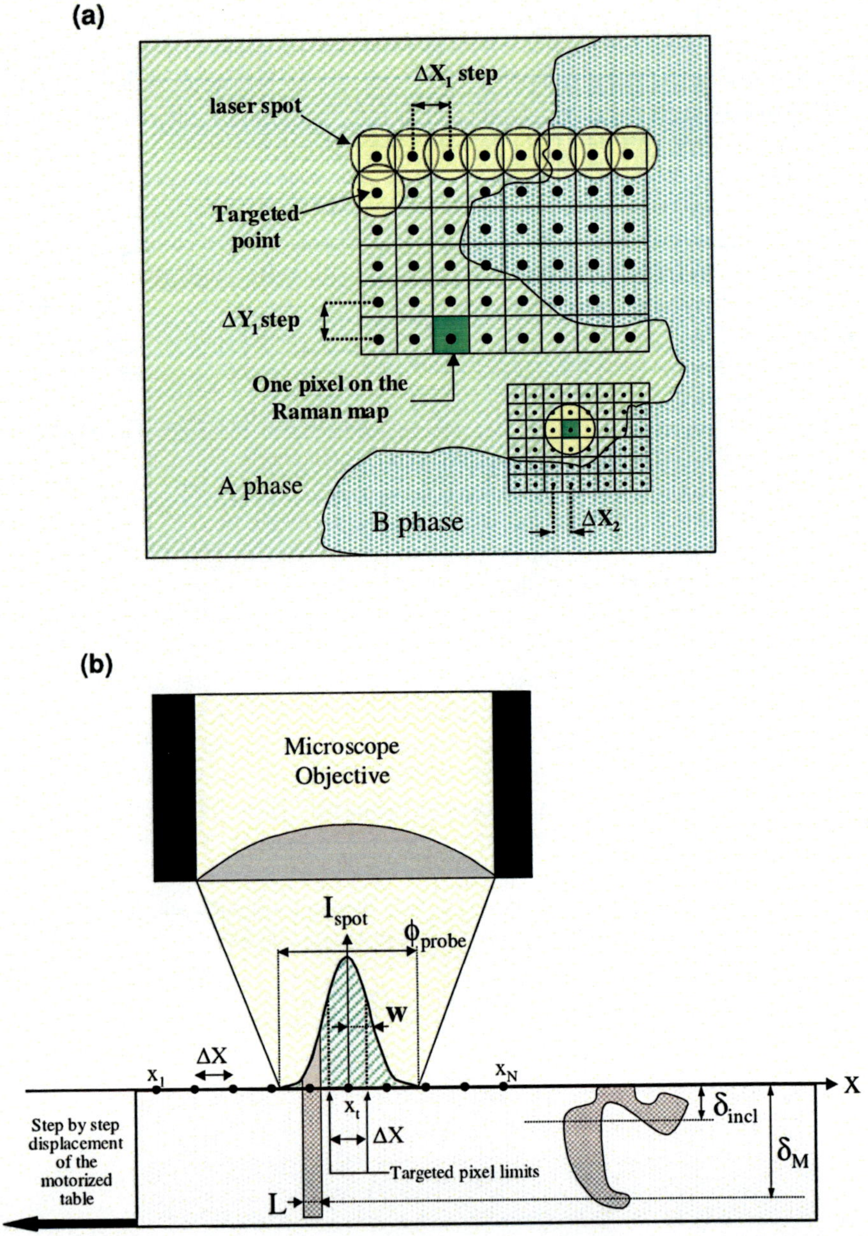

Fig. 3.7 **a** Schematic of two 8 × 6 raster mappings with square *pixels* (displacement steps $\Delta X = \Delta Y$); **b** schematic of the in-line mapping (ΔX step) of material M (laser penetration δ_M) with one L-wide and one irregular inclusions (laser penetration δ_{incl}). Each of the N spectra is attached to a ΔX-wide *pixel*. I_{spot} gives the Gaussian distribution (half width w) of the laser intensity across the experimental spot diameter (ϕ_{probe}) and all the scattering collected from the illuminated region contributes to the targeted pixel (x_t). The Raman contribution of the L-wide inclusion is proportional to the gray area below the laser intensity curve

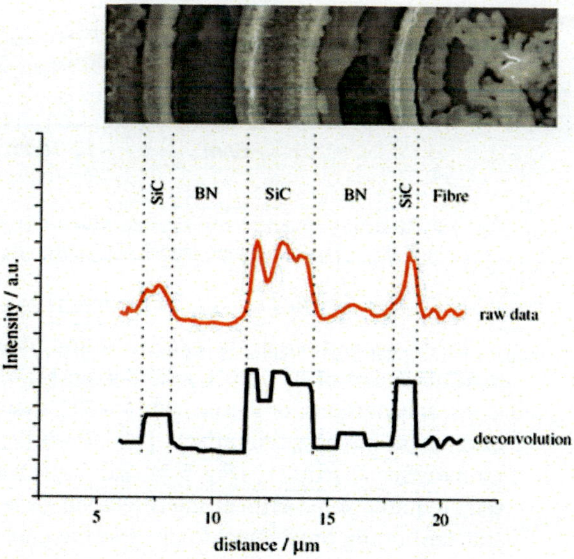

Fig. 3.8 The laser spot lineshape can be deconvoluted from the in-line Rayleigh map along the radius of a BN/SiC-covered SiC fiber in order to enhance the lateral resolution. The same procedure would apply to Raman mapping. Reprinted with permission from Elsevier [1].

Eq. (3.11). Because of the square root, increasing C (decreasing ΔX) for a given mapping time t_{rec} will involve strong concessions on the map dimension (cutting ΔX by two divides A_{map} by four; Fig. 3.7a) and/or the signal to noise ratio (reduced t_{spec}).

3.3 Data Processing

3.3.1 Spectra Simulation

The scattering of light by identical vibrators (wavevector $\bar{v}_c$) should result in a lorentzian-shaped band (half width γ), but most experimental lineshapes are best fitted by a convolution with a Gaussian profile (half width: $\sigma\sqrt{(2ln2)}$) representing the distribution of vibrators in slightly different states:

$$I\left(\bar{v}_c+\bar{v}\right) \propto \int_{-\infty}^{+\infty} \frac{\gamma}{\sigma} \times \frac{e^{-\left(\frac{\bar{v}'^2}{2\sigma^2}\right)} d\bar{v}'}{(\bar{v}-\bar{v}')^2+\gamma^2} \tag{3.12}$$

The use of simple Gaussian and Lorentzian profiles is often preferred to the Voigt profile described by Eq. (3.12) as they depend on three parameters instead of four (many fitting modules misleadingly name Voigt profile the simple sum of one Lorentzian and one Gaussian).

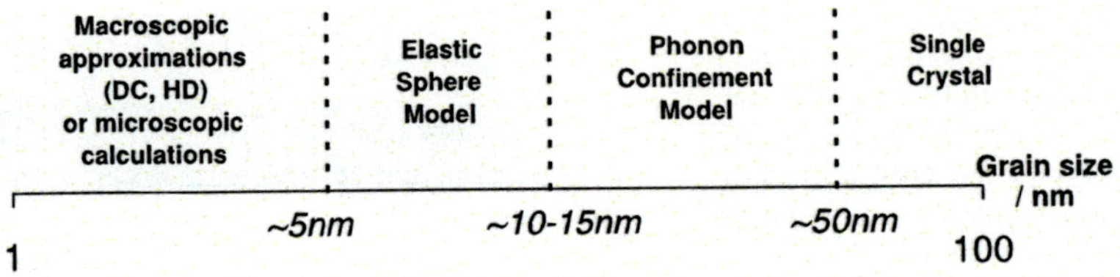

Fig. 3.9 Modeling spherical nanocrystals vibrations as a function of the grain size (Dielectric Continuum (DC); Hydrodynamic model (HD). After Ref. [1]

To the second order, the exact position, intensity, width, and lineshape of each band depend on many different parameters [1, 45] such as neighboring defects (inclusions, substitutions, or vacancies) [46–51], grains size and shape [25, 52–61], phase orientation (polarization effects [11, 62–66]), or thermomechanical stress (anharmonic effects) [67–71] (Fig. 3.1). Taking full benefit of the available information thus requires some critical interpretation of the raw spectra and an adaptation of standard fitting procedures to the specific case under focus. For instance, different models may be necessary to correlate Raman spectra to grain size (Fig. 3.9):

- In grains much larger than the wavelength, phonons propagate almost in the same way as in perfect "infinite" crystals.
- When the grain size falls in the tens of nanometer range, the Phonon Confinement Model (PCM; see Sect. 3.3.2) accounts for the phonon coherence length limitation by a weighed integration of the optical dispersion curves from Brillouin zone center outwards [1, 72, 73].
- Below a certain size, the very notion of collective vibrations disappears and the Elastic Sphere Model (ESM) takes over, using first principle description of low wavenumber vibrations in a "free-standing" sphere of a homogeneous (constant density) elastic medium [1, 74–78]. In this scheme, the Raman wavenumber values are inversely proportional to nanoparticles diameter [1, 77, 79, 80].
- In semiconductor quantum structures too small to be seen as homogeneous, macroscopic approximations taking the geometry of the sample into account by mechanical and/or electrostatic boundary conditions (the so-called dielectric continuum and hydrodynamic models, with all their modified versions) can be used [56, 81–86].

3.3.2 "Blind" Data Processing

In the (unfavorable but frequent) case when one has no preconceived idea about the Raman parameters most relevant to a specific issue, the comparison of spectra recorded on different samples and/or for different conditions is necessary. If nothing appears at first sight, a chemometric analysis might prove very useful. The main

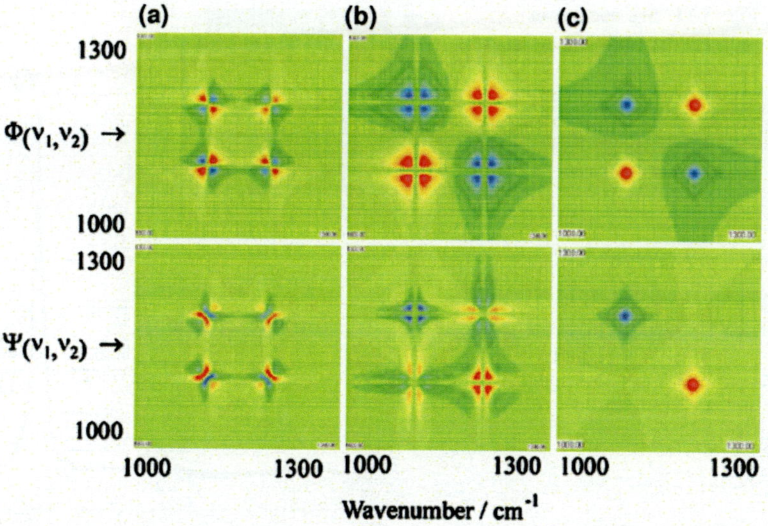

Fig. 3.10 Synchronous (*top*) and asynchronous (*bottom*) 2D-correlation for "model" $20\,\text{cm}^{-1}$-wide bands at 1100 and $1200\,\text{cm}^{-1}$ having opposite wavenumber (**a**), width (**b**), or intensity (**c**) variations. The $1100\,\text{cm}^{-1}$ peak shifts up by $10\,\text{cm}^{-1}$, widens by $2\,\text{cm}^{-1}$ and rises in intensity by 50 %. Reprinted with permission from Elsevier [92].

option is to use statistical multivariate analysis (principal components analysis, factor analysis, ... [87]) to isolate the linear combinations of spectral features changing the most from one spectrum to another (the so-called *principal components*). These well-documented methods are very valuable to tell whether different samples should be connected or not but two-dimensional correlation (C2D) is a good alternative if one is interested in the effect of a given physical parameter p (which can be time but also temperature, pressure, voltage, ...) on a given sample [88, 89]. Supposing that spectra I($\bar{\nu}$,p) are recorded from $p_{\min}$ to $p_{\max}$, the synchronous and asynchronous components of the spectra bidimensional correlation (Fig. 3.10), respectively $\Phi\,(\bar{\nu}_1, \bar{\nu}_2)$ and $\psi\,(\bar{\nu}_1, \bar{\nu}_2)$, are obtained through the following calculation:

$$\Phi\,(\bar{\nu}_1, \bar{\nu}_2) + i\psi\,(\bar{\nu}_1, \bar{\nu}_2) = \frac{\int_0^{\infty} \mathrm{FT}\,[I\,(\bar{\nu}_1, p)] \times \mathrm{FT}^*\,[I\,(\bar{\nu}_2, p)]\,\mathrm{d}P}{\pi\,(p_{\max} - p_{\min})} \tag{3.13}$$

In Eq. (3.13), FT designate Fourier transform, P being the conjugate variable of p parameter. The level curves of $\Phi\,(\bar{\nu}_1, \bar{\nu}_2)$ and $\psi\,(\bar{\nu}_1, \bar{\nu}_2)$ give a bi-dimensional representation of the correlated spectral features: Any local maximum on ϕ diagonal ($\bar{\nu}_1 = \bar{\nu}_2$) indicates a p-driven band shift, whereas off-diagonal extrema ($\bar{\nu}_1 \neq \bar{\nu}_2$) indicate coupled variations in $\bar{\nu}_1$ and $\bar{\nu}_2$ (a positive (negative) sign corresponding to variations proceeding in the same (opposite) direction). If $\Phi\,(\bar{\nu}_1, \bar{\nu}_2) \times \psi\,(\bar{\nu}_1, \bar{\nu}_2)$ is positive, then the signal starts changing in $\bar{\nu}_1$ "before" (i.e. at lower p value) it does in $\bar{\nu}_2$ (and conversely).

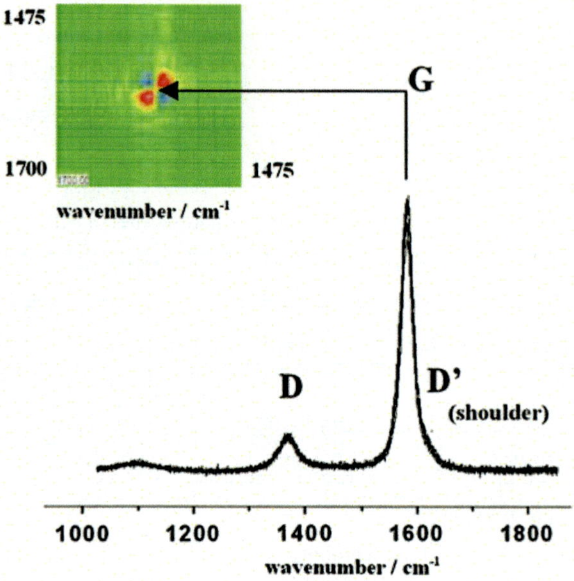

Fig. 3.11 Six spectra of an FT700 carbon fiber strained from 0.093 to 0.533 % (λ_{laser} = 457.9 nm) and the corresponding synchronous component of a 2D-correlation treatment. After [92]

2D-correlation should in theory apply to as-recorded spectra. Yet, baseline fluctuations and noise variations generate typical "streak-like" features and a baseline subtraction improves the results. The interpretation becomes even simpler if spectra are normalized prior to the calculations [90, 91]. Figure 3.10 shows the C2D spectra for two "model" bands at 1100 and 1200 cm^{-1} having opposite variations in wavenumber, width, or intensity. Intensity fluctuations give circular "halos" whereas width changes give "butterfly wings"-like features. As for wavenumber fluctuations, they can easily be recognized by crossing "egg timers" (of opposite signs) on the synchronous component and crescent-shaped features on the asynchronous one. The main limitation of C2D is that unless p is applied harmonically, there is no quantitative information associated with the values of Φ $(\bar{\nu}_1, \bar{\nu}_2)$ and ψ $(\bar{\nu}_1, \bar{\nu}_2)$. Yet, the technique is very sensitive to the smallest variations. A very didactic illustration is provided with the spectra of a carbon fiber strained uniaxially (Fig. 3.11). There is little apparent effect of straining on the wavenumbers but the synchronous C2D function clearly shows a shift of G band.

3.3.3 "LADIR-PARADIS" Program

The proprietary softwares of Raman spectrometers manufacturers allow for hyperspectral maps mathematical treatment but many do not allow for the use of distinct models depending on the region of the map under consideration. Yet, the goal of

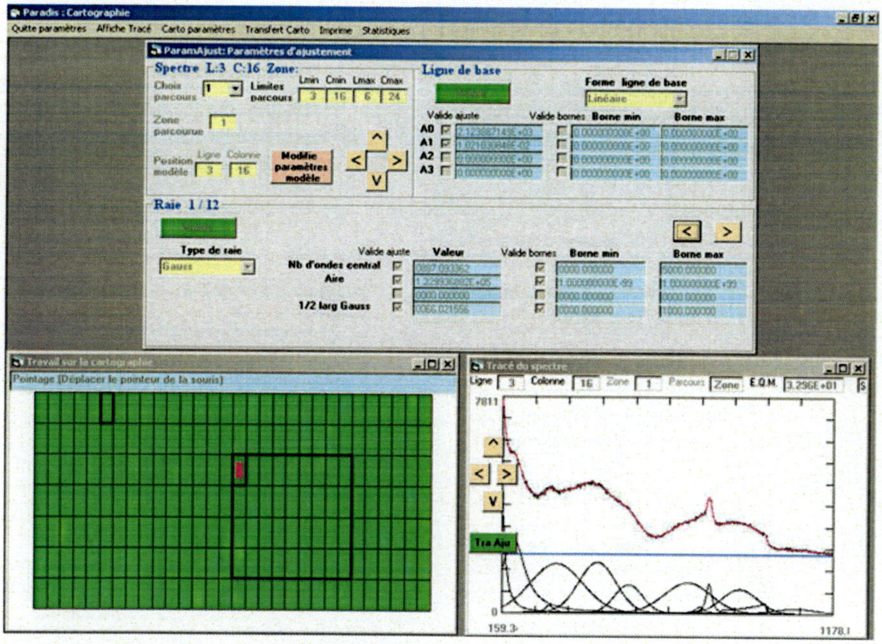

Fig. 3.12 The mapping module in "LADIR-PARADIS" program. Each pixel in the *lower left* window corresponds to one spectrum. The other two windows show the mathematical adjustment (*lower right*) and band parameters (*top*) of the active spectrum (in *pink*). Spectra in the 1 × 1 and 9 × 4 boxes were fitted with specific models

Raman mapping being to point spectral variations, a predefined model may not apply equally well to all spectra, especially if different phases are probed.

This led us to write the "LADIR-PARADIS" program (the French word for paradise is an acronym for "Programme d'Ajustement de RAies Destiné à l'Imagerie de Surfaces"—"Peak adjustment program devoted to surface mapping").[2] The opening window is very similar to what most programs offer, with the possibility to choose a baseline form and input as many bands as wanted for a spectrum adjustment. The parameters of the bands (lineshape, starting width, starting height, and starting center) are set individually and a fit using the Levenberg-Marquardt algorithm can then be launched (there is a possibility to restrict the variations of each spectral parameter between preset values). If a fitted spectrum has been chosen from a map, it can be saved as a model for the global treatment of any hyperspectral map. One specificity of "LADIR-PARADIS" is to include phonon confinement-based fits (see further; Sect. 3.3.2) but its true originality lies within the mapping window (Fig. 3.12):

- the pixels take different colors depending on the correlation factor between the adjusted and experimental spectra (green = good adjustment, red = unsatisfactory adjustment, black = aborted adjustment)

[2] This program is available from the authors.

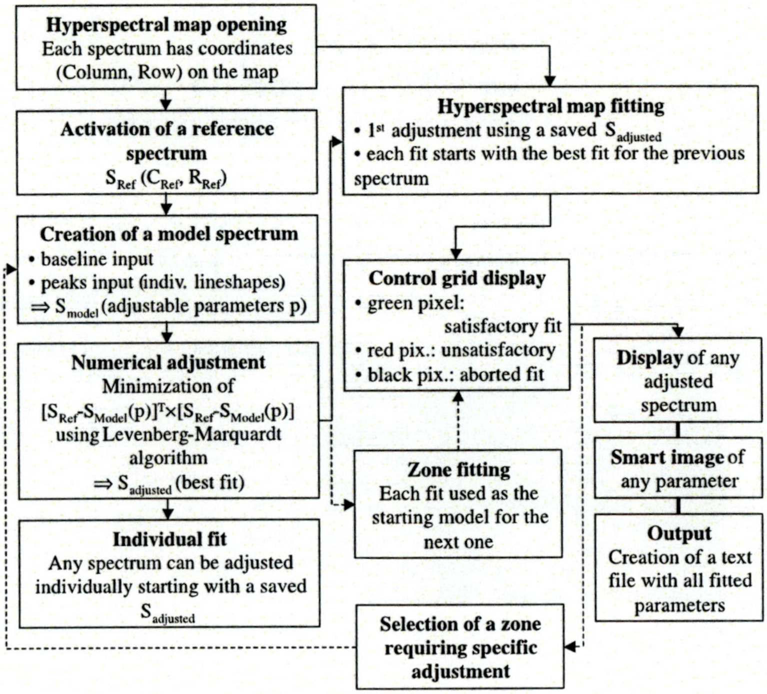

Fig. 3.13 LADIR-PARADIS program working chart

- it is possible to readjust all spectra from any area using a "local" model file (Figs. 3.12 and 3.13).

3.4 Case Studies

3.4.1 Raman Mapping of Carbon in SiC Fibers

Most natural and synthetic fibers exhibit radial structural and compositional variability. Hence the need for cross-sectional mapping which, considering a typical diameter in the range of tens of micrometers, can be easily done by μRS.

Graphite (sp^2 hybridization) has doubly degenerate E_{2g} modes at 42 and 1582 cm^{-1}. The latter is referred to as G band and corresponds to vibrations in the graphene planes. Additional modes called D and D′ (the letter stands for "disorder") appear in the Raman spectra whenever flaws are created, grain size is reduced or graphene planes are curved (Fig. 3.11). Thomsen et al. showed that only a double resonance mechanism can rigorously account for the D band [93] but an empirical relation linking its intensity to the size of graphitic domains (L_g) has been used since 1970 [94]:

$$\frac{I_{D1350\text{cm}^{-1}}}{I_{G1580\text{cm}^{-1}}} = \frac{C(\lambda)}{L_g\,(\text{nm})}; \quad C(\lambda = 514.5\,\text{nm}) = 44 \tag{3.14}$$

Figure 3.14 gives an example where Eq. (3.14) was used to measure the size of carbon moieties (at the limit of the validity range [95]) on the cross-section of a silicon carbide fiber.

3.4.2 Raman Mapping of SiC in SiC Fibers

The Phonon Confinement Model (PCM) was originally proposed to describe the selection rule breaking for phonons confined in isotropic nanospheres of diameter L [72, 73]. It was later adapted to specific cases such as non isotropy [47, 96], size distribution [25, 47], or non spherical shapes (for instance in nanowires and slabs [52, 54, 55, 97]) but we shall present the original approach.

Taking k_{BZ} as the edge of Brillouin zone, q as the reduced wavevector ($0 \leq q = k/k_{BZ} \leq 1$), and assuming isotropic mode dispersion, the Raman intensity is modeled with a weighed integration of dispersion curves $\bar{\nu}_{\text{disp}}(\text{q})$:

$$\text{I}(\bar{\nu}) \propto \int_{q=0}^{q=1} dq \times \frac{\text{e}^{-\text{C}_0 \text{k}_{\text{BZ}}^2 (\text{q}-\text{q}_0)^2 \text{L}^2}}{\left[\bar{\nu} - \bar{\nu}_{\text{disp}}(\text{q})\right]^2 + \left(\frac{\Gamma_0}{2}\right)^2} \tag{3.15}$$

In Eq. (3.15), Γ_0 is the natural Raman linewidth and C_0 governs the confinement strength [52].

The PCM does not apply to acoustic modes (their energy is nil at Brillouin zone center) and is very seldom used with Transverse Optical (TO) modes (on account of their low dispersion). Because the Longitudinal Optical (LO) branches usually reach their maximum frequency at the Brillouin zone center, the integration in Eq. (3.15) introduces contributions on the low frequency side of Raman modes and the resulting peaks are asymmetrical (Fig. 3.1).

"LADIR-PARADIS" program offers the possibility to adjust the Raman bands using Eq. (3.15). The fit returns q_0 and L values based on dispersion curves of the form given in Eq. (3.16) [73], but the program could easily be adapted to alternative expressions [54, 55, 97, 98].

$$\bar{\nu}_{\text{disp}}(\text{q}) = \sqrt{A + \sqrt{A^2 - B\,(1 - \cos \pi q)}} \tag{3.16}$$

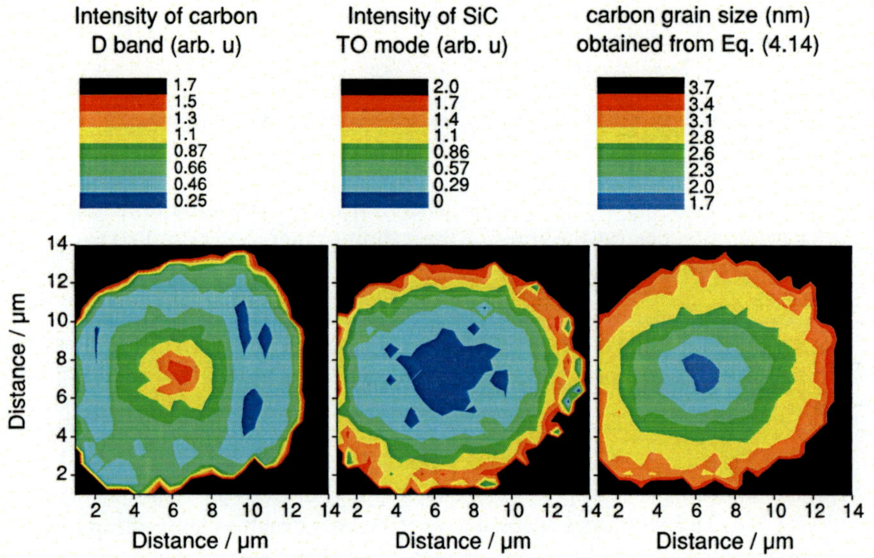

Fig. 3.14 Smart Raman maps (25 × 25 spectra) of the cross-section of an SA3 SiC fiber after 10 h-annealing at 1600 °C in reducing atmosphere (see Refs. [99, 100] for details)

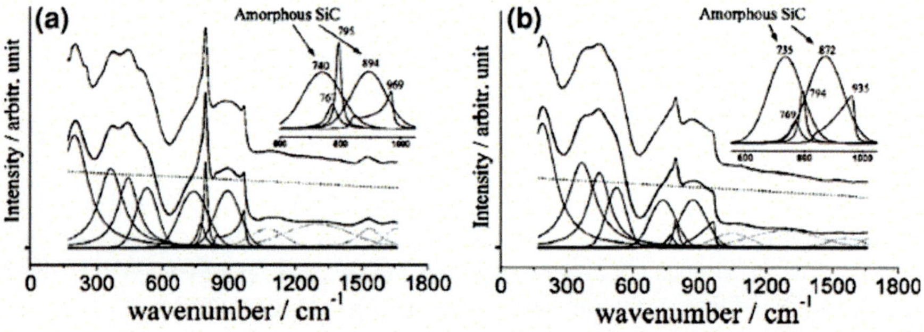

Fig. 3.15 Examples of SiC spectra fitted with a combination of Lorentz, Gauss, and "PCM" bands. The spectra were recorded at the inner (**a**) and outer (**b**) limits of the "pure SiC" region visible in Fig. 3.2. "With permission from Springer Science+Business Media: [101], figure N°6"

Figure 3.15 shows the frontier spectra of the "pure SiC" region pointed in Fig. 3.2. Their TO and LO contributions at ~795 and 970 cm^{-1} were fitted with the PCM. Note that a significant contribution from amorphous SiC was also present.

The physical interpretation of L parameter is a key element of the PCM as the phonon coherence length may either correspond to an actual grain size [24, 72, 102, 103] or to the size of any homogeneous domain limited by defects/impurities [48, 73, 104–110]. Carles et al. even showed in Ge_xC_{1-x} films that L corresponded to the actual particle size for diameters above ~20 nm but characterized domains where alloying had taken place in smaller ones [96].

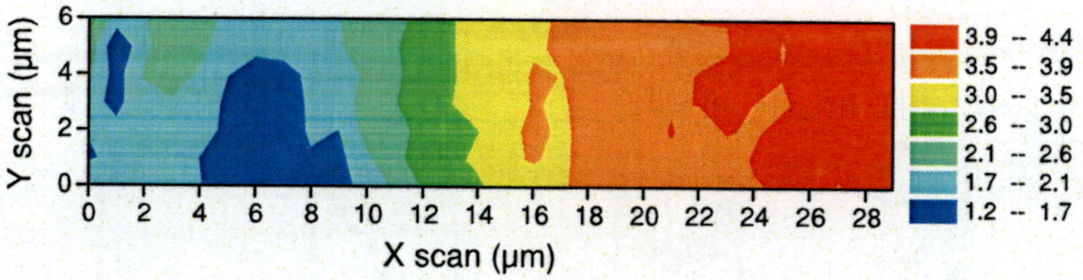

Fig. 3.16 Smart map of the Phonon Confinement Model L parameter (in nm unit) after mathematical adjustment (Eqs. 3.15 and 3.16) of the rectangular mapping from Fig. 3.2 in the SiC-rich region of SCS-6 fiber. Adapted from Ref. [101]

The L values calculated using the PCM in Fig. 3.16 are almost two orders of magnitude smaller than the diameters of the nanocrystals observed by TEM in the same region of the SCS-6 fiber [111]. The phonons appear to be confined by some free carbon in this slightly non stoichiometric region [111, 112]. In the SA-3 fiber however, the calculated L value corresponds to SiC domains extension and is unaffected by the many stacking faults appearing in TEM [113].

Disorder and nanometric dimensions often contribute to phonon confinement and their relative contributions can theoretically be isolated whenever $q_0 \neq 0$ is allowed in Eq. 3.15:

- Disorder disturbs short range order and results in Brillouin zone folding. q_0 should then stand away from Brillouin zone center and L will be inversely proportional to the density of defects.
- If the confinement takes place in perfectly ordered nanocrystals, then the activated modes should be centered at the Γ point ($q_0 = 0$) and L will represent the grain size.

3.4.3 *In situ* Micro-Raman Extensometry of Fibers and Fiber-Reinforced Composites

Because bonds vibrations have some anharmonic character, Raman bands have a tendency to shift in materials submitted to external stress ([1] and references therein). In perfect crystals, symmetry considerations make Raman bandshifts simple linear combinations of the stress tensor components [114–116]. Conversely, if several Raman bands are observed simultaneously and/or simplifying assumptions are made on the stress state, the latter can be measured from experimentally observed band shifts (Fig. 3.17) [41, 42, 69, 117–119].

In nano/polycrystalline or amorphous materials, there is no direct relationship between bonds straining, which commands Raman shifts, and the macroscopic stress field. Yet, if one (or several) direction is made particular by the geometry of the sample, empirical calibrations of stress-induced shifts remain possible. This is the

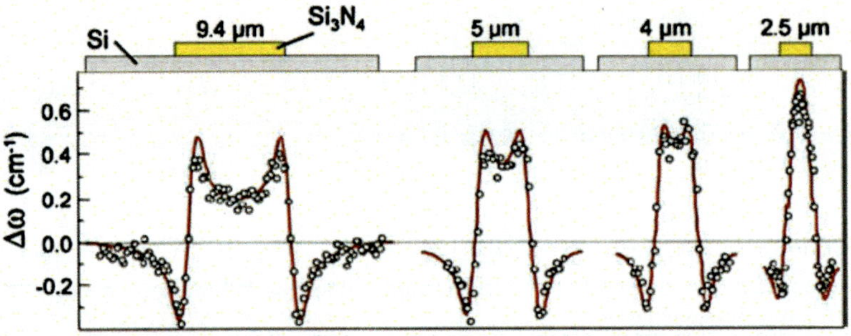

Fig. 3.17 The experimental measurement of Raman shifts in silicon (*points*) can be fitted with a stress-parameterized model (*lines*) taking into account the compression induced by silicon nitride, the laser probe diameter, the penetration depth, etc. Reproduced with permission from Wiley-Blackwell [119]

principle of Raman microextensometry (μRE) technique, which was mainly applied to fiber-reinforced composite materials. In these composites, the fibers are expected to resist axial tensions transferred from the surrounding matrix through interfacial shear stress. Their mechanical characterization therefore calls for a non destructive *in situ* measurement of the fiber axial stress [67, 68].

The first step is to calibrate the S^{ε} and S^{σ} factors linking the wavenumber shift of any specific band to, respectively, the axial strain (ε_f) and axial stress (σ_f) applied to a free-standing fiber (Young's modulus E_f; radius r_f):

$$S^{\varepsilon} = \frac{\bar{\nu}_{\mathrm{vib}} - \bar{\nu}_{\mathrm{vib}}^{\mathrm{no\ strain}}}{\varepsilon_{\mathrm{f}}} = E_{\mathrm{f}} \times \frac{\bar{\nu}_{\mathrm{vib}} - \bar{\nu}_{\mathrm{vib}}^{\mathrm{no\ strain}}}{\sigma_{\mathrm{f}}} = E_{\mathrm{f}} \times S^{\sigma} \qquad (3.17)$$

The in situ analysis of a similar fiber through a polymer matrix (or a thin ceramic matrix layer [120, 121]) can then be used to estimate the axial stress in an embedded fiber. A mapping along the fiber axis gives a strain profile ε_f (l) from which the Interfacial Shear Stress τ (ISS), a key element in the damage tolerance of composites, can be derived :

$$\tau = -E_{\mathrm{f}}\frac{\mathrm{r_f}}{2}\left[\frac{\mathrm{d}\varepsilon_{\mathrm{f}}(l)}{\mathrm{d}l}\right] = -\frac{\mathrm{r_f}}{2}\left[\frac{\mathrm{d}\sigma_{\mathrm{f}}(l)}{\mathrm{d}l}\right] \qquad (3.18)$$

The microdroplet test (where a matrix droplet is deposited and dried on a fiber) provides the simplest way of actually preparing a sample and measuring the interfacial shear stress for a given fiber/matrix system [122–124] (Fig. 3.18a). The interfacial to axial stress ratio measures how well the stress will transfer to the reinforcing fiber in a true composite. Its diminution for increasing strains results from the degradation of the fiber/matrix interface. A number of "in line" Raman mappings have been performed to measure axial stress and ISS in polymer-embedded fibers [68, 125–127, 129–132]. Figure 3.18b shows an example for the in-line mapping of an aramid fiber reinforcing an epoxy matrix strained to 1.7 % [131]. The horizontal

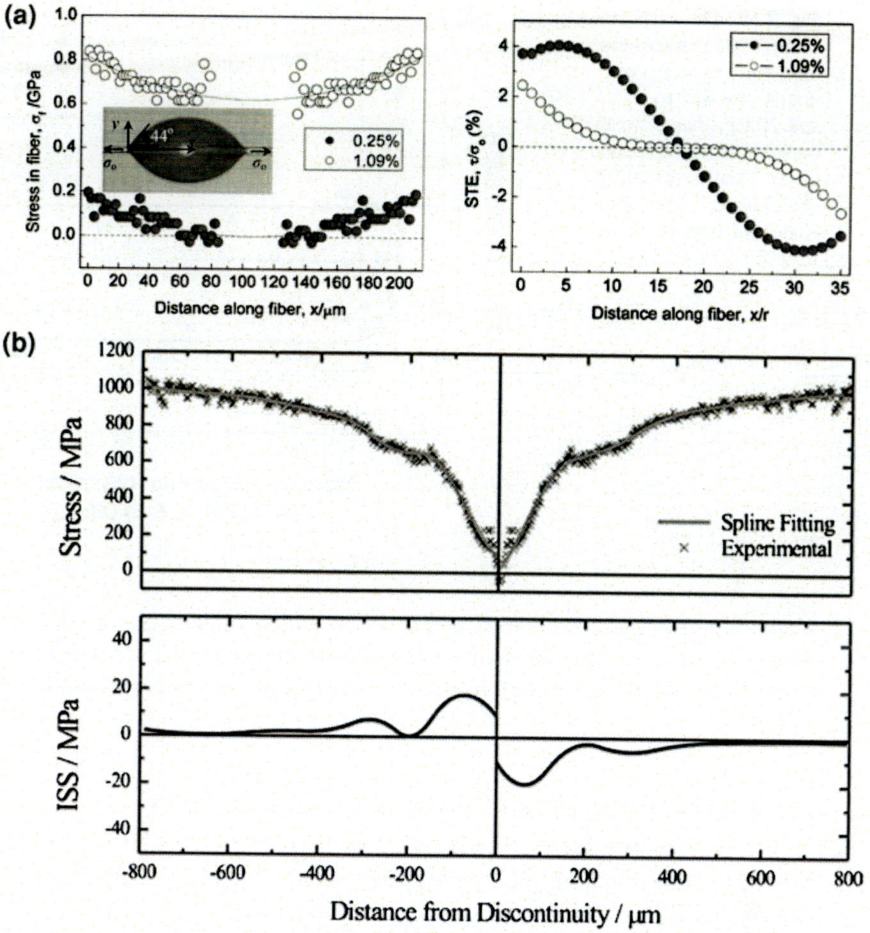

Fig. 3.18 The wavenumber shift of Kevlar-29 fiber 1610 cm^{-1} Raman peak was measured through **(a)** epoxy droplets (at 0.25 or 1.09 % fiber strain) and **(b)** the epoxy matrix of a 1.7 %-strained uniaxial composite. The fiber axial stress (σ_f) was then derived from preliminary calibrations and converted into the interfacial shear stress (ISS or τ) using Eq. 3.18. The τ/σ_f ratio measures the stress transfer efficiency (STE). Reprinted with permission from Elsevier [124] and [131]

axis origin corresponds to a fiber break where the nil stress results from full debonding. The ineffective length, over which the stress supported by the fiber builds up from zero (at a fiber tip or a matrix crack) to its maximum value, is around 400 μm.

Because the Raman signal of Carbon NanoTubes (CNTs) shifts under axial stress [133], these nanotubes were investigated as possible small dimension embedded stress gauges in the vicinity of weakly scattering reinforcing fibers (Fig. 3.19) [134–136] or in composites directly reinforced with CNTs [136, 137]. More generally, Raman spectroscopy is a lead technique for the analysis of CNTs [1, 138, 139] because a CNT is formally obtained by rolling the $n\mathbf{a}_1 + m\mathbf{a}_2$ vector of a graphene

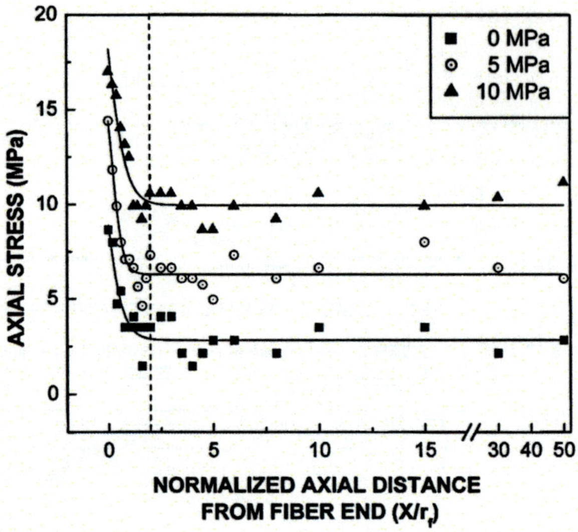

Fig. 3.19 The axial stress along a glass fiber was measured in an epoxy matrix stressed to 0, 5, and 10 MPa based on the band shifts of embedded CNTs Raman scattering (nil abscissa = fiber end). Reprinted with permission from Elsevier [135]

plane (n and m two integers; $\mathbf{a}_1$ and $\mathbf{a}_2$ basal lattice vectors) and each $\{n, m\}$ set, which relates to properties such as the tube diameter or the electrical properties (metallic vs. semiconducting), resonates for a specific excitation wavelength.

3.4.4 Raman Mapping of Micro-Indented Samples

Microindentation load-displacement curves are used to measure mechanical properties such as hardness (defined according to the indenter geometry), Young's modulus, strength, plasticity, etc. The stress can reach very high values locally (as evidenced through wavenumber shifts) and the Raman study of micro/nano-indentations is a possible way of monitoring phase transitions under non hydrostatic stress [140, 141]. Mapping capabilities are a great asset as the pressure conditions vary from one point to another (the imprints dimensions are in the range of a few tens of micrometers) because of the imperfect geometry of the indenters, the possibility of intragranular versus intergranular indentation, crack propagation at the imprints corners, materials pileup at the imprint periphery, etc., [38, 44, 141–148].

Figure 3.20 illustrates the Raman mapping of a Vickers microindentation in ZnSe. The indenter pyramid shape is easily recognized from the intensity variations (Fig. 3.20c; the deeper the imprint, the lower the intensity collected with a focus on the sample surface plane) but the wavenumber shifts are random, indicating a very inhomogeneous stress repartition in the imprint [149].

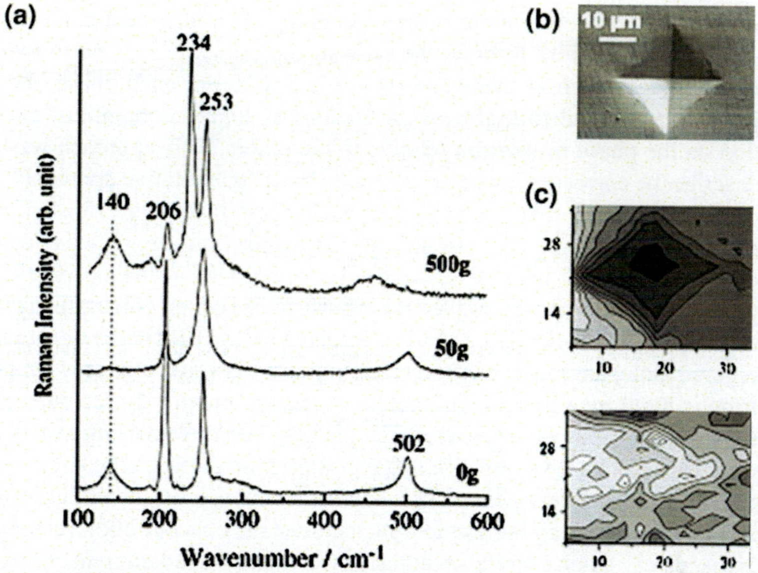

Fig. 3.20 **a** Raman spectra of ZnSe indented with a Vickers pyramid-shaped diamond micro-indenter ($\lambda_{laser} = 632.8$ nm); **b** Micrograph of the indenter imprint after 50 g loading; **c** Intensity variation (*top*) and wavenumber shifts (*bottom*) of the 206 cm^{-1} peak on the map of (**b**) imprint (X- and Y- scales in μm). Adapted from Ref [149]

3.4.5 Metal Corrosion Layers

Raman spectroscopy is interesting for the study of corrosion because the technique has a strong sensitivity to metal-corroding anions bonds. Raman scattering allows for a phase selective microscopic scale characterization [150, 151], with the additional advantages of easy implementation, minimal sample preparation and imaging possibilities [152]. Corrosion spots may even be observed through coatings such as paint.

The observation and characterization of ancient artefact corrosion products is a key to judging the validity of long-term corrosion models since laboratory accelerated corrosion testing does not simulate long-term corrosion properly [153]. The nature and extent of long-term corrosion processes in metal-based cultural heritage objects and monuments is strongly influenced by the environment (usually soils or the atmosphere) they have stayed in [154] and a better understanding of the mechanisms at play will lead to new corrosion diagnostics tools, to improved restoration treatments, or to propose alternative conservation strategies. The possibility to predict corrosion progress will also be very helpful in view of considering iron for nuclear waste storage overcontainers.

In the specific frame of a study on the indoor atmospheric corrosion of iron, a "qualitative" Raman-based localization of corrosion phases was achieved [155].

Each phase was detected either directly through the integrated intensity in a specific ROI or, on occasion, through the ratio of intensities integrated in two ROIs. This led to images such as those shown in Fig. 3.21b. For each phase, the color levels were user-adjusted to highlight the regions of higher content but gave no indication on the phase proportion relative to the others. This approach was sufficient to describe the corrosion layers organization but a "quantitative approach" was needed, in addition to the characterization of each phase reactivity [156, 157], in order to propose quantitative diagnosis indexes [158] and confront predictive corrosion models. "LADIR-CAT" (for "Corrosion ATmosphérique"—"ATmospheric Corrosion") program was thus developed for the treatment of Raman corrosion layer hyperspectral maps [159], following a strategy initiated at LADIR for the quantitative study of organic molecule redox conformers [160] and wool dye mixtures [161]: each spectrum is fitted by a linear combination of spectra previously recorded under strictly similar conditions on pure reference phases. The different intensities of the references thus account for the difference in Raman scattering efficiency and light penetration of the different phases. This approach circumvents the difficulty (specially for solids) of calibrating the Raman intensity versus concentration relationship, which is mandatory when spectra are adjusted from normalized references [162]. However, it implicitly assumes the phase distribution is invariant along the whole investigated thickness (see Sect. 3.2.2 and discussion of Eq. 3.9 in Sect. 3.2.3).

Figure 3.22 gives an example of "LADIR-CAT" fitting and Fig. 3.23 presents the general working scheme of the program, which returns two kinds of data:

- One matrix for each phase, with as many columns and rows as there are respectively X and Y pixels in the Raman map. The number in each cell represents the calculated phase percentage for the corresponding pixel. Each matrix can be opened in a commercial software for a 2D contour fill plotting (Fig. 3.21c).
- A table giving the average percent content of each phase over the whole map, which can be presented in the form of a sector diagram (Fig. 3.21a).

LADIR-CAT was successfully used to map nine samples from the reinforcing chains of the Amiens cathedral (fifteenth century) and propose a quantitative estimation of phase composition in the corrosion system [158].

3.4.6 Raman Mapping of Biological Samples

When too many scatterers contribute to micro-Raman spectra for any one of them to be tracked, which is typically the case in samples of biological interest (lipids, proteins, carbohydrates, nucleic acids, etc., are probed simultaneously), a selectivity can be achieved with specific signal enhancers (SERS). A Raman mapping then becomes possible, as illustrated in Fig. 3.24 and in-vivo analysis possibilities were even demonstrated in small animals [163]. Because of the signal complexity in biological samples, a multivariate statistical analysis might prove useful for the analysis of Raman maps [164].

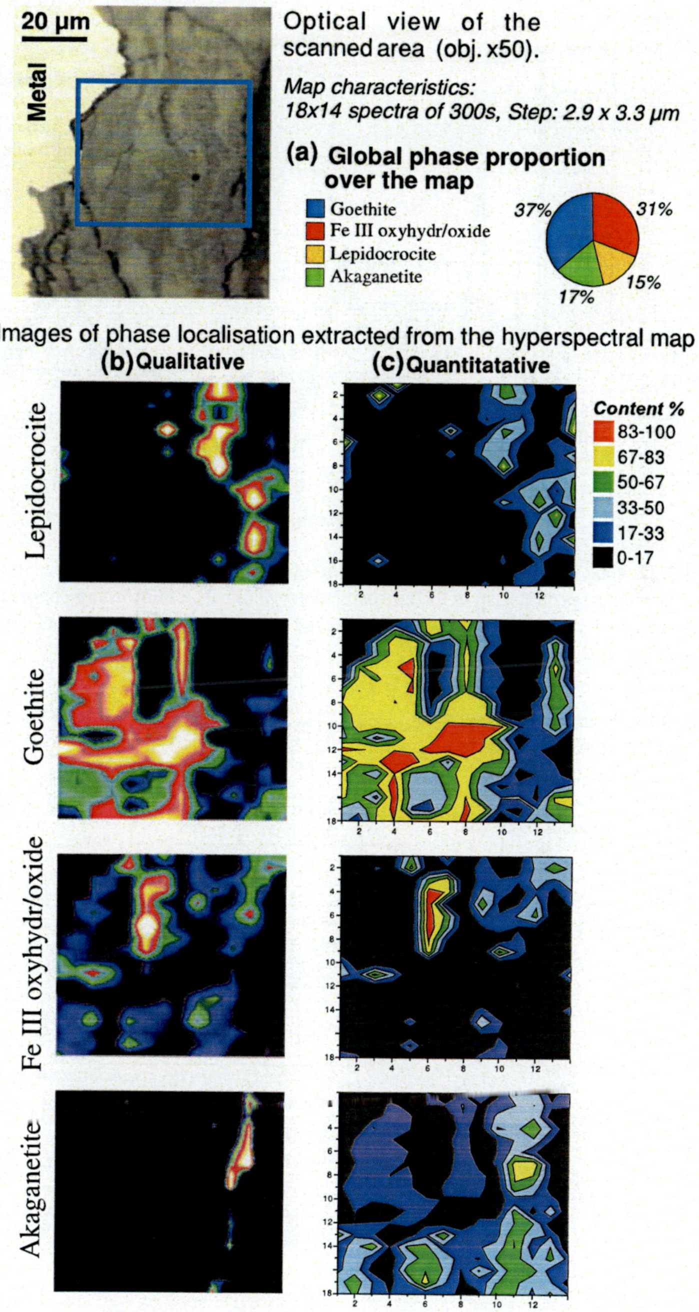

Fig. 3.21 Raman structural mapping of the indoor atmospheric corrosion layer of a sample from Amiens Cathedral. **a** the overall phase repartition, **b** "qualitative" phase mapping (based on regions of interest), and **c** quantitative results obtained from spectral decomposition by "LADIR-CAT" program

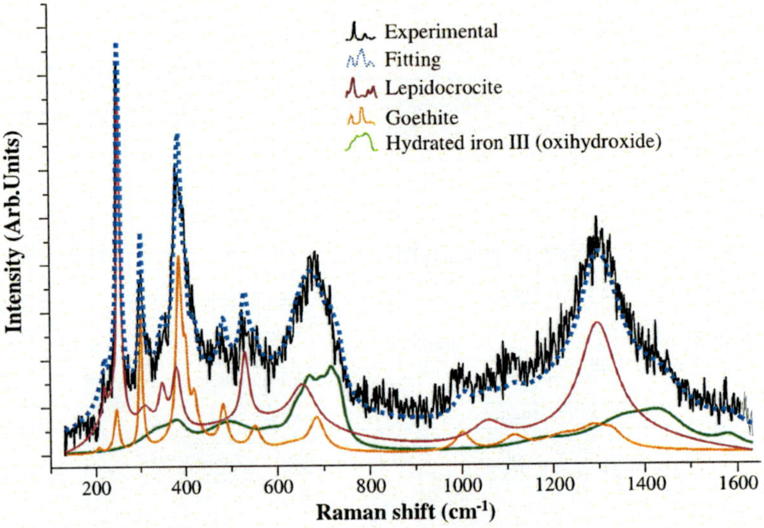

Fig. 3.22 One spectrum ($\lambda_{laser} = 514.5$ nm, objective ×100, laser power = 100 μW, 300 s) taken from the hyperspectral map of the corrosion layer of one iron reinforcing chain in Amiens cathedral. "LADIR-CAT" program was run to adjust the data with a combination of three reference phase spectra

Fig. 3.23 Working scheme of "LADIR-CAT" application for automated quantitative treatment of Raman maps by the decomposition of the spectra set

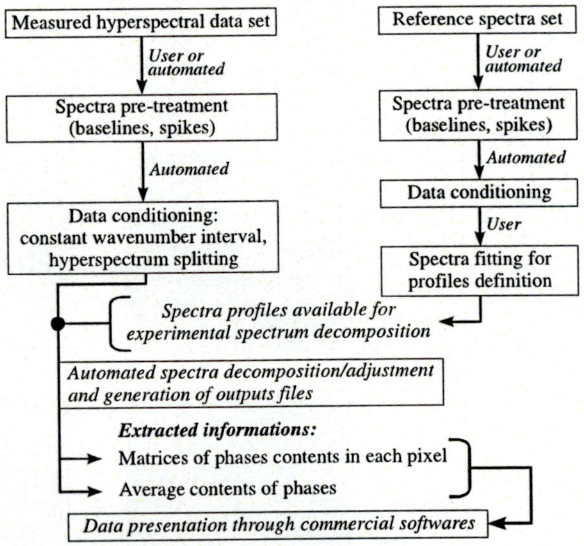

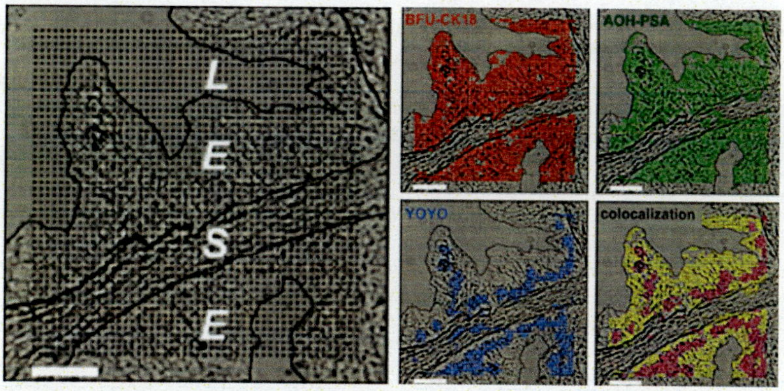

Fig. 3.24 532 nm-excited Raman scattering was recorded in XY-raster pattern (50 × 50 micrometers) over a tissue section spanning some stromal tissue (S) and prostate glands epithelia (E) or lumen (L). The sample had been stained with **i)** CK18 and PSA antibodies-conjugated Composite Organic–Inorganic Nanoparticles (COINS) associating silver (for resonance enhancement) with either basic fushin (BFU) or acridin orange (AOH) Raman labels; **ii)** DNA-specific YOYO® fluorescent dye. Localization images were obtained after applying thresholds to each probe weight in a spectral deconvolution based on reference spectra. The colocalization image identifies epithelial nuclei (*magenta*) and coexpression of CK18 and PSA specifically in the epithelium (*yellow*). Scale bars, 10 microns. Adapted, with permission from American Chemical Society [165]

3.5 Perspectives

Raman micro-spectroscopy with mapping capabilities offers great promise for fast and non-destructive analysis of various nanophased samples. From a technical point of view, the lateral resolution (about 1 square micrometer) will probably increase in the future with new objectives (×200 magnification) and tip-enhancement heads. However, the automated recording of hyperspectral maps is advantageous only if the large amount of data generated can be processed at once. Existing software packages provide a number of data treatment options but the possibility for users to write their own procedures would probably greatly improve the versatility of the method.

References

1. G. Gouadec, Ph Colomban, Raman spectroscopy of nanomaterials: how spectra relate to disorder, particle size and mechanical properties. Prog. Cryst. Growth Charact. Mater. **53**, 1–56 (2007)
2. W. Kiefer, (ed.), G. Gouadec, (Guest ed.), Ph. Colomban (Guest ed.), Special issue on the Raman study of nanomaterials. J. Raman Spectr. **38**, 597–796 (2007)
3. G. Turrell, Chap 4: Raman Imaging, in *Raman Microscopy Developments and Applications*, ed. by G. Turrell, J. Corset (Academic Press, London, 1996)
4. Ph. Colomban, Raman analyses and "Smart Imaging" of nanophases and nanosized materials. Spectrosc. Eur. **15**, 8–16 (2003)

5. M. Havel, Ph. Colomban, Smart Raman and Rayleigh Spectroscopy for the analysis of nanomaterials. Microsc. Anal. **20**, 13–15 (2006)
6. Ph. Colomban, Raman Microspectrometry and Imaging of Ceramic Fibers in CMCs and MMCs, in *Advances in Ceramic Matrix Composites V*, ed. by N.P. Bansal, J.P. Singh, E. Ustundag (The American Ceramic Society, Westerville, 2000)
7. M. Baranska, L.M. Proniewicz, Raman mapping of caffeine alkaloid. Vibr. Spectr. **48**, 153–157 (2008)
8. D.W. Piston, Choosing objective lenses: the importance of numerical aperture and magnification in digital optical microscopy. Biol. Bull. **195**, 1–4 (1998)
9. M. Born, E. Wolf, *Principles of Optics* (Pergamon Press, Oxford, 1985)
10. E. Abbe, Archiv. Mikroscopische Anat. **9**, 413 (1873)
11. P.J. Pauzauskie, D. Talaga, K. Seo et al., Polarized Raman confocal microscopy of single gallium nitride nanowires. J. Am. Chem. Soc. **127**, 17146–17147 (2005)
12. D.A. Long, *Raman Spectroscopy* (McGraw-Hill, New York, 1977)
13. M. Delhaye, J. Barbillat, J. Aubard et al., Chap 3: Instrumentation, in *Raman Microscopy Developments and Applications*, ed. by G. Turrell, J. Corset (Academic Press, London, 1996)
14. I. De Wolf, Micro-Raman spectroscopy to study local mechanical stress in silicon integrated circuits. Semicond. Sci. Technol. **11**, 139–154 (1996)
15. G. Turrell, M. Delhaye, P. Dhamelincourt, Chap 2: Characteristics of Raman Microscopy. in *Raman Microscopy Developments and Applications*. (Academic Press, London, 1996)
16. C.J.R. Sheppard, T. Wilson, Depth of field in the scanning microscope. Optics Lett. **3**, 115–117 (1978)
17. C.-B. Juang, L. Finzi, C.J. Bustamante, Design and application of a computer-controlled confocal scanning differential polarization microscope. Rev. Sci. Instrum. **59**, 2399–2408 (1988)
18. V.K. Ramshesh, J.J. Lemasters, Pinhole shifting lifetime imaging microscopy. J. Biomed. Opt. **13**, 064001 (2008)
19. K. Kneipp, M. Moskovits, H. Kneipp, *Surface-Enhanced Raman Scattering*. (Springer, Berlin, 2006)
20. N.P.W. Pieczonka, R.F. Aroca, Single molecule analysis by surface-enhanced Raman scattering. Chem. Soc. Rev. **37**, 946–954 (2008)
21. M. Montagna, R. Dusi, Raman scattering from small spherical particles. Phys. Rev. B **52**, 10080–10089 (1995)
22. A. Roy, A.K. Sood, Growth of CdS_xSe_{1-x} nanoparticles in glass matrix by isochronal thermal annealing: confined acoustic phonons and optical absorptions studies. Solid State Comm. **97**, 97–102 (1996)
23. M. Ferrari, F. Gonella, M. Montagna et al., Detection and size determination of Ag nanoclusters in ion-exchanged soda-lime glasses by waveguided Raman spectroscopy. J. Appl. Phys. **79**, 2055–2059 (1996)
24. V. Paillard, P. Puech, M.A. Laguna et al., Improved one-phonon confinement model for an accurate size determination of silicon nanocrystals. J. Appl. Phys. **86**, 1921–1924 (1999)
25. W.K. Choi, Y.W. Ho, V. Ng, Effect of size of Ge nanocrystals embedded in SiO_2 on Raman spectra. Mater. Phys. Mech. **4**, 46–50 (2001)
26. M. Ivanda, K. Babocsi, C. Dem et al., Low wavenumber Raman scattering from CdS_xSe_{1-x} quantum dots embedded in a glass matrix. Phys. Rev. B **67**, 235329 (2003)
27. K.W. Adu, Q. Xiong, H.R. Gutierrez et al., Raman scattering as a probe of phonon confinement and surface optical modes in semiconducting nanowires. Appl. Phys. **A85**, 287 (2006)
28. M. Ivanda, A. Hohl, M. Montagna et al., Raman scattering of acoustical modes of silicon nanoparticles embedded in silica matrix. J. Raman Spectrosc. **37**, 161 (2006)
29. M. Ivanda, K. Furic, S. Music et al., Low wavenumber Raman scattering of nanoparticle and nanocomposite materials. J. Raman Spectr. **38**, 647–659 (2007)
30. C. Pinghini, D. Aymes, N. Millot et al., Low-frequency Raman characterization of size-controlled anatase TiO_2 nanopowders prepared by continuous hydrothermal syntheses. J. Nanopart. Res. **9**, 309–315 (2007)

31. S.K. Ram, M.N. Islam, S. Kumar et al., Evidence of bimodal crystallite size distribution in μc-Si: H films. Mater. Sci. Eng. A **159–160**, 34–37 (2009)
32. S.K. Gupta, R. Desai, P.K. Jha et al., Titanium dioxide synthesized using Titanium chloride: size effect study using Raman spectroscopy and photoluminescence. J. Raman Spectr. **41**, 350–355 (2010)
33. D. Richards, Near-field microscopy: throwing light on the nanoworld. Phil. Trans. R. Soc. Lond. A **361**, 2843–2857 (2003)
34. E. Bailo, V. Deckert, Tip-enhanced Raman scattering. Chem. Soc. Rev. **37**, 921–930 (2008)
35. B.-S. Yeo, J. Stadler, T. Schmid et al., Tip-enhanced Raman spectroscopy—Its status, challenges and future directions. Chem. Phys. Lett. **472**, 1–13 (2009)
36. N.J. Everall, Confocal Raman microscopy: why the depth resolution and spatial accuracy can be much worse than you think. Appl. Spectr. **54**, 1515–1520 (2000)
37. J.P. Tomba, J.M. Pastor, Confocal Raman microspectroscopy with dry objectives: a depth profiling study on polymer films. Vibr. Spectr. **44**, 62–68 (2007)
38. T. Jawhari, J.C. Merino, J.M. Pastor, Micro-Raman spectroscopy study of the process of microindentation in polymers. J. Mater. Sci. **27**, 2231–2242 (1992)
39. S. Schuckler, S.W. Huffman, I.W. Levin, Vibrational microspectroscopic imaging: spatial resolution enhancement. Proc. SPIE **5321**, 157 (2004)
40. J. Kasim, Y. Ting, Y.Y. Meng et al., Near-field Raman imaging using optically trapped dielectric microsphere. Opt. Express **16**, 7976 (2008)
41. A. Atkinson, S.C. Jain, Spatially resolved stress analysis using Raman spectroscopy. J. Raman Spectrosc. **30**, 885–891 (1999)
42. B. Dietrich, K.F. Dombrowski, Experimental challenges of stress measurements with resonant Micro-Raman spectroscopy. J. Raman Spectrosc. **30**, 893–897 (1999)
43. K. Chikama, K. Matsubara, S. Oyama et al., Three-dimensional confocal Raman imaging of volume holograms formed in ZrO_2 nanoparticle-photopolymer composite materials. J. Appl. Phys. **103**, 113108 (2008)
44. L.-H. He, E.A. Carter, M.V. Swain, Characterization of nanoindentation-induced residual stresses in human enamel by Raman microspectroscopy. Anal. Bioanal. Chem. **389**, 1185–1192 (2007)
45. W.B. White, The structure of particles and the structure of crystals: information from vibrational spectroscopy. J. Ceram. Process. Res. **6**, 1–9 (2005)
46. Ph Colomban, F. Romain, A. Neiman et al., Double Perovskites with oxygen structural vacancies: Raman spectra, conductivity and water uptake. Solid State Ionics **145**, 339–347 (2001)
47. J.E. Spanier, R.D. Robinson, F. Zhang et al., Size-dependent properties of CeO_{2-y} nanoparticles as studied by Raman scattering. Phys. Rev. B **64**, 2450407 (2001)
48. I. Kosacki, T. Suzuki, V. Petrovsky et al., Raman scattering and lattice defects in nanocrystalline CeO_2 thin films. Solid State Ionics **149**, 99–105 (2002)
49. A.V. Gomonnai, Y.M. Azhniuk, V.O. Yukhymchuk et al., Confinement-, surface- and disorder-related effects in the resonant Raman spectra of nanometric $CdS_{1-x}Se_x$ crystals. Phys. Status Solidi B **239**, 490–499 (2003)
50. D. Chiriu, P.C. Ricci, C.M. Carbonaro, Vibrational properties of mixed $(Y_3Al_5O_{12})_x$-$(Y_3Sc_2Ga_3O_{12})_{1-x}$crystals. J. Appl. Phys. **100**, 033101–033105 (2006)
51. Ph Colomban, A. Tournié, L. Bellot-Gurlet, Raman identification of glassy silicates used in ceramic, glass and jewellry: a tentative differentiation guide. J. Raman Spectrosc. **37**, 841–852 (2006)
52. I.H. Campbell, P.M. Fauchet, The effects of mycrocrystal size and shape on the one phonon Raman spectra of crystalline semiconductors. Solid State Comm. **58**, 739–741 (1986)
53. K.R. Patton, M.R. Geller, Phonons in a nanoparticle mechanically coupled to a substrate. Phys. Rev. B **67**(155), 418 (2003)
54. S. Hofmann, C. Ducati, R.J. Neil et al., Gold catalysed growth of silicon nanowires by plasma enhanced chemical vapor deposition. J. Appl. Phys. **94**, 6005–6012 (2003)
55. Q. Xiong, R. Gupta, K.W. Adu et al., Raman spectroscopy and structure of crystalline gallium phosphide nanowires. J. Nanosci. Nanotech. **3**, 335–339 (2003)

56. C. Trallero-Giner, Optical phonons and resonant Raman scattering in II–VI spheroidal quantum dots. Phys. Status Solidi B **241**, 572–578 (2004)
57. A.K. Arora, M. Rajalakshmi, T.R. Ravindran, Phonon Confinement in Nanostructured Materials, in *Encyclopedia of Nanoscience and Nanotechnology*, ed. by H.S. Nalwa (American Scientific Publishers, Valencia, 2004)
58. L. Saviot, D.B. Murray, M.D.C. Marco de Lucas, Vibrations of free and embedded anisotropic elastic spheres: application to low-frequency Raman scattering of silicon nanoparticles in silica. Phys. Rev. B **69**, 113402 (2004)
59. K.-Y. Lee, J.-R. Lim, H. Rho et al., Evolution of optical phonons in CdS nanowires, nanobelts, and nanosheets. Appl. Phys. Lett. **91**, 201901 (2007)
60. A.G. Rolo, M.I. Vasilevskiy, Raman spectroscopy of optical phonons confined in semiconductor quantum dots and nanocrystals. J. Raman Spectr. **38**, 618–633 (2007)
61. Y.-T. Nien, B. Zaman, J. Ouyang et al., Raman scattering for the size of CdSe and CdS nanocrystals and comparison with other techniques. Mater. Lett. **62**, 4522–4524 (2008)
62. S. Pojprapai, J.L. Jones, M. Hoffman, Determination of domain orientation in Lead Zirconate Titanate ceramics by Raman spectroscopy. Appl. Phys. Lett. **88**, 162903 (2006)
63. M. Deluca, T. Sakashita, G. Pezzotti, Polarized Raman scattering of domain structures in polycrystaline Lead Zirconate Titanate. Appl. Phys. Lett. **90**, 051919 (2007)
64. M. Becker, H. Scheel, S. Christiansen et al., Grain orientation, texture, and internal stress optically evaluated by micro-Raman spectroscopy. J. Appl. Phys. **101**, 063531 (2007)
65. D.B. Murray, C.H. Netting, R.B. Mercer et al., Polarizability calculation of vibrating nanoparticles for intensity of low wavenumber Raman scattering. J. Raman Spectr. **38**, 770–779 (2007)
66. H.M. Fan, X.F. Fan, Z.H. Ni et al., Orientation-dependant Raman spectroscopy of single Wurtzite CdS nanowires. J. Phys. Chem. C **112**, 1865–1870 (2008)
67. C. Galiotis, Laser Raman spectroscopy, a new stress/strain measurement technique for the remote and online non-destructive inspection of fiber reinforced polymer composites. Mater. Technol. **8**, 203–209 (1993)
68. L.S. Schadler, C. Galiotis, Fundamentals and applications of micro Raman spectroscopy to strain measurements in fibre reinforced composites. Inter. Mater. Rev. **40**, 116–134 (1995)
69. E. Anastassakis, Selection rules of Raman scattering by optical phonons in strained cubic crystals. J. Appl. Phys. **82**, 1582–1591 (1997)
70. G. Lucazeau, Effect of pressure and temperature on Raman spectra of solids: anharmonicity. J. Raman Spectrosc. **34**, 478–496 (2003)
71. D. Bollas, J. Parthenios, C. Galiotis, Effect of stress and temperature on the optical phonons of aramid fibers. Phys. Rev. B **73**, 094103 (2006)
72. H. Richter, Z.P. Wang, L. Ley, The one phonon Raman spectrum in microcrystalline silicon. Solid State Comm. **39**, 625–629 (1981)
73. P. Parayanthal, F.H. Pollak, Raman scattering in alloy semiconductors: "spatial correlation" model. Phys. Rev. Lett. **52**, 1822–1825 (1984)
74. H. Lamb, On the vibrations of an elastic sphere. Proc. London Math. Soc. **13**, 189–212 (1882)
75. A. Tamura, K. Higeta, T. Ichinokawa, Lattice vibrations and specific heat of a small particle. J. Phys. C: Solid State Phys. **15**, 4975–4991 (1982)
76. E. Duval, Far-Infrared and Raman vibrational transitions of a solid sphere: selection rules. Phys. Rev. B **46**, 5795–5797 (1992)
77. A. Tanaka, S. Onari, T. Arai, Low-frequency Raman scattering from CdS microcrystals embedded in a Germanium Dioxide glass matrix. Phys. Rev. B **47**, 1237–1243 (1993)
78. L. Saviot, D.B. Murray, The connection between elastic scattering cross sections and acoustic vibrations of an embedded nanoparticle. Phys. Status Solidi C **1**, 2634–2637 (2004)
79. E. Duval, A. Boukenter, B. Champagnon, Vibration eigenmodes and size of microcrystallites in glass: observation by very-low-frequency Raman scattering. Phys. Rev. Lett. **56**, 2052–2055 (1986)
80. P. Verma, W. Cordts, G. Irmer et al., Acoustic vibrations of semiconductor nanocrystals in doped glasses. Phys. Rev. B **60**, 5778–5785 (1999)

81. M.C. Klein, F. Hache, D. Ricard et al., size dependence of electron–phonon coupling in semiconductor nanospheres: the case of CdSe. Phys. Rev. B **42**, 11123–11132 (1990)
82. F. Comas, C. Trallero-Giner, N. Studart et al., Interface optical phonons in spheroidal dots: Raman selection rules. Phys. Rev. B **65**, 073303 (2002)
83. E.P. Pokatilov, S. Klimin, V.M. Fomin et al., Multi-phonon Raman scattering in semiconductor nanocrystals: importance of non-adiabatic transitions. Phys. Rev. B **65**, 075316 (2002)
84. M.I. Vasilevskiy, Dipolar vibrational modes in spherical semiconductor quantum dots. Phys. Rev. B **66**, 195326 (2002)
85. V.A. Fonoberov, A.A. Balandin, Polar optical phonons in Wurtzite spheroidal quantum dots: theory and application to ZnO and ZnO/MgZnO nanostructures. J. Phys. Cond. Matter **17**, 1085–1097 (2005)
86. Q.H. Zhong, Optical phonon modes in a free-standing quantum wire with ring geometry. Phys. Lett. A **372**, 5932–5937 (2008)
87. H. Shinzawa, K. Awa, W. Kanematsu et al., Multivariate data analysis for Raman spectroscopic imaging. J. Raman Spectr. **40**, 1720–1725 (2009)
88. I. Noda, Generalized two-dimensional correlation method applicable to infrared, Raman, and other types of spectroscopy. Appl. Spectrosc. **47**, 1329–1336 (1993)
89. Y.M. Jung, I. Noda, New approaches to generalized two-dimensional correlation spectroscopy and its applications. Appl. Spectr. Rev. **41**, 515–547 (2006)
90. M.A. Czarnecki, Interpretation of two-dimensional correlation spectra: science or art? Appl. Spectrosc. **52**, 1583–1590 (1998)
91. M.A. Czarnecki, Two-dimensional correlation spectroscopy: effect of normalization of the dynamic spectra. Appl. Spectrosc. **53**, 1392–1397 (1999)
92. G. Gouadec, J.-P. Forgerit, Ph Colomban, Choice of the working conditions for Raman extensometry of carbon and SiC fibers by 2D correlation. Comp. Sci. Technol. **62**, 505–511 (2002)
93. C. Thomsen, S. Reich, Double resonant Raman scattering in graphite. Phys. Rev. Lett. **85**, 5214–5217 (2000)
94. F. Tuinstra, J.L. Koenig, Raman spectrum of graphite. J. Chem. Phys. **53**, 1126–1130 (1970)
95. A.C. Ferrari, J. Robertson, Interpretation of Raman spectra of disordered and amorphous carbon. Phys. Rev. B **61**, 14095–14107 (2000)
96. R. Carles, A. Mlayah, M.B. Amjoud et al., Structural characterization of Ge microcrystals in Ge_xC_{1-x} films. Jap. J. Appl. Phys. **31**(pt 1), 3511–3514 (1992)
97. Y.K. Kim, H.M. Jang, Raman lineshape analysis of nano-structural evolution in cation—ordered $ZrTiO_4$—based dielectrics. Solid State Comm. **127**, 433–437 (2003)
98. A.K. Arora, M. Rajalakshmi, T.R. Ravindran et al., Raman spectroscopy of optical phonon confinement in nanostructured materials. J. Raman Spectr. **38**, 604–617 (2007)
99. M. Havel, Ph Colomban, Skin/Bulk nanostructure and corrosion of SiC based fibres. A surface Rayleigh and Raman study. J. Raman Spectrosc. **34**, 786–794 (2003)
100. M. Havel, Ph. Colomban, Rayleigh and Raman images of the bulk/surface nanostructure of SiC based fibres. Compos. Part B. **35**, 139–147 (2004)
101. M. Havel, D. Baron, Ph Colomban, Smart Raman/Rayleigh imaging of nanosized SiC materials using the spatial correlation model. J. Mater. Sci. **39**, 6183–6190 (2004)
102. A. Tanaka, S. Onari, T. Arai, Raman scattering from CdS microcrystals embedded in a Germanate glass matrix. Phys. Rev. B **45**, 6587–6592 (1992)
103. J. Zuo, C. Xu, Y. Liu et al., Crystallite size effects on the Raman spectra of Mn_3O_4. Nanostruct. Mater. **10**, 1331–1335 (1998)
104. K.K. Tiong, P.M. Amirtharaj, F.H. Pollak et al., Effects of As^+ ion implantation on the Raman spectra of GaAs: "Spatial Correlation" interpretation. Appl. Phys. Lett. **44**, 122–124 (1984)
105. J. Zuo, C. Xu, B. Hou et al., Raman spectra of nanophase Cr_2O_3. J. Raman Spectrosc. **27**, 921–923 (1996)
106. A. Fischer, L. Anthony, A.D. Compaan, Raman analysis of short-range clustering in Laser—Deposited CdS_xTe_{1-x} Films. Appl. Phys. Lett. **72**, 2559–2561 (1998)
107. W.F. Zhang, Y.L. He, M.S. Zhang et al., Raman scattering study on anatase TiO_2 nanocrystals. J. Phys. D **33**, 912–916 (2000)

108. Y.K. Kim, H.M. Jang, Polarization leakage and asymmetric Raman line broadening in microwave dielectric $ZrTiO_4$. J. Phys. Chem. Solids **64**, 1271–1278 (2003)
109. H.M. Jang, T.-Y. Kim, I.-W. Park, Nano-sized clusters with tetragonal symmetry in $PbTiO_3$—based relaxor ferroelectrics. Sol. State Comm. **127**, 645–648 (2003)
110. R.S. Chen, C.C. Chen, Y.S. Huang et al., A comparative study of microstructure of RuO_2 nanorods via Raman scattering and field emission scanning electron microscopy. Sol. State Comm. **131**, 349–353 (2004)
111. X.J. Ning, P. Pirouz, The microstructure of SCS-6 SiC fibre. J. Mater. Res. **6**, 2234–2248 (1991)
112. A.B. Mann, M. Balooch, J.H. Kinney et al., Radial variations in modulus and hardness in SCS-6 silicon carbide fibers. J. Am. Ceram. Soc. **82**, 111–116 (1999)
113. M. Havel, D. Baron, L. Mazerolles et al., Phonon confinement in SiC nanocrystals: comparison of the size determination using transmission electron microscopy and Raman spectroscopy. Appl. Spectrosc. **61**, 855–859 (2007)
114. S. Ganesan, A.A. Maradudin, J. Oitmaa, A lattice theory of morphic effects in crystals of the diamond structure. Ann. Phys. **56**, 556–594 (1970)
115. E. Anastassakis, A. Pinczuk, E. Burnstein et al., Effect of static uniaxial stress on the Raman spectrum of silicon. Solid State Comm. **8**, 133–138 (1970)
116. E. Anastassakis, Inelastic light scattering in the presence of uniaxial stress. J. Raman Spectr. **10**, 64–76 (1981)
117. I. De Wolf, H.E. Maes, S.K. Jones, Stress measurements in silicon devices through Raman spectroscopy: bridging the gap between theory and experiment. J. Appl. Phys. **79**, 7148–7156 (1996)
118. I. De Wolf, Stress measurements in Si microelectronics devices using Raman spectroscopy. J. Raman Spectrosc. **30**, 877–883 (1999)
119. I. De Wolf, Raman spectroscopy: about chips and stress. Spectroscopy Europe **15**, 6–13 (2002)
120. J. Wu, Ph Colomban, Raman spectroscopy study on the stress distribution in the continuous Fibre-reinforced CMC. J. Raman Spectrosc. **28**, 523–529 (1997)
121. G. Gouadec, S. Karlin, Ph. Colomban, Raman Extensometry Study of NLM202 and Hi-Nicalon SiC Fibres. Compos. Part B **29**, 251–261 (1998)
122. B. Mottershead, S.J. Eichhorn, Deformation micromechanics of model regenerated cellulose fibre-epoxy/polyester composites. Comp. Sci. & Tech. **67**, 2150–2159 (2007)
123. W.T.Y. Tze, S.C. O'Neill, C.P. Tripp et al., Evaluation of load transfer in the cellulosic–fiber/polymer interphase using a micro-Raman tensile test. Wood and Fiber Sci. **39**, 184–195 (2007)
124. L. Zhenkun, Q. Wei, K. Yilan et al., Stress transfer of single fiber/microdroplet tensile test studied by micro-Raman spectroscopy. Compos. Part A **39**, 113–118 (2008)
125. C. Galiotis, A. Paipetis, C. Marston, Unification of fibre/matrix interfacial measurements with Raman microscopy. J. Raman Spectrosc. **30**, 899–912 (1999)
126. A. Paipetis, C. Galiotis, Y.C. Liu et al., Stress-transfer from the matrix to the fibre in a fragmentation test: Raman experiments and analytical modeling. J. Comp. Mater. **33**, 377–399 (1999)
127. E. Pisanova, S. Zhandarov, E. Mäder et al., Three techniques of interfacial bond strength estimation from direct observation of crack initiation and propagation in polymer–fibre systems. Compos. Part A **32**, 435–443 (2001)
128. R.J. Young, C. Thongpin, J.L Stanford et al., Fragmentation analysis of glass fibres in model composites through the use of Raman spectroscopy. Compos. Part A **32**, 253–269 (2001)
129. P.I. Gonzalez-Chi, R.J. Young, Deformation micromechanics of a thermoplastic–thermoset interphase of epoxy composites reinforced with polyethylene fiber. J. Mater. Sci. **39**, 7049–7059 (2004)
130. G. Anagnostopoulos, A.G. Andreopoulos, J. Parthenios et al., Global method for measuring stress in polymer fibers at elevated temperatures. Appl. Phys. Lett. **87**, 131910 (2005)
131. G. Anagnostopoulos, J. Parthenios, A.G. Andreopoulos et al., An experimental and theoretical study of the stress transfer problem in fibrous composites. Acta Mater. **53**, 4173–4183 (2005)

132. A.B. Coffey, C.M. O'Bradaigh, R.J. Young, Interfacial stress transfer in an aramid reinforced thermoplastic elastomer. J. Mater. Sci. **42**, 8053–8061 (2006)
133. R. Kumar, S.B. Cronin, Optical properties of carbon nanotubes under axial strain. J. Nanosci. Nanotech. **8**, 1–9 (2007)
134. J.R. Wood, Q. Zhao, M.D. Frogley et al., Carbon nanotubes: from molecular to macroscopic sensors. Phys. Rev. B **62**, 7571–7575 (2000)
135. Q. Zhao, M.D. Frogley, H.D. Wagner, The use of carbon nanotubes to sense matrix stress around a single glass fiber. Comp. Sci. Technol. **61**, 2139–2143 (2001)
136. Q. Zhao, M.D. Frogley, H.D. Wagner, Direction-sensitive stress measurements with carbon nanotubes sensors. Polym. Adv. Techn. **13**, 759–764 (2002)
137. C.A. Cooper, R.J. Young, M. Halsall, Investigation into the deformation of carbon nanotubes and their composites through the use of Raman spectroscopy. Compos. Part A **32**, 401–411 (2001)
138. M.S. Dresselhaus, G. Dresselhaus, R. Saito et al., Raman spectroscopy of carbon nanotubes. Phys. Rep. **409**, 47–99 (2005)
139. C. Thomsen, S. Reich, Chap 3: Raman Scattering in Carbon Nanotubes. in *Light Scattering in Solids IX: Novel Materials and Techniques*, ed. by M. Cardona, R. Merlin (Springer, Heidelberg, 2006)
140. V. Domnich, Y. Aratyn, W.M. Kriven et al., Temperature Dependance of silicon hardness: experimental evidence of phase transformations. Rev. Adv. Mater. Sci **17**, 33–41 (2008)
141. C.R. Das, H.C. Hsu, S. Dhara et al., A complete Raman mapping of phase transitions in Si under indentation. J. Raman Spectr. **41**, 334–339 (2010)
142. S. Kouteva-Arguirova, V. Orlov, W. Seifert et al., Raman investigation of stress and phase transformation induced in silicon by indentation at high temperatures. Eur. Phys. J. Appl. Phys. **27**, 279–283 (2004)
143. P. Puech, F. Demangeot, J. Frandon et al., GaN nanoindentation: a micro-Raman spectroscopy study of local strain fields. J. Appl. Phys. **96**, 2853–2856 (2004)
144. N. Orlovskaya, D. Steinmetz, S. Yarmolenko et al., Detection of temperature- and stress-induced modifications of $LaCoO_3$ by micro-Raman spectroscopy. Phys. Rev. B **72**, 014122 (2005)
145. T. Wermelinger, C. Borgia, C. Solenthaler et al., 3-D Raman spectroscopy measurements of the symmetry of residual stress fields in plastically deformed sapphire crystals. Acta Mater. **55**, 4657–4665 (2007)
146. C.R. Das, S. Dhara, H.C. Hsu et al., The mechanism of the recrystallization process in epitaxial GaN under dynamic stress field: atomistic origin of planar defect formation. J. Raman Spectr. **40**, 1881–1884 (2009)
147. Y.B. Gerbig, S.J. Stranick, D.J. Morris et al., Effect of crystallographic orientation on phase transformations during indentation of silicon. J. Mater. Res. **24**, 1172–1183 (2009)
148. T. Wermelinger, R. Spolenak, Correlating Raman peak shifts with phase transformation and defect densities: a comprehensive TEM and Raman study on silicon. J. Raman Spectr. **40**, 679–686 (2009)
149. Ph Colomban, M. Havel, Raman imaging of stress-induced phase transformation in transparent ZnSe ceramics and Sapphire single crystal. J. Raman Spectrosc. **33**, 789–795 (2002)
150. D. Neff, S SR, L. Bellot-Gurlet et al., Structural characterization of corrosion products on archaeological iron: an integrated analytical approach to establish corrosion forms. J. Raman Spectrosc. **35**, 739–745 (2004)
151. S. Réguer, D. Neff, L. Bellot-Gurlet et al., Deterioration of iron archaeological artefacts: micro-Raman investigation on Cl-containing corrosion products. J. Raman Spectrosc. **38**, 389–397 (2007)
152. L. Bellot-Gurlet, D. Neff, S. Réguer et al., Raman studies of corrosion layers formed on archaeological irons in various media. J. Nano Res. **8**, 147–156 (2009)
153. Ph Colomban, S. Cherifi, G. Dexpert, Raman identification of corrosion products on automotive galvanized steel sheets. J. Raman Spectr. **39**, 881–886 (2008)

154. P. Dillmann, G. Béranger, P. Piccardo et al. (eds). *Corrosion of Metallic Heritage Artefacts. Investigation, Conservation and Prediction for Long-Term Behaviour*, European federation of corrosion publications-EFC, vol. 48 (Woodhead, Cambridge, 2007)
155. D. Neff, L. Bellot-Gurlet, P. Dillmann et al., Raman imaging of ancient rust scales on archaeological iron artefacts for long-term atmospheric corrosion mechanisms study. J. Raman Spectrosc. **37**, 1228–1237 (2006)
156. H. Antony, S. Perrin, P. Dillmann et al., Electrochemical study of indoor atmospheric corrosion layers formed on ancient iron artefacts. Electrochimica Acta **52**, 7754–7759 (2007)
157. Monnier J (2008) Corrosion Atmosphérique sous Abri d'Alliages Ferreux Historiques. Caractérisation du Système, Mécanismes et Apport à la Modélisation, Ph.D Thesis, Université Paris-Est (http://tel.archives-ouvertes.fr/tel-00369510/fr/)
158. J. Monnier, L. Legrand, L. Bellot-Gurlet et al., Study of archaeological artefacts to refine the model of iron long-term indoor atmospheric corrosion. Journal of Nuclear Materials **379**, 105–111 (2008)
159. J. Monnier, L. Bellot-Gurlet, D. Baron et al., A methodology for Raman structural quantification imaging and its application to iron indoor atmospheric corrosion products. J. Raman Spectrosc. **42**, 773–781 (2011)
160. S. Bernad, T. Soulimane, S. Lecomte, Redox and conformational equilibria of cytochrome c_{552} from thermus thermophilus adsorbed on chemically modified silver electrode probed by SERS. J. Raman Spectrosc. **35**, 47–54 (2004)
161. F. Salpin, F. Trivier, S. Lecomte et al., A new quantitative method: non-destructive study by Raman spectroscopy of dyes fixed on wool fibres. J. Raman Spectrosc. **37**, 1403–1410 (2006)
162. F. Dubois, C. Mendibide, T. Pagnier et al., Raman mapping of corrosion products formed onto spring steels during salt spray experiments. A correlation between the scale composition and the corrosion resistance. Corros. Sci. **50**, 3401–3409 (2008)
163. S. Keren, C. Zavaleta, Z. Cheng et al., Noninvasive molecular imaging of small living subjects using Raman spectroscopy. PNAS **105**, 5844–5849 (2008)
164. A. Beljebbar, O. Bouché, M.D. Diébold et al., Identification of Raman spectroscopic markers for the characterization of normal and adenocarcinomatous colonic tissues. Crit. Rev. Oncol. Hematol. **72**, 255–264 (2009)
165. B.R. Lutz, C.E. Dentinger, L.N. Nguyen et al., Spectral analysis of multiplex Raman probe signatures. ACS Nano **2**, 2306–2314 (2008)

Chapter 4
Raman Imaging of Micro- and Nano-Structured Materials

Valérie Guieu and François Lagugné-Labarthet

Abstract Raman microscopy has emerged as a powerful technique to characterize anisotropic materials with a submicrometer resolution. The use of polarized light allows one to get precise information about the local organization of the relevant molecular groups through the determination of the most probable distribution function. Such a polarized analysis can be conducted under a confocal microscope but caution must be considered because of the use of high numerical aperture objectives. The molecular orientation can be effectively correlated with the topography of the sample when atomic force microscopy experiments are conducted on the same object. In this chapter, we present 2D and 3D Raman imaging results that have been conducted on mesostructured polymer surfaces and on functionalized metallic surfaces that can serve as SERS platforms. In addition, we compare the results obtained on semiconductor nanowires in confocal Raman conditions or benefiting from local tip enhancement.

4.1 Introduction

The constant development of optical microscopy since the sixteenth century has always been driven by the desire to improve the spatial resolution and image contrast aiming at achieving a better characterization of smaller and fainter specimens. Numerous techniques such as confocal, dark-field, phase contrast, Brewster angle, and polarization microscopy have emerged as improvements of conventional

V. Guieu · F. Lagugné-Labarthet
Institut des Sciences moléculaires, UMR 5255, Université Bordeaux 1,
351 Cours de la Libération, 33405 Talence Cedex, France

F. Lagugné-Labarthet (✉)
University Of Western Ontario, Department of Chemistry, 1151 Richmond Street,
London, ON N65B7, Canada
e-mail: flagugne@uwo.ca

A. Zoubir (ed.), *Raman Imaging*, Springer Series in Optical Sciences 168,
DOI: 10.1007/978-3-642-28252-2_4, © Springer-Verlag Berlin Heidelberg 2012

optical microscopy. Being a pure imaging tool, optical microscopy suffers, however, from its low physical and chemical specificity. This can be remedied by combining it with optical spectroscopy techniques such as fluorescence, infrared, Raman, or nonlinear optical measurements such as vibrational sum frequency generation (VSFG) [1], coherent anti-stokes Raman scattering (CARS) [2], or two-photon excited fluorescence [3]. Such developments have been successfully applied to a wide range of materials with good spatial and spectral resolution. However, their spatial resolution is restricted by the diffraction limit imposed by the wavelength and the numerical aperture of the microscope objective. In infrared microscopy, for instance, the lateral resolution is a few micrometers which is insufficient to resolve submicron structures. In addition, two-dimensional infrared detectors are restricted in terms of their individual pixel sizes and number which is limiting for imaging purposes. Vibrational nonlinear optical techniques offer numerous advantages over linear optical spectroscopy but they often require complex instrumentation and have access to a limited spectral range [3]. Fluorescence spectroscopy is being widely developed either using continuous lasers or pulsed lasers and has succeeded to obtain subdiffraction spatial resolution [4, 5]. However, fluorescence techniques still rely on the use of exogenous labeling agents.

In this context, Raman microscopy presents some advantages to probe materials or biomaterials [6–10] offering molecular information through the vibrational fingerprint of the sample. Raman microscopy also present the tremendous advantage to characterize anisotropic materials when polarized light is used [11]. It allows one to get precise information about the local organization of the relevant molecular groups through the determination of the most probable distribution function. Such polarized analyses can be conducted under a confocal microscope and performed on surfaces as well as small-scale anisotropic materials. Herein, we show a few examples of Raman imaging using polarized light on semiconductor nanowires with various crystallographic orientations.

Although Raman is particularly well suited for materials with large scattering cross-sections, the acquisition time is often too long for diluted systems, monolayers, and in situ measurements of biological materials. For such samples, significant enhancement of the Raman signal can be achieved with metallic nanoparticles or metallic structures that are designed to enhance the electromagnetic field locally through a surface enhanced Raman scattering (SERS) effect [12–14]. The SERS effect can be generated from a variety of approaches ranging from the synthesis of colloidal solution of silver or gold nanoparticles with defined shapes [15–18], electrochemical methods with the use of a template [19–22], nano-imprint lithography [23, 24], or advanced lithography techniques to fabricate plasmonic platforms with high reproducibility and enhancement factors ranging from 10^3 to 10^6 [25–29]. This enhancement is of great interest since it compensates for the extremely small Raman scattering cross-sections which are typically of 10^{-30}cm^2/molecule/steradian. When combined with electronic resonance of specific molecules (dye in general), surface enhanced resonance Raman scattering (SERRS) studies have demonstrated single molecule detection level [30, 31]. In this chapter we describe examples of 2D and 3D array platforms for local Raman enhancement of monolayers under a microscope.

Beyond confocal imaging, tip enhanced Raman spectroscopy (TERS), which is exhaustively described in this book by experts in the field [32–35], presents unique capabilities in terms of image spatial resolution, correlation between the spectral signal and the topography, as well as for the study of single molecules [36]. Such developments are of interest in particular for nanomaterials characterization and in biology for the detection of ultra low concentration samples [37–39]. Nevertheless, the potential of TERS and future also depend of the versatility of its instrumentation for different types of samples (transparent or opaque) as well as the reproducibility of the results.

This chapter is organized as follows: in the first section, polarized Raman imaging of GaN crystalline nanowires is investigated using a standard confocal microscope. TERS imaging of GaN nanowires is also presented providing both topography and the optical signal. In the second section, anisotropic polymer thin films are investigated using polarized light. More specifically, we present the complementarity of second harmonic microscopy and Raman microscopy to determine the odd and even order parameters in a poled azopolymer thin film. Finally, in the last section, we present some results on the mapping of 2D and 3D SERS structures using a confocal microscope.

4.2 Raman Imaging of Single Nanostructures

In this section we review the application of Raman confocal imaging and tip enhanced Raman spectroscopy applied to single GaN nanowires.

4.2.1 Polarized Confocal Imaging of Single GaN Nanowires

One-dimensional semiconductor nanowires of gallium nitride (GaN) are prime candidates for future nanoscale devices such as blue laser and high-power/high-temperature electronics and sensors for biological applications [40–42]. Recent advances in the field of nanostructure synthesis allow one to control not only the size but also the crystallographic orientations of the nanostructure by properly choosing the substrate with best matching lattice constants. For example, GaN wires grown on (100)-$LiAlO_2$ substrate have their $\vec{c}$ axis perpendicular to the growth direction and display a triangular cross-section, while wires grown on (111)-MgO have their growth axis along the [001] direction with an hexagonal Wurzite-type lattice. Raman scattering can thus be very helpful to probe the crystallinity, mechanical stress, temperature effects, the confinement of phonons modes, or the lattice dynamics in nanocrystals, allowing one to correlate the optical measurements with the anisotropy of the nanowire. In addition, these optical techniques could be implemented in synthesis reactors to monitor in situ and in real time the growth process of the nanoobjects.

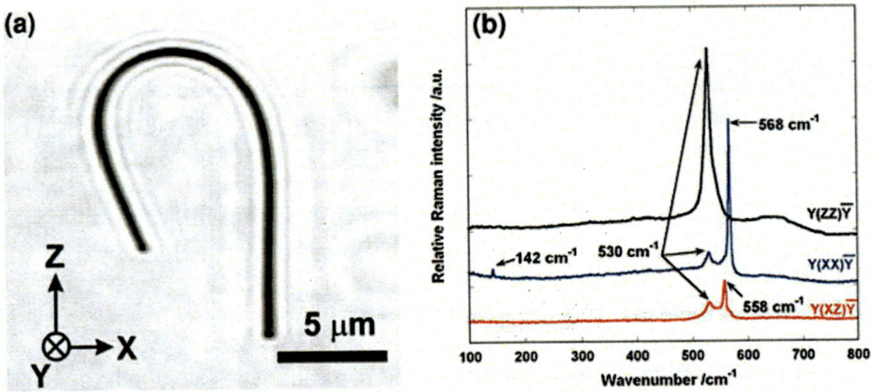

Fig. 4.1 **a** Video image and **b** polarized Raman spectra of the GaN nanowire in the $Y(ZZ)\bar{Y}$, $Y(XX)\bar{Y}$ and $Y(XZ)\bar{Y}$polarization configurations, respectively [43]

We here report polarization imaging of an individual GaN nanowire deposited onto a fused silica surface.

As reported by Pauzauskie et al. [43]. single [001] nanowire with a Wurtzite-like hexagonal cross-section was first cut and positioned on a microscope glass slide (Fig. 4.1a). Atomic force microscopy (AFM) reveals a diameter of about 170 nm and a length of 41 µm. The Raman setup is based on an inverted microscope combined with a spectrometer (LabRAM INV, grating 600 grooves/mm, resolution 4 cm^{-1}). The sample is scanned with a piezoelectric stage with an intrinsic accuracy of 1 nm in the lateral directions. The light beam input excitation wavelength was fixed at $\lambda = 514.5$ nm from an Ar^+ ion laser and the input polarization was selected with a half-waveplate. Focussing on the sample was done with a 100X objective (N.A. = 0.9) and the Raman signal was collected in the backscattering geometry. A polarizer serves as an analyzer just in front of the entrance slit of the spectrometer. The laser intensity was fixed to 10 mW at the focal point. The depolarization ratios were checked with a reference sample of liquid CCl_4. The pinhole of the confocal microscope was closed at about 140 micron, which corresponded to a focal volume of about 1 μm^3, limiting significantly the background signal of the glass substrate. The detector was a liquid nitrogen- cooled CCD camera and acquisition time was fixed to 1 s/spectrum in the imaging mode while an acquisition time of 20 s/spectrum was used to record most spectra shown in the present work. Raman spectra in the polarization configurations $Y(ZZ)\bar{Y}$,$Y(XX)\bar{Y}$ and $Y(XZ)\bar{Y}$ are shown in Fig. 4.1b. The nanowire was aligned so that the Z polarization direction matches the growth direction (i.e., the $\vec{c}$ axis) of the nanowire. Y and $\bar{Y}$ are the directions of the input excitation and the backscattered Raman signal, respectively.

As reported by various authors for thin films [44, 45] and for bundles of nanowires [46, 47], GaN that crystallizes in the hexagonal Wurtzite-type structure belongs to the C_{6v}^4 space group and has two formula units per primitive cell. According to the factor group analysis, the Raman active modes are $1A_1(x^2 + y^2, z^2)$+$1E_1(xz, yz)$+

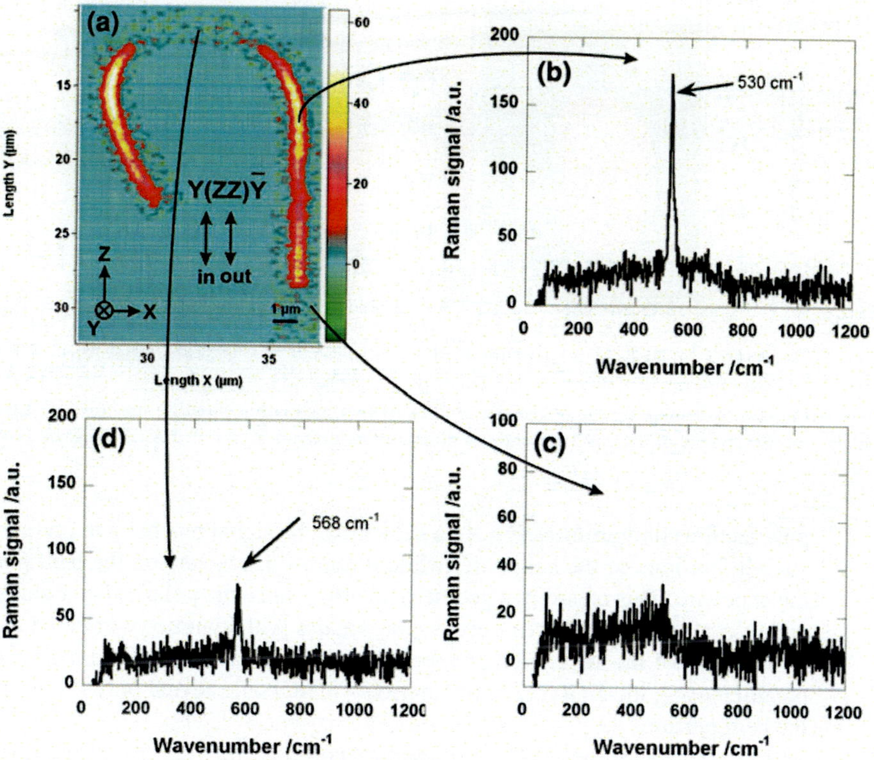

Fig. 4.2 **a** Polarized Raman image generated by integration of the (509–552 cm^{-1}) spectral range (A_1(TO) mode) for $Y(ZZ)\bar{Y}$ polarization configuration. **b–d** Raman spectra in the different parts of the image

$2E_2(x^2 - y^2, xy)$ while $2B_2$ modes are silent. Since GaN wurtzite structure is non-centrosymmetric, both A_1 and E_1 modes can be split into longitudinal optical (LO) and transverse optical (TO) components. For GaN single nanowire, four main signals are observed at 142, 530, 557, and 568 cm^{-1} and they can be assigned to E_2(low), A_1(TO), E_1(TO), and E_2(high) symmetry type modes, respectively. In the $Y(XZ)Y$ configuration, only E_1(TO) mode is expected, and the peak at 530 cm^{-1} [(A_1 (TO)] denotes a polarization leak. The polarized spectra shown in Fig. 4.5 allows an unambiguous assignment which is in good agreement with previous work on GaN thin films.

Mapping of the nanowire was performed by recording step-spectra at every 200 nm with an integration time of 1 s. Such short acquisition time was possible in order to rule out any thermal effect due to the high laser power under the sample. By intensity integration of the (509–552 cm^{-1}) spectral domain around the A_1(TO) mode (530 cm^{-1}), the variations of the Raman signal over the full nanowire can be represented (Fig. 4.2a). The images show a spatial lateral resolution better that 1 µm, with a reasonable acquisition time of about 1 h for a polarized image. For the $Y(ZZ)\bar{Y}$

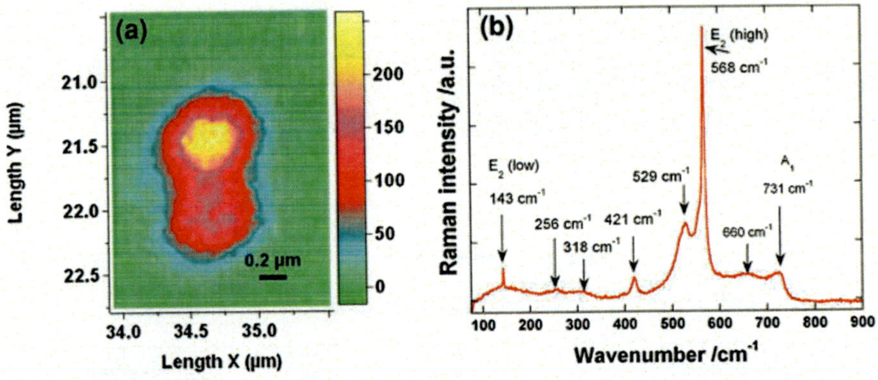

Fig. 4.3 **a** Raman image of the extremities of two nanowires standing up on MgO substrate. **b** Raman spectra of the end face without polarization analysis $Z(YY + YX)\bar{Z}$ integrated over 20 s

polarization configuration, Fig. 4.2a shows that the signal reaches a maximum on the straight portions of the nanowire, while it almost disappears on the bent portion of the nanowire. This is purely a geometrical effect since the polarization configuration and selected integrated area only display modes with symmetry oriented along the growth axis of the wire. Complementary polarized spectroscopic image would be obtained using the $Y(XX)\bar{Y}$configuration which would reveal only the A_1 mode in the bent portion.

To complement these polarized spectra, non-polarized spectra of the end face of a standing up GaN nanowire grown on MgO were recorded. Actually two nanowires stuck together can be seen on Fig. 4.3 and the corresponding Raman image integrated over the most intense peak at 568 cm^{-1}. A similar spectrum was observed for a single nanowire, which excludes any coupling effects between both nanowires. In addition to the phonon modes observed earlier, additional broader peaks are detected at 256, 318, 421, 660, and 731 cm^{-1}. The latter can be assigned to the LO-type mode of A_1 symmetry. The origin of the peaks at 318 and 421 cm^{-1} arises from second-order Raman effects providing information on the vibrational states over the Brillouin zone; they are assigned to overtone processes of acoustic phonons. As suggested in previous experimental and theoretical studies the 318 cm^{-1} band belongs to A_1 symmetry class, while the peak at 421 cm^{-1} belongs to E_2 symmetry. The peak at 256 cm^{-1} can be assigned to a zone-boundary phonon activated by surface disorder that occurs at the end of the nanowire. From lattice dynamic calculations and phonon-dispersion relations for hexagonal GaN, the E_2 optical branch produces a peak in the phonon density of states around 250 cm^{-1}. It is noteworthy that this peak is observed only from the extremity of a single nanowire but never when the nanowire is flat on the surface. This allows one to conclude that finite-size effects, defects, and phonon confinement phenomena can only be observed at the end(s) of the nanowires. The broad peak at 660 cm^{-1} is observed for both the nanowire extremities and also for the $Y(ZZ)\bar{Y}$ polarization.

As reported by Liu et al. this broad peak is presumably due to a defect state with an energy close to the A_1 (LO) or it can be tentatively assigned to the first overtone of a B_1 silent mode. In their studies, Pauzauskie and co-authors have shown that it was possible to discriminate nanowires grown with different crystallinities. Wires with [001] or $[1\bar{1}0]$orientations or GaN nanotubes are expected to have distinct Raman polarized signals. This represents a valuable asset of Raman polarized microscopy to determine the crystalline orientation of an unknown nanowire. The large scattering cross-section of the studied materials was certainly favorable to perform these studies at the single wire level [43, 48].

The high spatial resolution of modern Raman confocal instruments together with a piezoelectric stage demonstrates unambiguously the possibility to image the optical properties of nano-objects with spatial resolution limited by the Abbe' criterion to about 0.5 μm, keeping the full control of input and output polarization. With the help of a higher resolution spectrometer, stress-induced shifts and crystallinity modifications at various growth temperatures can be monitored at the scale of a single nanowire or nanodevice [49].

4.2.2 Single GaN Nanowire Imaging Using TERS

Tip enhanced Raman scattering (TERS) has been applied successfully to the study of nanomaterials such as single-walled nanotubes [33] with sub-15 nm spatial resolution [50, 51].

Herein, tip enhanced Raman imaging of a single GaN nanowire with $[1\bar{1}0]$ orientation is conducted. The nanowire as grown on (100)-$LiAlO_2$ had its $\vec{c}$ axis oriented perpendicular to the growth direction and had a diameter of about 170 nm. The wire was deposited at the surface of a glass slide. The sample was studied in transmission mode using the combination of an AFM (Bioscope II, Veeco) operating in non-contact mode and an inverted confocal microscope connected to a high resolution spectrometer (LabRAM HR, HORIBA Jobin Yvon). Communication between the two instruments was enabled. The tip was coated with a silver layer (30 nm) and a wavelength of 514.5 nm was used as the excitation wavelength.

For each (x, y) position of the tip, an order was sent to the Raman microscope to acquire a signal during a selected duration. The height of the tip could be selected to be either close or far from the sample while the feedback was maintained (Fig. 4.4).

Due to the very weak signal of the nanowire, only a line scan was performed. For each (x, y) point a Raman spectra was acquired with the tip in feedback while a second one was acquired with the tip retracted to 2 μm above the sample. As shown in Fig. 4.5a, the E_2(high) and E_1(TO) modes were observed at 568 and 558 cm^{-1}, respectively. The difference of the Raman signals (Tip close minus Tip far) can be related to the tip enhancement excluding the far field contribution. Shown in Fig. 4.5b, the E_1 mode at 558 cm^{-1} is only observed. This was observed by Matsui et al. [52] and is presumably due to the orientation of the E_1 mode that has a component along the tip axis while the E_2 mode has an in-plane contribution only. The variation of the intensity

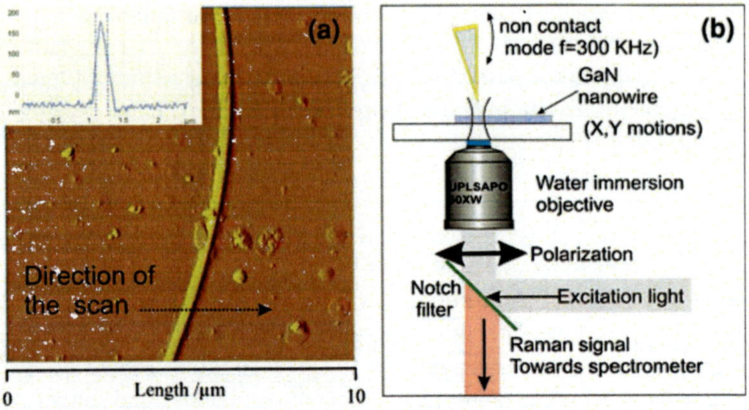

Fig. 4.4 **a** AFM image of a GaN $[1\bar{1}0]$ wire deposited at the surface of a fused silica microscope coverslip. **b** Experimental setup

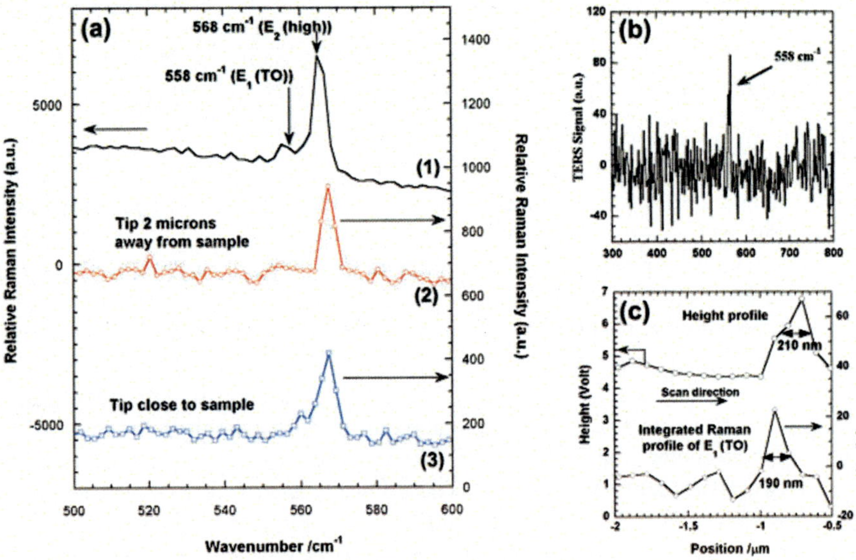

Fig. 4.5 **a** Raman spectra of a $[1\bar{1}0]$ GaN wire either in confocal mode *1* or with the interaction of an AFM tip located close to *3* or far from the sample *2*. **b** Difference of the Raman spectra. TERS = (tip close to sample-tip far from sample). **c** Variation of the Raman signal of the 558 cm^{-1} mode along the nanowire cross-section

of the E_1 mode is shown in Fig. 4.5c together with the variation of its topography. The cross-section corresponds to the wire diameter (~170 nm) and the optical signal shows a full width at half-maximum (FWHM) of about 190 nm. These results confirm that the TERS geometry improves the spatial resolution of the hyperspectral images but the enhancement depends on the symmetry of the vibrational modes.

4.3 Mesostructured Polymer Surfaces

Azobenzene polymer thin films have attracted much attention in the past decade for their potential as photonic devices and due to the possibility to change its optical properties through cycles of Trans-Cis isomerization [53]. It is possible to modulate the surface of such azopolymer thin films upon irradiation with an interference pattern, leading to the formation of surface relief diffraction gratings with high diffraction efficiency [54]. In the present study, we have fabricated a grating with 25 % diffraction efficiency on the first order using an interference pattern. This surface relief grating was subsequently electrically poled using an intense DC electric field. The surface relief amplitude of the grating was about 390 nm as measured by atomic force microscopy which corresponds nearly to the initial thickness of the film. The poling experiment was conducted to generate nonlinear optical properties of the thin grating by creating a non-centrosymmetric arrangement of the chromophores within the grating [55, 56]. These properties were conducted to fabricate a device that has both functions: diffraction and generation of second harmonic light [57].

The grating was first inscribed onto a thin film of pDR1M (poly(4'(((2 -(methacryloyloxy)ethyl)ethyl)-amino)-4-nitroazobenzene), Fig. 4.6b) (thickness = 410nm), using the (p+p) polarization configuration [56, 58]. In such polarization configuration, the resulting optical field from the interference pattern is always polarized along the same direction, i.e., perpendicular to the grating grooves, which implies that the resulting molecular reorientation of the chromophores would always be in the same direction after $\text{Trans}_{//}$–Cis–$\text{Trans}_{\perp}$ cycles. Assuming a uniaxial photo-induced reorientation of the rod-like molecules in the perpendicular direction with respect to the actinic optical field, the azobenzene entities will have the same probability to orient in the in-plane (Y) and in the out-of-plane (Z) directions that are equivalent in the angular hole burning model (Fig. 4.6a) [59]. For such uniaxial orientation, the molecular orientation can be described by a unique angle θ between the molecular director and the reference axis which in the present case is the out-of-plane direction (Z, Fig. 4.6a). In this work, Z direction was chosen as the reference because the electrical poling field (3 KV) was induced perpendicular to the thin film surface, i.e., along the Z direction.

Assuming this uniaxial orientation and cylindrical symmetry (the molecules have a revolution symmetry along their long molecular axis), the angular distribution function $f(\theta)$ of the chromophores along the out-of-plane Z direction can be described on the basis of the Legendre's polynomials using the information entropy theory [60]. In this general method, the distribution function is obtained by maximizing the information entropy, and the Lagrange method of undetermined multipliers leads to the normalized function $f(\theta)$which includes the second, $\langle P_2 \rangle$, and fourth,$\langle P_4 \rangle$, order parameters.

$$f(\theta) = \frac{\exp\left(\lambda_2 P_2(\cos\theta) + \lambda_4 P_4(\cos\theta)\right)}{\int_{-1}^{+1} \exp\left(\lambda_2 P_2(\cos\theta) + \lambda_4 P_4(\cos\theta)\right) \mathrm{d}(\cos\theta)} \quad (4.1)$$

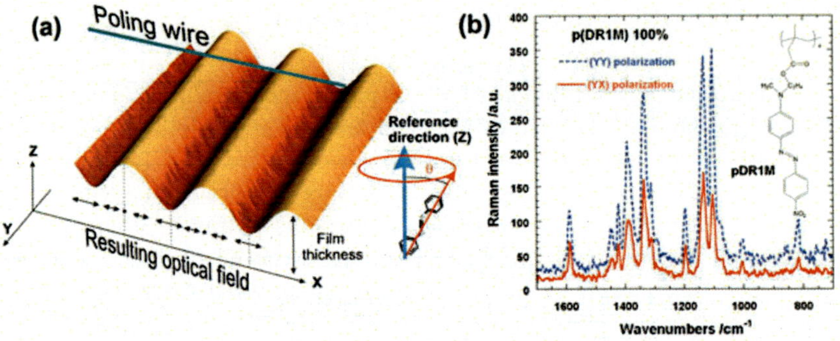

Fig. 4.6 **a** Components of the incident optical field upon grating inscription using the (p+p) polarization configuration of the interfering beams and angular distribution of the azo chromophores relative to the reference direction. **b** Polarized *YY* and *YX* polarized Raman spectra of a pDR1M polymer thin film in the 700–1700 cm^{-1} spectral range

and where the coefficients $\langle P_2 \rangle$ and $\langle P_4 \rangle$ are:

$$\begin{cases} \langle P_2 \rangle = \int\limits_{-1}^{+1} P_2 (\cos\theta) f(\theta) \, \mathrm{d}(\cos\theta) \\ \langle P_4 \rangle = \int\limits_{-1}^{+1} P_4 (\cos\theta) f(\theta) \, \mathrm{d}(\cos\theta) \end{cases} \tag{4.2}$$

with

$$\begin{cases} P_2 (\cos\theta) = \frac{1}{2} \left(3\cos^2\theta - 1\right) \\ P_4 (\cos\theta) = \frac{1}{8} \left(35\cos^4\theta - 30\cos^2\theta + 3\right) \end{cases} \tag{4.3}$$

The resulting distribution function can be derived knowing the experimental $\langle P_2 \rangle$ and $\langle P_4 \rangle$. The Lagrange multipliers λ_2, λ_4 are numerically calculated using equations (4.1) and (4.2). The normalized distribution function, $f(\theta)$ can be determined for the various areas of the gratings where $\langle P_2 \rangle$ and $\langle P_4 \rangle$ have been determined.

The theoretical expressions of polarized Raman intensities taking into consideration the use of a large N.A. microscope objective have been exhaustively detailed for gratings inscribed with (p+p) interference pattern [60, 61]. We remind here the essential expressions to obtain $\langle P_2(\cos\theta) \rangle$ and $\langle P_4(\cos\theta) \rangle$. In previous studies the first two even order parameters were determined relative to the actinic field polarization which is along X for (p+p) interference pattern. However, to compare these parameters with nonlinear optical measurements, a common axis which corresponds to the *Z* or *Y* directions is selected as a proper reference (Fig. 4.6a)

The polarized Raman intensities depend on the averaging over all orientations in space:

$$\boldsymbol{I}_{IJ} \propto \boldsymbol{I}_0 \left\langle \boldsymbol{\alpha}_{IJ}^2 \boldsymbol{n}_T \right\rangle \tag{4.4}$$

where i, j denote the input and analyzer polarization directions, respectively, and α_{IJ} the molecular polarizability tensor elements, and where the quantities:

$$\left\langle \alpha_{IJ}^2 n_T \right\rangle = \int_0^{2\pi} \mathrm{d}\psi \int_0^{2\pi} \mathrm{d}\varphi \int_{-1}^{+1} n_T(\theta) \left[T(\theta, \varphi, \psi) \overleftrightarrow{\alpha} T^t(\theta, \varphi, \psi) \right]^2 \mathrm{d}(\cos\theta) \qquad (4.5)$$

are calculated using the normal T and transposed T^t classical Euler matrices. Since the integral involves only the angle θ, the averaged quantities $\langle \alpha_{IJ}^2 \rangle$ can be expressed in terms of $\langle \cos^2\theta \rangle$ and $\langle \cos^4\theta \rangle$, thus $\langle P_2 \rangle$ and $\langle P_4 \rangle$. In the rigorous treatment, the use of high numerical aperture objective (N.A. = 0.9) must be taken into account. The polarized intensities can be expressed of selected polarization input and output and are function of the tensor elements as well as the parameters of the microscope objectives used for the study. The input polarization of the excitation light (i term) as well as the polarization of the analyzer (j term) are leading to the theoretical expression of the general polarized Raman intensities $I_{ij} = f\left(\langle \alpha_{ij}^2 \rangle, A, B, C_0, C_1, C_2 \right)$, namely:

$$\begin{cases} I_{YX} = \left[\langle \alpha_{YX}^2 \rangle A + \langle \alpha_{YZ}^2 \rangle B \right] (2C_0 + C_2) + \left[\langle \alpha_{ZX}^2 \rangle A + \langle \alpha_{ZZ}^2 \rangle B \right] (4C_1) \\ \qquad + \left[\langle \alpha_{XX}^2 \rangle A + \langle \alpha_{XZ}^2 \rangle B \right] C_2 \\ I_{YY} = \left[\langle \alpha_{YY}^2 \rangle A + \langle \alpha_{YZ}^2 \rangle B \right] (2C_0 + C_2) + \left[\langle \alpha_{ZY}^2 \rangle A + \langle \alpha_{ZZ}^2 \rangle B \right] (4C_1) \\ \qquad + \left[\langle \alpha_{XY}^2 \rangle A + \langle \alpha_{XZ}^2 \rangle B \right] C_2 \end{cases} \qquad (4.6)$$

In (4.6), I_{XY} and I_{YY} are obtained when the grooves of the grating are oriented along the Y direction.

In the second set of experiments, the grating is rotated under the microscope with its grooves oriented in the direction perpendicular to the input polarization. One gets the equation set (4.7):

$$\begin{cases} I_{XY} = \left[\langle \alpha_{XY}^2 \rangle A + \langle \alpha_{XZ}^2 \rangle B \right] (2C_0 + C_2) + \left[\langle \alpha_{ZY}^2 \rangle A + \langle \alpha_{ZZ}^2 \rangle B \right] (4C_1) \\ \qquad + \left[\langle \alpha_{YY}^2 \rangle A + \langle \alpha_{YZ}^2 \rangle B \right] C_2 \\ I_{XX} = \left[\langle \alpha_{XX}^2 \rangle A + \langle \alpha_{XZ}^2 \rangle B \right] (2C_0 + C_2) + \left[\langle \alpha_{ZX}^2 \rangle A + \langle \alpha_{ZZ}^2 \rangle B \right] (4C_1) \\ \qquad + \left[\langle \alpha_{YX}^2 \rangle A + \langle \alpha_{YZ}^2 \rangle B \right] C_2 \end{cases} \qquad (4.7)$$

With

$$\begin{cases} A = \pi^2 \int_0^{\theta m} (\cos^2\theta + 1) \sin\theta \mathrm{d}\theta = \pi^2 \left(\frac{4}{3} - \cos\theta_m - \frac{1}{3}\cos^3\theta_m \right) \\ B = 2\pi^2 \left(\frac{2}{3} - \cos\theta_m + \frac{1}{3}\cos^3\theta_m \right) \end{cases} \qquad (4.8)$$

The quantities A and B come from integrations of the squares of the electric vector components over the scattering cone and the constants C_0, C_1, C_2 are related to the focalization efficiency, and they are calculated via integration of the incident electric field vector components over the effective irradiated volume [62–64]. All these parameters are dependent on the angular semi-aperture of the objective ($\theta_m = 64.16°$ for the 100X, N.A. $= 0.9$ microscope objective) and the refractive index of the sample n ($n = 1.645$).

Finally, knowing the intensities I_{ij} in the different parts of the grating (δ), and their ratio $R_1(\delta) = I_{XY}/I_{XX}$ and $R_2(\delta) = I_{YX}/I_{YY}$, the first two even order parameters, namely $\langle P_2 \rangle$ and $\langle P_4 \rangle$, can be estimated.

Experimentally, from the Raman polarized spectra measured ($\lambda_{excitation} = 752.5$ nm) in the [700–1700 cm^{-1}] spectral domain (Fig. 4.6b), we integrate a particular vibrational mode, the orientation of which is well defined. The $\upsilon_s NO_2$ mode is oriented along the long molecular axis of the trans-shaped azobenzene chromophore and is thus a good probe to estimate the orientation of the rod-shaped azobenzene chromophore. The other modes at 1104 ($\upsilon\Phi$-N, υ18a), 1134 and 1196 (δCH ring, υ9a and υ9b), 1392 (υN $=$ N) and 1589 cm^{-1} (ωC $=$ C, 8a and 8b) are also characteristic of the trans azobenzene chromophores but their directions are less unidirectional than $\upsilon_s NO_2$. In the following we have arbitrarily selected the whole 1044–1180 cm^{-1} spectral range to calculate the polarized intensities, I_{ij} their ratio, $R_1(\delta)$ and $R_2(\delta)$ and finally the order parameters, $\langle P_2 \rangle$ and $\langle P_4 \rangle$. As illustrated in Fig. 4.7a, b, we have thus obtained the polarized Raman images over a (5µm)^2 area for *YY* and *YX* polarizations, respectively. Under these conditions, the variations over several grating periods of the average polarized Raman intensities are plotted in Fig. 4.7c for the full set of input/output polarizations. The variation of the final order parameters $\langle P_2 \rangle$, $\langle P_4 \rangle$ along the position on the grating are reported in Fig. 4.7d.

The $\langle P_2 \rangle$ order parameter varies from +0.228 (bottoms) to +0.266 (tops). As expected, these values are positive, but they denote an orientational ordering along the Z reference direction which is far from being perfect. For a perfect non-centrosymmetric alignment along Z one should expect $\langle P_2 \rangle = 1.0$. In contrast, the parameter $\langle P_4 \rangle$ is always negative and varies markedly from −0.067 (bottoms) to −0.230 (tops): the influence of $\langle P_4 \rangle$ is thus preponderant on the shape of the corresponding distribution functions [56]. Actually, using only the sets of $\langle P_2 \rangle$, $\langle P_4 \rangle$, the polar distribution functions are asymmetric and display maxima at ±34° (bottoms), ±39° (intermediates), and ±42°(tops) with respect to the direction of reference [56]. Also, the distribution function in the top regions is significantly narrower than in the bottoms regions (the FWHM varies from ±28 to ±60°). These observations suggest the existence of increasing force and pressure constraints during the peak relief formation and during the related chromophore diffusion mechanisms. Because of the large force gradients during the isomerization cycles and the angular reorientations, the chromophore orientation distributions appear significantly perturbed by the polymer migration from areas of high field intensity (that correspond to the bottoms of the grating) to area of lower field intensity (tops of the grating).

As a demonstration of high resolution Raman confocal imaging on mesostructured polymer thin films, both Atomic Force microscopy surface topography and Raman images were conducted on superimposed surface relief gratings with a periodicity of ∼1µm (Fig. 4.8). Raman measurements were recorded with a LabRAM

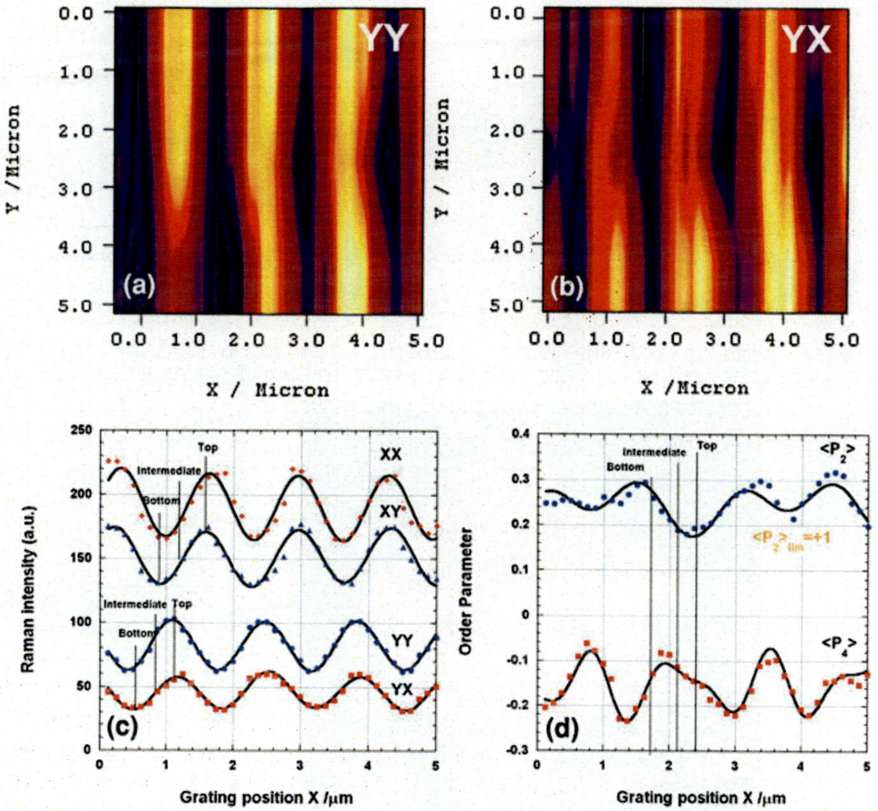

Fig. 4.7 **a, b** Raman polarized images of a mesostructured polymer in the *YY* and *YX* polarization configurations, respectively. **c** Variation of the polarized Raman measurements *YY*, *XX*, *YX* and *XY* along the grating position *X*. **d** $\langle P_2 \rangle$ and $\langle P_4 \rangle$ order parameter along the grating vector

HR 800 (HORIBA Jobin Yvon) spectrometer combined with an inverted microscope (Olympus IX 71) and a *X*, *Y* Piezoelectric stage to position the sample with the best accuracy and repeatability. A 100X, 0.9 NA microscope objective was used. A Raman spectrum (Fig. 4.8c) was recorded every 100 nm along the *X* and *Y* directions with an integration time of 5 s per spectrum ($\lambda_{excitation} = 752.5$ nm). The variation of the Raman signal was integrated over the (1060–1180) cm^{-1} spectral domain (Fig. 4.8d). It must be pointed out that details on the Raman profile are obtained with a spatial resolution better than 500 nm (FWHM of the Raman profiles in Fig. 4.8e). Such high spatial measurements were made possible due to the use of reproducible positioning piezoelectric stage.

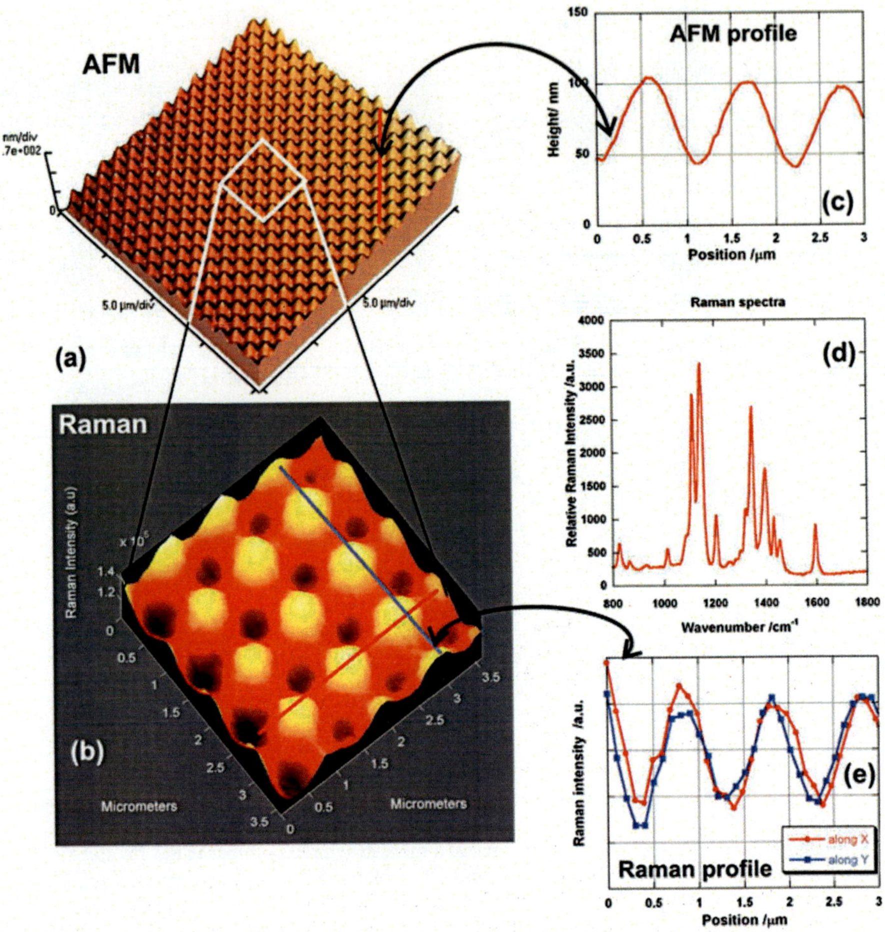

Fig. 4.8 **a** Atomic force microscopy image and **b** Raman intensity variation of a vibrational mode. ($\nu_s NO_2$) of an azobenzene-containing polymer thin that is periodically structured (periodicity of $\sim 1 \mu m$). **d** Raman Spectra of azobenzene polymer **c** with a 5 s acquisition time. **c** and **e** Variation of the AFM topography and the Raman intensity over the *X* and *Y* directions, respectively

4.4 Raman Microscopy Using SERS Platforms

Herein, we describe the fabrication and the characterization of 2D and 3D surface enhanced Raman scattering (SERS) platforms for the detection of monolayers adsorbed onto their surfaces. The 2D structures are designed to be used in transmission mode while the 3D structures are conceived for potential remote measurements through an optical fiber.

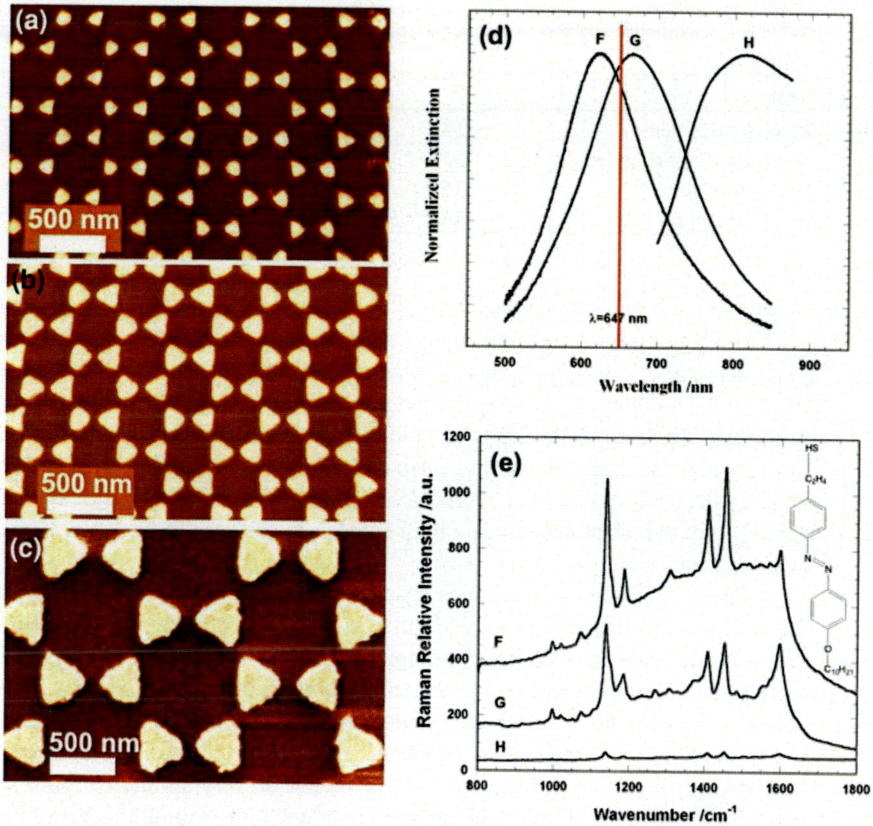

Fig. 4.9 **a–c** Fischer patterns inscribed on glass by e-beam lithography. The sizes of the triangles are 145 nm (**a**), 185 nm (**b**) and 385 nm (**c**). **d** Microabsorption spectra of the patterns with 145 nm (F), 185 nm (G), 385 nm (H) triangle sizes. **e** Maximum Raman enhancement was obtained for the e-beam pattern with the smallest structures and with plasmon frequency maximum located at $\lambda = 600$ nm. The irradiation source was $\lambda = 647$ nm and with intensity of 220μW measured at the sample. Spectra were averaged over 10 different spots on the patterns

4.4.1 Electron-Beam Lithography 2D SERS Platforms

In this work, we have investigated the possibility to use SERS platforms prepared by electron beam lithography together with a Raman confocal microscope. Due to the quasi-perfect reproducibility of the electron-beam (e-beam) lithography method, the absorption band of the plasmonic devices can be tailored precisely by adjusting the size of the features and associated gaps which is of tremendous interest to tune surface enhanced Raman effects [65–67].

Using these SERS platforms, we have specifically investigated the adsorption of azobenzene monolayers organized at the surface of nanostructured gold.

Table 4.1 Geometric and spectroscopic parameters of the substrates as prepared by e-beam lithography

Sample	F	G	H
Triangles Size /nm	145	185	385
Maximum Extinction /nm	622	667	813
Gold coverage	16 %	25 %	57 %
Enhancement Factor	1.07×10^5	3.93×10^4	1.49×10^3

The substrates were prepared using conventional e-beam lithography approach on glass substrates. ZEP 520 photoresist was used in conjunction with a LEO 1530 SEM microscope equipped with e-beam lithography capabilities. A standard development procedure was used. A thin layer of gold (40 nm) was deposited using electron beam evaporation. The SERS substrates (also called Fischer patterns) were then functionalized with azobenzene thiol molecules (10^{-4} M solution for 12 h) and washed to remove any excess of azobenzene molecules or aggregates that were noncovalently adsorbed onto gold.

Shown in Fig. 4.9d are the transmission spectra performed for samples with triangle sizes varying from 145 nm to 385 nm (Fig. 4.9a–c). The transmission spectra were redshifted for patterns with larger triangular sizes. The Raman spectra were recorded on the e-beam samples and significant enhancements were observed for samples with plasmon frequencies in resonance with the excitation wavelength (Fig. 4.9e). The excitation laser wavelength ($\lambda = 647$ nm) being in the middle of the plasmons bands of platforms F and G, the Raman spectra should be very similar for both samples before normalization to the gold surface coverage. However, in this set of samples, the Raman spectra were slightly more intense for F sample and since the gold coverage of F is smaller than for G, this results in a higher Raman enhancement as shown in Table 4.1.

More precisely, one has to consider that in SERS, the enhancement factor at each molecule varies with the geometry, the size, and the nature of the metallic substrate but depends also of the matching of the electric field of both the excitation source and the Raman signal with the plasmon band [19, 68]. The electromagnetic enhancement factor of SERS signal is thus given by [68]:

$$F = |E(v_{\text{excitation}})|^2 \, |E(v_{\text{Raman}})|^2 \tag{4.9}$$

where $E(v_{\text{excitation}})$ and $E(v_{\text{Raman}})$ are the local electric-field enhancement factors at the incident frequency ($v_{\text{excitation}}$) and at the Raman Stokes frequency(v_{Raman}), respectively. However, since the plasmon width is large compared to the Raman stokes shift, both electromagnetic field are often in resonance and/or preresonance with the plasmon band. Therefore, an approximation is to assume that $E(v_{\text{excitation}})$ and $E(v_{\text{Raman}})$ are the same leading to an enhancement factor proportional to $E = |E(v_{\text{excitation}})|^4$. In such approximation, the matching of the excitation laser line with the plasmon frequency is an essential condition to obtain efficient Raman surface enhancements.

In order to establish the relationship between the nanostructured surface and the SERS activity of the platform, the enhancement factors have been determined. The determination of the Raman Enhancement Factor (REF) in SERS is a prerequisite to quantify the enhancement factor of the Raman signal. The REF can be estimated by comparing the measured SERS intensities (I_{SERS}) with the non-enhanced Raman scattering intensities (I_{NE}) using (4.10) [25, 69, 70]:

$$REF = \frac{I_{SERS}}{I_{bulk}} \frac{N_{bulk}}{N_{SERS}} \tag{4.10}$$

where N_{bulk} and N_{SERS} are the number of molecules under laser illumination for the reference sample (liquid) and for the functionalized surface patterns. Despite an inherent experimental difficulty in precisely measuring the REF, it is necessary to have a good knowledge of the metallic surface area to estimate the number of scattering molecules that participate in the Raman signal. In the present approach we have considered the following points to precisely estimate the enhancement factors obtained with our optical configuration:

First, a Raman reference sample was measured in solution. A 10^{-2} M azobenzene solution was prepared in chloroform and probed with our microscope apparatus using a 60X, 0.7 N.A. objective. The volume of the focal point in the solution was first measured by monitoring the change of intensity of the 520 cm^{-1} band of a silicon substrate immersed in the same solvent. By varying the focus of the laser on the immersed silicon sample, the intensity variation of the Si peak shows a Gaussian type variation as shown in Fig. 4.10b. The width at half height gives the waist length of the focal volume in the considered liquid. The transverse dimension (diameter of the spot) is determined with the numerical aperture of the objective used for a specific wavelength. For the liquid Raman reference measurement, the waist of the beam is about 14.5μm (Fig. 4.10b) and the diameter of the beam with a 0.7 N.A objective is about $1.22\lambda/N.A. = 1.13\mu m$, resulting in a cylindrical volume of $\approx 14.5 \mu m^3 (\pm 10\,\%)$.

Second, knowing the concentration of the reference solution, we can estimate the number of molecules at the focus, N_{bulk}, which is about 8.73×10^7 molecules.

Third, for the SERS platforms, we assume that our gold patterns are homogeneously covered with a monolayer of azobenzene moieties with molecular area of 0.3 nm^2 per molecule. The surface of gold can be measured with image analysis software from the SEM images of our patterns. Such surfaces vary with the size of the inscribed features. Assuming that a monolayer is formed on gold, and that we know the total gold surface as well as the molecular area, one can determine the number of scattering moieties under the focalized beam.

I_{bulk} and I_{SERS} are the Raman intensities of a given vibrational mode of the azobenzene chromophore. Intensities of the laser irradiation source as well as acquisition times were normalized for comparison of the Raman intensities.

Summarized in Table 4.1 are the enhancement factors calculated for the 1140 cm^{-1} Raman peak (δ(CH) ring+ν_{9a}) for the e-beam substrates. For all the samples it is

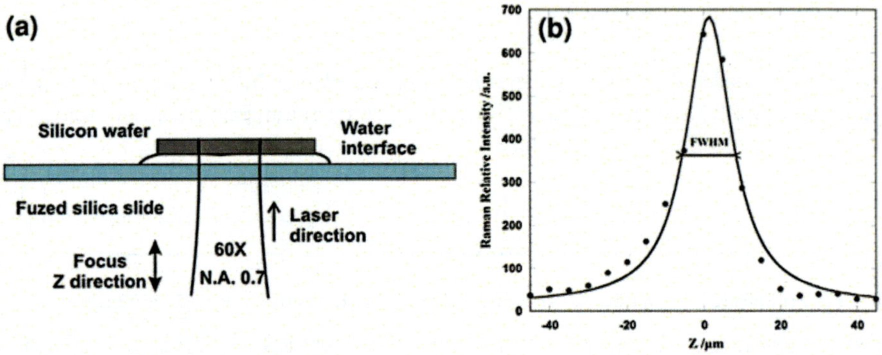

Fig. 4.10 **a** Setup to measure the longitudinal dimension of the beam waist in a liquid sample. **b** variation of the Raman signal of silicon (520 cm^{-1})in function of the Z position of the objective. The FWHM is ~14.5 microns

noteworthy that the magnitude of the enhancement varies between 10^3 and 10^5. The maximum enhancements were observed for measurements where the excitation source matches the plasmon frequency of the array. When such conditions were fulfilled, the Raman enhancement factors were about 3.7×10^4 and 1.7×10^5. In addition, the e-beam substrates are more versatile since the gap and the size of the structures are well controlled. Even in simple structures like the Fischer's patterns, this permits one to precisely tune the plasmon frequency of the surface according to the Raman excitation line. In the case of a functionalization with a self-assembled monolayer, the Raman spectra can be recorded in short times (seconds) while no signal was recorded on a flat gold surface as prepared by e-beam evaporation. This emphasizes the tremendous interest of developing e-beam patterns on transparent substrates (glass coverslips, fused silica) for Raman measurements in a transmission geometry setup. Further polarized measurements on such structures will also be of interest to map the symmetry of the plasmon modes [71].

4.4.2 3D SERS Substrate for Raman Imaging

The discovery of optical fibers has had a significant impact especially in telecommunication technologies as an efficient material for light transmission. Analytical chemists have also benefited from such systems because they can be coupled to various spectroscopic instruments, thus opening new applications, such as remote analysis or sampling in hostile environment [72, 73]. Besides UV–visible and fluorescence detection techniques [74, 75], vibrational spectroscopy is also gaining benefits from the use of optical fibers: remote Raman spectroscopy provides rapid identification of molecules through the spectral analysis of their chemical fingerprints and therefore is a powerful approach for analyzing and imaging of molecular species

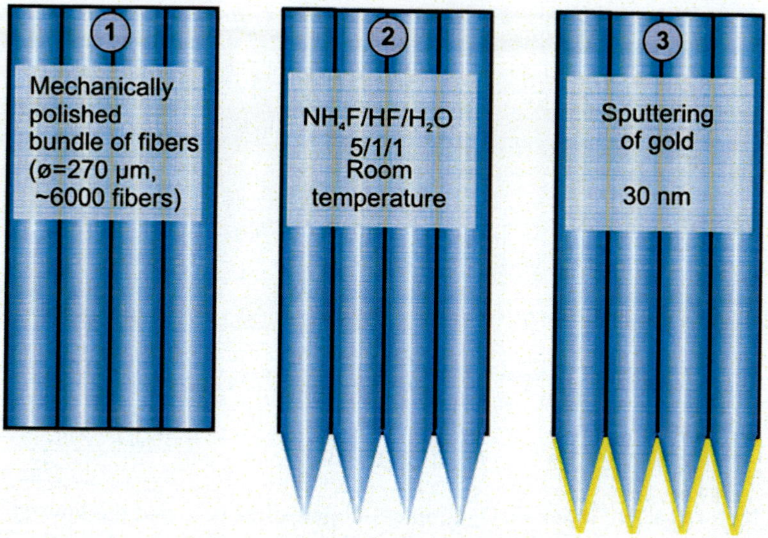

Fig. 4.11 Fabrication of a tip array onto a fiber bundle

in complex or sensitive environments. However, the optical loss of optical fibers is one major drawback limiting its applications. The development of new optical fiber sensors to perform remote analysis benefiting from surface enhanced Raman effect is of tremendous interest for many applications which require in situ or in vivo spatially resolved measurements [76–81].

We have investigated [82, 83] the design and fabrication of sharp metallic nanostructured arrays resulting from the chemical etching of an optical fibers bundle followed by the deposition of a thin gold layer (30 nm) over the tips formed by etching. Such a 3D substrate demonstrated a high SERS activity for reference molecules and Raman confocal imaging showed large Raman enhancement localized at the apex of the tips. This approach introduces an easy, versatile, and reproducible experimental method to design and fabricate three-dimensional SERS-active micrometer-sized platforms. As the substrates are portable, robust and stable sensors, their durability allows safe handling and use in vivo.

Silica imaging fibers with a diameter of 270 μm embedding 6000 individually cladded 3–4 μm diameter optical fibers were polished with lapping films and then etched in a buffer of hydrofluoric acid solution (HF).

The etching of the fiber array was accomplished through the combination of several methods and resulted in the formation of conical tips whose shape, length, and angle were dependent on experimental conditions (time and concentration) [84]. When the optical fibers bundle was left for 90 min in an HF aqueous solution consisting of 40 % aqueous NH_4F and 48 % HF in deionized water in a proportion of 5/1/1, conical tips were formed on the core of each individual fiber (Figs. 4.11 and 4.12). This selective etching was achieved due to the different etching rates between the

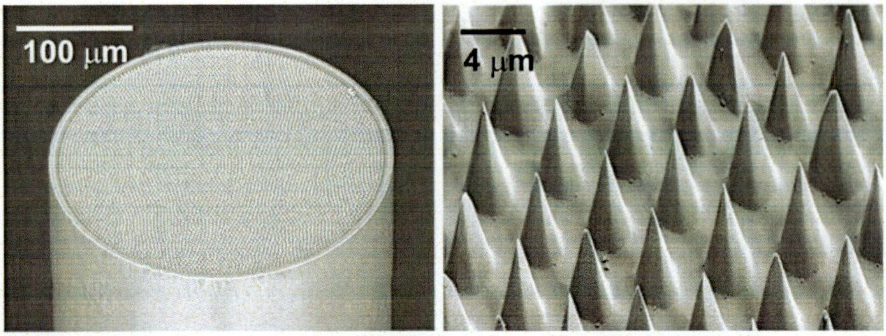

Fig. 4.12 Scanning electron microscopy (SEM) images at different magnifications of the optical-fiber bundle etched in 40 % aqueous NH_4F and 48 % HF in deionized water in a proportion of 5/1/1 for 90 min

high-refractive index GeO_2 gradient-doped silica core and the low-refractive index fluorine doped cladding. The etched fiber was then mounted on a rigid holder and the etched face was sputter coated with a gold film with a 30 nm thickness. Other structures (not shown here) were formed when a different composition of etchant was used [82].

The tip array was functionalized with benzenethiol (BT) because of their well-known Raman and Raman-enhanced spectra. Moreover, immersion of the gold surface in a BT solution followed by intense rinsing results in a monolayer of benzene rings linked to the golden fiber optics array through a covalent bond, allowing a more accurate evaluation of the number of molecules probed into the laser beam and thus a better order of magnitude for the Raman Enhancement Factor.

We have measured the SERS effect on the fiber optics array using a confocal microspectrometer equipped with a precision motorized piezoelectric *XYZ* sample stage. The 647 nm excitation line of a Kr^+ laser was chosen to avoid luminescence background from the optical fiber. The images presented below (Fig. 4.13a) were obtained by investigating the enhancement potential of conical tips (Fig. 4.12).

The spectrum obtained from the BT monolayer adsorbed on the nanotip array appeared significantly enhanced as compared to that of the reference solution. Moreover, it has to be noted that no Raman signal was detected for a BT monolayer deposited on flat gold (i.e., with a roughness <1 nm).

The large intensity Raman spectra of the BT monolayer was observed at the extremities of the tips by recording the variation of the Raman enhancement intensity as a function of the position of the sample (step 0.1 μm for X and Y positions and 1 μm for the Z position for 3D imaging, see Fig. 4.13c). 2D and 3D images demonstrated a surface enhanced Raman scattering (SERS) effect induced by the strong curvature of the metallic tip apexes.

The Raman enhancement factor of the nanostructure has been calculated as described in Sect. 4.1 taking into account not only the number of probed molecules in the pure BT solution and onto the gold nanostructured surface (0.22 nm^{2}per mole-

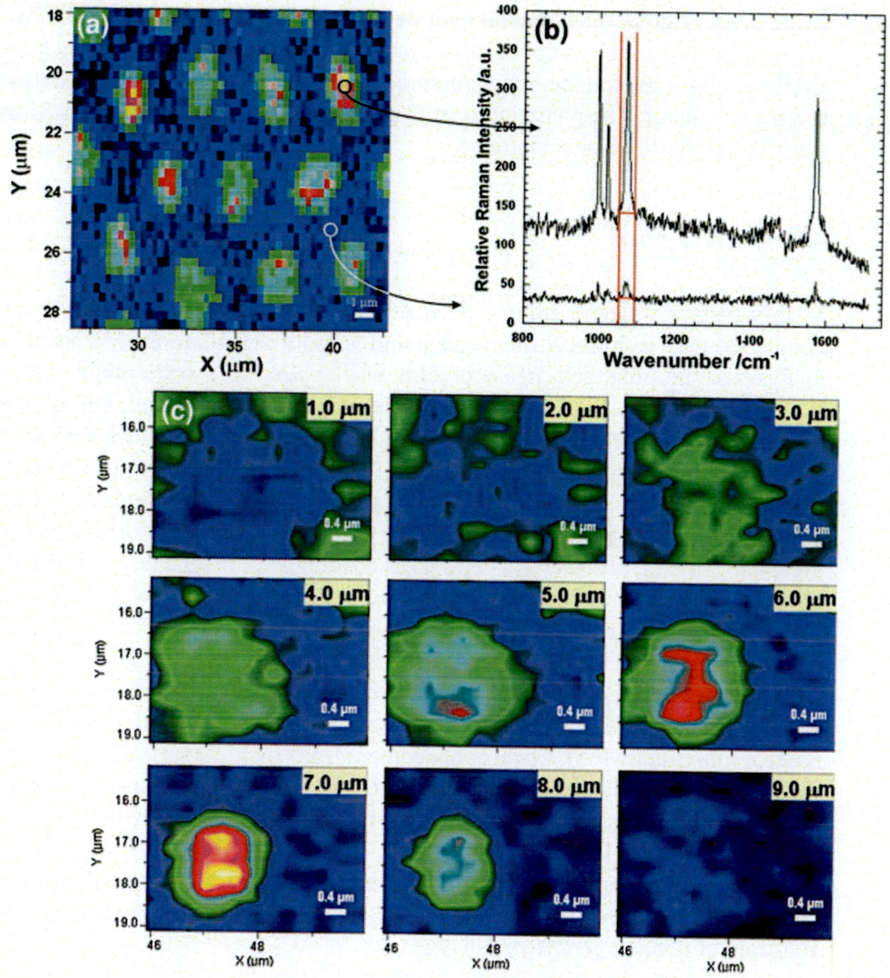

Fig. 4.13 2D **a** and 3D **c** Raman mapping of BT adsorbed on the optical fiber array using a 100X, 0.9 N.A. microscope objective. The variation of integrated Raman intensity of the υ_1 phenyl mode vibration at $1072\,\text{cm}^{-1}$ is plotted. **b** Typical Raman spectra at the maximum and minimum Raman enhancement locations. The acquisition time was set to 1 s. per spectrum

cule [85]), but also the experimental conditions such as the laser power. Although not very large according to the standards of SERS enhancement, such a Raman Enhancement Factor of $\sim 3 \times 10^4$ emphasizes the success of the use of an fiber optics array of conical sharp tips for Raman enhancement, which allows one to probe dilute solutions or to reduce acquisition times. We have reported the fabrication and the optical characterization of an array of gold metalized nanotips at the surface of a fiber bundle. Confocal Raman microscopy measurements clearly demonstrated that the Raman signal was enhanced at the apices of the tips once the tips were covered with a thin layer of gold. Such optical fiber devices present an interesting potential for in vivo applications such as endoscopic probes. In such case, the fiber array guides the excitation light toward the nanostructured face and collects the backscattering light toward a detector. The potential of such nanotips-array-terminated fibers for remote sensing has yet to be demonstrated.

4.5 Conclusions

In this article, we have reviewed recent work of Raman imaging measurements conducted with a confocal microscope and with its combination with an AFM. The analysis of the polarized measurements was applied to a wide range of materials such as inorganic nanostructures or anisotropic polymer thin films. Order parameter and associated distribution functions can be defined and measured precisely using polarized light in conjunction with an optical microscope. Beyond measurements spatially limited by diffraction, the integration of Raman measurements with AFM trough the TERS approach is a challenging endeavor, but recent works have shown its great capabilities, provided that one takes advantage of resonance conditions and/or enhancements of the field from a metallic tip. We present here a few preliminary results conducted on crystalline nanowires.

Raman imaging is also benefiting from recent advances in the plasmonic field where enhancement of the Raman signal is reproducible and can be tuned accordingly with the SERS platforms geometries using advanced nanofabrication methods such as e-beam lithography [65]. Local enhancement from such 2D surfaces can be obtained with a good reproducibility. Plasmons are also governing enhancements in TERS setup and the enhancement is shown to be strongly dependent of the tip-sample separation distance [86–88].

The combination of such SERS platforms with TERS setup could be an effective controllable way to benefit from tunable antenna effects between the tip and a metallic substrate of defined geometry [89–92].

Acknowledgments The authors would like to warmly acknowledge all the collaborators who have participated in experiments, data analysis, and interpretations of the numerous laboratory results in this manuscript in particular, C.Sourisseau, N.Marquestaut, N. Sojic, J. L. Bruneel and D. Talaga (Institut des Sciences Moléculaires, Université Bordeaux 1, France), P. Yang and P.Pauzauskie (Chemistry Department, University of California Berkeley, USA). The authors are also grateful for the financial support from the CNRS (Chemistry Department), the Agence Nationale pour la Recherche (ANR), the Région Aquitaine, and the Canadian discovery grant as well as the Canadian Research Chair program (FLL).

References

1. K. Cimatu, S. Baldelli, J. Am. Chem. Soc. **128**, 16016 (2006)
2. E. Conor, X.S. Xie, Ann. Rev. Anal. Chem. **1**, 883 (2008)
3. F. Lagugné-Labarthet, Y.R. Shen, in *Optical Imaging and Microscopy*, ed. by P. Török, F.-J. Kao (Springer, New York, 2007), Vol. 87, p. 237

4. E. Betzig, G.H. Patterson, R. Sougrat, O.W. Lindwasser, S. Olenych, J.S. Bonifacino, M.W. Davidson, J. Lippincott-Schwartz, H.F. Hess, Science **313**, 1642 (2006)
5. M.J. Rust, M. Bates, X. Zhuang, Nat Methods **3**, 793 (2006)
6. C. Matthäus, T. Chernenko, J.A. Newmark, C.M. Warner, M. Diem, Biophys. J. **93**, 668 (2007)
7. C. Matthäus, Boydston-White, M. Miljkovic, M. Romeo, M. Diem. Appl. Spectrosc. **60**, 1 (2006)
8. T. Lefèvre, M.-E. Rousseau, M. Pézolet, Appl. Spectrosc. **60**, 841 (2006)
9. M.-E. Rousseau, L. Beaulieu, T. Lefèvre, J. Paradis, T. Asakura, M. Pézolet, Biomacromolecules **7**, 2512 (2006)
10. M.-E. Rousseau, T. Lefèvre, L. Beaulieu, T. Asakura, M. Pézolet, Biomacromolecules **5**, 2245 (2004)
11. P.R. Carey, Annu. Rev. Phys. Chem. **57**, 527 (2006)
12. D.L. Jeanmaire, R.P. Van Duyne, J. Electroanal. Chem. **84**, 1 (1977)
13. M. Fleischmann, P.J. Hendra, A.J. McQuillan, Chem. Phys. Lett. **26**, 163 (1974)
14. M. Moskovits, J. Raman Spectrosc. **36**, 485 (2005)
15. Y. Li, W. Cai, G. Duan, Chem. Mater. **20**, 615 (2008)
16. C.E. Talley, J.B. Jackson, C. Oubre, N.K. Grady, C.W. Hollars, S.M. Lane, T.R. Huser, P. Nordlander, N.J. Halas, Nano Lett. **5**, 1569 (2005)
17. N. Anderson, P. Anger, A. Hartschuh, L. Novotny, Nano Lett. **6**, 744 (2006)
18. A.R. Tao, P. Yang, J. Phys. Chem. B **109**, 15687 (2005)
19. J.J. Baumberg, T.A. Kelf, Y. Sugawara, S. Cintra, M.E. Abdelsalam, P.N. Bartlett, A.E. Russell, Nano Lett. **5**, 2262 (2005)
20. J. Henzie, K.L. Shuford, E.-S. Kwak, G.C. Schatz, T.W. Odom, J. Phys. Chem. B **110**, 14028 (2006)
21. Y. Lu, G.L. Liu, J. Kim, Y.X. Mejia, L.P. Lee, Nano Lett. **5**, 119 (2005)
22. S.J. Oldenburg, J.B. Jackson, S.L. Wescott, N.J. Halas, J. Appl. Phys. Lett. **75**, 2897 (1999)
23. R.A. Alvarez-Puebal, B. Cui, J.-P. Bravo-Vasquez, T. Veres, H. Fenniri, J. Phys. Chem. C **111**, 6720 (2007)
24. B. Cui, T. Veres, Microelectron. Eng. **84**, 1544 (2007)
25. N. Félidj, J. Aubard, G. Lévi, J. R. Krenn, M. Salermo, G. Schider, B. Lamprecht, A. Leitner, F. R. Aussenegg: Phys. Rev. B **65**, 075419(1) (2002)
26. A.G. Brolo, E. Arctander, R. Gordon, B. Leathem, K.L. Kavanagh, Nano Lett. **4**, 2015 (2004)
27. M.A.D. Jesus, K.S. Giesfeldt, J.M. Oran, N.A. Abu-Hatab, N.V. Lavrik, M.J. Sepaniak, Appl. Spectrosc. **59**, 1501 (2005)
28. B. Cui, L. Clime, K. Li, T. Veres, Nanotechnology **19**, 145302/1 (2008)
29. C.L. Haynes, A.D. McFarland, R.P. Van Duyne, Anal. Chem. **77**, 338A (2005)
30. S. Nie, D.R. Emory, Science **275**, 1102 (1997)
31. K. Kneipp, H. Kneipp Appl, Spectrosc. **60**, 322 (2006)
32. A. Rasmussen, V. Deckert, J. Raman Spectrosc. **37**, 311 (2006)
33. Y. Saito, N. Hayazama, H. Kataura, T. Murakami, K. Tsukagoshi, Y. Inouye, S. Kawata, Chem. Phys. Lett. **136**, (2005)
34. B. Pettinger, G. Picardi, R. Schuster, G. Ertl, Single Mol. **5**, 285 (2002)
35. A. Hartschuh, Angew. Chem. Int. Ed. **47**, 8178 (2008)
36. C.C. Neacsu, J. Dreyer, N. Behr, M.B. Raschke, Phys. Rev. B **73**, 193406/1 (2006)
37. F. Wei, D. Zhang, N.J. Halas, J.D. Hartgerink, J. Phys. Chem. B **113**, 9158 (2008)
38. C. Dana, T. Deckert-Gaudig, C. Budich, M. Laue, R. Moeller, D. Naumann, V. Deckert, J. Popp J. Raman Spectrosc. **40**, 240 (2009)
39. W. Zhang, B.S. Yeo, T. Schmid, R. Zenobi, J. Phys. Chem. C **111**, 1733 (2007)
40. P. Pauzauskie, T. Kuykendall, Y. Zhuang, J. Goldberger, D. Suirbully, J. Denlinger, P. Yang, Nat. Mater. **3**, 524 (2004)
41. H. Choi, J. Johnson, R. He, S.-K. Lee, F. Kim, P. Pauzauskie, J. Goldberger, R. Saykally, P. Yang, J. Phys. Chem. B **107**, 8721 (2003)
42. Y. Huan, X. Duan, Y. Cui, C.M. Lieber, Nano Lett. **2**, 101 (2002)

43. P.J. Pauzauskie, D. Talaga, K. Seo, P. Yang, F. Lagugné-Labarthet, J. Am. Chem. Soc. **127**, 17146 (2005)
44. H. Harima, J. Phys. Condens Matter. **14**, R967 (2002)
45. T. Azuhata, T. Sota, K. Suzuki, S. Nakamura, J. Phys. Condens Matter. **7**, L129 (1995)
46. J. Liu, X.-M. Meng, Y. Jiang, C.-S. Lee, I. Bello, S.-T. Lee, Appl. Phys. Lett. **83**, 4241 (2003)
47. H.-L. Liu, C.-C. Chen, C.-T. Chia, C.-C. Yeh, C.-H. Chen, M.-Y. Yu, S. Keller, S.P. DenBaars, Chem. Phys. Lett. **345**, 245 (2001)
48. T. Livneh, J. Zhang, G. Cheng, M. Moskovits, Phys. Rev. B **74**, 035320/1 (2006)
49. V. Poborchii, T. Tada, T. Kanayama, Appl. Phys. Lett. **89**, 233505/1 (2006)
50. A. Hartschuh, E.J. Sanchez, X.S. Xie, L. Novotny, Phys. Rev. Lett. **90**, 095503 (2003)
51. A. Hartschuh, H. Qian, A.J. Meixner, N. Anderson, L. Novotny, Surf. Interface Anal. **38**, 1472 (2006)
52. R. Matsui, P. Verma, T. Ichimura, Y. Inouye, S. Kawata, Appl. Phys. Lett. **90**, 061906 (2007)
53. J. Delaire, K. Nakatani, Chem. Rev. **100**, 1817 (2000)
54. A. Natansohn, P. Rochon, Chem. Rev. **102**, 4139 (2002)
55. R.D. Schaller, R. Saykally, Y.R. Shen, F. Lagugné-Labarthet, Opt. lett. **28**, 1296 (2003)
56. F. Lagugné-Labarthet, C. Sourisseau, R.D. Schaller, R.J. Saykally, P. Rochon, J. Phys. Chem. B **108**, 17059 (2004)
57. F. Lagugné-Labarthet, F. Adamietz, V. Rodriguez, C. Sourisseau, J. Phys. Chem. B. **110**, 13689 (2006)
58. F. Lagugné-Labarthet, T. Buffeteau, C. Sourisseau, J. Phys. Chem. B. **103**, 6690 (1999)
59. Z. Sekkat, M. Dumont, Synth. Met. **54**, 373 (1993)
60. C. Sourisseau, Chem. Rev. **104**, 3851 (2004)
61. C. Sourisseau, D. Talaga, Appl. Spectrosc. **60**, 1368 (2006)
62. C. Brémard, P. Dhamelincourt, J. Laurens, G. Turell, Appl. Spectrosc. **39**, 1036 (1985)
63. C. Brémard, J. Laureyns, J.C. Merlin, G. Turell, J. Raman Spectrosc. **18**, 305 (1987)
64. G. Turrell, J. Raman Spectrosc. **15**, 103 (1984)
65. S. Lal, N.K. Grady, J. Kundu, C.S. Levin, J.B. Lassiter, N.J. Halas, Chem. Soc. Rev. **37**, 898 (2008)
66. N. Marquestaut, A. Martin, D. Talaga, L. Servant, S. Ravaine, S. Reculusa, D.M. Bassani, E. Gillies, F. Lagugné-Labarthet, Langmuir **24**, 11313 (2008)
67. C.L. Haynes, R.P. Van Duyne, J. Phys. Chem. B **105**, 5599 (2001)
68. G.C. Schatz, M.A. Young, R.P. Van Duyne, in *Surface-Enhanced Raman Scattering*, ed. by K. Kneipp, M. Moskovits, H. Kneipp (Springer, Berlin, 2006) Vol. 103, p 19
69. N. Félidj, S. Truong, J. Aubard, G. Lévi, J.R. Krenn, A. Hohenau, A. Leitner, F.R. Aussenegg, J. Chem. Phys. **120**, 7141 (2004)
70. E.C. Le Ru, E. Blackie, M. Meyer, P.G. Etchegoin, J. Phys. Chem. C **111**, 13794 (2007)
71. O. Schubert, J. Becker, L. Carbone, Y. Khalavka, T. Provalska, I. Zins, C. Sönnichsen, Nano Lett. **8**, 2345 (2008)
72. D.S. Hubbard, M.P. Houlne, G.E. Kiefer, K. McMillan, D.J. Bornhop, Bioimaging **6**, 63 (1998)
73. Y. Kiomacho, H. Sato, K. Aizawa, H. Tashiro, Appl. Opt. **44**, 4722 (2005)
74. J.M. Song, P.M. Kasili, G.D. FGriffin, T. Vo-Dinh, Anal. Chem. **76**, 2591 (2004)
75. C. Amatore, A. Chauvin, P. Garrigue, L. Servant, N. Sojic, S. Szunerits, L. Thouin, Anal. Chem. **76**, 7202 (2004)
76. E.J. Smythe, M.D. Dickey, J. Bao, G.M. Whitesides, F. Capasso, Nano Lett. **9**, 1132 (2009)
77. D.J. White, A.P. Mazzolini, P.R. Stoddart, J. Raman Spectrosc. **38**, 377 (2007)
78. D.J. White, P.R. Stoddart, Opt. Lett. **30**, 598 (2005)
79. M.E. Hankus, H. Li, G.J. Gibson, B.M. Cullum, Anal. Chem. **78**, 7535 (2006)
80. T. Murphy, S. Lucht, H. Schmidt, H.-D. Kronfeldt, J. Raman Spectrosc. **31**, 943 (2000)
81. E. Polwart, R.L. Keir, C.M. Davidson, W.E. Smith, D.A. Sadler, Appl. Spectrosc. **54**, 522 (2000)
82. V. Guieu, F. Lagugné-Labarthet, L. Servant, D. Talaga, N. Sojic, Small **4**, 96 (2008)
83. V. Guieu, D. Talaga, L. Servant, N. Sojic, F. Lagugné-Labarthet, J. Phys. Chem. C **113**, 874 (2009)

84. P. Pantano, D.R. Walt, Rev. Sci. Instrum. **68**, 1357 (1997)
85. J. Engelking, M. Witteman, M. Rehahn, H. Menzel, Langmuir **16**, 3407 (2000)
86. R.M. Roth, N.C. Panoiu, M.M. Adams, R.M. Osgood, C.C. Neacsu, M.B. Raschke, Opt. Express **14**, 2922 (2006)
87. C.C. Neacsu, G.A. Steudle, M.B. Raschke, App. Phys. B. **80**, 295 (2005)
88. A.L. Demming, F. Festy, D. Richards, J. Chem. Phys. **122**, 184716 (2005)
89. D.P. Fromm, A. Sundaramurthy, P.J. Schuck, G. Kino, W.E. Moerner, Nano Lett. **4**, 957 (2004)
90. D.P. Fromm, A. Sundaramurthy, A. Kinkhabwala, P.J. Schuck, G.S. Kino, W.E. Moerner, J. Chem. Phys. **124**, 061101 (2006)
91. L. Novotny, Nature **455**, 887 (2008)
92. T. Deckert-Gaudig, V. Deckert, Small **5**, 432 (2009)

Chapter 5
Application of Raman-based Images in the Earth Sciences

Lutz Nasdala, Olivier Beyssac, J. William Schopf and Bernd Bleisteiner

5.1 Preface

An increasing fraction of the research done on geological samples deals with the questions how, and to which extent, minerals deviate from their ideal structural and chemical composition. Such deviations from perfectness are commonly referred to as *real structures* of minerals. There exist many types of real structures, including point defects such as substitutions of an atomic species by another element, atomic vacancies, or interstitials (i.e., atoms at lattice sites that are normally unoccupied in the ideal structure), dislocations, domain or twin boundaries, stacking faults, volume units occupied by other phases (such as inclusions), strain, and many more; for an overview see [122].

If such real-structure properties are not (more or less) homogeneously distributed within minerals but show a marked heterogeneity pattern, this pattern is described as *internal texture*. Note that they were, and are, also called "internal structures",

L. Nasdala
Institut für Mineralogie und Kristallographie, Universität Wien,
Althanstr. 14, A-1090 Wien, Austria
e-mail: lutz.nasdala@univie.ac.at

O. Beyssac
Institut de Minéralogie et de Physique des Milieux Condensés (IMPMC), UMR 7590
CNRS-UPMC, Campus Jussieu, Case Courrier 115, 4 place Jussieu, F-75005 Paris, France

J. W. Schopf
Center for the Study of Evolution and the Origin of Life (Institute of Geophysics and Planetary Physics), Department of Earth and Space Sciences, and Molecular Biology Institute, University of California, 595 Charles E. Young Drive East, Los Angeles, CA 90095, USA

J. W. Schopf
Pennsylvania State University Astrobiology Research Center, 435 Deike Building, University Park, PA 16802, USA

B. Bleisteiner
HORIBA Jobin Yvon GmbH, Neuhofstrasse 9, D-64625 Bensheim, Germany

A. Zoubir (ed.), *Raman Imaging*, Springer Series in Optical Sciences 168,
DOI: 10.1007/978-3-642-28252-2_5,

especially in earlier papers. In view of the definitions of the two terms, however, heterogeneity patterns should be described more correctly as textures [22]. Internal textures of minerals include, for instance, growth zoning, sector zoning, diffusion zoning, patchy or other alteration patterns, stress and strain textures, exsolution textures, and many more. The detection and visualization of such patterns of heterogeneity therefore provide a wealth of information about the formation, the secondary alteration, and eventually the destruction or decomposition of crystals and minerals.

In view of internal textures, analytical and imaging techniques are currently developed and applied virtually at two different levels. First, an increasingly broad range of techniques is used to merely reveal the internal textures, to assist other analyses or applications. To give an example, bias of electron probe micro analyser (EPMA) or ion microprobe results due to point measurements having straddled the boundary between different growth generations or zones, can be avoided by placing measurements carefully within single zones. For this, knowledge of the location, size, and extension of zones is needed, whereas knowledge of their chemical and/or structural differences is not necessary here. This description may, for instance, still be applicable to cathodoluminescence (CL) images of minerals which have proven to be most useful but whose reliable interpretation is in many cases still pending. Second, image generation techniques are developed toward a more quantitative level. The understanding of what exactly varies among zones, domains, or other interior regions assists the reconstruction of the growth conditions and post-growth history of minerals and their host rocks, especially if appropriate calibrations for the respective heterogeneity feature exist.

In this chapter, we describe internal textures of crystals, minerals, and mineral assemblages as detected by Raman spectroscopy. As an introduction, and to underline the motivation of applying Raman-based heterogeneity images in the Earth sciences, overviews of the different internal textures, and the broad range of techniques to visualize them are given. These overviews should be considered rather as selections of typical examples; they do not claim to be complete. Second, a number of Raman experimental details are discussed; however, only those which are of particular relevance to mineralogical applications. In particular, the different imaging concepts are compared, and it is argued why Earth scientists prefer to apply hyperspectral Raman maps whereas direct images, if applied at all, are rarely used in this field. Also, the reader is advised of number of common (and often published) artefacts related to the Raman analysis of minerals. For a more complete and more detailed overview of Raman experimental aspects, however, the reader is referred to Chap. 1 of this book. The main part of this chapter deals with Raman-based images applied in the Earth sciences. After a brief general overview, which again does not claim to be complete, examples for applications in three selected areas of research are presented in more detail. These include applications in high-pressure petrology and mineralogy, research addressing phenomena of radiation damage, and the study of fossils and bio-minerals.

5.2 Introduction: An Overview of Internal Textures of Minerals

The study of textural relationships among crystals/grains of minerals in rocks is one of the main tools in petrology and has a long tradition; this is especially done by petrologists using transmitted light optical microscopy of thin sections (see [25, 121], and references therein). This chapter cannot cover this topic; we just present two selected, excellent examples to underline the value of rock textures in interpreting rocks (Fig. 5.1). For instance, the occurrence of pronounced, wing-like fringes of calcite and quartz on both sides of (comparably hard) pyrite crystals (Fig. 5.1a) indicates a metamorphic growth process under conditions of strong shear stress [75]. In this example, phase identification was easily possible from the optical properties observed in cross-polarised transmitted light. The same observation technique did not provide much information in the second example (Fig. 5.1b, left part): A silica sample from the famous Permian "petrified forest" location in Chemnitz, Saxony, was merely found to consist of numerous quartz grains. However, the corresponding CL image reveals minute variations in optically active defects in the quartz lattice. Even the cell structure of the petrified wood can be recognized here (Fig. 5.1b, right part). Applying the CL technique, [52] found that the primary quartz, which typically shows yellow CL, was partially overprinted by low-temperature hydrothermal fluids, resulting in short-lived blue CL (cf. [53]); and they concluded that the initial silicification must have occurred at elevated temperatures.

In the study of single-crystals of minerals, the understanding of their internal textures is of enormous importance for geoscientists. As an example, the wide range of information on the growth and post-growth history of zircon ($ZrSiO_4$) that is potentially extracted from internal textures of individual crystals of this mineral, was presented and discussed in detail by [22]. Correspondingly, more and more techniques are applied to visualize internal textures. A number of examples for the application of direct imaging techniques are presented in Fig. 5.2, and a few examples of colour-coded images and maps are shown in Fig. 5.3 (the principal difference between direct imaging and mapping techniques will be discussed in Sect. 5.3 below).

Our suite of examples for direct images of minerals starts with the common phenomenon of inclusions of other phases. In the case of gases, fluids, or other minerals that were trapped inside the host mineral during its primary growth, the composition of the inclusions may allow one to conclude about the formation conditions of the host mineral. In the example shown in Fig. 5.2a, the detection of kaolinite filling fibrous or tube-like inclusions in Thai gem garnet [120] indicated a secondary alteration history. Similarly, zircon crystals from a potassic granulite from the Plešovice quarry, southern Bohemian Massif, show brownish phases (interpreted as iron oxides or hydroxides) at large fractures, which was interpreted by [137] to indicate a low-temperature alteration history. These zircon crystals (Fig. 5.2b) are also characterized by marked primary growth zoning (i.e., growth zones with alternating low and high levels of incorporated trace uranium) and sector zoning (i.e., preferred U incorporation at pyramid compared to prism faces). This complex zoning is recognized from

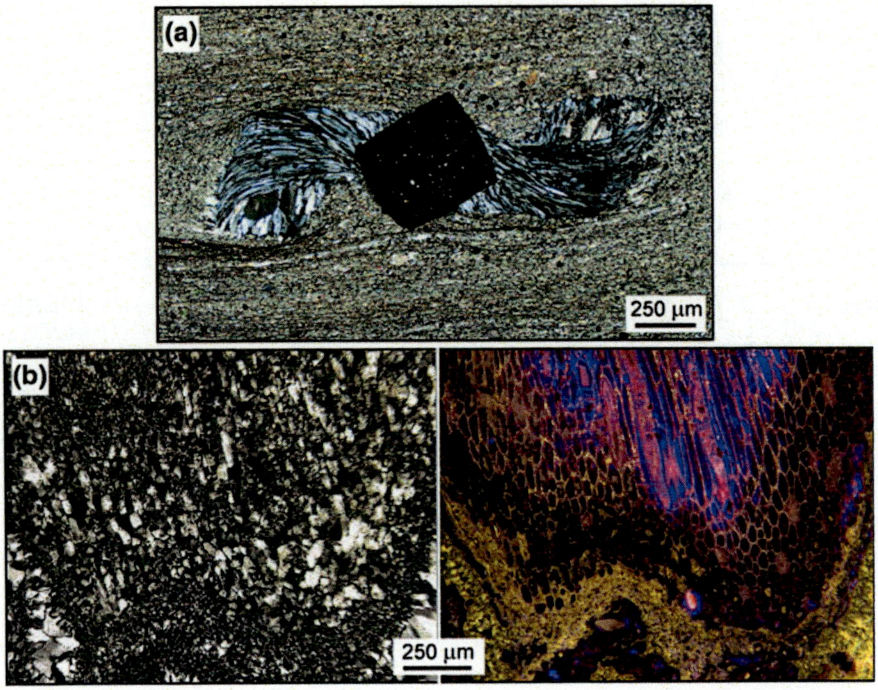

Fig. 5.1 Images of rock samples showing genetically informative textures. **a** Transmitted light photomicrograph (*cross-polarised light*) of a rock thin-section of a fine-grained, metamorphosed schist from Lourdes, France. The central, isometric pyrite crystal (opaque, *black*) is bracketed by wing-like, fibrous strain fringes consisting of quartz and calcite. Sample courtesy D. Köhn (Universität Mainz, Germany). **b** Pair of cross-polarised light (*left*) and CL (*right*) images of a silicified wood (*Psaronius sp.*) specimen from Chemnitz, Saxony. Image courtesy J. Götze (TU Bergakademie Freiberg, Germany)

the variably irradiation-lowered interference colors in zones with elevated U concentrations (in more detail discussed in Sect. 5.4.3 below). The next example belongs to an utterly different type. A photoluminescence (PL) image of a polished sample mount containing synthetic monazite crystals (Fig. 5.2c) that were grown using the Li–Mo flux technique [55], does not show notable internal textures of these crystals. However, the surrounding epoxy exhibits, as an analytical artifact, extensive electron-beam damage. Each single imaging scan in the scanning electron microscope (SEM) is seen as rectangular area, and slower (and correspondingly longer) scans are recognized from the higher intensity of the defect luminescence. Particularly, pronounced internal zoning is seen in a CL image of a cut and polished diamond single-crystal from the Guaniamo area, Venezuela (for sample description cf. [132]). This image is now well-known as the "Picasso diamond". The observation of three types of interior regions enabled [21] to assign variations of δ^{13}C values to three growth stages of this diamond crystal. The transmission electron microprobe (TEM) brightfield image of the thin (i.e., electron-beam transparent) diamond foil presented in Fig. 5.2e shows numerous straight, parallel dislocations. This observation indicates mechanical deformation that was not followed by notable annealing upon uplift, which may suggest that the diamond was deformed during its uplift (R. Wirth, pers. comm.). An effect of mechanical stress as an artifact of the preparation procedure is shown in Fig. 5.2f. The back-scattered electron (BSE) image of a chip of reference zircon #91500 [160] shows comparably weak growth zoning (Fig. 5.2f, left image). The corresponding CL image (Fig. 5.2f, right image) shows fan-shaped striations that are due to near-surface strain introduced during the mechanical preparation process. This image pair reconfirms again that luminescence imaging is most sensitive to very minor distortions and changes in electronic levels (cf. also [51]). Another effect of stress, here again of natural origin, is the so-called planar deformation features (PDFs) that develop in brittle minerals as a result of a shock-wave, as it typically occurs in the impact of meteorites to the Earth surface. In suevites from the Ries crater, Germany, extremely high pressures of $\geq$30 GPa (or $\geq$300 kbar) during the short but intense impact shock are indicated by the occurrence of stishovite (a very-high pressure silica polymorph), with the neighboring diaplectic quartz showing numerous PDFs (Fig. 5.2g; [141]). Our final direct imaging example (Fig. 5.2h) was again obtained under the optical microscope; it represents a decidedly different type of textures, namely those resulting from natural ion-irradiation processes. The sample consists of fluorite with scattered inclusions containing actinide-bearing mineral phases. In these inclusions, U and Th, and their instable daughter nuclei, may undergo alpha-event decays, which results to the ejection of alpha particles (i.e., ^{4}He cores with energies in the range 3.9–8.8 MeV) in random directions. Due to the well-defined alpha energies and the uniform masses of ejected helium cores, trajectories of all alpha particles belonging to the same type of decay event have the same length, which eventually results in the formation of radiation-damaged areas that have the shape of concentric hollow spheres. These are typically observed in thin sections as circles (more or less blurred, depending on the dimension of the source of the radioactivity; [46]). Their presence indicates that the sample has spent extended periods of time under rather "cold" conditions near the Earth's surface (because otherwise thermal annealing would have removed the radiation damage).

Mapping of geological samples is commonly done in the EPMA, to visualize distribution patterns of major and trace elements within single-crystals; two examples are presented for this. The first one (Fig. 5.3a) shows primary zoning due to a rhythmic growth process [48], through minor variations in the incorporation of Ba^{2+} in the structure of sanidine feldspar ($KAlSi_3O_8$). Secondary zoning, here caused by low-temperature "wet" alteration near the Earth surface, is indicated in Fig. 5.3b: Crystals of intensely weathered ilmenite [$(Fe,Mg)TiO_3$] show extended surficial regions where the primary constituents (including Mg^{2+} substituting for Fe^{2+}) are depleted whereas elements that are not incorporated in the primary growth of this mineral (such as Al^{3+}) are strongly enriched [142]. Primary growth zoning is also seen in the hyperspectral PL map of a synthetic zircon crystal doped with holmium (see [55]). The map visualizes colour-coded the intensity distribution of the 540 nm emission of Ho^{3+} (which is due to the $^5S_2 \rightarrow {}^5I_8$ electronic transition; [43]) and shows that the Ho incorporation decreased appreciably during progressing crystal

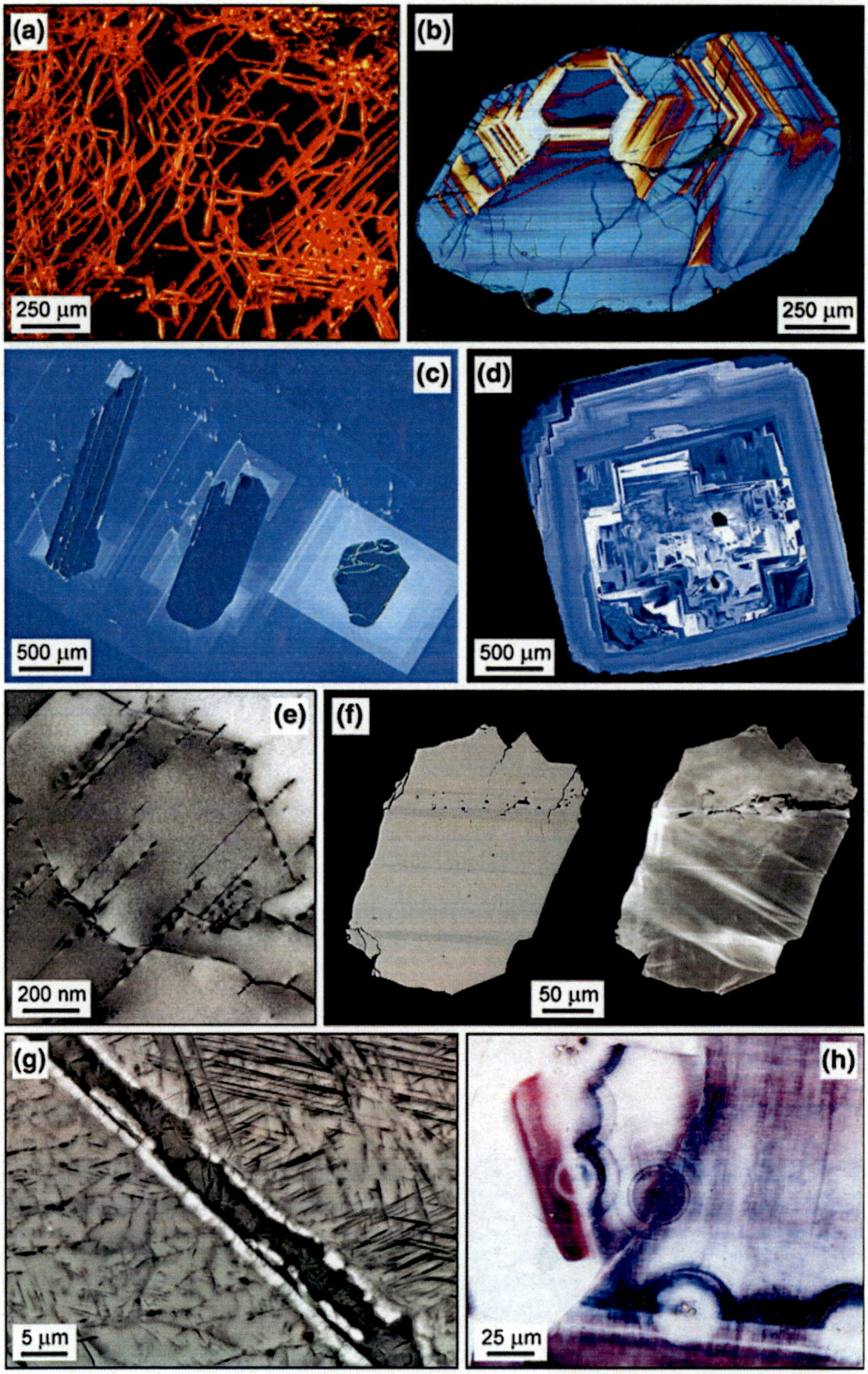
(a)
250 µm
(b)
250 µm
(c)
500 µm
(d)
500 µm
(e)
200 nm
(f)
50 µm
(g)
5 µm
(h)
25 µm

◄ **Fig 5.2** Various images of internal textures of minerals. **a** Photomicrograph of a pyrope garnet specimen from Ban Nong Bon, Trat Province, Thailand. Fibrous inclusions are filled with secondary kaolinite. Image courtesy B. Phichaikamjornwut (Chiang Mai University, Thailand). **b** Transmitted light photograph (cross-polarised light) of a thin section of a zoned zircon single-crystal from Plešovice, Czech Republic. Sample courtesy J. Košler (University of Bergen, Norway). **c** PL image (UV illumination) of polished, Sm-doped $CePO_4$ crystals embedded in araldite epoxy. The epoxy is affected by electron beam damage during the SEM imaging. Sample courtesy K. Ruschel (Universität Wien, Austria). **d** CL image of the famous "Picasso diamond", revealing three growth generations. Image courtesy D. Schulze, A. Dias (University of Toronto, Canada), and N. Cayser (University of Edinburgh, U.K.). **e** TEM brightfield image of a placer diamond from the Ukraine, showing numerous straight dislocations. Image courtesy R. Wirth (Helmholtz Zentrum / GFZ Potsdam, Germany). **f** Pair of BSE (*left*) and CL (*right*) images of a chip of reference zircon #91500 originating from Kuehl Lake, Ontario, Canada. The CL image shows stress striations caused by the mechanical preparation procedure. Sample courtesy M. Wiedenbeck (Helmholtz Zentrum / GFZ Potsdam, Germany); image courtesy A. Kronz (Universität Göttignen, Germany). **g** BSE image of a suevite clast from the Ries impact crater, Germany, showing a pseudotrachylite vein with stishovite (high BSE intensity). The neighbouring diaplectic quartz (*gray* BSE), especially on the *upper right side*, is characterized by numerous PDFs (planar deformation features) caused by the shock-impact. Image courtesy V. Stähle (Universität Heidelberg, Germany). **h** Transmitted light photograph (*plane-polarised light*) of fluorite from Wölsendorf, Bavaria, Germany. Small radioactive inclusions are surrounded by so-called radiohaloes, i.e., concentric rings that originate from alpha particle damage. Sample courtesy M.A. Götzinger (Universität Wien, Austria)

growth. Also, somewhat more holmium was incorporated at prism faces than at pyramid faces, resulting in additional weak sector zoning. This sector zoning is in contrast to the incorporation of other trace elements, such as Dy or U, which are more often preferentially incorporated in zircon at pyramid faces ([101, 137]; cf. Fig. 5.2b above). In the MetriPol birefringence image (for the technique cf. [62, 64]) shown in Fig. 5.3d, a radial stress field can be recognized in the host diamond from the birefringence halo surrounding an olivine inclusion. This observation indicates that the inclusion remains under significant remnant pressure within the diamond host. Finally, a largely different example is presented in Fig. 5.3e; showing the lateral distribution of a structural parameter: This electron back-scatter diffraction (EBSD) map reveals the variation of the crystallographic orientation of fibrous calcite crystals within aggregates in stalagmites [112].

More and more imaging and mapping techniques are developed and used successfully to study internal textures of single-crystals, and textures of mineral assemblages. These techniques include, for instance, fluorescence lifetime imaging microscopy (FLIM; [124]), hyperspectral CL imaging [29, 85], Nomarski differential interference contrast (DIC) microscopy [2, 50], time-of-flight secondary ion mass spectrometry (TOF-SIMS) mapping [16, 57, 91], ion microprobe mapping [21, 66], and many more. This list, and the above selection of examples on how textural relationships among minerals, and internal textures of minerals, can be visualized, is not claimed to be complete; neither in terms of the information content of textures, nor in terms of the analytical techniques that are applied. The range and variety covered by our examples, however, is meant to underline the wealth of valuable information that can be obtained from the detection of heterogeneity in different sorts of images and/or textural distribution patterns.

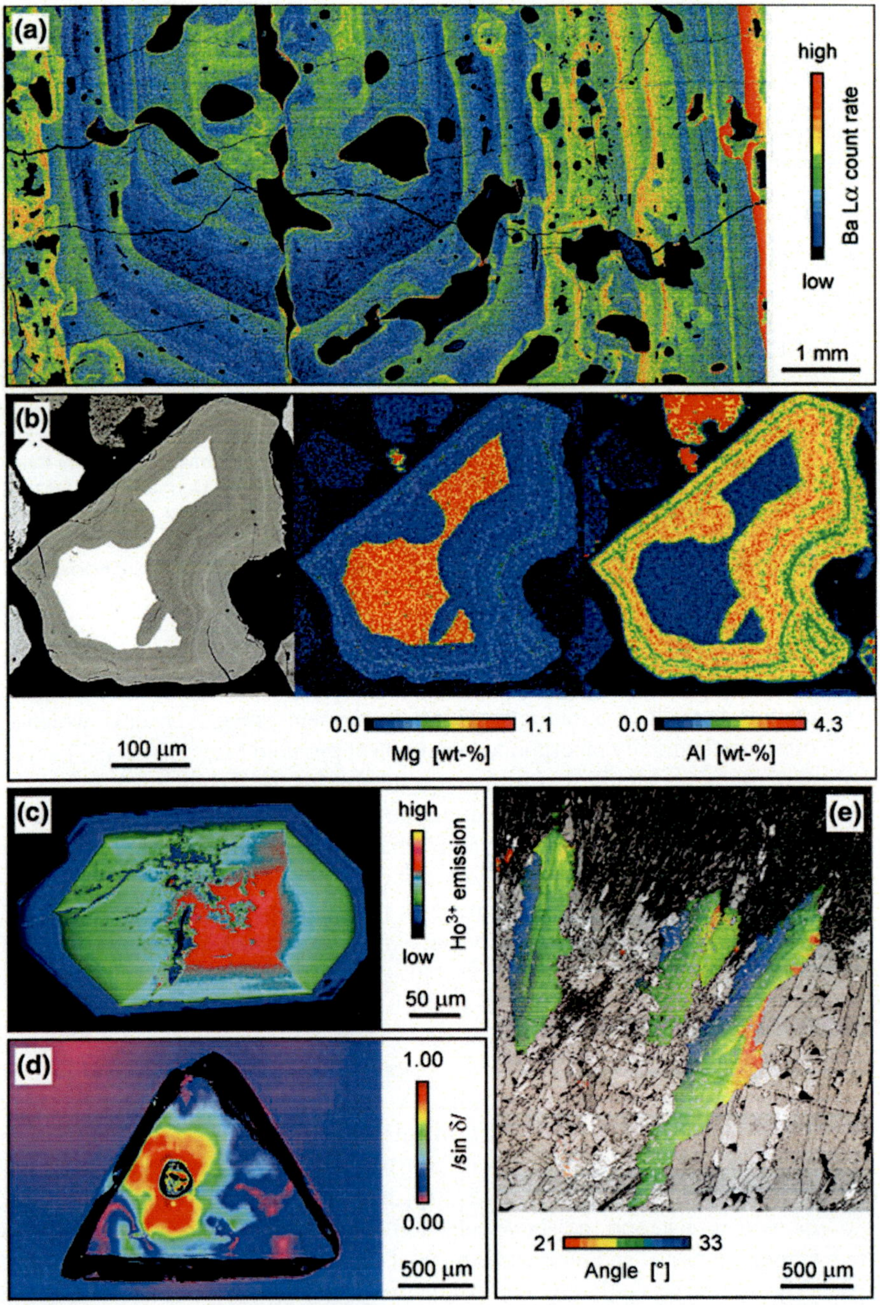
(a)
high
Ba Lα count rate
low
1 mm
(b)
0.0
1.1
Mg [wt-%]
0.0
4.3
Al [wt-%]
100 μm
(c)
high
Ho3+ emission
low
50 μm
(d)
1.00
/sin δ/
0.00
500 μm
(e)
21
33
Angle [°]
500 μm

◂ **Fig 5.3** Colour-coded images and maps of internal textures of minerals. **a** Barium distribution within a sanidine feldspar crystal from the Taapaca volcano, Chile. Glass, voids, and mineral inclusions appear black. Image courtesy C. Ginibre, A. Kronz, and G. Wörner (Universität Göttingen, Germany). **b** Altered ilmenite crystal from Rakovnik, Central Bohemia, Czech Republic. Left, BSE image; middle, Mg distribution map; right, Al distribution map. Image courtesy R. Škoda (Masaryk University, Brno, Czech Republic). **c** Map of the laser-induced PL emission of a synthetic, Ho-doped $ZrSiO_4$ crystal, showing pronounced growth zoning. Sample courtesy J.M. Hanchar (Memorial University of Newfoundland, St. John's, Canada). **d** MetriPol birefringence image of a diamond crystal containing an olivine inclusion, from Shandong Province, China. Sample courtesy H.J. Milledge (University College, London, U.K.); image courtesy D.Howell (Macquarie University, Sydney, Australia). **e** EBSD map of radiaxial fibrous calcite in a stalagmite from a cave near Iserlohn, Germany, showing the angle of the c-axis of calcite crystals with respect to the N-S direction of the image (measured clockwise). Image courtesy R.D. Neuser (Universität Bochum, Germany)

5.3 Experimental

In this section, we provide a brief outline of technical aspects related to the generation of Raman-based images and their particular application in the Earth sciences. The Raman analysis of minerals may be particularly challenging, when compared to the analysis of other materials. One major aspect is that the minerals have normally formed in natural, virtually "polluted" environments whose conditions may deviate appreciably from the controlled growth of, for instance, semiconductors or pharmaceutical products. Correspondingly, mineral samples consist rarely of structurally and chemically well-defined crystals; rather, they are commonly multiply doped with formula elements and characterized by a range of structural defects. This results in common analytical problems such as the co-excitation of PL (due to the presence of optically active, element-related or defect centres) by the incident laser beam. Such and other aspects are discussed in the following, and special care will be taken in the discussion of artifacts that may potentially affect the results of Raman analyses of minerals (cf. also [104]). More technical details relevant for the generation of Raman-based images, can for example be found in Chap. 1 of this book.

5.3.1 Methodology: Spectrometers and Image-Generation Techniques

5.3.1.1 Spectrometers

Most systems used for studies of geologic materials involving Raman-based images are commercial dispersive, single-stage spectrometers equipped with Rayleigh rejection filters. There also exist a range of commercial Fourier-transform (FT) Raman systems; however, these have not become widely accepted in the Earth sciences community. Their apparent advantage of the reduction of laser-induced luminescence with near-infrared (NIR) compared to visible or ultraviolet (UV) excitation cannot make up for the many disadvantages of FT systems, including the lower scattering intensity at longer excitation wavelengths and the lower spatial resolution performance (which is due to the dependence of the resolution on the wavelength and the lack of confocality of such systems). Among the dispersive systems, single-stage configurations are clearly preferred by Earth scientists because of their higher signal efficiency ("optical throughput") compared to that of double or triple monochromator systems. This is an important advantage for Raman studies of geological samples because minerals are often poor Raman scatterers. The need for high signal efficiency is enhanced further by the fact that many minerals tend to be sensitive to light absorption, which results in the necessity to use comparably low incipient laser power. Double or triple monochromators have the analytical advantage of a much better dispersion, i.e., their spectral resolution is higher and bands close to the laser line, with wavenumbers well below the cut off of conventional Rayleigh rejection filters, can be measured. These abilities, however, are less relevant for most geoscientific applications, compared to the need to have high signal efficiency. For these reasons, double or triple monochromators are only used occasionally in the Earth sciences.

One of the major advantages of Raman spectroscopy is its ability to analyze samples on a micrometre scale, which is of particular relevance to the Earth scientists because of the commonly small sizes of crystals and their narrow intergrowth within the samples to be analyzed. The microscope is therefore a key element of the Raman spectrometer, enabling one to localize the mineral phases of interest in a matrix with complex micro-texture. Most systems are now equipped with high-grade research microscopes that are optimized for high magnification objectives (50× to 100×) with high numerical aperture (NA; up to 0.9). However, high spatial resolution cannot be achieved alone by coupling a spectrometer system with a powerful optical microscope. High lateral and especially depth resolution requires a confocal arrangement of the optical pathway. Raman systems are therefore often either equipped with a true confocal aperture [144], or pseudo-confocality is achieved by a combination of a narrow spectrometer entrance slit in one direction with (software-controlled) pixel masking on the charge-coupled device (CCD) detector in the other direction [70, 162]. This allows one to virtually cut off the illuminated sample area, which has the shape of a truncated cone, at its broad base; resulting in a genuine lateral resolution not much larger than the focal-spot diameter (Fig. 5.4, left part).

In evaluating Raman maps, it needs to be considered that their real resolution is controlled and limited by two independent factors, namely, the step size (i.e., the inter-distance between neighboring single analyses across the mapped area) and the lateral resolution of each single analysis. Note, however, that there exist different definitions of what the term "lateral resolution" (and analogously "depth resolution") describes. The spatial resolution performance of an analysis system is commonly described by the size of the analyzed volume (this is mostly used to define the spatial resolution of point analyzes; see Fig. 5.4a). Second, it is possible to define spatial resolution by the minimum distance between two objects that can just be resolved (this is mostly used to define the spatial resolution of images).

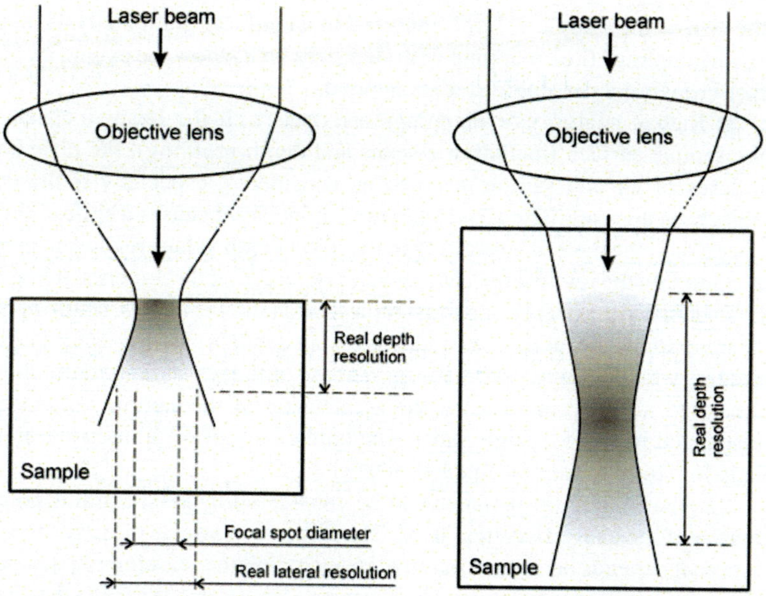

Fig. 5.4 Simplified sketches (not to scale) of the beam path of a confocal Raman spectrometer. The analyzed volume areas are shown gray. Left, beam focused at the sample surface; right, beam focused inside the sample. Note that in the latter case the focus quality is decreased appreciably, resulting in an enlarged focal-spot diameter and dramatically decreased depth resolution

The lateral resolution (in the sense of the second definition above) that is obtainable with a confocal microscope setup can be estimated using the formula $\Delta d_{min} = F\lambda / NA$, where Δd_{min} is the minimum distance between two objects just resolved, F is a constant (depending on the optical system of the microscope, the nature of the sample, and the criterion applied in the definition of resolution), and λ is the wavelength of the exciting laser light. For the Rayleigh criterion, the theoretical minimum distance is expressed by $\Delta d_{min} = 1.22\lambda / 2\ NA$ ($\sim 0.6\lambda / NA$). When the Sparrow criterion is used, values of slightly below $0.6\ \lambda / NA$ are calculated. Note that both criteria yield only the appropriate estimates in the case of flat and highly reflective samples, whereas sample-dependent features such as surface roughness may introduce notable uncertainty. A more realistic evaluation of the spatial resolution (in the sense of the first definition above) uses expressions for the focal volume of the laser beam based on $d_w = 1.22\lambda / NA$ (i.e., the laser beam waist) and d.o.f. $= 4\ \lambda / NA^2$ (i.e., the laser depth of focus). However, the real lateral resolution exceeds notably the diameter of the laser beam waist (Fig. 5.4). Experimental evaluation of the lateral and axial resolution may therefore yield performance values which are notably different from estimates based on the above formulae; this depends on the surface quality and roughness, refraction, and many other factors [13, 30]. Applying the second definition above leads consequently to notably smaller estimates of the spatial resolution: It is well possible to visualize objects in Raman-based images whose size is much smaller than the size of the sample volume analyzed in a confocal point analysis, in particular if "oversampling" (i.e., mapping with step sizes well below the extension of the volume that is analysed per single step) is applied.

If Raman analyses or mappings are done with the beam focused well below the sample surface (including z-scans and depth profiling), the different refractive indexes of air and sample need to be considered, especially if the common dry objectives are used. The air-sample optical interface causes a shift of the laser focus deeper into the material, resulting in incorrect depth values read from the mechanical z-control on the microscope. Moreover, rays entering the material at large angles will be focussed more deeply into the sample than shallower angles, resulting in increasing degradation of the laser focus with increasing depth [4, 31–33, 146]. The loss of focus quality, which is associated with an increase of the waist diameter of the illuminated area, and a strong increase of the z-extension of the sample volume analyzed, is visualized in Fig. 5.4 (right part). The application of oil immersion objectives may help to reduce these unwanted effects.

Another important parameter to be considered is the spectral dispersion performance of the analysis system used. The dispersion affects directly the spectral resolution; it depends on several parameters including the wavelength, the focal length of the system (distance between the focussing element and the CCD detector), the grating groove density, and the pixel size of the CCD. When the entrance slit is smaller than the CCD pixel element, the spectral resolution will be nearly equivalent to the dispersion on two pixels. Depending on these parameters the spectral resolution can vary appreciably; typical values lie in the range between several wavenumbers and a few tenths of wavenumbers. The spectral resolution with which a measurement was done is of importance for all applications in which band fitting is applied and FWHMs (full width at half maximum) of bands are used. Examples for such applications are the estimation of the degree of crystallinity, or the crystal size, of mineral phases based on band FWHMs. Depending on the spectral resolution, however, each spectral signal will be recorded with too broad FWHM, for which reason mathematical band width correction needs to be applied. Of course, such correction is more reliable if the artificial band broadening (i.e., the contribution of the limited resolution of the system) is small compared to the signal to be detected; quantitative FWHM measurements should therefore be done with a system with sufficiently high spectral resolution. A mathematic correction procedure has, for instance, been published by [155]. Note, however, that the reliability of any mathematical correction will depend on how accurately the spectral resolution performance of the respective experimental setup has been determined by the operator.

Another aspect to be considered is the wavenumber accuracy of a system and, connected with that, the wavenumber calibration. Proper wavenumber calibration is needed especially for quantitative applications beyond the simple phase identification; for instance, the determination of remnant pressures or compressive strain from the band up-shift. Raman spectrometers are therefore calibrated with atomic calibration lamps (in particular neon lamps) or with reference samples for the absolute wavenumber calibration. To avoid the necessity of frequent recalibration, spectrometers with a stable optical setup are needed. This is especially the case for successful imaging (e.g. of stress phenomena in minerals), in particular when mapping over large areas, which may sometimes take several hours to days.

5.3.1.2 Strategies for the Generation of Raman-Based Images

There are two principal ways of generating Raman images, namely, direct imaging (also called global illumination imaging) and hyperspectral imaging (also called Raman mapping). For a more detailed discussion of the two principles, the reader is referred to [150] see also [28, 60]. Direct images are obtained based on the simultaneous and global illumination of the area to be imaged, during a single measurement at a specific wavenumber. Such a direct image then represents the total intensity distribution (in the case of polychromatic images), or the intensity distribution of the chosen wavenumber (in the case of monochromatic images), across the imaged area. Monochromatic images are, for instance, obtained using tunable bandpass filters; ideally, they show the intensity distribution pattern of a certain, pre-selected Raman band. The advantage of this technique is that direct images are obtained in very short times (typically seconds). However, because global imaging cannot discriminate between Raman and background signals (such as luminescence, which may easily be much more intense than the Raman signal itself), the direct imaging method is mostly suitable for samples with almost no background signal. In the Earth sciences, where samples often show a high stray background or luminescence, global imaging is therefore rarely used.

Hyperspectral images or maps, in contrast, are produced by recording a full Raman spectrum at each x, y location of the area to be imaged (with each analysis corresponding to one pixel of the future image). They are obtained by scanning the sample with either a laser spot or line. Recording such maps requires an automated xyz stage to move the sample relative to the fixed laser (however, there also exist a few designs of spectrometers with oscillating mirrors, in which the focused laser scans the area to be sampled). In point-by-point mapping, the sample region illuminated by the laser spot is stepped, or raster-scanned, in two spatial dimensions (x and y) with a spectrum being recorded at each x, y position, and the CCD detector is entirely used to obtain the spectral information. In static line-by-line mapping, a laser line is raster-scanned along one of the sample axes (x or y; the dimension that is parallel to the entrance slit of the spectrometer) and the two-dimensional CCD detector is utilized to acquire both spatial and spectral information. One part of the spatial information (i.e., the x or y coordinate of the respective spectrum) is hence probed and obtained by line-scanning the sample in a direction perpendicular to the line focus direction. The second half of the spectral information (i.e., the y or x coordinate), in contrast, is registered on the CCD detector in the dimension parallel to the spectrometer's entrance slit. The other dimension of the CCD detector is then used to record the spectral information. Recently, a number of dynamic line-by-line mapping techniques have been developed, which decrease appreciably the total acquisition time. One of them is based on the synchronization of the sample movement/scanning with the light collection and data readout on the CCD detector (for applications on geological objects see

[8]). Second, average spectra may be obtained through scan moving the laser focus across a rectangular area (1–300 μm size), which is done by two mirrors based on piezoelectric elements. The latter technique is especially helpful in finding small crystals or inclusions in samples, which sometimes are difficult to find by simple inspection under an optical binocular. Moreover, the piezoelectric-driven mirrors can also be used for classical point mapping with very fine steps down to several tens of nanometres without the necessity to move the sample. The latter includes the ability to resolve very fine textures in minerals.

Well-resolved maps may be generated from hyperspectral data sets consisting of 10^4–10^5 single spectra. In spite of the technical advantages summarized above, it takes a comparably large amount of time to obtain such huge data sets. The disadvantage of Raman mapping being a time-consuming technique, however, is more than compensated for by several advantages, mostly resulting from the possibility of reliable spectral treatment. For instance, spectra can be corrected for broad band background features, and after appropriate band fitting, the distribution pattern of almost every spectral parameter (band intensities, intensity ratios, band positions, band FWHMs, etc.) can be visualized in colour-coded images. In view of this, it appears most understandable that in the Earth sciences, hyperspectral mapping has become the Raman-based imaging method of choice.

5.3.2 Common Analytical Artifacts

Owing to the comparably complex chemical and structural composition of geological samples, the interpretation of Raman spectra and, correspondingly, of Raman-based images generated from them, may be particularly complex and challenging. Apart from a range of common Raman analytical artifacts that are either due to the sample itself or connected with the analytical procedure, there is another aspect that needs to be considered: There is still a clear mismatch, on the international scale, between the enormously increased interest of Earth scientists in applying vibrational spectroscopic techniques, and the still limited training opportunities in this field. A significant portion of Raman studies in international peer-reviewed journal is hence currently published by self-taught Raman spectroscopists. This, and the many challenges in interpreting reliably Raman spectra of geological materials, has resulted in many Raman analytical artifacts being published in recent years, even in leading international journals. In the following, a number of such (unfortunately, common) artifacts are discussed. By doing this, we would like to encourage the readers to be critical in reading and using published Raman analytical results, and to take great care in the interpretation of their own Raman spectra and images.

One first problem arises from the fact that minerals are often richly colored (i.e., intensely light-absorbing) and sensitive to thermal effects. This complicates the analysis especially if high lateral resolution is needed, i.e., if the laser beam is focused to a very small focal spot. The resulting high laser power density in the focal-spot area may then easily cause heavy local temperature increase, which in turn may result in local sample alteration or degradation. Three examples for this are shown in Fig. 5.5. Due to moderate temperature increase caused by the laser beam (8 mW), the radiation damage in a radiohalo in the silicate mineral cordierite was partially annealed, which is seen from the permanent strong reduction of the radio-colouration across the mapped area (Fig. 5.5a). As the sample experienced partial structural recovery during the analysis, it must be considered that band parameters derived from such spectra do not represent the initial structural state anymore; and based on such experimental results the degree of structural damage must be under-estimated. Even more dramatic changes are seen in Fig. 5.5b; heavy light absorption caused by Fe^{2+} in this biotite mica has resulted in local disintegration of the sample, seen as crater-like "burned" spots. Similarly, a few mW of laser light focused to a $\sim 1\,\mu m^2$ area resulted in a power density that was high enough to cause oxidation of Fe^{2+} in the spinel-structured magnetite (Fe_3O_4), to form hematite (Fe_2O_3; Fig. 5.5c). In all three examples, the artifact was easily recognizable by thorough sample inspection after the analysis; note that this is not always the case. Comparably, moderate light absorption that does not result in permanent damage but only in temporary temperature increase, may also affect notably the Raman results. For example, strong temperature increase in the focal-spot area may affect down-shift and broadening of Raman bands (for the example of diamond documented in detail in [94], for graphite in [68]). The potential occurrence of photo-chemical reactions under the impact of the laser beam should also be considered. All these aspects underline that the laser energy (and wavelength) has to be chosen carefully, to remain well below the threshold of any permanent or temporary changes of the sample. However, the use of very low excitation energies may hamper appreciably the analysis, because (especially if poorly Raman-scattering minerals are to be studied) long accumulation times per single analysis are needed. This, in turn, may make it impractical to apply hyperspectral mapping. In the case of sample materials that are particularly sensitive to light absorption and related effects, line-by-line mapping appears advantageous compared to point-by-point mapping, because of the distribution of the beam energy over a larger area.

Another source of artifacts may be the sample preparation, for instance, near-surface strain introduced by the mechanical preparation (cf. Fig. 5.2f above). An excellent example has been published by [115], showing extensive up-shift and broadening of pyrite (FeS_2) Raman bands resulting from the mechanical impact during the sample grinding and polishing procedure (Fig. 5.6a). Similar effects limit severely the reliability of Raman measurements of graphitic carbon in polished rock thin-sections. Preparation-induced surficial structural damage results in strongly broadened Raman bands, hence faking apparent high disorder and limited crystallinity of the bulk material studied [9, 104]. Here, an alternative to the cautious application of low-impact preparation techniques can be the analysis of included graphite flakes below the sample surface that were "shielded" from the mechanical impact by an overlying (transparent) grain (compare [118]).

Aspects of the confocality, and the resulting high lateral and depth resolution, have already been discussed above. Analytical artifacts may arise from either the underestimation or the overestimation of the real spatial resolution performance of

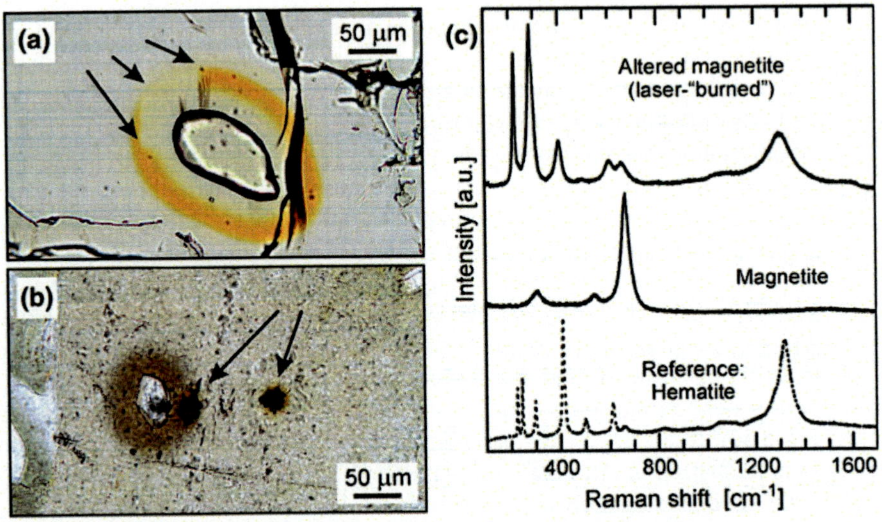

Fig. 5.5 Analytical artefacts I. **a** Transmitted-light microphotograph of a radiohalo in cordierite around a large monazite inclusion from the Strangways Metamorphic Complex, central Australia. The yellow, irradiation-induced colour of the halo appears bleached in a rectangular area where Raman mapping (Ar^+ 514.5 nm excitation) was done (*arrows*). Sample courtesy A. Möller (University of Kansas, Lawrence, U.S.A.). **b** Transmitted-light microphotograph of a Fe-rich mica (biotite). Two Raman analysis spots (He–Ne 632.8 nm excitation) are seen as "burned" circles (*arrows*). **c** Raman spectra of a Fe^{2+}–Fe^{3+} spinel (magnetite) obtained with different laser energies. In an analysis with low laser power (He–Ne 632.8 nm excitation, 0.3 mW; spectrum in the *middle*), the magnetite spectum was obtained. Analysis of the same sample with higher laser power (3 mW) yielded a spectrum that is reminiscent of that of hematite, indicating oxidation under the laser beam (*upper* spectrum)

a Raman system. In any case, accurate experimental determination of the real lateral and depth resolution of the system used, for instance, by "oversampling" line scans across the boundary of two neighboring, transparent crystals, helps to avoid misestimates of the spatial resolution performance, and hence potentially associated misinterpretations. Underestimation of the spatial resolution performance comes into the picture when randomly placed point analyzes is wrongly assumed to represent bulk samples. For instance, even the thorough characterization of reference samples with bulk techniques such as X-ray diffraction does not ensure that the obtained Raman spectrum is indeed that of the bulk material; often surficial layers of weathering products are analyzed in the Raman spectra instead (quite a number of examples for this can be found in the various online spectra databases). A more complex example is presented in Fig. 5.6b: The Raman spectrum obtained from an (obviously insufficiently cleaned) rock thin-section consists of an overlap of bands of the sectioned spessartine garnet, a blue dye originating from a former label applied with a felt-tip pen, and tiny remnants of the silicon carbide polishing powder. Examples for overestimates of the spatial resolution are presented by [95] and in the focused ion beam (FIB) study of [110]. In these two contributions, cases are discussed where surficial, transparent layers of only 0.1 and 1.5 μm thickness were attempted to be analyzed with conventional Raman spectrometers whose real spatial resolution were probably on the order of $\geq$3 μm. Correspondingly, the Raman signal of the underlying bulk was co-detected and wrongly assigned to the surficial layers, which had led to far-reaching but incorrect conclusions. A similar but rather simple example is presented in Fig. 5.6c: The Raman spectrum of a small ($\leq$5 μm sized) chip of zircon in a microprobe mount is strongly affected by bands of the neighbouring araldite epoxy as soon as the spectrum is obtained without maximum confocality.

Probably, the most commonly published artifact related to the Raman spectroscopic investigation of geological materials is the misinterpretation of laser-induced PL bands as Raman bands. It is generally assumed that luminescence emissions are easily recognized because they are much broader than Raman bands; however, this "rule of thumb" has many exceptions. In particular, problems occur if samples contain notable amounts of elements with a 4*f* electronic configuration, especially three-valent REE (rare earth element) ions, or optically active centers with 3*d* configuration yielding narrow-band PL (especially Cr^{3+}). The widths of PL emission bands of such ions (or, more precisely, widths of individual sub-bands that arise from the crystal-field splitting affecting the respective electronic transitions) may be as narrow as 0.1–0.3 nm, which is equivalent to FWHMs of only a few cm^{-1}. For instance, the 514.5 nm excitation Raman spectrum of a radiation-damaged and probably heat-treated zircon sample of Zhang et al. [165], consisting of crystalline ZrO_2 and amorphous SiO_2, showed bands of the ZrO_2 plus even narrower bands with apparent Raman shifts in the range 1,000–1,100 cm^{-1}. The latter were assigned to Raman bands of crystalline $ZrSiO_4$ and SiO_2 by Zhang et al. [165]. Nasdala et al. [104] presented a Raman spectrum of the very same sample with 632.8 nm excitation, only showing the ZrO_2 Raman bands; and they discussed that the additional bands in the 514.5 nm excitation spectrum must have been green (wavelength near 545 nm) PL emissions. The analogous misinterpretation, caused by the use of an inappropriate excitation wavelength, was published by Xian et al. [164] who suggested using the 1,461 cm^{-1} "Raman band" of zircon for a genetic assignment of this mineral. However, [96] showed, *inter alia* with the example of a narrowly zoned crystal from a gneiss from the Saxonian Erzgebirge (for sample description see Tichomirowa et al. [147]) that this band (actually a sub-band of the $^4S_{3/2} \rightarrow {}^4I_{15/2}$ electronic transition of Er^{3+} at 556.4 nm) was an artifact, overlaying the Raman spectrum of zircon at 514.5 nm excitation (Fig. 5.6d).

Another more recent example for mistaking REE emissions as Raman bands is found in [27]. These authors presented "Raman" spectra (795 nm excitation) of fluorite, however in the range 400–2,400 cm^{-1}, which is first questionable insofar as the only first-order Raman band of fluorite lies outside this range (322 cm^{-1}; [44]). Second, the spectral range chosen (with 795 nm excitation corresponding to 820–980 nm wavelength) is strongly affected by PL emissions related to Er^{3+} [44] and the $^4F_{3/2} \rightarrow {}^4I_{9/2}$ electronic transition of Nd^{3+}. Strong effects of REE emissions, affecting the Raman spectrum of xenotime (YPO_4), are presented in Figs. 5.6e. Xenotime is known to contain commonly extended quantities of lanthanide elements,

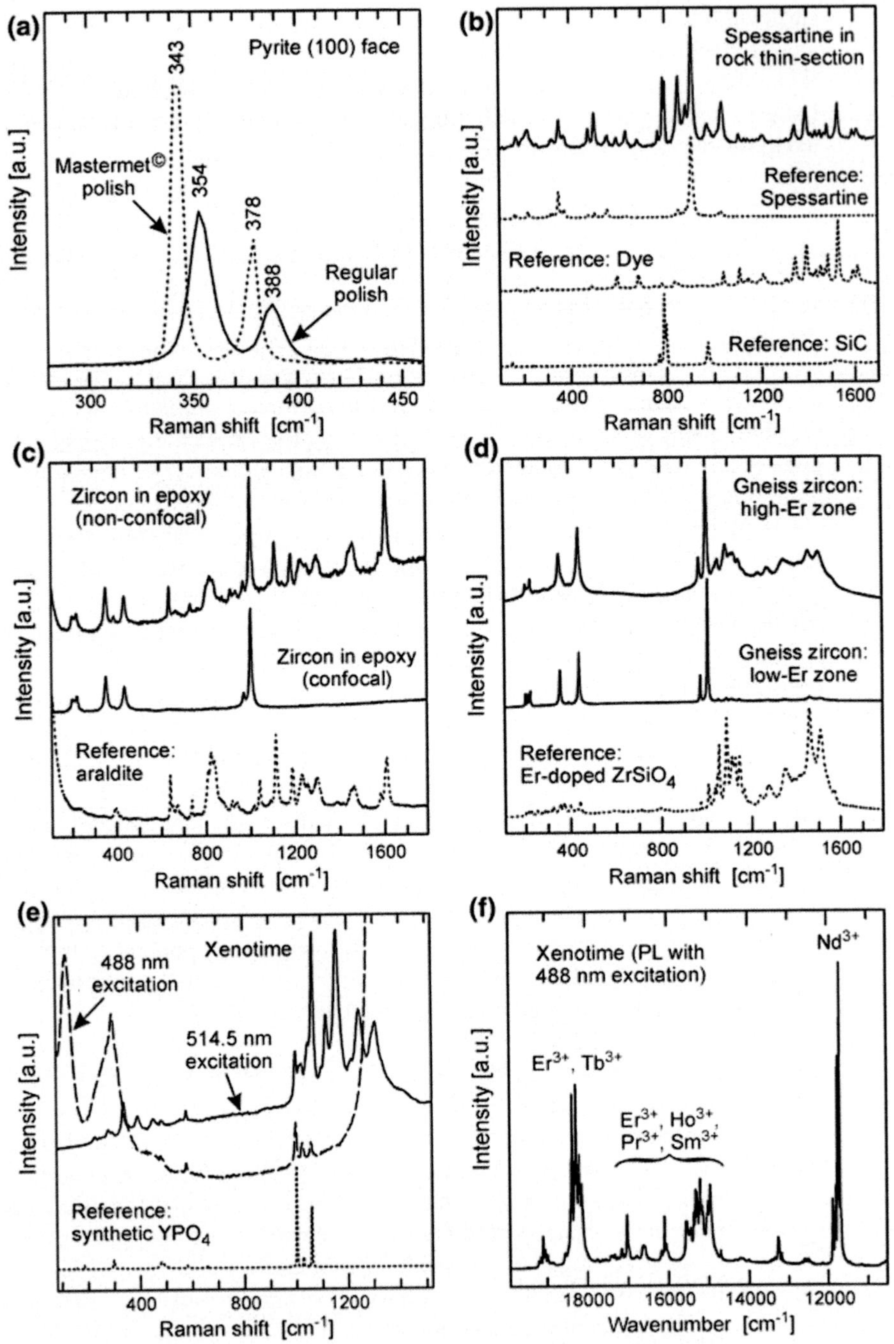
(a)
Pyrite (100) face
343
354
378
388
Mastermet© polish
Regular polish
Intensity [a.u.]
300 350 400 450
Raman shift [cm-1]
(b)
Spessartine in rock thin-section
Reference: Spessartine
Reference: Dye
Reference: SiC
Intensity [a.u.]
400 800 1200 1600
Raman shift [cm-1]
(c)
Zircon in epoxy (non-confocal)
Zircon in epoxy (confocal)
Reference: araldite
Intensity [a.u.]
400 800 1200 1600
Raman shift [cm-1]
(d)
Gneiss zircon: high-Er zone
Gneiss zircon: low-Er zone
Reference: Er-doped ZrSiO4
Intensity [a.u.]
400 800 1200 1600
Raman shift [cm-1]
(e)
Xenotime
488 nm excitation
514.5 nm excitation
Reference: synthetic YPO4
Intensity [a.u.]
400 800 1200
Raman shift [cm-1]
(f)
Xenotime (PL with 488 nm excitation)
Nd3+
Er3+, Tb3+
Er3+, Ho3+, Pr3+, Sm3+
Intensity [a.u.]
18000 16000 14000 12000
Wavenumber [cm-1]

◀ **Fig 5.6** Analytical artefacts II. **a** Raman spectra (Ar^+ 488 nm excitation) of pyrite from the Habachtal, Austria. Strain induced by the manual sample preparation results in notable broadening and shifts of Raman bands (dotted spectrum) whereas the spectrum (solid graph) of a sample which was polished using Mastermet© is equivalent to that of an unprepared pyrite crystal face. Spectra courtesy E. Libowitzky (Universität Wien, Austria). **b** Raman spectrum of a garnet grain (spessartine) in a thin section. Due to insufficient cleaning of the section prior to the analysis, the spectrum shows bands of minute remnants of the polishing powder and a blue dye originating from the felt-tip pen used to label the section. **c** Confocal and non-confocal Raman spectra of a 3–5 μm sized zircon chip embedded in araldite epoxy. **d** The Raman spectrum of natural zircon may be affected by laser-induced Er emission (540–560 nm) when obtained with 514.5 nm excitation. **e** Depending on the excitation wavelength, the Raman spectrum of natural xenotime is often obscured by intense, laser-induced REE emissions. **f** Laser-induced photoluminescence spectrum of the same xenotime, showing several groups of narrow REE emission bands

and in studying natural xenotime samples it is therefore most difficult to find an excitation wavelength with which the Raman spectrum obtained is not affected by laser-induced REE emission bands (see PL spectrum in Fig. 5.6f; cf. also xenotime Raman spectra in the various online databases). The main Raman bands of xenotime, assigned to stretching vibrations of PO_4 tetrahedrons, lie at 1001, 1026, and 1059 cm^{-1}. With 514.5 nm (corresponding to 19,436 cm^{-1}) excitation, these Raman bands are recorded at 18,435–18,337 cm^{-1}. This "collides" with the group of strong emission bands in this spectral region (Fig. 5.6f), resulting in Raman bands being fully obscured by PL emission bands in the 514.5 nm excitation spectrum (Fig. 5.6e, solid line). With 488 nm (corresponding to 20,492 cm^{-1}) excitation, in contrast, the three main phosphate stretching vibrations are recorded in the range 19,491–19,433 cm^{-1}, which lies virtually in the "gap" between two groups of PL emission bands (Fig. 5.6e, dashed spectrum).

A reliable way to distinguish between Raman and PL bands is to obtain spectra at different excitations: Raman bands will have the same Raman shift independent of the excitation (with the exception of few materials whose bands show dispersion effects, such as defect bands in graphite; [156, 157]). Luminescence emissions, in contrast, will appear with different apparent Raman shifts, because they are controlled by energetic differences among electronic levels and hence characterized by certain frequencies. The main goal for a Raman analyst, however, will be to choose an excitation wavelength at which laser-induced PL emissions are avoided as much as possible. For instance, natural corundum (Al_2O_3) samples contain mostly trace or minor amounts of Cr^{3+} substituting for Al^{3+}. In addition to the most intense *R* lines of the Cr^{3+} emission center ($^2E \rightarrow {}^4A_2$ electronic transition; [111]), there are extended "sidebands" (main transitions modulated by vibrations [123]) in the range 13,500–15,000 cm^{-1}. Note that the red "Fe impurity fluorescence" of ruby described by Aminzadeh [1] is actually Cr^{3+} sidebands, too. Even though these sidebands are (at room temperature) more than one order of magnitude lower in intensity than the *R* lines, they may, depending on the Cr^{3+} concentration of the respective sample, still be several orders of magnitude more intense than the Raman bands of corundum. Therefore, red (i.e., He–Ne 632.8 nm, or Kr^+ 647.1 nm or 676.4 nm) excitation would be a most unfortunate choice for the Raman analysis of natural corundum samples.

The above list of artifacts is, unfortunately, incomplete; there exist several more problems and uncertainties (for all of which published examples can be found in the literature, too). Artifacts in Raman-based images may relate to the imperfect surface topography of samples, for which reason the application of auto-focus system may be advantageous. Failure to consider the strong orientation dependence of Raman spectra (normally described by the Porto notation; [23]) may lead to wrong conclusions, if apparently different spectra are not recognized to be simply due to different orientations of the same sample material with respect to the polarization of the laser (and, if applicable, the recorded laser light). One such problem was published by Furuichi et al. ([42]; discussed by [103]). Improper wavenumber calibration may cause apparently shifted Raman bands, which in turn may result in uncertain quantitative estimates of pressure and strain or, more simply, hamper phase identification through searches in spectra databases. Analogously, failure to consider the experimental broadening of bands in evaluating measured band FWHMs, or improper determination of the real apparatus function, may result in incorrect estimates of grain sizes [5] or the degree of disorder. The same may be caused by the application of questionable background-correction models. Finally, many published misinterpretations were simply caused by incorrect assignments of bands or spectra. The latter is handicapped by the still existing limits in the availability of really reliable reference spectra. Several groups worldwide have undertaken tremendous efforts to solve this problem; however, a sufficiently comprehensive *and* fully correct database of Raman spectra of minerals is not available to date.

5.4 Applications in the Earth Sciences

5.4.1 Generalities

Raman-based images (or, more precisely, hyperspectral Raman maps) have been applied successfully in virtually all sub-disciplines of the Earth sciences. However, whereas the application of Raman analyzes has a long tradition in the Earth sciences (starting as early as 1928 when Landsberg and Mandelstam [80] analyzed α-quartz and calcite), the use of Raman-based images of geological samples has become prevalent only in the past few years, certainly favorably affected by the recent methodological progress. In the following, we will present a number of brief examples, followed by the more detailed discussion of applications in three selected fields. This compilation, however, must remain incomplete; it mainly aims at demonstrating the diversity of possible applications based on the particular experimental advantages of the Raman technique. For detailed and extensive overviews of the applications of Raman point analysis in the Earth sciences, the reader is referred to Smith and Carabatos-Nédelec [138] and Nasdala et al. [104].

The many applications of Raman spectroscopy to the investigation of gem materials (see [20, 73]) are favored by the ability of this technique to study samples nondestructively. Raman analyzes can be done without any need for sample

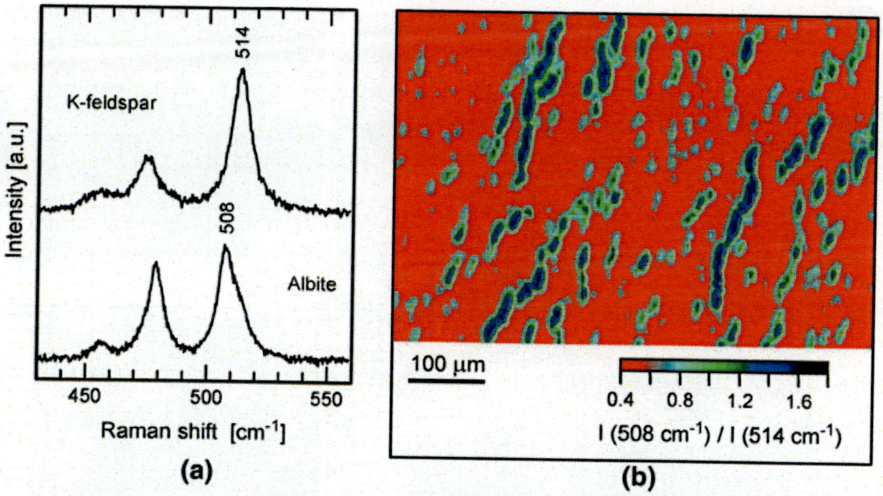

Fig. 5.7 **a** Two Raman spectra taken from a cabochon-cut gem feldspar (cat's eye moonstone) from Tanzania. **b** Raman map (step width 5 μm) of the same specimen. The sample shows a perthitic texture, with Na-feldspar lamellae in a K-feldspar matrix. Sample courtesy T. Häger (Universität Mainz, Germany)

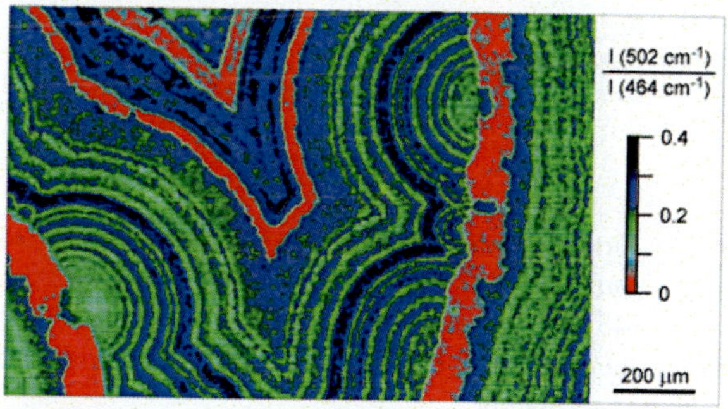

Fig. 5.8 Raman map (step width 6 μm) of an agate section from St. Egidien, Saxony, Germany. The intensity ratio of the main bands of moganite (502 cm^{-1}) and α-quartz (464 cm^{-1}) reveals strong, rhythmic growth zoning. Sample courtesy J. Götze (TU Bergakademie Freiberg, Germany)

preparation (which is mostly connected with permanent changes or damages to the sample) and even without separation of the material to be analyzed; this is also applicable for Raman mappings. An example is shown in Fig. 5.7: Detection of a perthitic texture (i.e., exsolution of K and Na feldspar upon slow cooling) in a gem "moonstone" revealed a slow cooling history, which in turn is valuable information in provenance assignment. Raman mapping without the need (and possibility) to separate the object to be analyzed includes the study of fluid inclusions or the remnant pressures acting on solid inclusions; these two examples will be discussed in more detail below. All of these applications have in common that distribution patterns of phases (e.g., exsolutions, alterations) or physical parameters, (e.g., stress) bear more information than single spectra on a gemstone's, or other sample's, formation and post-formation history. An example for how Raman maps may reveal formation conditions and their changes during the growth of minerals on the basis of phases

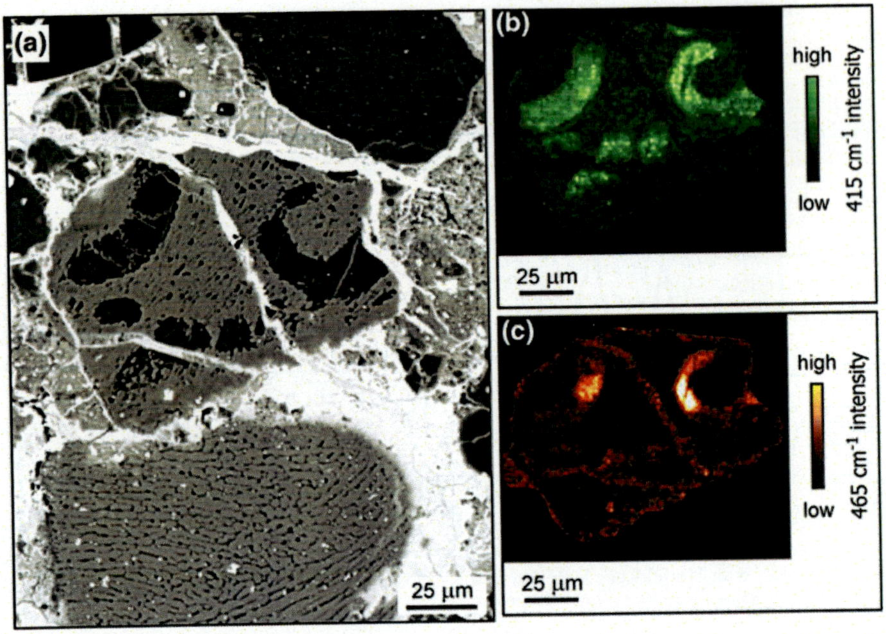

Fig. 5.9 Silica-rich condrule in meteorite Acfer 182, which belongs to the unequilibrated CH condrites. **a** BSE image. **b** Distribution image generated from the intensity of the main Raman band of the silica polymorph cristobalite. **c** Analogous intensity distribution for the main band of α-quartz. Sample courtesy D.C. Hezel (Natural History Museum, London, U.K.)

formed is presented in Fig. 5.8. Raman maps visualizing primary zoning due to different levels of incorporated nonformula elements were presented by Geisler et al. [45] and Váczi [151], and Hövelmann et al. [61] published Raman maps visualizing the sensitivity of this technique to changes of the isotopic composition in alteration experiments.

Another field in which Raman mapping is increasingly applied is the study of cosmic materials, including meteorites [40, 143] or dust of extraterrestrial origin [14]. For instance, Hezel et al. [58, 59] studied silica-rich chondrules in a large suite of chondritic meteorites, and they found them to contain occasionally several co-existing silica polymorphs, including glassy SiO_2, tridymite, cristobalite, and α-quartz. The distribution of the two latter polymorphs in a pyroxene-rich chondrule is shown in Fig. 5.9. It can be seen clearly that the two colour-coded images reveal different patterns, which documents that the SiO_2 phase composition varies appreciably even on a scale of within single chondrules.

Raman mapping is also increasingly used to study minerals and related solids in experimental run products. Applications cover the wide range from the industrial cementation from substrates prior to electrolysis [10] to experiments addressing the chemical exchange processes in mineral alteration processes [113]. Our example (Fig. 5.10) visualizes the Ca carbonate distribution in an experimentally carbonated

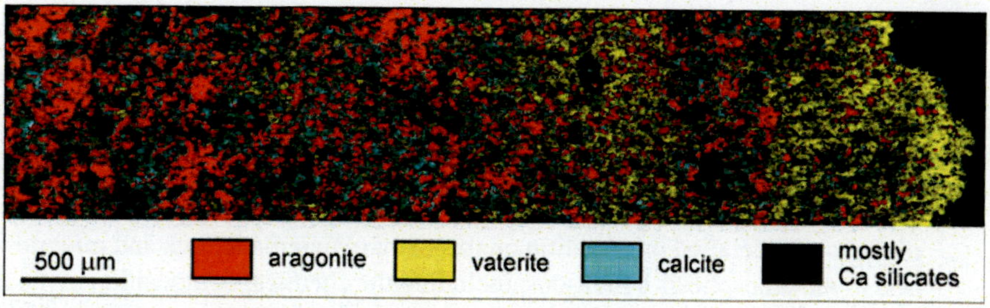

Fig. 5.10 Compositional Raman map (step width 5.9 μm) of an experimentally carbonated Portland cement core (carbonation "flow" direction from the left towards the right side of the image). For details see Corvisier et al. [19]. Note the high concentration of vaterite near the carbonation front whereas the carbonated zone is mostly composed of aragonite and calcite

Portland cement core. Such materials are of enormous interest for estimates of the underground CO_2 storage, either in deep saline aquifers or in depleted oil and gas fields. The CO_2 storage efficiency relies sensitively on the well bore integrity. Recent experimental studies as well as natural cases showed that Portland cement, which was used as sealing material around the well, reacted under down-hole conditions when exposed to CO_2-rich fluids [19]. This alteration, which is mainly controlled by carbonation processes, could affect severely the hydro-mechanical properties of the well bore ensemble [36]. It is known that metastable carbonate polymorphs (vaterite and/or aragonite) can form and even be preserved during progressing alteration. Their spatial distribution within the sample, however, is hardly accessible with common chemical characterization methods. The Raman map shown in Fig. 5.10 was obtained by Corvisier et al. [19] from a cement core that was experimentally carbonated in a high-pressure, high-temperature vessel. It demonstrates that vaterite is the first $CaCO_3$ polymorph formed. Using numerical models, and considering the width of the vaterite front in their Raman maps, Corvisier et al. [19] draw quantitative conclusions on kinetics and dissolution parameters. They found that, in general, the presence of the different $CaCO_3$ polymorphs in carbonated Portland cement shows that local equilibrium cannot be assumed and that kinetics plays an important role in the carbonation process.

Further, mineralogical applications of Raman-based images fall into the field of art history and archeology (e.g., [6, 82]). This topic, however, is covered in a separate chapter in this book and will therefore not be discussed here. In the following, applications in the fields of high-pressure petrology, radiation damage, and mineralized fossils are presented and discussed in more detail.

5.4.2 Applications in High-Pressure Petrology and Mineralogy

High-pressure minerals are remnants of, and hence bearer of information on, geological processes that have occurred at considerable depths of the Earth's interior, down to the transition zone, and even the lower mantle (i.e., at depths of more than about 410–660 km below the Earth's surface; [3]). In contrast, direct sampling of the Earth's interior material through cores from deep boreholes is limited to very few tens of kilometres. Therefore, the study of material from deeper levels of the Earth that was transported by geological processes to the surface is of major interest. Such material (ultra-high pressure rocks, rock xenoliths, and mineral inclusions) may provide detailed information on the primary magmatic rock and mineral formation or metamorphosis conditions, including the chemical composition in the Earth's interior (for example, see [12]).

However, a presumption is that the material originating at deep levels has not undergone substantial chemical and structural changes upon transportation to the surface, which includes the necessity of some shielding or encapsulation, and normally requires relatively rapid uplift. These preconditions are ideally fulfilled for inclusions trapped inside diamond crystals or in other, relatively rigid host minerals (e.g., zircon, garnet, kyanite). For instance, the presence of syngenetic wadsleyite (β-$MgSiO_4$) inclusions in diamond crystals reveals diamond formation at the depths of at least about 400–550 km below the Earth's surface, whereas ringwoodite (γ-$MgSiO_4$) is only to be expected in super-deep diamonds that were formed in source regions of about 600 km depth or more (cf. [24]). Another example is the study of (mostly high-density) fluid inclusions in high-pressure minerals. Such fluids trapped inside diamond crystals yield information on both the formation of diamond and the composition and evolution of fluids in the mantle [74, 159]. Similarly, the study of H_2O-CO_2 fluids (for the assignment of CO_2 bands see [65]) trapped in garnet crystals from highly deformed paragneisses from the Nestos Shear Zone, Xanthi, Greece (see Fig. 5.11) allowed Krenn et al. [77] to draw detailed conclusions on both the fluid and metamorphic evaluation along the metamorphic host rock's pressure–temperature path after its metamorphism at ca. 1.2 GPa and 730 °C. These authors concluded from detailed studies of the fluid inclusions, geobarometric data and textural relationships that the rock must have undergone initial rapid decompression at high temperatures above 600 °C, and further slow cooling at pressures below 0.2 GPa.

5.4.2.1 Multi-Phase Inclusions in High-Pressure Metamorphic Rocks

Aragonite is a high-pressure polymorph of $CaCO_3$ which may form from calcite during high-pressure metamorphism. Aragonite may then be trapped by a surrounding host mineral and hence be preserved during the pressure release after the metamorphic event. The identification of aragonite in mineral inclusions, therefore, indicates relatively high-grade metamorphic conditions exceeding 0.7 GPa at 300 °C (blueschist facies) to <2 GPa at 700–800 °C (eclogite facies); for the stability fields see Johannes and Puhan [67]. In fact, aragonite is very rarely observed in high-pressure metasediments except in some blueschist-facies terranes with a very low geothermal gradient such as the Franciscan Complex in California or Western Crete in Greece. This is considered as a kinetic issue as the back-transformation of aragonite to calcite is extremely fast. Preservation of aragonite therefore requires an anti-clockwise

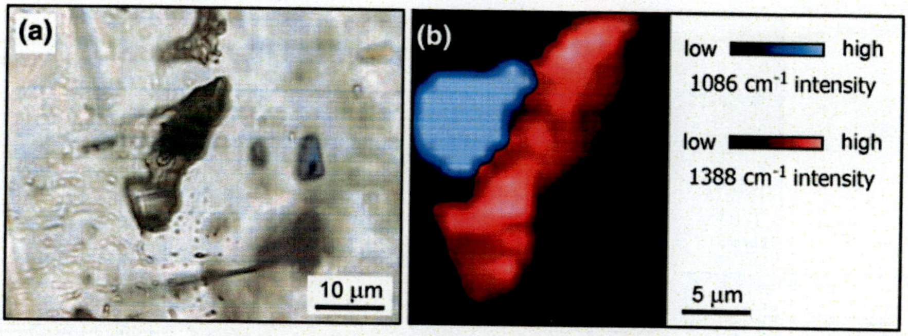

Fig. 5.11 Two-phase (solid and gas) inclusion in a garnet crystal from a high-pressure rock from the Rhodope Mountains, Greece. **a** Transmitted light photomicrograph, plane-polarised light. **b** Color-coded Raman map (step width 1 µm) of the central area, showing an overlay of the intensity distribution patterns of the main bands of calcium carbonate [$\nu_3(CO_3)$ at 1,086 cm^{-1}] and carbon dioxide [$\nu_+(CO_2)$ at 1,388 cm^{-1}]. Image courtesy R. Kaindl (Universität Innsbruck, Austria)

pressure–temperature path for the host rock, or relatively fast decompression accompanied by continuous cooling.

Recently, Chopin et al. [18] reported a new occurrence of aragonite in blueschist metamorphic rocks from Alpine Corsica (France). The host rocks are siliceous marbles which were found at the direct contact of serpentinites from the meta-ophiolitic ensemble of Cape Corsica. In the study of Chopin et al. [18], Raman micro-spectroscopy was the key tool for distinguishing unambiguously aragonite from calcite on a micrometre scale, and Raman-based imaging has been used to decipher the unusual and spectacular textural behavior of aragonite in these rocks. Aragonite is actually preserved as inclusions in Ca-rich garnets, whereas in the quartz-carbonate matrix only calcite is observed. Raman maps (Fig. 5.12) show that aragonite is present as fibers in garnets which are perpendicular to the faces of the garnet, resulting in a sectoral (instead of a radial) arrangement of the fibers. In addition, calcite and quartz are present as equidimensional inclusions in the centers of the garnet grains; carbonaceous material is concentrated along the margin of garnet and calcite-quartz assemblage in the matrix. The simultaneous presence of aragonite and garnet in the same rock may appear paradoxical as garnet indicates that the rock was actually submitted to significant metamorphic temperatures, whereas aragonite may back-transform to calcite extremely and rapidly at such temperatures. This was interpreted by Chopin et al. [18] as prograde intergrowths during subduction metamorphism. Garnet nucleated in a calcite-quartz matrix and followed its growth in an aragonite-quartz matrix. During exhumation, the aragonite matrix transformed back to calcite, whereas the carbonates included in garnet were mechanically shielded and preserved their original structure.

Similarly, detailed information on metamorphic conditions and the effective pressure–temperature path was drawn by O'Brien and Ziemann [114] from detailed observations on multi-phase inclusions in garnet crystals in a carbonate-bearing

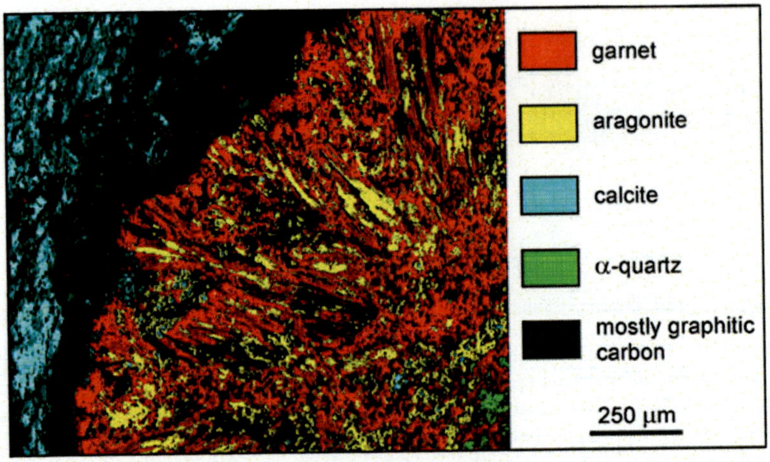

Fig. 5.12 Raman map (3.3 μm step width) of a garnet crystal containing α-quartz and calcite in its core, and fibrous aragonite in the rim. The rock matrix is mainly composed of calcite; the section came from a garnet-bearing high-pressure marble from the Cape Corse area, North-Eastern Corsica, France

eclogite from the Saidenbach reservoir, Saxonian Erzgebirge, Germany (Fig. 5.13). The Saidenbach area is famous for the occurrence of ultra-high pressure rocks (e.g., eclogites and high-pressure gneisses). The discovery of quartz pseudomorphs after coesite [126] and later the detection of preserved coesite [87, 88, 104] showed that—based on the stability fields of Mirwald and Massonne [89]—the host rocks must have experienced pressures of at least 2–2.5 GPa. Moreover, the discovery of euhedral microdiamond crystals included in the same host minerals [86, 97] even indicated peak pressures of at least 4 GPa (at about 900 °C), or more. The latter estimate was made based on the carbon stability diagram of Kennedy and Kennedy [71]; however, it does not consider the hypothetical possibility of metastable microdiamond formation at lower pressures.

Raman mapping has been applied to study the distribution patterns of coesite and α-quartz in high-pressure SiO_2 inclusions by Nasdala et al. [104] and Korsakov et al. [76]. In their study, O'Brien and Ziemann [114] found high-pressure inclusions in which these SiO_2 polymorphs are associated with additional mineral phases. In addition to estimates of peak metamorphic conditions, the Raman maps of O'Brien and Ziemann [114] yielded valuable information on the post-metamorphic history of the metamorphic host rock. First, these authors found that coesite is not always surrounded by a rim of palisade quartz (see lower left rim of the grain shown in Fig. 5.13a), whose presence indicates initial coesite → α-quartz transformation upon pressure release; and they concluded that the SiO_2 inclusions were overgrown by garnet at conditions well into the coesite stability field and hence trapped and shielded. Second, O'Brien and Ziemann [114] found aragonite relics (Fig. 5.13b). From the above observation, they concluded that fracturing of the host garnet (see Fig. 5.13a, c), which allows for significant volume expansion of the inclusion and hence trans-

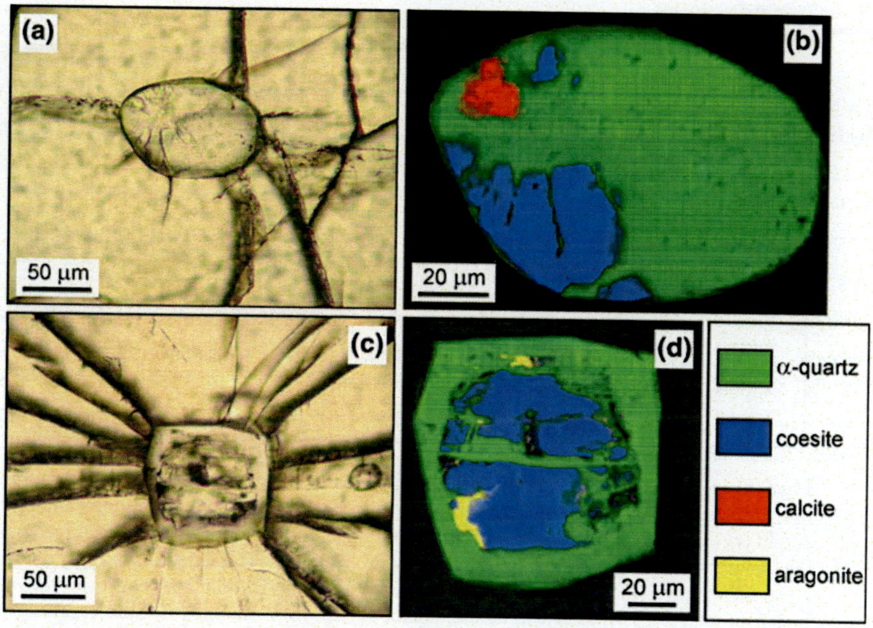

Fig. 5.13 Two SiO_2 inclusions in garnet from a high-pressure gneiss occurring near the Saidenbach reservoir, Saxonian Erzgebirge, Germany. **a** and **c** Photomicrographs (plane-polarised transmitted light) of the 30 μm thick thin section. **b** and **d** The corresponding colour-coded Raman maps show the phase distributions of coesite remnants as well as of $CaCO_3$ polymorphs. Image courtesy M. Ziemann (Universität Potsdam, Germany)

formation into the more volume-expanded low-pressure polymorphs, cannot have occurred at temperatures higher than 300 °C (because otherwise the aragonite would have transformed to calcite).

5.4.2.2 Patterns of Internal Stress in Diamond Crystals

The *in situ* point-by-point mapping technique with a confocal Raman system has proven extremely useful in the diamond research. In addition to highly resolved point analyzes, this technique provides a nondestructive means to study not only the mineralogical composition of inclusions, but also their inter-growth [102]. This in turn may provide information about both the formation environment of the diamond crystals and their post-growth history. For instance, the discovery of serpentinization of olivine inclusions may indicate an alteration history. Second, Raman maps have successfully been applied to reveal internal growth textures of diamond crystals. The various growth zones are characterized by minute variations in the width of the main first-order lattice vibration of diamond (LO = TO phonon near $\sim$1,332 cm^{-1}; cf. [140]), which are assigned to small variations in trace elements and/or structural

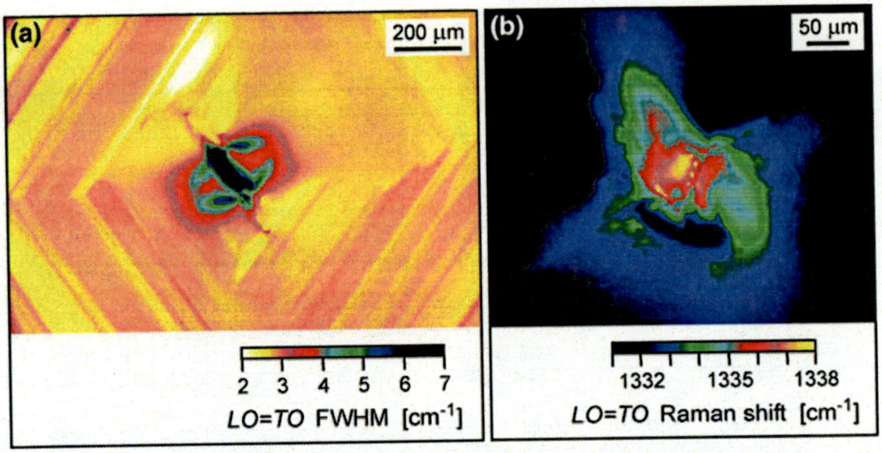

Fig. 5.14 Examples for internal textures of diamond single crystals. **a** Raman map (step width 5 μm) of a large diamond crystal from the Panda kimberlite, N.W. Territories, Canada, containing a graphite single-crystal inclusion. The inclusion itself is surrounded by a strain halo (which is recognised from the strong broadening of the main diamond mode). The bulk of the diamond crystal yields minute zonal variations in the width of the Raman band, revealing the diamond's primary growth zoning. **b** Raman map (step width 4 μm) of a diamond crystal from Kankan, Guinea, revealing a halo of compressive strain around a larnite inclusion. The up-shift of the main diamond mode corresponds to pressures as high as 2.7 GPa. Sample courtesy F.E. Brenker (Universität Frankfurt, Germany), J.W. Harris (University of Glasgow, U.K.), and De Beers Consolidated Mines Ltd.

defects among zones. Maps revealing the zoning are of particular importance as the visualization of textural relationship of inclusions and growth zones may, for instance, assist in assigning included phases as proto-, syn-, or epigenetic in nature. To give an example, Nasdala et al. [105] found euhedral graphite single-crystal inclusions in diamond crystals from the Panda district, Canada, to be evenly surrounded by growth zones of their host diamond (Fig. 5.14a). As this is characteristic of protogenetic inclusions, authors concluded that in case of the Panda specimens, graphite must have been the primary phase, and was overgrown by diamond.

Much stronger broadening of the LO = TO phonon, accompanied by notable band shift, is observed at elevated pressures. Consequently, the primary diamond zoning is easily overprinted if diamond is affected by stress/strain. Calibrations for the anisotropic increase of the Raman shift with compressive strain (ranging between 0.7 and 2.2 cm^{-1} per GPa) have been provided by Grimsditch et al. [54], Hanfland et al. [56], and Sharma et al. [134]. Compressive strain values determined from the diamond LO = TO position extracted from Raman maps at locations in close proximity to inclusions, were found to agree very well with remnant pressures of the inclusions themselves: Nasdala et al. [102, 105] determined 2.6 and 2.2 GPa for single-crystal graphite inclusions in two Ekati diamond crystals (cf. Fig. 5.14a), whereas [49] calculated 2.4 and 2.1 GPa from the lattice parameters of these same graphite inclusions. Nasdala et al. [102, 105] showed that Raman maps obtained from unprepared diamond crystals allow one to estimate also the lateral extension (several hundred μm depending on the type and size of the included phase) of haloes of compressive strain around inclusions (Fig. 5.14b). These haloes of compressive strain (also referred to as fossilised pressures, remnant pressures, or overpressures) are due to the heterogeneous volume expansion of the diamond-inclusion couples upon pressure release during uplift; they are formed provided the strain is not released through fracturing. Near the ends of such fractures, notable dilative strain can also be observed. Complementary to strain birefringence techniques [63, 64, 81], Raman mapping can therefore be used to evaluate the internal strain of the diamond crystals. This is of interest in reconstructing the diamond genesis as well as for technical applications (such as the selection of stress-free diamonds for anvil cells).

5.4.3 Patterns of Internal Radiation Damage

The structural order in nominally crystalline minerals may be decreased, up to complete amorphization, due to the impact of radioactivity. The accumulation of structural radiation damage results in marked changes of physical properties and decreases the chemical resistance of radiation-damaged minerals, when compared with their well-crystalline analogues [34]. Better understanding of irradiation effects is first of relevance for the Earth sciences, for instance, to support the interpretation of U-Pb geochronology results (radiation damage may bias such results, by enhancing the susceptibility of minerals to be dated to secondary Pb loss; [136]), or for an interpretation of the thermal history of samples. Second, one of the main motivations for studying radiation-damaged minerals arises from the need to find alternative radioactive waste forms, for the safe long-term disposal of spent fuel from nuclear power stations or highly radioactive Pu from disarmed nuclear weapons [35].

The degree of radiation damage in a given mineral can, in principle, be estimated using any method of structural analysis, provided an appropriate calibration is available. Several methods have been used for this, including birefringence determination in cross-polarized light [15], X-ray powder diffraction [93], X-ray absorption [37], nuclear magnetic resonance spectroscopy [38], infrared spectroscopy [163], and TEM [158]. The use of Raman spectroscopy to quantify radiation damage in zircon was introduced by Nasdala et al. [98]. This technique has been applied in many zircon studies since (see [101], and references therein); and similar Raman investigations were done on other radiation-damaged minerals such as biotite [99], monazite [133], cordierite [106], fergusonite [149], Gieré et al. [47], diamond [108], and quartz [78]).

Many of these studies have also applied Raman-based images, motivated by the common fine-scale heterogeneity of radiation-damaged specimens. Raman line scans and maps of radiohaloes [i.e., damage haloes in host minerals caused by the impact of alpha particles (high-energy helium cores) originating from neighboring, radioactive phases] helped to estimate the extent and extension of the alpha particle damage [106, 107]; (Fig. 5.15) and to assign causes of observed property changes. For instance, the typical ring-like pattern of radiocoloration and/or radioluminescence

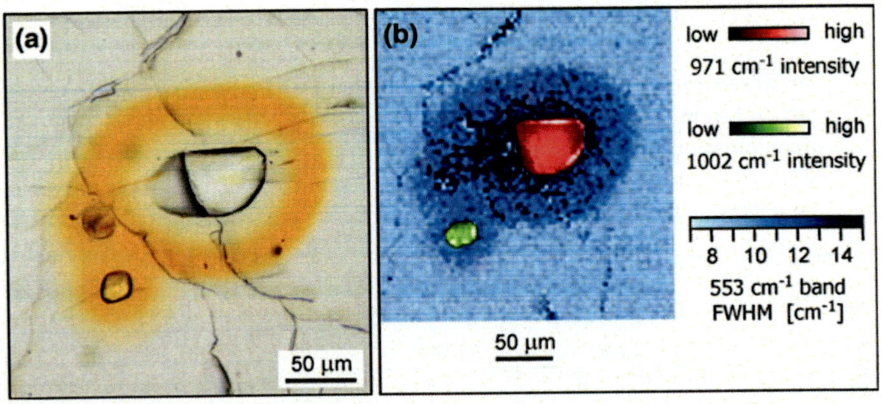

Fig. 5.15 Radiohaloes (i.e., alpha-particle damage haloes) in cordierite. **a** Photomicrograph (*plane-polarised light*) of a yellow coloured radiohalo in cordierite around (colourless) monazite and zircon inclusions. The sample came from a garnet-cordierite gneiss from the Strangways Metamorphic Complex, central Australia. Sample courtesy A. Möller (University of Kansas, Lawrence, U.S.A.). **b** Composite Raman map, showing the distribution patterns of monazite (*red*) and zircon (*green*) as well as the gradual broadening of the 553 cm^{-1} cordierite band (*blue*)

in such haloes (see again Fig. 5.2h) corresponds very well to the distinct maxima of the structural radiation damage as predicted by Monte Carlo simulations (for the theoretical background cf. [167]) and observed by Raman spectroscopy (Fig. 5.16b); however, observed patterns do not correspond to the rather broad ionization distribution curves. This contradicts earlier assignments of the observed changes in optical activity to ionization effects [90] and suggests that defect centers are the major cause of irradiation-induced changes in color and/or emission behavior [99, 106, 107].

In the case of U- and Th-bearing and hence self-irradiating minerals, the majority of the structural damage to the host lattice is due to alpha recoils, i.e., recoil impulses of heavy daughter nuclei upon emission of an alpha particle, which form tiny damage clusters of a few tens of nanometres in size [26, 135, 158]. Due to the short range of alpha recoils, radiation damage and actinide content in self-irradiating minerals correlate often on a submicrometre scale, provided the radiation damage has not been annealed through thermal annealing or heterogeneous chemical alteration processes. In an annealing study of cut-in-half crystals, [107] showed that the band broadening in Raman maps of zircon (being mainly controlled by radiation damage) correlates very well with the observed BSE intensity, whereas the chemical composition (i.e., the average atomic number) has very minor effects on the BSE of this mineral. Another example consists of zoned, complexly altered zircon crystals from Bancroft, Ontario (for a description of the location see [125]). Neighboring zones with different uranium concentrations have experienced different degrees of radiation damage and hence different degrees of volume expansion, which led to intense fracturing of the less damaged (and less expanded) zones [83]. These fractures were then perfect pathways for alteration fluids; alteration occurred preferentially at the ends of fractures, which

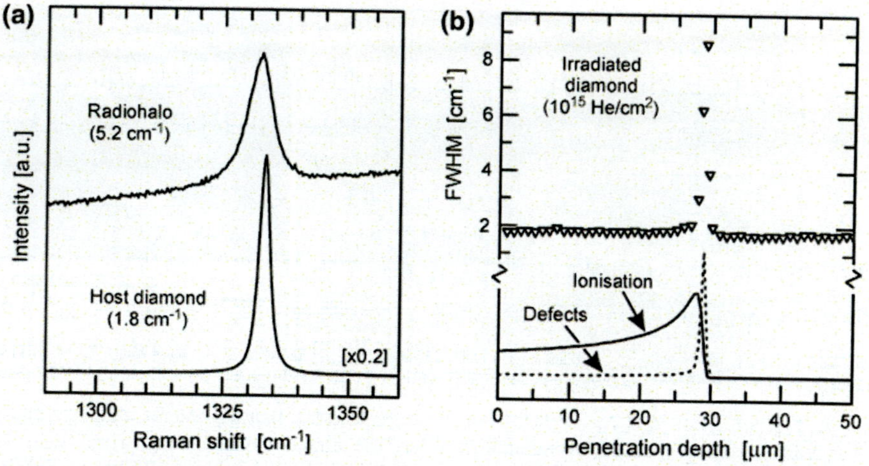

Fig. 5.16 Radiation damage in diamond. **a** Two example spectra of a brown spotted diamond from Namaqualand, RSA. The surficial brown spots are radiohaloes that were generated through the bombardment with alpha particles emitted from external radioactive sources. Note the notable broadening of the diamond LO = TO phonon in these haloes, indicating significant disorder. Sample courtesy J.W. Harris (University of Glasgow, U.K.) and De Beers Consolidated Mines Ltd. **b** Plot of the FWHM of the diamond LO = TO phonon versus depth below surface (small triangles). This Raman line-scan was done on a polished diamond plate that was irradiated with 8.8 MeV helium ions in a Tandem accelerator. The observed FWHM distribution corresponds well with the distribution pattern of point defects as predicted by Monte Carlo simulation (shown below; *dashed*)

protrude into the more damaged zones ([79, 109]; Fig. 5.17a). The alteration resulted in the formation of new (and, therefore, crystalline) zircon, which is the reason for the lesser radiation damage observed in the altered areas (Fig. 5.17b).

Finally, Raman maps were also used successfully to visualize structural artifacts caused by the EPMA analysis of radiation-damaged minerals. In contrast to earlier assumptions, according to which the impact of the high-energy electron beam would cause additional structural damage, Nasdala et al. [100, 101] detected notable annealing of the pre-existing radiation damage within the EPMA spots. Raman maps (Fig. 5.18) allowed them to get relatively precise estimates of the extent of the recovery and its lateral extension. In conclusion, these authors suggested performing Raman and other spectroscopic analyzes always before the EPMA analysis, or at least to place Raman analyzes far enough away from EPMA spots, to avoid any bias of quantitative spectroscopic results.

5.4.4 Investigation of Fossils and Biominerals

Fossils are preserved remnants, including preserved traces or marks, of organisms from the remote past. Their preservation is often associated with mineralization

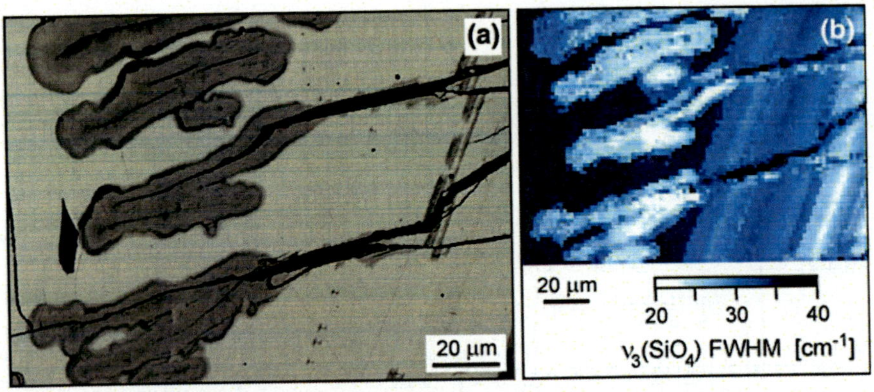

Fig. 5.17 Zoned zircon crystal from Bancroft, Ontario, showing intense chemical and physical alteration that emanated from large fractures. The alteration reaction was driven by low-T fluids, and the large fractures have served as "fast pathways" for these fluids. **a** BSE image. The finger-like altered areas are characterised by low BSE intensity. **b** Raman map (step width 1.5 µm), revealing variations in the crystallinity, both among the primary zones and between the host and the (less radiation-damaged) altered patches. Sample courtesy D. Moser (Royal Ontario Museum, Toronto, Canada) and J.M. Hanchar (Memorial University of Newfoundland, St. John's, Canada)

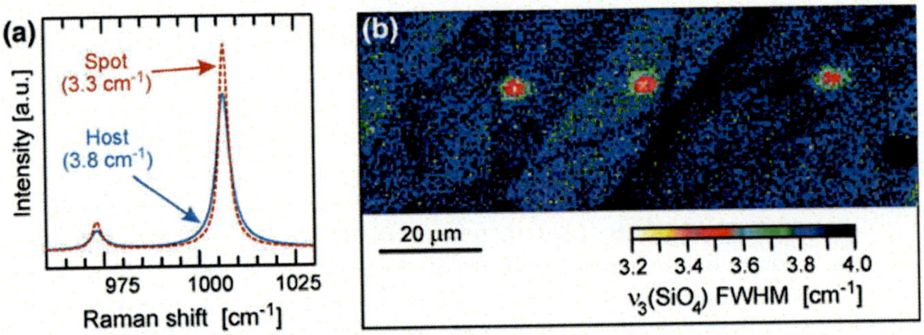

Fig. 5.18 Recovery of the self-irradiation damage in reference zircon #91500 (see Fig. 5.2f) resulting from the impact of the electron beam during electron microprobe analysis. **a** Raman spectra of the mildly radiation-damaged host crystal (*blue*) and the partially reconstituted zircon inside an EPMA analysis spot (*red*). Band FWHM values are given in brackets. **b** Raman map (step width 0.5 µm) of a polished chip. Locations and lateral extensions of three EPMA spots can be recognised from the FWHM decrease. The *blue-black* background pattern reflects weak zoning and polishing effects. Sample courtesy M. Wiedenbeck (Helmholtz Zentrum / GFZ Potsdam, Germany)

processes (see, for instance, the petrified wood specimen in Fig. 5.1b). Some rocks (such as limestone) and natural resources (such as coal) are mainly composed of fossils, and fossils are also important as biomarkers used in stratigraphy. There are general, enormous efforts to improve the knowledge of the Earth's geological history through the investigation of its fossil record.

Among the most exciting topics in the current study of fossils are still the questions, when the first forms of life appeared on Earth and as to which degree these have been fossilized? The search for traces of life in the very primitive Earth has excited geologists and paleontologists for decades; it has even resulted in some kind of quest for the oldest fossil on Earth. Schopf [128] described mineralized, cellular microbial filaments from a ~3.5 Ga old Apex chert in Western Australia; these were later reinvestigated, including Raman spectroscopy and mapping, by Schopf et al. [130, 131] and discussed critically, by Brasier et al. [11]. Similar in age are micrometer-scale mineralized tubes in Archean pillow lavas from the Barberton Greenstone Belt, South Africa (Fig. 5.19a), which were interpreted by Furnes et al. [41] as traces of microbial activity in these ~3.5 Ga old rocks, i.e., drillholes of micro-organisms. This proposed new type of biomarker, too, was discussed critically for instance by M. Brasier and colleagues, who provided an alternative interpretation in stating that "decomposition—possibly of organic matter—produced fluids that drove a mineral grain into the glass as chemical reactions at the grain ate into the glass, sort of like a corrosive-tipped pile driver" (cited from [72]). Unfortunately, Raman maps did not yield any evidence of graphitic carbon (whose presence could have been a key argument supporting the biogenicity of the microtubules); the apparent tube fillings were found to consist entirely of titanite ($CaTiSiO_5$; Fig. 5.19b). The oldest traces of life thus far were claimed to be aggregates of graphitic carbon occurring in the ~3.8 Ga old Isua Supracrustal Belt, southern West Greenland, interpreted to be remnants of a vast microbial ecosystem in the early Archaean [92, 127]. The biogenic origin of this graphite, however, was later questioned [39, 84]. Van Zuilen et al. [153] assigned the graphite formation to siderite ($FeCO_3$) decomposition, which contrasts the initial assignment as micro-fossils. The graphitic inclusions from the Isua Supracrustal Belt were recently reinvestigated in detail by Papineau et al. [116, 117] who employed a complex of modern micrometre- and nanometre-scale techniques, including FIB, TEM, Synchrotron-based X-ray microscopy, NanoSIMS, and Raman spectroscopy and mapping. These authors found heterogeneous internal patterns of the isotopic composition and structural features within the graphite, possibly pointing to multiple formation process through deposition from CO_2- and CH_4-bearing fluids. A definite assignment to either biogenic or abiogenic formation, however, was still not possible.

The above three examples may underline that one crucial aspect in the research addressing the earliest developments of life on Earth is whether or not the objects studied are fossils, i.e., of biogenic origin indeed (see discussion by [119]). Investigating the textural relationships between organic compounds and minerals in fossils (or, more generally, bio-mineralizations) is a key to assess their biogenicity and the pathway for their preservation over geological periods of time. For a summary of studies of microfossils using Raman-based images (including three-dimensional techniques) see Schopf and Kudryavtsev [129], Schopf et al. [131], and Taylor et al. [145]. Two examples involving putative fossils and biomineralization are presented in more detail in the following; one perhaps representing one of the most primitive forms of animal on Earth, and the other microfossil occurring in a high-grade metamorphic (i.e., deeply buried) rock. However, Raman spectroscopy and mapping have found much more versatile applications; these techniques are used to study various sorts of fossils and recently formed biominerals. Examples for the latter include

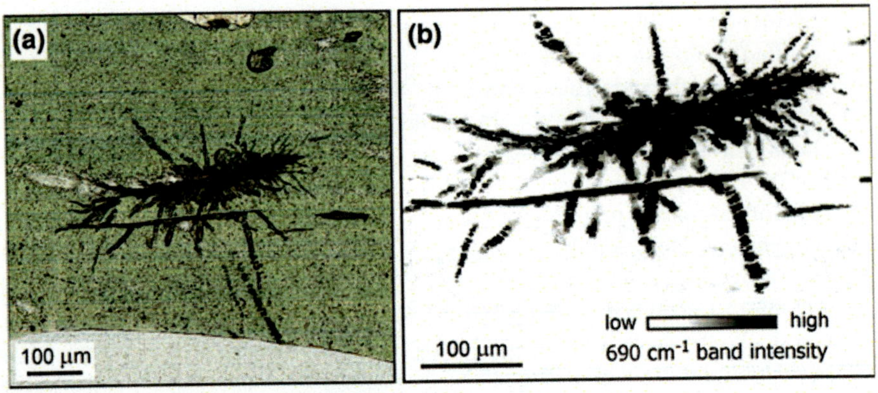

Fig. 5.19 Pillow lava from the Hooggenoeg Formation, Barberton Greenstone Belt, South Africa. Sample courtesy H. Furnes (University of Bergen, Norway). **a** Photomicrograph (plane-polarised light) of volcanic glass replaced by chlorite (*green*), containing an assemblage of *dark brownish*, tubular textures. **b** Raman map (step width 1.5 μm) showing the intensity distribution of the main titanite band

studies of apatitic bone tissue (e.g., [69, 148, 152]) and dentin materials [161], and the visualization of distribution patterns of the various $CaCO_3$ polymorphs in pearls [139] or fish otoliths [154, 166].

5.4.4.1 Lowest Cambrian Ctenophore ("Comb Jelly") Embryo

Geologic time is divided into two vast eons, namely, the Precambrian [extending from the formation of the Earth some 4,500 myr. (millions of years) ago to the first appearance of small shelly invertebrate animals about 543 myr. ago] and the Phanerozoic. The latter, i.e., the more recent of the two main eons, includes the familiar evolutionary sequence from invertebrates to fish, then reptiles, birds and mammals, and from spore-plants to naked seed plants to flowering plants. Animals, with their complex nerve systems and multicellular tissues, are first known in ~580 Ma old strata of the Ediacaran Period of the Neoproterozoic Era, the very latest unit of the Precambrian. During the earliest (Cambrian) period of the Phanerozoic, soon after animals first appear in the fossil record, the various lineages of animal life rapidly diversified (during the so-called "Cambrian Explosion of Life") to give rise to virtually all of the major animal body-plans that have survived to the present. Among the most primitive of the successful animal pioneers were coelenterates such as jelly fish, animals that have a rather simple organization composed of only two layers of cells (unlike the three-layered organization of more advanced animals known as bilaterians). And perhaps the earliest-evolved of such coelenterate-grade organisms were members of the Ctenophora ("comb jellies"), a phylum composed mostly of gelatinous, planktonic, spheroidal to oblate voracious predators that is represented in modern oceans by fewer than 100 living species.

Despite the seemingly fragile nature of these gelatin-like soft-bodied animals, fossils of the group have been reported from Phanerozoic strata of China, Canada, and Germany. The most anatomically informative of these, exhibiting the eight comb rows (ctenes) diagnostic of the group, are beautifully preserved adult ctenophores (*Maotianoascus octonarius*) known from the Middle Cambrian (~530 Ma old) Chengjiang Fauna of Yunnan, China. But the most ancient ctenophore and the only embryonic comb jelly known from the fossil record is that shown here from the lowermost Cambrian (~540 Ma old) Kuanchuanpu Formation of Ningqiang, in Shaanxi Province, south China [17]. This apatite-permineralized comb jelly egg—enclosing an embryo in late development, before hatching—well illustrates the combined use of confocal laser-scanning microscopy (CLSM) and Raman mapping. The CLSM image (Fig. 5.20a) shows the presence of eight embryonic comb rows and provides high-resolution morphologic information unavailable by any other means, whereas the Raman maps elucidate the mineralization events by which it was fossilized: Fig. 5.20b shows that the originally soft organic tissues of the fossil, now apatite-infused, are preserved as carbonaceous kerogen; and Fig. 5.20c demonstrates that the void spaces in such tissues were secondarily infilled by calcite. Thus, the complementary information provided by CLSM and Raman establishes the biological affinities of this exquisitely preserved ctenophore embryo, documents its molecular-structural composition, elucidates its mode of preservation, and demonstrates that this primitive lineage dates from a very early stage in the Cambrian Explosion of Life.

5.4.4.2 Studies on Fossilized Spores in a High-Pressure Metamorphic Rock

Fossils are generally considered as features of the sedimentary domain, whereas metamorphism is supposed to erase the fossil record due to deformation and/or chemical transformations. Traces or fragments of fossils have been described, nonetheless, in some metamorphic rocks, but rare are those conserving an organic heritage from the living organism. More generally, studying fossilization processes during advanced diagenesis and metamorphism may open up new possibilities for tracking biogenicity in the geological record, especially in the quest of the early life as the rocks from this period have generally suffered a burial history. Recently, Bernard et al. [7] reported remarkably well preserved plant spores from high-pressure metamorphic rocks from the Alps; these rocks were buried to nearly 35 km depth during the Alpine collision. Bernard et al. [7] carried out a systematic characterization study involving Raman imaging together with nanometre-scale imaging techniques, including TEM.

Raman mapping has been used to visualize the distribution of carbonaceous material and minerals in the spore as well as in the surrounding rock matrix. Because these spores were found in carbonated septaria, most mineral phases involved are carbonates, in particular calcite ($CaCO_3$) and ankerite [$(Fe, Mg)Ca(CO_3)_2$]. Raman spectroscopy is actually very powerful to distinguish calcite from ankerite, as the difference in chemical composition induces slight but significant shifts of both the internal CO_3 stretching modes and the external vibrational modes (with the extents of

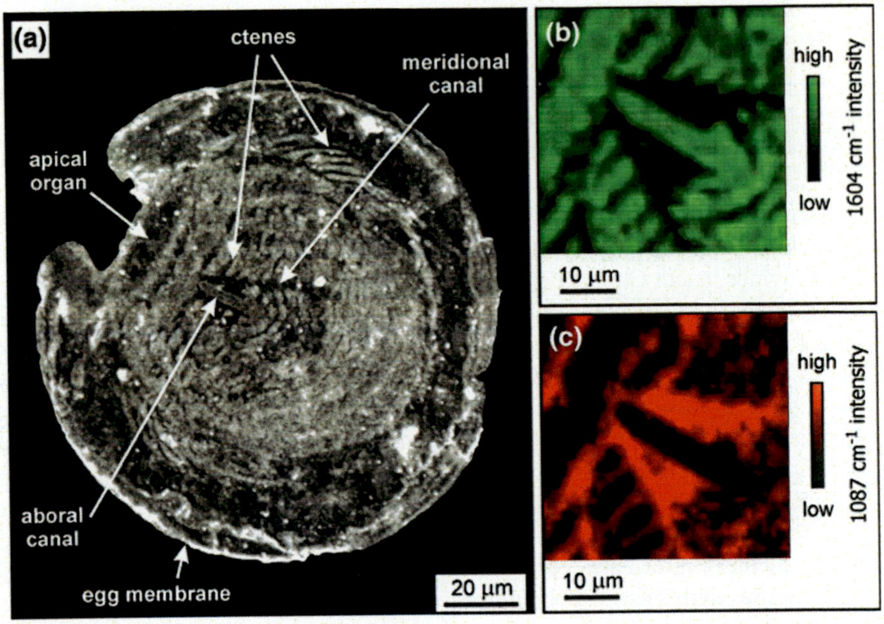

Fig. 5.20 Ctenophore embryo from the Lower Cambrian Kuanchuanpu Formation, Ningqiang, Shaanxi Province, China. **a** CLSM image. **b** Raman map of the central part, showing the intensity of a major band of carbonaceous kerogen. **c** Raman map of the same area, showing the intensity of the main calcite band

the band shifts being controlled by the extent of the cation substitution). Figure 5.21 depicts a Raman image obtained on one spore from these metamorphic rocks whose morphology is perfectly preserved. The wall of the spore section is composed of disordered but graphitic carbonaceous material, and some of the original ornamentation of the spore wall is still clearly observable on the outer side of this carbonaceous wall. The rock matrix is mainly composed of calcite which also fills much of the spore. Ankerite is always observed in contact with biogenic carbonaceous material, either as small aggregates disseminated in the matrix or as a few micrometres thick coronae lying along the internal side of the spore wall. Bernard et al. [7] recorded similar Raman image on other spores section, and they observed in all cases that (i) the spore wall is composed of organic material, (ii) an ankerite coronae is present along the inner side of the spore walls, and (iii) spores may be filled by calcite or other minerals such as phengites. Based on the Raman-based textural patterns (with a resolution on the micrometre scale) and geochemical information with nanometre-scale resolution, Bernard et al. [7] proposed a fossilization model for these metamorphic spores. They interpreted the carbonaceous spore wall as a remnant of the original sporopollenin-rich outer wall of the spore, whereas the ankerite coronae were assigned to the mineralization of the original inner cellulosic wall. This replacement of the inner carbonaceous part of the spore wall might have occurred very early

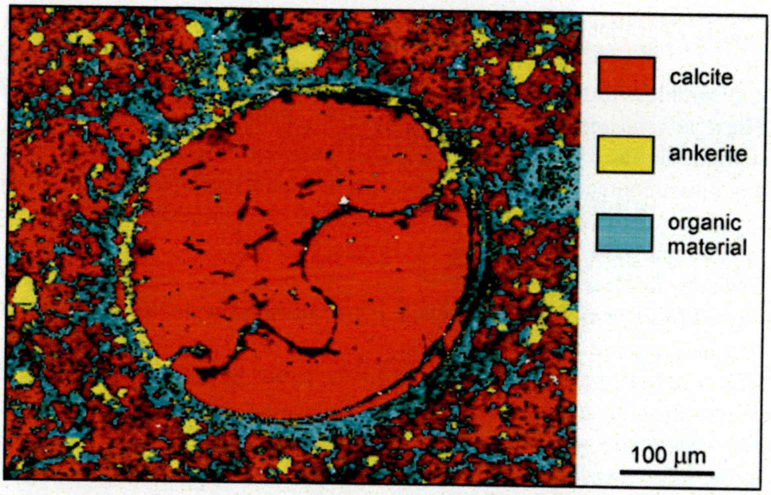

Fig. 5.21 Compositional Raman map (step width 2.2 μm) of a lycophyte megaspore in a thin section of a Triassic septaria, from the Vanoisse massif, Western Alps, France. This fossil has been remarkably preserved although it experienced high-pressure metamorphism during the Alpine orogeny

during diagenesis, with replacement of products being preserved during the whole metamorphic history.

5.5 Conclusions and Outlook

In this chapter, we attempted to summarize the use of Raman-based images in current Earth sciences research. Perhaps more than in other scientific fields and disciplines, Earth scientists may extract valuable information from the study of internal textures and distribution patterns of all sorts of properties or parameters within their samples. The examples presented in this chapter range from rock-forming and accessory minerals through gemstones, meteorites, solid or fluid inclusions, and microfossils, to synthetic analogues of minerals in technical products. The features whose heterogeneity patterns are visualized in our examples show similar diversity; they include the distribution of individual phases within assemblages, chemical zoning and heterogeneity, patterns of chemical alteration or disintegration, variations in the distribution of polymorphs of the same chemical phase due to either primary growth zoning or phase transformations, stress/strain distribution, and patterns of natural and artificial irradiation damage. Neither of these diverse ranges are claimed to be complete. Already, until now, Raman-based imaging has proven to be most flexible in its possible applicability to the study of minerals and related objects, with virtually unlimited opportunities in the detailed study of mineral growth and post-growth alteration processes.

It is clear that Raman-based imaging is only one out of many routine techniques to choose from for visualizing such textures and distribution patterns within geological samples. Raman images or maps, however, provide some unique advantages. These advantages include the ability to study samples nondestructively and often without the need for any special sample preparation. Analyzes are possible without any direct contact with the sample, and with an excellent spatial resolution down to the μm^3 range. Sir Chandrasekhara Venkata Raman predicted the many-sidedness of Raman spectroscopy already in his Nobel lecture, presented in Stockholm, Sweden, in December, 1930, in which he said "*The universality of the phenomenon, the convenience of the experimental technique and the simplicity of the spectra obtained enable the effect to be used as an experimental aid to the solution of a wide range of problems in physics and chemistry. ... It follows that the new field of spectroscopy has practically unrestricted scope in the study of problems relating to the structure of matter.*" This description of versatility is in particular applicable to the implementation of Raman-based images in Earth sciences research. It can be foreseen that in the not too far distance future, Raman-based images will be used routinely at universities and geoscience research institutions.

Acknowledgments The lead author of this chapter is most grateful to G. Irmer (Freiberg), W. Hofmeister (Mainz), R.J. Lehnert (Reutlingen), and I. Reese (Hamburg) for many stimulating discussions on various Raman analytical problems. O.B. is indebted to M. Belleil (Marne-La-Vallée) for his invaluable help with line-scan imaging techniques. J.W.S. is thankful to A. Kudryavtsev for many years of fruitful cooperation. We wish to thank S. Bernard and B. Goffé (both Paris), N. Cayser (Edinburgh), A. Dias, D. Schulze (both Toronto), C. Ginibre, A. Kronz, G. Wörner (all Göttingen), J. Götze (Freiberg), D. Howell (Sydney), R. Kaindl (Innsbruck), E. Libowitzky (Wien), R.D. Neuser (Bochum), B. Phichaikamjornwut (Chaing Mai), R. Škoda (Brno), V. Stähle (Heidelberg), R. Wirth, and M. Ziemann (both Potsdam) for providing spectra, maps, and images. Samples were kindly made available by F.E. Brenker (Frankfurt), H. Furnes, J. Košler (both Bergen), J. Götze, M. Tichomirowa (both Freiberg), M.A. Götzinger, K. Ruschel (both Wien), T. Häger, D. Köhn (both Mainz), J.M. Hanchar (St. John's), J.W. Harris (Glasgow), D.C. Hezel (London), H.J. Milledge (London), A. Möller (Lawrence), D. Moser (Toronto), M. Wiedenbeck (Potsdam), and De Beers Consolidated Mines Ltd. Thanks are due to A. Wagner for sample preparation and T. Váczi for editorial help. A part of the results presented in this chapter were obtained in projects funded by the European Commission (contract MEXC- CT- 2005- 024878, to L.N.), the Austrian Science Fund (FWF; grant P20028-N10, to L.N.), and the French Agence Nationale pour la Recherche (project GeoCARBONS, to O.B.).

References

1. A. Aminzadeh, Appl. Spectrosc. **51**, 817–819 (1997)
2. A.T. Anderson, Am. Mineral. **68**, 125–129 (1983)
3. D.L. Anderson, *Theory of the Earth* (Blackwell Scientific Publications, Boston, 1989)
4. K.J. Baldwin, D.N. Batchelder, Appl. Spectrosc. **55**, 517–524 (2001)
5. L. Bayarjargal, T.G. Shumilova, A. Friedrich, B. Winkler, Eur. J. Mineral. **22**, 29–34 (2010)
6. L. Bellot-Gurlet, D. Neff, S. Reguer, J. Monnier, M. Saheb, P. Dillmann, J. Nano. Res. **8**, 147–156 (2009)
7. S. Bernard, K. Benzerara, O. Beyssac, N. Menguy, F. Guyot, G.E. Brown Jr, B. Goffe, Earth Planet Sci. Lett. **262**, 257–272 (2007)

8. S. Bernard, O. Beyssac, K. Benzerara, Appl. Spectrosc. **62**, 1180–1188 (2008)
9. O. Beyssac, B. Goffé, J.P. Petitet, E. Froigneux, J.N. Rouzaud, Spectrochim. Acta A **59**, 2267–2276 (2003)
10. O. Bøckman, T. Østvold, G.A. Voyiatsis, G.N. Papatheodorou, Hydrometallurgy **55**, 93–105 (2000)
11. M.D. Brasier, O.R. Green, A.P. Jephcoat, A.K. Kleppe, M.J. Van Kranendonk, J.F. Lindsay, A. Steele, N.V. Grassineau, Nature **416**, 76–81 (2002)
12. F.E. Brenker, L. Vincze, B. Vekemans, L. Nasdala, T. Stachel, C. Vollmer, M. Kersten, A. Somogyi, F. Adams, W. Joswig, J.W. Harris, Earth Plan Sci. Lett. **236**, 579–587 (2005)
13. J.L. Bruneel, J.C. Lassègues, C. Sourisseau, J. Raman Spectrosc. **33**, 815–828 (2002)
14. M.J. Burchell, J. Mann, J.A. Creighton, A.T. Kearsley, G. Graham, I.A. Franchi, Meteorit. Planet. Sci. **41**, 217–232 (2006)
15. B.C. Chakoumakos, T. Murakami, G.R. Lumpkin, R.C. Ewing, Science **236**, 1556–1559 (1987)
16. E. Chatzitheodoridis, D. Kostopoulos, I. Lyon, T. Henkel, N. Cornelius, E. Baltatsis, T. Reischmann, 17th Annual V.M. Goldschmidt Conference, Cologne, Germany, August 2007. Geochim. Cosmochim. Ac. **71**, A163 (2007)
17. J.Y. Chen, J.W. Schopf, D.J. Bottjer, C.Y. Zhang, A.B. Kudryavtsev, A.B. Tripathi, X.Q. Wang, Y.H. Yang, X. Gao, Y. Yang, Proc. Natl. Acad. Sci. **64**, 6289–6292 (2007)
18. C. Chopin, O. Beyssac, S. Bernard, J. Malavieille, Eur. J. Mineral. **20**, 857–865 (2008)
19. J. Corvisier, F. Brunet, A. Fabbri, S. Bernard, N. Findling, G. Rimele, V. Barlet-Gouédard, O. Beyssac, B. Goffe, Eur. J. Mineral. **22**, 63–74 (2010)
20. C. Coupry, D. Brissaud, in *Raman microscopy, developments and applications*, eds. by G. Turell, J. Corset (Academic Press, London, 1996), pp. 421–453
21. J. Craven, B. Harte, D. Fisher, D.J. Schulze, Miner. Mag. **73**, 193–200 (2009)
22. F. Corfu, J.M. Hanchar, P.W.O. Hoskin, P. Kinny, in *Zircon*, eds. by J.M. Hanchar, P.W.O. Hoskin, Reviews in Mineralogy and Geochemistry vol. 53, Mineralogical Society of America, Washington DC, (2003), pp. 469–500
23. T.C. Damen, S.P.S. Porto, B. Tell, Phys. Rev. **142**, 570–574 (1966)
24. G.F. Davies, *Dynamic Earth* (Cambridge University Press, Cambridge, 1999)
25. W.A. Deer, R.A. Howie, J. Zussman, *An Introduction to the Rock-Forming Minerals*, 2nd edn. (Pearson Education Ltd., Essex, 1992)
26. R. Devanathan, L.R. Corrales, W.J. Weber, A. Chartier, C. Meis, Nucl. Instr. Meth. Phys. Res. B **228**, 299–303 (2005)
27. H. Dill, B. Weber, Ore. Geol. Rev. **37**, 65–86 (2010)
28. T. Dörfer, W. Schumacher, N. Tarcea, M. Schmitt, J. Popp, J. Raman Spectrosc. **41**, 684–689 (2009)
29. J. England, M. Cusack, N.W. Paterson, P. Edwards, M.R. Lee, R. Martin, J. Geophys. Res. **111**, G03001 (2006)
30. N.J. Everall, Appl. Spectrosc. **54**, 773–782 (2000)
31. N.J. Everall, Appl. Spectrosc. **54**, 1515–1520 (2000)
32. N.J. Everall, Appl. Spectrosc. **62**, 591–598 (2008)
33. N.J. Everall, J. Lapham, F. Adar, A. Whitley, E. Lee, S. Mamedov, Appl. Spectrosc. **61**, 251–259 (2007)
34. R.C. Ewing, Nucl. Instr. Meth. Phys. Res. B **91**, 22–29 (1993)
35. R.C. Ewing, W. Lutze, W.J. Weber, J. Mater. Res. **10**, 243–246 (1995)
36. A. Fabbri, J. Corvisier, A. Schubnel, F. Brunet, B. Goffe, V. Barlet-Gouédard, G. Rimmele, in 3rd International Symposium on GeoProc'2008, Villeneuve d'Ascq, France, June 2008, Proceedings, eds. by J.-F. Shao, N. Burlion (Wiley Press, New York, 2008), pp. 613–621
37. F. Farges, Phys. Chem. Miner. **20**, 504–514 (1994)
38. I. Farnan, H. Cho, W.J. Weber, Nature **445**, 190–193 (2007)
39. C.M. Fedo, M.J. Whitehouse, Science **296**, 1448–1452 (2002)
40. T. Ferroir, P. Beck, B. Van de Moortèle, M. Bohn, B. Reynard, A. Simionovici, A. El Goresy, P. Gillet, Earth Plan. Sci. Lett. **275**, 26–31 (2008)

41. H. Furnes, N.R. Banerjee, K. Muehlenbachs, H. Staudigel, M. de Wit, Science **304**, 578–581 (2004)
42. H. Furuichi, N. Fujii, E. Ito, Y. Kanno, S. Watanabe, H. Tanaka, Am. Mineral. **88**, 926–928 (2003)
43. M. Gaft, R. Reisfeld, G. Panczer, *Modern Luminescence Spectroscopy of Minerals and Materials* (Springer, Berlin, 2005)
44. A.R. Gee, D.C. O'Shea, H.Z. Cummins, Solid State Commun. **4**, 43–46 (1965)
45. T. Geisler, B.E. Burakov, V. Zirlin, L. Nikolaeva, P. Pöml, Eur. J. Mineral. **17**, 883–894 (2005)
46. R.V. Gentry, Science **184**, 62–66 (1974)
47. R. Gieré, C.T. Williams, R. Wirth, K. Ruschel, Chem. Geol. **261**, 333–345 (2009)
48. C. Ginibre, G. Wörner, A. Kronz, Elements **3**, 261–266 (2007)
49. J. Glinnemann, K. Kusaka, J.W. Harris, Z. Kristallogr. **218**, 733–739 (2003)
50. J. Götze, Mater. Characterizat. **60**, 594–602 (2009)
51. J. Götze, U. Kempe, Miner. Mag. **72**, 909–924 (2008)
52. J. Götze, R. Rößler, Veröff. Mus. Naturkunde Chemnitz **23**, 35–50 (2000)
53. J. Götze, M. Plötze, D. Habermann, Mineral. Petrol. **71**, 225–250 (2001)
54. M.H. Grimsditch, E. Anastassakis, M. Cardona, Phys. Rev. B **18**, 901–904 (1978)
55. J.M. Hanchar, R.J. Finch, P.W.O. Hoskin, E.B. Watson, D.J. Cherniak, A.N. Mariano, Am. Mineral. **86**, 667–680 (2001)
56. M. Hanfland, K. Syassen, S. Fahy, S.G. Louie, M.L. Cohen, Phys. Rev. B **31**, 6896–6899 (1985)
57. R.M. Hazen, Am. Mineral. **91**, 1715–1729 (2006)
58. D.C. Hezel, H. Palme, F.E. Brenker, L. Nasdala, Meteor. Planet. Sci. **38**, 1199–1215 (2003)
59. D.C. Hezel, H. Palme, L. Nasdala, F.E. Brenker, Geochim. Cosmochim. Ac. **70**, 1548–1564 (2006)
60. G.A. Hope, R. Woods, C.G. Munce, Miner. Eng. **14**, 1565–1577 (2001)
61. J. Hövelmann, A. Putnis, T. Geisler, B.C. Schmidt, U. Golla-Schindler, Contrib. Mineral. Petrol. **159**, 43–59 (2010)
62. D. Howell, Quantifying stress and strain in diamond. PhD thesis, University of College, London, U.K, 2009
63. D. Howell, L. Nasdala, Aust. J. Earth Sci. **55**, 1175–1178 (2008)
64. D. Howell, I. Wood, D. Dobson, A. Jones, L. Nasdala, J.W. Harris, Contrib. Mineral. Petrol. **160**, 705–717 (2010)
65. G. Irmer, T. Graupner, GeoRaman 2002, Prague, Czech Republic, June 2002. Acta Univ. Carol. Geol. **46**, 34 (2002)
66. M. Ito, S. Messenger, Appl. Surf. Sci. **255**, 1446–1450 (2008)
67. W. Johannes, D. Puhan, Contrib. Mineral. Petrol. **31**, 28–38 (1971)
68. H. Kagi, I. Tsuchida, M. Wakatsuki, K. Takahashi, N. Kamimura, K. Iuchi, H. Wada, Geochim. Cosmochim. Ac. **58**, 3527–3530 (1994)
69. M. Kazanci, H.D. Wagner, N.I. Manjubala, H.S. Gupta, E. Paschalis, P. Roschger, P. Fratzl, Bone **41**, 456–461 (2007)
70. I. Keen, L. Rintoul, P.M. Fredericks, 14h European Symposium on Polymer Spectroscopy (ESOPS 14), Dresden, Germany, September 2001. Macromolecular Symposia, vol. 184 (2002), pp. 287–298
71. C.S. Kennedy, G.C. Kennedy, J. Geophys. Res. **81**, 2467–2470 (1976)
72. R.A. Kerr, Science **304**, 503 (2004)
73. L. Kiefert, H.A. Hänni, T. Ostertag, in *Handbook of Raman spectroscopy*, eds. by I.R. Lewis, H.G.M. Edwards (CRC Press, New York, 2001), pp. 469–489
74. O. Klein-BenDavid, E.S. Izraeli, E. Hauri, O. Navon, Lithos **77**, 243–253 (2004)
75. D. Koehn, D.G.A.M. Aerden, P.D. Bons, C.W. Passchier, J. Metamorph. Geol. **19**, 217–232 (2001)
76. A.V. Korsakov, D. Hutsebaut, D. Theunissen, P. Vandenabeele, A.S. Stepanov, Spectrochim. Acta A **68**, 1046–1052 (2007)
77. K. Krenn, C. Bauer, A. Proyer, E. Mposkos, G. Hoinkes, Lithos **104**, 33–53 (2008)

78. R. Krickl, L. Nasdala, J. Götze, D. Grambole, R. Wirth, Eur. J. Mineral. **20**, 517–522 (2008)
79. T.E. Krogh, G.L. Davis, Carnegie Inst Wash. Year B. **74**, 619–623 (1975)
80. G. Landsberg, L. Mandelstam, Naturwissenschaften **16**, 557–558 (1928)
81. A.R. Lang, Nature **213**, 248–251 (1967)
82. D. Lau, C. Villis, S. Furman, M. Livett, Anal. Chim. Acta **610**, 15–24 (2008)
83. J.K.W. Lee, J. Tromp, J. Geophys. Res.-Sol. Ea. **100**, 17753–17770 (1995)
84. A. Lepland, G. Arrhenius, D. Cornell, Precambrian Res. **118**, 221–241 (2002)
85. C.M. MacRae, N.C. Wilson, S.A. Johnson, P.L. Phillips, M. Otsuki, Microsc. Res. Techniq. **67**, 271–277 (2005)
86. H.-J. Massonne, *VIIth Intl Kimberlite Conf, Cape Town, RSA, April 1998. Proceedings, PH Nixon Volume, 1999* (1999), pp. 533–539
87. H.-J. Massonne, Eur. J. Mineral. **13**, 565–570 (2001)
88. H.-J. Massonne, L. Nasdala, Am. Mineral. **88**, 883–889 (2003)
89. P. Mirwald, H.-J. Massonne, J. Geophys. Res. **85**, 6983–6990 (1980)
90. C. Moazed, R. Overbey, R.M. Spector, Nature **258**, 315–317 (1975)
91. D.W. Mogk, E.A. Mathez, Geochem. Geophys. Geost. **1**, 1047 (2000)
92. S.J. Mojzsis, G. Arrhenius, K.D. McKeegan, T.M. Harrison, A.P. Nutman, C.R.L. Friend, Nature **384**, 55–59 (1996)
93. T. Murakami, B.C. Chakoumakos, R.C. Ewing, G.R. Lumpkin, W.J. Weber, Am. Mineral. **76**, 1510–1532 (1991)
94. T.A. Nachalnaya, V.D. Andreyev, E.V. Gabrusenok, Diam. Relat. Mater. **3**, 1325–1328 (1994)
95. L. Nasdala, Am. Mineral. **94**, 853–855 (2009)
96. L. Nasdala, J.M. Hanchar, Anal. Lett. **38**, 727–734 (2005)
97. L. Nasdala, H.-J. Massonne, Eur. J. Mineral. **12**, 495–498 (2000)
98. L. Nasdala, G. Irmer, D. Wolf, Eur. J. Mineral. **7**, 471–478 (1995)
99. L. Nasdala, M. Wenzel, M. Andrut, R. Wirth, P. Blaum, Am. Mineral. **86**, 498–512 (2001)
100. L. Nasdala, C.L. Lengauer, J.M. Hanchar, A. Kronz, R. Wirth, P. Blanc, A.K. Kennedy, A.-M. Seydoux-Guillaume, Chem. Geol. **191**, 121–140 (2002)
101. L. Nasdala, M. Zhang, U. Kempe, G. Panczer, M. Gaft, M. Andrut, M. Plötze, in *Zircon*, eds. by J.M. Hanchar, P.W.O. Hoskin, Reviews in Mineralogy and Geochemistry vol. 53, Mineralogical Society of America, Washington DC (2003a), pp. 427–467
102. L. Nasdala, F.E. Brenker, J. Glinnemann, W. Hofmeister, T. Gasparik, J.W. Harris, T. Stachel, I. Reese, Eur. J. Mineral. **15**, 931–935 (2003)
103. L. Nasdala, B. Wopenka, C.L. Lengauer, Am. Mineral. **89**, 912–913 (2004)
104. L. Nasdala, D.C. Smith, R. Kaindl, M. Ziemann, in *Spectroscopic methods in mineralogy*, eds. by A. Beran, E. Libowitzky, EMU Notes in Mineralogy 6, European Mineralogical Union (2004), pp. 281–343
105. L. Nasdala, W. Hofmeister, J.W. Harris, J. Glinnemann, Am. Mineral. **90**, 745–748 (2005)
106. L. Nasdala, M. Wildner, R. Wirth, N. Groschopf, D.C. Pal, A. Möller, Mineral. Petrol. **86**, 1–27 (2006)
107. L. Nasdala, A. Kronz, J.M. Hanchar, M. Tichomirowa, D.D. Davis, W. Hofmeister, Am. Mineral. **91**, 1738–1746 (2006)
108. L. Nasdala, D. Grambole, J.W. Harris, M. Wildner, W. Hofmeister, in *GeoRaman '08 Conference, Ghent, Belgium, June 2008. Book of Abstracts*, eds. by P. Vandenabeele, L. Moens (2008), p. 52
109. L. Nasdala, J.M. Hanchar, D. Rhede, A.K. Kennedy, T. Váczi, Chem. Geol. **269**, 290–300 (2009)
110. L. Nasdala, R. Grötzschel, S. Probst, B. Bleisteiner, Can. Mineral. **48**, 351–359 (2010)
111. D.F. Nelson, M.D. Sturge, Phys. Rev. **137**, A1117–A1130 (1965)
112. R.D. Neuser, D.K. Richter, Sediment. Geol. **194**, 149–154 (2007)
113. D.R.D. Niedermeier, A. Putnis, T. Geisler, U. Golla-Schindler, C.V. Putnis, Contrib. Mineral. Petrol. **157**, 65–76 (2009)
114. P.J. O'Brien, M.A. Ziemann, Eur. J. Mineral. **20**, 827–834 (2008)

115. A. Pačevski, E. Libowitzky, P. Žinković, R. Dimitrijević, L. Cvetković, Can. Mineral. **46**, 249–261 (2008)
116. D. Papineau, B.T. De Gregorio, G.D. Cody, M.D. Fries, S.D. Mojzsis, A. Steele, R.M. Stroud, M.L. Fogel, Geochim. Cosmochim. Ac. **74**, 5862–5883 (2010)
117. D. Papineau, B.T. De Gregorio, R.M. Stroud, A. Steele, E. Pecoits, K. Konhauser, J. Wang, M.L. Fogel, Geochim. Cosmochim. Ac. **74**, 5884–5905 (2010)
118. J.D. Pasteris, Appl. Spectrosc. **43**, 567–570 (1989)
119. J.D. Pasteris, B. Wopenka, Astrobiology **3**, 727–738 (2003)
120. B. Phichaikamjornwut, P. Ounchanum, A. Boonsoong, P. Limtrakun, in 6th European Conference on Mineralogy and Spectroscopy, Stockholm, Sweden, September 2007. Programme with Abstracts, p. 70 (2007)
121. H. Pichler, C. Schmitt-Riegraf, *Gesteinsbildende Minerale im Dünnschliff* (Ferdinand Enke Verlag, Stuttgart, 1987)
122. A. Putnis, *Introduction to Mineral Sciences* (Cambridge University Press, Sydney, 1992)
123. U. Rothamel, J. Heber, W. Grill, J. Phys. Condens. Matter. **50**, 297–304 (1983)
124. A.G. Ryder, M.A. Przyjalgowski, M. Feely, B. Szczupak, T.J. Glynn, Appl. Spectrosc. **58**, 1106–1115 (2004)
125. J. Satterly, Ontario Dept of Mines, Annual Report 1956, vol. 65 (part 6), pp. 108–115 (1957)
126. E. Schmädicke, Eur. J. Mineral. **3**, 231–238 (1991)
127. M. Schidlowski, Nature **333**, 313–318 (1988)
128. J.W. Schopf, Science **260**, 640–646 (1993)
129. J.W. Schopf, A.B. Kudryavtsev, Geobiology **3**, 1–12 (2005)
130. J.W. Schopf, A.B. Kudryavtsev, D.G. Agresti, T.J. Wdowiak, A.D. Czaja, Nature **416**, 73–76 (2002)
131. J.W. Schopf, A.B. Kudryavtsev, A.D. Czaja, A.B. Tripathi, Precam. Res. **158**, 141–155 (2007)
132. D. Schulze, B. Harte, J.W. Valley, J.M. Brenan, D.M.D. Channer, Nature **423**, 68–70 (2003)
133. A.M. Seydoux-Guillaume, R. Wirth, L. Nasdala, M. Gottschalk, J.M. Montel, W. Heinrich, Phys. Chem. Miner. **29**, 240–253 (2002)
134. S.K. Sharma, H.K. Mao, P.M. Bell, J.A. Xu, J. Raman Spectrosc. **16**, 350–352 (1985)
135. A.P. Shpak, O.Y. Grechanovsky, A.S. Litovchenko, S.Y. Sayenko, Probl. At. Sci. Tech. **2**, 29–33 (2007)
136. L.T. Silver, S. Deutsch, J. Geol. **71**, 721–758 (1963)
137. J. Sláma, J. Košler, D.J. Condon, J.L. Crowley, A. Gerdes, J.M. Hanchar, M.S.A. Horstwood, G.A. Morris, L. Nasdala, N. Norberg, U. Schaltegger, M.N. Tubrett, M.J. Whitehouse, Chem. Geol. **249**, 1–35 (2008)
138. D.C. Smith, C. Carabatos-Nédelec, in *Handbook of Raman spectroscopy*, ed. by I.R. Lewis, H.G.M. Edwards (CRC Press, New York, 2001), pp. 349–422
139. A.L. Soldati, D.E. Jacob, U. Wehrmeister, T. Häger, W. Hofmeister, J. Raman Spectrosc. **39**, 525–536 (2008)
140. S.A. Solin, A.K. Ramdas, Phys. Rev. B **1**, 1687–1698 (1970)
141. V. Stähle, R. Altherr, M. Koch, L. Nasdala, Contrib. Mineral. Petrol. **155**, 457–472 (2008)
142. M. Stárková, T. Sidorinová, Z. Táborský, J. Malec, 2nd CEMC, Szklarska Poreba, Poland, Sept 2008. Mineralogia Special Papers 32 (2008), p. 150
143. A. Steele, M.D. Fries, H.E.F. Amundsen, B.O. Mysen, M.L. Fogel, M. Schweizer, N.Z. Boctor, Meteorit. Planet Sci. **42**, 1549–1566 (2007)
144. R. Tabaksblat, R.J. Meier, B.J. Kip, Appl. Spectrosc. **46**, 60–68 (1992)
145. P.D. Taylor, A.B. Kudryavtsev, J.W. Schopf, Invetrebrate Biol. **127**, 87–97 (2008)
146. A. Tfayli, O. Piot, M. Manfait, J. Biophoton. **1**, 140–153 (2008)
147. M. Tichomirowa, H.-J. Berger, E.A. Koch, B.V. Belyatski, J. Götze, U. Kempe, L. Nasdala, U. Schaltegger, Lithos **56**, 303–332 (2001)
148. J.A. Timlin, A. Carden, M.D. Morris, J.F. Bonadio, C.E. Hoffler, K.M. Kozloff, S.A. Goldstein, J. Biomed. Opt. **4**, 28–34 (1999)
149. N. Tomašić, Ć.A. Gajovi, V. Bermanec, D.S. Su, Rajić Linarić M, Ntaflos T, Schlögl R. Phys Chem Minerals **33**, 145–159 (2006)

150. P.J. Treado, M.P. Nelson, in *Handbook of Raman spectroscopy*, eds. by I.R. Lewis, H.G.M. Edwards (CRC Press, New York, 2001), pp. 191–249
151. T. Váczi, Mineralogical studies on zircon. PhD thesis, University of Wien, Austria, 2009
152. A.A. Van Apeldoorn, Y. Aksenov, M. Stigter, I. Hofland, J.D. de Bruijn, H.K. Koerten, C. Otto, J. Greve, C.A. van Blitterswijk, J. R. Soc. Interface **2**, 39–45 (2005)
153. M.A. van Zuilen, A. Lepland, G. Arrhenius, Nature **418**, 627–630 (2002)
154. G.I. Veinott, T.R. Porter, L. Nasdala, Trans. Am. Fish. Soc. **138**, 1157–1165 (2009)
155. P. Verma, S.C. Abbi, K.P. Jain, Phys. Rev. B **51**, 16660–16667 (1995)
156. R.P. Vidano, D.B. Fischbach, L.J. Willis, T.M. Loehr, Solid State Commun. **39**, 341–344 (1981)
157. Z. Wang, X. Huang, R. Xue, L. Chen, J. Appl. Phys. **84**, 227–231 (1998)
158. W.J. Weber, R.C. Ewing, L.-M. Wang, J. Mater. Res. **9**, 688–698 (1994)
159. Y. Weiss, R. Kessel, W.L. Griffin, I. Kiflawi, O. Klein-BenDavid, D.R. Bell, J.W. Harris, O. Navon, Lithos **112**, 660–674 (2009)
160. M. Wiedenbeck, J.M. Hanchar, W.H. Peck, P. Sylvester, J. Valley, M. Whitehouse, A. Kronz, Y. Morishita, L. Nasdala, J. Fiebig, I. Franchi, J.P. Girard, R.C. Greenwood, R. Hinton, N. Kita, P.R.D. Mason, M. Norman, M. Ogasawara, R. Piccoli, D. Rhede, H. Satoh, B. Schulz-Dobrick, O. Skar, M.J. Spicuzza, K. Terada, A. Tindle, S. Togashi, T. Vennemann, Q. Xie, Y.F. Zheng, Geostand. Geoanal. Res. **28**, 9–39 (2004)
161. D.M. Wieliczka, P. Spencer, M.B. Kruger, Appl. Spectrosc. **50**, 1500–1504 (1996)
162. K.P.J. Williams, G.D. Pitt, D.N. Batchelder, B.J. Kip, Appl. Spectrosc. **48**, 232–235 (1994)
163. J.A. Woodhead, G.R. Rossman, L.T. Silver, Am. Mineral. **76**, 74–82 (1991)
164. W.S. Xian, M. Sun, J. Malpas, G.C. Zhao, M.F. Zhou, K. Ye, J.B. Liu, D.I. Phillips, Anal. Lett. **37**, 119–130 (2004)
165. M. Zhang, E.K.H. Salje, I. Farnan, A. Graeme-Barber, P. Daniel, R.C. Ewing, A.M. Clark, H. Lennox, J. Phys. Condens. Mat. **12**, 1915–1925 (2000)
166. F. Zhang, W. Cai, Z. Sun, J. Zhang, Anal. Bioanal. Chem. **390**, 777–782 (2008)
167. J.F. Ziegler, J.P. Biersack, U. Littmark, *The Stopping and Range of Ions in Solids* (Pergamon Press, New York, 1985)

Chapter 6
Raman Mapping in the Scientific Investigations of Works of Art

Polonca Ropret, Costanza Miliani and Silvia A. Centeno

Abstract Raman mapping in works of art has traditionally been performed using a motorized xy stage that moves a small artwork or a sample taken from the object in a step-by-step manner in two directions under the microscope objective. This configuration allows to obtain important information about the objects in a non-invasive manner and, in the case of samples, on the stratigraphy and the composition of all layers. Three case studies are presented to discuss the capabilities and limitations of this approach. In the first case, the distribution of crystalline and amorphous phases in a sample from a historical glaze were mapped and the resulting composition was related to the production technology of the decoration. In the second case, Raman mapping was performed non-destructively in a ceramic fragment to contribute to elucidate the complex process involved in the lustre decoration formation. And in the third case, the composition of the different paint layers in a sample cross-section removed from a polychromed wooden sculpture was mapped in order to help differentiate original paint layers from those resulting from restoration interventions. Among the limitations of the mapping approach using a motorized xy stage are the impossibility to use it with art objects that do not fit on the stage, under the

P. Ropret (✉)
Conservation Centre, Institute for the Protection of the Cultural Heritage of Slovenia, Poljanska 40, 1000 Ljubljana, Slovenia
e-mail: polona.ropret@rescen.si

P. Ropret
Museum Conservation Institute, Smithsonian Institution, 4210 Silver Hill Rd., Suitland, MD 20746, USA

C. Miliani
c/o Dipartimento di Chimica, Istituto di Scienze e Tecnologie Molecolari del CNR (CNR-ISTM), Universita' di Perugia, Via Elce di Sotto 8, 06123 Perugia, Italy

S. A. Centeno
Department of Scientific Research, The Metropolitan Museum of Art, New York, NY 10028, USA

A. Zoubir (ed.), *Raman Imaging*, Springer Series in Optical Sciences 168,
DOI: 10.1007/978-3-642-28252-2_6, © Springer-Verlag Berlin Heidelberg 2012

microscope objective, and the time required for the automatic optimization of the focus distance at each point. Examples of in situ non-invasive mapping experiments in two contemporary paintings using a novel configuration in which a set of mirrors is placed in the horizontal exit of the microscope attached to the spectrometer are discussed. The advantages of this system are, in addition to allowing to map larger objects, its full confocality and the possibility to work with multiple laser excitations. The limitations of this approach are also discussed.

6.1 Introduction

Since the mid-1980s, Raman spectroscopy has turned into an indispensable technique for the non-destructive and non-invasive characterization of artists' materials and techniques, archaeological materials and the technological processes involved in their manufacture, and deterioration mechanisms. This is attested, for example, by the number articles published in the *Journal of Raman Spectroscopy*, other scientific journals, and specialized journals in cultural heritage, archaeology, and art conservation. The *Journal of Raman Spectroscopy* has so far produced four special issues dedicated to the application of the technique to these fields [1–4]. A review of the use of the technique in the art and archaeology fields is beyond the scope of this chapter; nonetheless it is worth mentioning that wide ranges of artistic media and of archaeological and historic materials have been studied by Raman spectroscopy, including manuscripts on parchment, stone sculpture, wall paintings, wooden artifacts, textiles, glass, jewelry, gemstones, ceramics, metalwork, rock art, works of art on paper, traditional oil paintings, photographic materials, and modern and contemporary works of art [1–15]. The reference libraries of Raman spectra of materials of interest to these fields that have so far been made available include synthetic and natural organic and inorganic pigments [14–21], artistic varnishes, resins, gums, waxes and other binding media [22], minerals [23], and fibers [9].

What makes the technique extremely powerful for the examination of art objects and artifacts, in addition to its molecular specificity and lateral resolution, is its non-invasive nature, that allows to examine precious objects without the removal of sample material. Non-invasive Raman mapping provides the possibility to correlate chemical data with information derived from complementary analytical techniques such as X-ray fluorescence spectroscopy (XRF), open architecture X-ray diffraction (XRD), UV-vis reflectance spectroscopy, and multispectral imaging (See, for example, [24]).

However, objects of cultural significance are generally hybrid organic-inorganic materials intrinsically characterized by both micro and macro-scale heterogeneities, due not only to the artists' techniques but also to the effects of ageing, exposure to the environment, and past conservation practices. Although several non-invasive techniques in addition to Raman spectroscopy are available to identify the various components in bulk samples, this type of analysis sometimes results in a loss of spatial information, often crucial to answer questions related to conservation and restoration. Therefore, with the aim to gain insight into the microscale structure of artworks' multilayered materials, such as those of paintings, minute stratigraphic samples (typically <1 mm^3) often need to be removed. Generally, the samples are embedded in a synthetic resin in a cross-section configuration, polished, and viewed under optical and/or scanning electron microscopes to study the layering structure. In this context, chemical mapping is obviously very significant as it allows for a detailed analysis of the materials distribution within the stratigraphy.

Traditionally, Raman mapping has been performed using a motorized xy stage that moves the object or microscopic sample in two directions in a step-by-step manner under the microscope attached to the spectrometer. This mapping approach has been used to study, for example, the deterioration mechanism of pigments in wall paintings [25]; the painting technique of 16th century artists [26]; rust scales on archaeological iron artifacts [27]; the spatial distribution of corrosion products in ancient Chinese artifacts [28]; and the composition of crusts in stained glass windows [29]. Data from Raman mapping in microsamples has been typically combined with information obtained by using complementary analytical techniques. Aze et al. used micro-Raman semiquantitative mapping in combination with backscattered electron imaging, elemental microprobe mapping, and in situ X-ray diffraction to study the fading mechanism of the red lead pigment in sample cross-sections from wall paintings [25]; Lau et al. combined Raman maps with elemental maps acquired using energy-dispersive X-ray microanalysis (EDS) to characterize paint layers in microsamples from 16th century paintings [26]; and McCann et al. used backscattered electron imaging to complement Raman mapping in the analysis of microscopic samples from corroded Chinese bronze money trees [28].

As with other analytical techniques, Raman mapping results may comprise a single value collected at each (x,y) point, or a full spectrum. When spectra are available, the data can be arranged in a hyperspectral cube where each (x_m,y_n) point has a spectrum associated with it, with its intensity in the z direction. Multivariate analysis may then be performed, for example, by principal component analysis (PCA) [26] (See Chap. 1).

Among the limitations of the mapping approach using a motorized xy stage are the impossibility to use it with objects that do not fit on the stage and, due to the fact that more often than not the surfaces of art objects are not flat, the time required for the automatic optimization of the focus distance at each point. The use of fiber optics coupled to the spectrometer also has limitations due to the lack of mapping capabilities, at least in commercial spectrometers, and also due to their lower optical throughput, lack of confocality, and their limited flexibility regarding the laser excitation.

A novel mapping mechanism, that uses a horizontal exit in the microscope attached to the spectrometer has been recently developed [30]. This system has multiple advantages, among them its full confocality and the possibility to use it with multiple excitation lines.

In the present article, experimental set ups are described for each mapping approach and examples of the application of both methods to the characterization of artists' materials and techniques in works of art are presented and discussed in the context of their capabilities and limitations.

6.2 Raman Mapping Using a Motorized xy Stage

6.2.1 Experimental Setup

Micro-Raman spectra were obtained using a JASCO Ventuno spectrophotometer equipped with a charge-coupled device (CCD) detector cooled to 50°C. Raman spectra were acquired using green radiation from a Nd:YAG laser. Backscattering illumination and collection of the scattered light were made through an Olympus confocal microscope (long-focus Olympus 5, 20, 50, 100x objectives). Working with a 1200 l/mm grating a spectral resolution of about 2 cm^{-1} was achieved. Calibration of the spectrometer was accomplished by using the Raman lines of a polystyrene standard.

In this instrument, the mapping capabilities are provided by an automated xy stage on which the sample is placed, and an autofocus system that searches for the z position. In-plane maps were acquired with 50x and/or 100x objectives. Maps of variable dimensions were obtained by collecting spectra with steps between 4 and 20 μm along the x and y axes. Spectral measurements were made with continuous scans in the range 50–2000 cm^{-1}, exposure times between 10 and 200 s, 4 to 10 accumulations, and a maximum 10 mW laser power at the sample.

Maps were created by visualizing different parameters: height or areas of peaks, shift of peaks, or peak's full width at half maximum (FWHM).

Micro-FTIR mapping measurements were carried out using a Jasco IMV4000 instrument equipped with a line array MCT detector (1 x 16 elements), an 16x Cassegrain objective, and an automated sample stage. Spectra were collected in the reflection mode, in the range 650–6000 cm^{-1}, with a 4 cm^{-1} spectral resolution.

Scanning electron microscopy (SEM) backscattered electron images were taken using a Philips XL30 with a LaB_6 source. EDS microanalyses were carried out using an EDAX/DX4 detector.

6.2.2 Case Studies

Three case studies were selected to exemplify the use of Raman mapping with a motorized xy stage. In the first example, the distributions of crystalline and amorphous phases in a sample cross-section from a historical glaze were mapped and the resulting composition was related to the production technology of the decoration. In the second case, Raman mapping was performed in a fragment of a ceramic shard to obtain complementary information to that gained using other analytical techniques in order to elucidate the complex process involved in the lustre decoration formation. In the third case, the composition of the different paint layers in a sample cross-section removed from a polychromed wooden sculpture with a complex stratigraphy was mapped in order to help differentiate original paint layers from those resulting from restoration interventions.

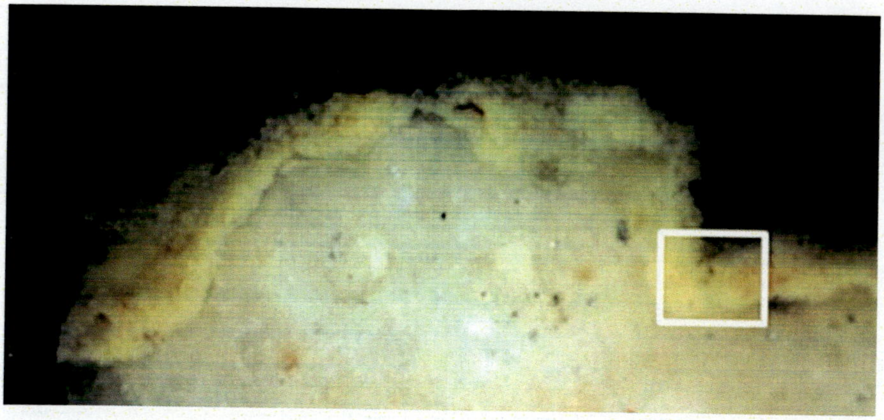

Fig. 6.1 Photomicrograph of a cross-section removed from a decoration in a Hotel Hungaria tile containing Naples yellow. Visible illumination. Original magnification 50x. Raman mapping measurements were carried out in a 100 × 72 µm area, indicated with a white rectangle at the right of the photomicrograph, with a step of 4 µm

6.2.2.1 Study on the Spatial Distribution of Modified Lead Antimonate Pigments in a Historical Glaze

Lead antimonate yellow ($Pb_2Sb_2O_7$), one of the oldest known synthetic pigments, was employed as a coloring agent for artistic glass production in Egyptian, Mesopotamian, Babylonian, and Assyrian workshops. In Western European art, lead antimonate yellow was used since the 16th century as a yellow enamel pigment in Italian *majolica* and, later, in paintings, adopting the name *luteolum Napolitanum* (Naples yellow) [31].

Interestingly, modified Naples yellows based on a ternary oxide of lead, antimony, and tin ($Pb_2Sb_{2-x}SnxO_{7-x/2}$) have been identified by X-ray diffraction and micro-Raman spectroscopy in Italian 16th century *majolica* [32], in 17th century paintings [33], and in Mid-European paintings dating from the 18th to the 19th centuries [34].

The modification of lead antimonate yellow, that has a cubic pyrochlore structure, with Sn^{4+} and also with Zn^{2+} and other suitable cations produces a distortion of the unit cell that strongly affects the Raman scattering features. The collapse of the intense band at 510 cm^{-1} (Sb-O mode) with respect to the 150 cm^{-1} scattering (Pb-O mode), and the appearance of a well resolved band at about 460 cm^{-1} have proved to be particularly reliable for the identification of the modified lead pyroantimonates [35].

A modified Naples yellow was also identified in later glazed tiles that decorate the Liberty façade of the Hotel Hungaria in Venice (*ca.* 1914). Figure 6.1 shows a photomicrograph taken with visible illumination of a cross-section removed from a yellow decoration in one of these tiles from the Hotel Hungaria containing Naples yellow. The area where Raman maps were acquired is also indicated in this photomicrograph.

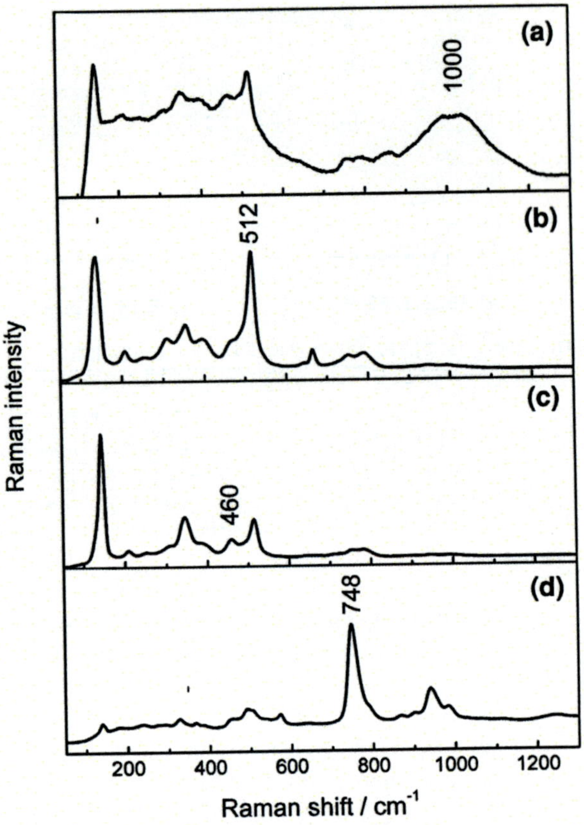

Fig. 6.2 Raman spectra recorded in the area indicated by a white rectangle in Fig. 6.1: glassy phase (**a**), Naples yellow (**b**), modified Naples yellow (**c**), Cr- doped malayaite (**d**), respectively. At each point, the Raman spectrum was recorded using a 40 s integration time, 4 accumulations, and a power of 10 mW

As a first step, 494 Raman spectra were acquired and comprehensively analyzed to determine the components that contributed to the detected signal. This revealed the presence of the following phases: silica glass, lead-antimonate, modified lead-antimonate and Cr-doped malayaite. Representative Raman spectra are shown in Fig. 6.2.

The Raman maps shown in Fig. 6.3b–f were produced using the features that overlap the least in the spectra shown in Fig. 6.2. The location of the glassy phase can be visualized in the chemical map shown in Fig. 6.3b, on the top right of the area. This image was acquired by mapping the FWHM of the Si-O stretching at ca. 1000 cm^{-1}, that is typically very broad.

The map for Naples yellow, presented in Fig. 6.3c, was acquired by recording the intensity of the scattering band of Sb-O at 510 cm^{-1}, the most intense in the pyrochloric structure. Naples yellow appears to be distributed all over the yellow layer, except in those areas where the glassy phase is present. By mapping the area at 460 cm^{-1}, the maximum of the diagnostic feature for the modified pyrochloric structure, it was possible to see that the compound is located in four small round spots at right and a bigger one in the upper left (Fig. 6.3d), which correspond to the

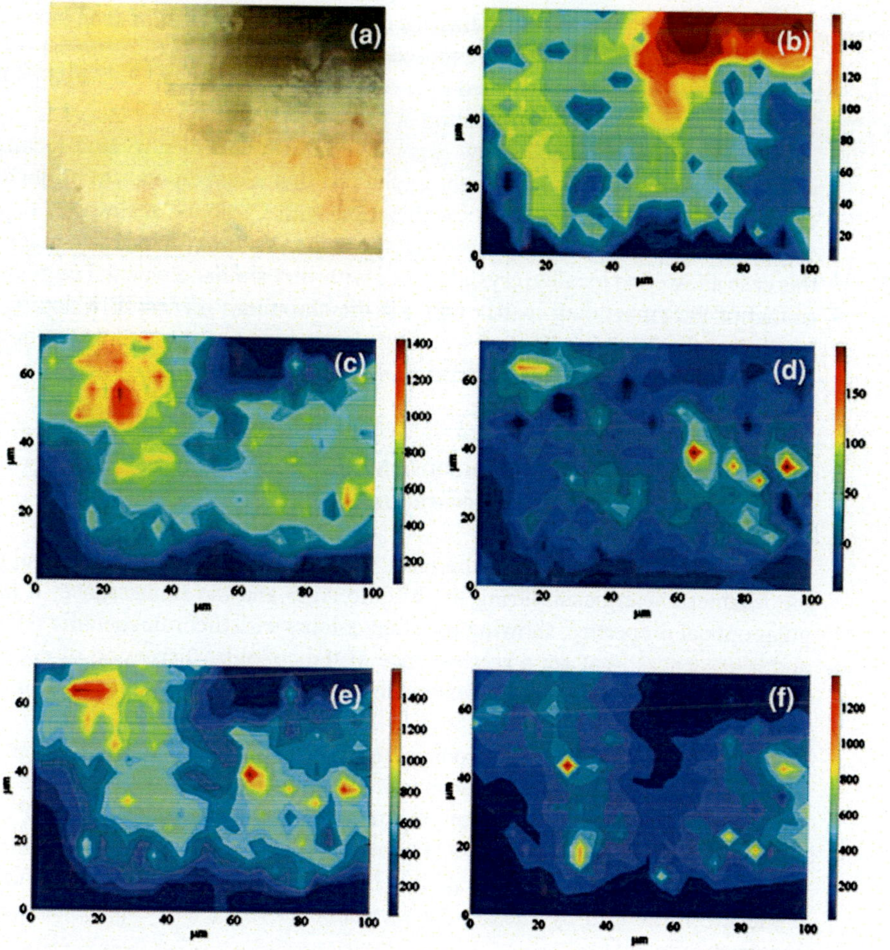

Fig. 6.3 Optical image and Raman maps of the area indicated by a white rectangle in the cross-section shown in Fig. 6.1: optical image, original magnification 100x (**a**), glassy phase, FWHM of the Si-O stretching at ca $1000\,cm^{-1}$ (**b**), Naples yellow, Sb-O mode at $510\,cm^{-1}$ (**c**), modified Naples yellow, $460\,cm^{-1}$ (**d**), modified Naples yellow, Pb-O stretching at $150\,cm^{-1}$ (**e**), and Cr- doped malayaite, Cr^{4+}-O mode at $748\,cm^{-1}$ (**f**), respectively. At each point, the Raman spectrum was recorded using a 40 s integration time, 4 accumulations, and a power of 10 mW

orange crystals visible in the optical image (Fig. 6.3a). Similar results were obtained by mapping the intensity of the Pb-O stretching at $150\,cm^{-1}$ (Fig. 6.3e) that increases in the modified Naples yellow with respect to the Sb-O mode, albeit the spots appear less defined than in the previous map since the band is not so specific for the modified structure as the $460\,cm^{-1}$ signal is. The darker crystals visible in the optical image contain Cr- doped malayaite that can be efficiently localized by mapping its resolved band at $748\,cm^{-1}$ (Fig. 6.3f). Cr- doped malayaite (Cr^{4+}- $CaSnOSiO_4$) is an important ceramic pigment, being the only one to have peculiar deep burgundy to pink shades that are stable when exposed to a wide range of temperatures [36]. This compound can be easily identified by a resonance-enhanced Cr^{4+}-O Raman line at 748 cm^{-1} using a Nd-YAG excitation at 532 nm [37].

Raman mapping of the yellow glaze layer allowed to identify and localize the modified lead-antimonate, highlighting the fact that it corresponds to bright orange crystals possibly introduced, in combination with Cr-doped malayaite, to give a warmer hue to the tiles. It is worth to stress the facts that molecular mapping in this case allowed to localize crystal phases with very similar elemental compositions to that of the embedding matrix and that the chromium present as a doping agent was below the detection limit of scanning electron microscopy coupled with energy dispersive X-ray spectrometry (SEM-EDS) analysis.

6.2.2.2 Mapping of Micro-Structural Modifications on a Glaze Surface Induced by a Lustre Decoration

Lustre was regarded as one of the most precious pottery decoration techniques in the Mediterranean basin during the Middle Ages and the Renaissance. It has singular optical properties, showing different colours with dichroic metallic reflections and iridescences, that are a consequence of the colloidal dispersion of copper and silver nanoparticles within the first layers of the glassy transparent medium of the glaze [38].

Lustre nanoparticles were produced by first putting a mixture of silver and/or copper salts and oxides over the surface of a previously glazed and decorated ceramic piece, then firing the whole system at a temperature between 500 and 700°C (third firing) in a reductive atmosphere. At temperatures above 500°C, the glaze softens and the Cu^{+} and/or Ag^{+} ions penetrate into the glaze where, in a reductive atmosphere produced in the kiln by smoke, the formation of silver and copper nanoparticles occurred.

The Medieval Persian ceramics that will be discussed in the present article as a case study have been dated to the 10th century and show gold and reddish lustre (Fig. 6.4) whose composition and structure have already been studied by UV-vis is reflectance spectroscopy, transmission electron microscopy (TEM), and X-ray absorption fine structure (XAFS) [39, 40]. It has been proved that the gold color is mainly due to Ag nanoparticles, though Ag^{+}, Cu^{+}, and Cu^{2+} are also dispersed within the glassy matrix. The red color is mainly due to Cu nanoparticles although some Ag nanoparticles, Ag^{+}, Cu^{+}, and Cu^{2+} ions may be present. These findings are in substantial agreement with results of studies on Italian Renaissance pottery previously reported [38, 41]. Yet, the presence of copper and silver ions in the glaze confirms that the lustre formation is mediated by copper- and silver- alkali ion exchange, followed by nucleation and growth of metal nanoparticles.

Micro-Raman spectroscopy, that is sensitive to both crystalline and amorphous phases, was applied to obtain complementary information to that obtained by TEM and XAFS to reconstruct the complex process involved in the lustre formation.

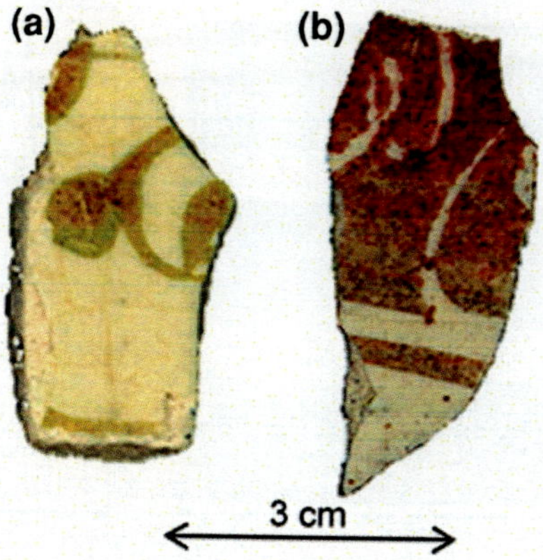

Fig. 6.4 Medieval Persian shards dated to the 10th century, belonging to the Islamic Museum of Tehran, decorated with gold lustre (**a**) and red lustre (**b**), respectively

On some of the samples studied, in the vicinity of the gold lustre decoration, micro-Raman spectroscopy revealed the presence of cuprite (Cu_2O). The complexity of the Raman spectra recorded made it difficult to immediately identify the cuprous oxide, however, through a micro-Raman mapping experiment it was possible to separate and distinguish the spectral features due to cuprite from those originating from the rest of the components of the matrix: lead oxides (149 cm^{-1}) and cassiterite (SnO_2, 636 cm^{-1}). These results are presented in Fig. 6.5. The distribution map of cuprite, obtained by recording the intensity of the signal at 227 cm^{-1}, revealed a constellation of small crystals similar to those visible in the SEM image taken on a different area of the same sample and shown in Fig. 6.6. The origin of the cuprite can be attributed to a residue from the original lustre mixture, indicating a non-perfect reducing atmosphere during the lustre production.

Interestingly, the glassy network structure resulted to be strongly altered by the lustre formation as revealed by the statistical comparison of the Si-O stretching Raman features of the white glaze and of the lustre glaze. Indeed, the comparison of the lustre and white glaze spectra showed a decrease in intensity and a shift to lower frequencies (lower connectivity) of the Si-O stretching of the glaze in correlation with the presence of lustre (Fig. 6.7).

In order to provide further evidence of the effect of lustre formation on the glassy phase structure, a Raman mapping study was conducted on the red lustre sample shown in Fig. 6.4b, for which an $80 \times 80\,\mu m^2$ area was mapped with a 20 μm spatial resolution. By mapping the area (Fig. 6.7b) or the FWHM (Fig. 6.7c) of the Si-O stretching band similar contour plots were obtained, both well describing the shape of the lustre decoration visible in the optical image (Fig. 6.7a). The spectral modifications observed are evidence that the lustre nanoparticles disrupt the surrounding glass structure during their nucleation and growth, leading to a less connected glaze, consistent with observations on copper-containing alkali borosilicate glass previously reported [42].

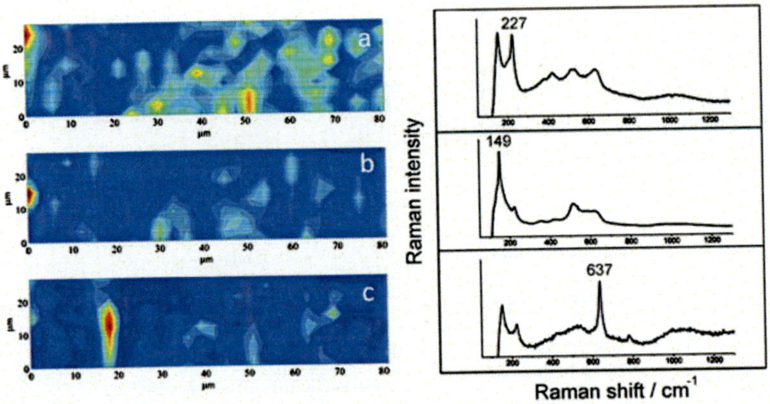

Fig. 6.5 Raman maps acquired in the gold lustre decoration on the shard shown in Fig. 6.4a: cuprite (**a**), lead oxide (**b**), and cassiterite (**c**), respectively. At each point, the Raman spectrum was recorded using a 40 s integration time, 4 accumulations, and a 10 mW power

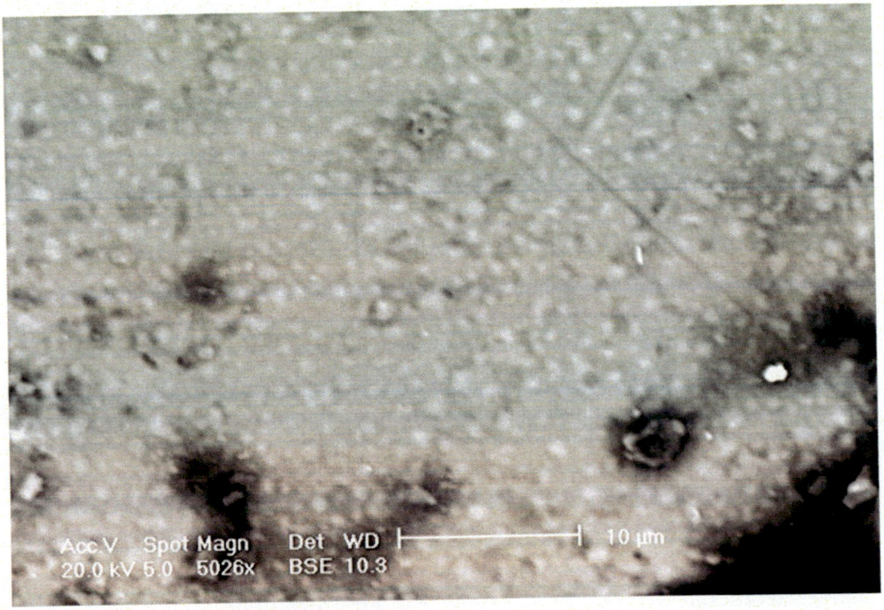

Fig. 6.6 SEM backscattered electron image of the surface of the shard shown in Fig. 6.4a. Cuprite crystals, a few microns in diameter, partially embedded in the glaze, are visible in this image

6.2.2.3 Mapping of the Pigment Distribution in a Layered Polychromy

Polychrome sculptures, more often than not, have been subjected to overpainting, exposed to physical contact with the viewer, and, therefore, to losses of their original surface, especially when in the form of religious statuary. Sculptures in churches

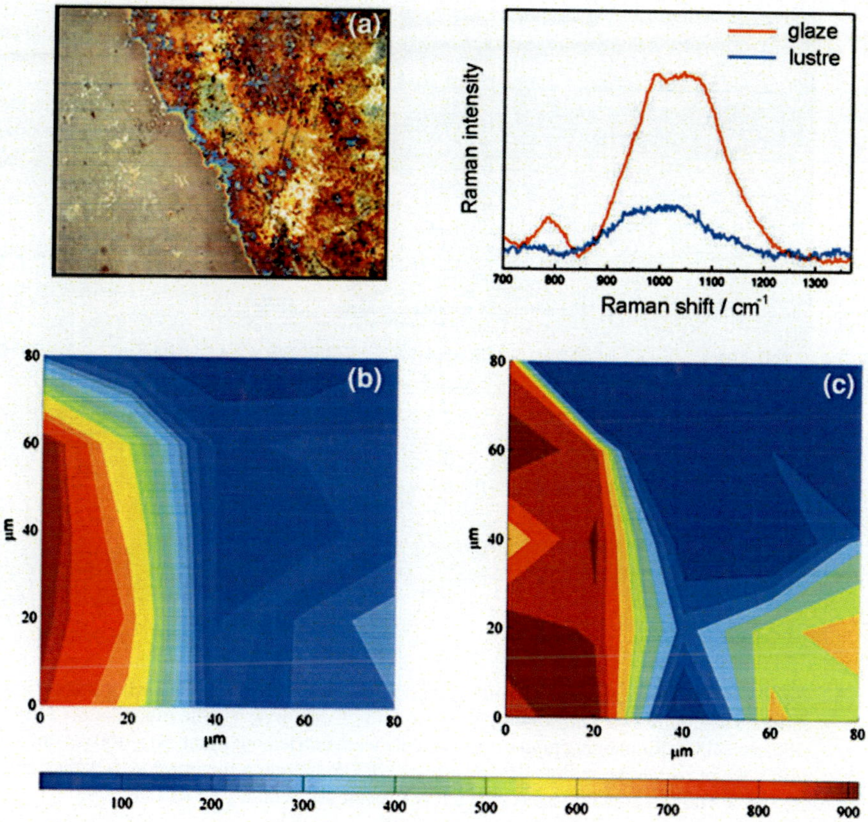

Fig. 6.7 Optical image and Raman maps of the sample shown in Fig. 6.4b, acquired at the interface between the red lustre and the glaze: optical image, 80 × 80 μm^2, 20 µm step size (**a**), contour plot obtained using the area of the Si-O stretching (**b**), and contour plot obtained using the FWHM of the Si-O stretching (**c**,) respectively. The insert at the upper right shows a comparison of the Raman spectra of the lustre and of the white glaze. At each point, the Raman spectrum was recorded using a 20 s integration time, 5 accumulations, and a 10 mW power

have been generally repainted not only when the original polychromy has been partially lost but also when fashion and taste changed. Previous to any conservation and restoration work on this kind of painted statuary it is essential to discriminate between the original and the restoration palettes and to characterize the original and the overpainting layers. In this context, the chemical mapping of a cross-section can be of great value to highlight the spatial distribution of the different molecular phases along the stratigraphy.

For the present article, an example was selected to highlight the strengths and the limitations of Raman mapping to study complex layered polychromies: an Italian wooden sculpture dated to the 17th century, representing *Our Lady of the Assumption*, that appeared to have been significantly altered from its original design.

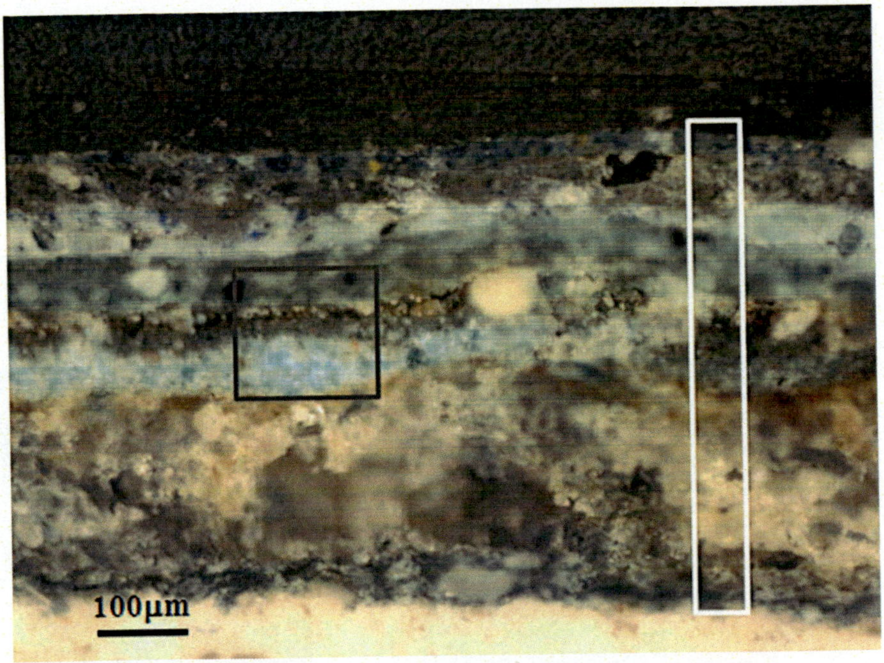

Fig. 6.8 Photomicrograph of a sample cross-section removed from the *Lady's* mantle in the polychromed sculpture *Our Lady of the Assumption* (17th century). Visible illumination. Original magnification 100x. Raman mapping measurements were carried out in the $50 \times 400\,\mu m$ area indicated with a white rectangle, with a $10\,\mu m$ step size. FTIR mapping measurements were carried out in the $125 \times 100\,\mu m$ area indicated with a black rectangle, with a $12.5\,\mu m$ step size

The photomicrograph of a cross-section taken from the Lady's mantle shown in Fig. 6.8 describes by itself the complex stratigraphy of the paint passage where several layers, at least seven, ranging in color from black to light blue, are visible in a thickness of about $400\,\mu m$.

Raman mapping was performed in a $50 \times 400\,\mu m$ area along the complete stratigraphy, with a $10\,\mu m$ step size, acquiring a total of 246 spectra (the spectral range was limited to $200\text{–}2000\,cm^{-1}$ for faster acquisition). Although many spectra were severely affected by a fluorescent background, possibly due to the oil medium used as a binder, the following pigments were identified by their characteristic Raman scattering features: indigo, a carbon-based black, calcite ($CaCO_3$), lead white (the basic $PbCO_3Pb(OH)_2$ and/or the neutral $PbCO_3$), massicot (orthorhombic PbO), and ultramarine blue (Fig. 6.9).

By mapping the area of the ν (C=O) band of indigo at $1690\,cm^{-1}$, it was possible to localize a thin layer that corresponds well with the first blue layer visible in the optical image, about $20\,\mu m$ thick, applied over the preparatory layer (Figs. 6.8 and 6.10a). Indigo is a colorant extracted from the leaves of various *Indigofera* and other plant species, *Isatis tinctoria* among them [43]. Indigo has been identified in manuscripts

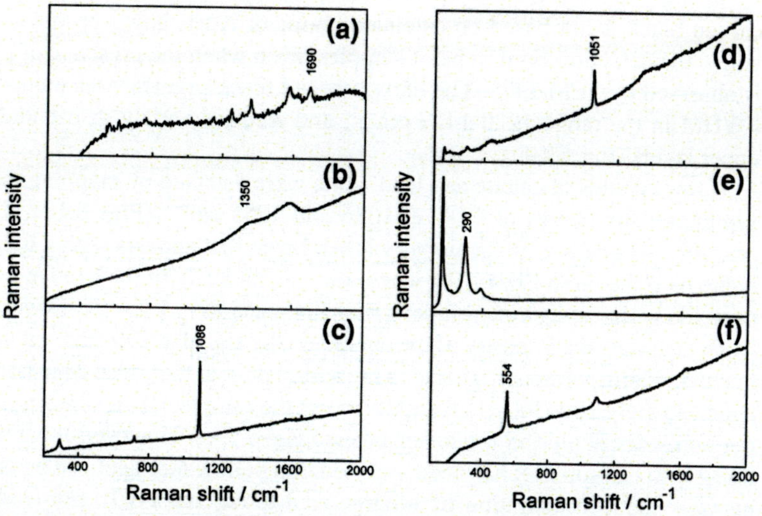

Fig. 6.9 Raman spectra acquired in the area indicated by a white rectangle in Fig. 6.8: indigo (**a**), carbon-based black (**b**), calcite (**c**), lead carbonate (**d**), massicot (**e**), and ultramarine blue (**f**), respectively. Spectra were recorded in the 100–2000 cm^{-1} spectral range, with a 2 cm^{-1} resolution, using a 50 s integration time, 3 accumulations, and a power of 4 mW

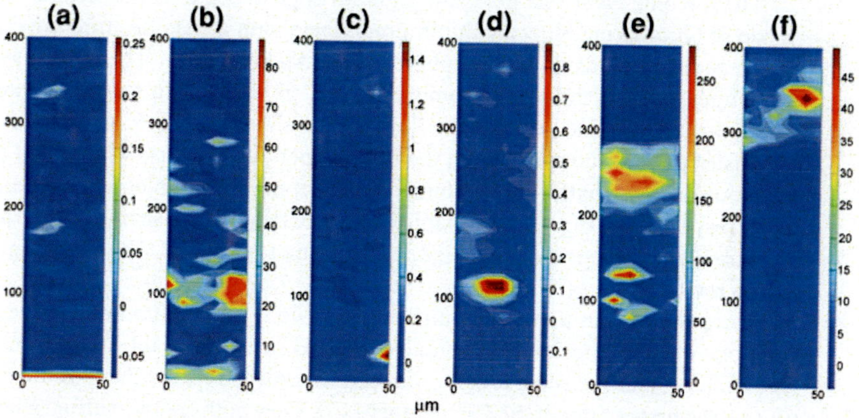

Fig. 6.10 Raman maps acquired in the area indicated by a white rectangle in Fig. 6.8. The contour plots were obtained by using the bands whose Raman shifts are indicated in Fig. 6.9: indigo (**a**), carbon-based black (**b**), calcite (**c**), lead carbonate (**d**), massicot (**e**), and ultramarine blue (**f**), respectively

on parchment [5], in oil paintings [43–45], in polychromed wooden objects [8], and in other artistic media, though it was mainly used for textile dyeing and printing.

In this case, indigo was mixed with a carbon-based black pigment, the latter with characteristic Raman bands due to amorphous carbon at ca. 1590 and 1330 cm^{-1}. The absence of a feature at 960 cm^{-1} due to the phosphate ion suggests that the carbon black is not based on calcinated bone or ivory, however this mode may be below the detection limit in paint samples, even when the presence of phosphorous is observed by SEM-EDS. The carbon-based black pigment was mapped using the FWHM in the range 1220-1437 cm^{-1}, and resulted widely distributed also in the upper layers (Fig. 6.10b).

A few crystals of calcite and lead white were detected by mapping the respective symmetric stretchings of CO_3^- at 1086 and 1051 cm^{-1} (Figs. 6.10c, d). Calcite is probably present also in the upper whitish layer, that appears with a light blue color in the map shown in Fig. 6.10c, where the low intensity of the band at 1086 cm^{-1} is most likely due to a strong hindering by fluorescence.

By mapping the intensity of the characteristic signal at 290 cm^{-1}, it was observed that the yellow pigment massicot is present mainly in the green-bluish layer, located approximately 200–300 μm from the top of the sample, albeit some crystals are also randomly distributed in the lower layers (Fig. 6.10e). However, this result must be taken with caution as it has been reported that yellow massicot can be easily formed by laser induced alteration of litharge (red tetragonal PbO), red lead or minium (Pb_3O_4), and the black plattnerite (PbO_2), especially when working with a green excitation [21]. All attempts to analyze the green-bluish layer with a low laser power only gave features due to massicot, therefore it was not possible to assess if another type of lead oxide is actually present in this layer.

The top blue layer that gives the external appearance to the Madonna's mantle is made of ultramarine blue, an aluminum-silicate with a sodalite structure containing sulfur radical anions as chromophores ($Na_8[Al_6SiO_{24}]S_n$). The totally symmetric stretching mode of the blue chromophore S^{3-} gives a strong scattering band at 554 cm^{-1} that was used to produce the Raman map shown in Fig. 6.10f. This map shows the distribution of ultramarine in the three upper paint layers, with a total thickness of about 100 μm, that in the photomicrograph appear bright blue, dark blue, and light blue, respectively (Fig. 6.8), probably due to the fact that the strong Raman scattering of ultramarine blue is masking signals from other black and white pigments present in the mixtures.

Moreover, it is important to point out that two types of ultramarine blue exist: the first one of natural origin, known also as lapis lazuli from the stone from which it is extracted, and the second one produced according to a synthetic route developed during the first quarter of the 19th century [43]. They both contain sulfur radical anions as chromophores entrapped in the sodalite cage and, therefore, give rise to exactly the same scattering signals, making the discrimination between the natural and the synthetic forms, that in this case imply original or modern pigment, impossible by Raman spectroscopy.

Complementary information was achieved by SEM-EDS. A backscattering SEM image taken on the left portion of the cross-section is shown in Fig. 6.11. EDS elemental analysis revealed that the white pigment mixed with ultramarine blue is zinc white (ZnO), a synthetic pigment first introduced around 1850 [43], a fact that consequently allows to identify the external blue layers, totaling 100 μm in thickness, as not original. ZnO is a weak Raman scatterer [46] and therefore masked by ultramarine blue in the mapping experiment.

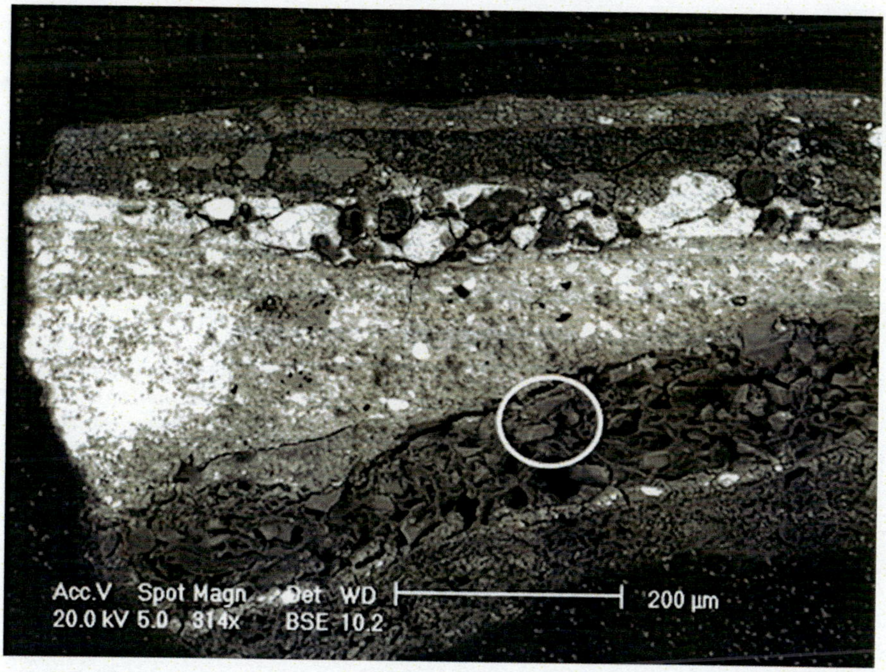

Fig. 6.11 SEM backscattered electron image of the sample shown in Fig. 6.8. The circle indicates two glassy particles showing typical conchoidal fractures and containing 1–3 wt% of cobalt (by EDS analysis)

In the SEM backscattered electron image, the second paint layer from the bottom appears to be made of particles showing the typical conchoidal fracture of a glassy material and containing 1–3 wt % of cobalt, therefore related to the use of smalt, a cobalt doped silica glass widely used as a pigment particularly after the 16th century [43].

Moreover, in the same green-bluish layer where massicot was identified by Raman mapping, EDS analysis showed the presence of about the 4% in weight of copper, along with a much larger amount of lead and minor amounts of other elements (Si, Fe). In order to clarify the nature of the copper-based compound, a micro-FTIR mapping experiment was conducted on a 125 × 100μm area, indicated in Fig. 6.8, with a 12.5 μm step size. The FTIR map presented in Fig. 6.12a was acquired by using the intensity of the typical combination band $\upsilon_1 + \upsilon_3$ of azurite ($2CuCO_3.Cu(OH)_2$) [47] and clearly shows that the green-bluish layer also contains this pigment. Again, as in the case of the zinc white and ultramarine blue pigments, the Raman bands of azurite in the range between 400 and 100 cm^{-1} were masked by the strong scattering of massicot in the Raman mapping measurements.

In the same FTIR mapping experiment, another blue pigment, Prussian blue ($Fe_4[Fe(CN)_6]_3$).14-16H_2O, was identified in the upper part of the stratigraphy by its characteristic strong CN stretching band at 2090 cm^{-1} (Fig. 6.12b). Prussian blue

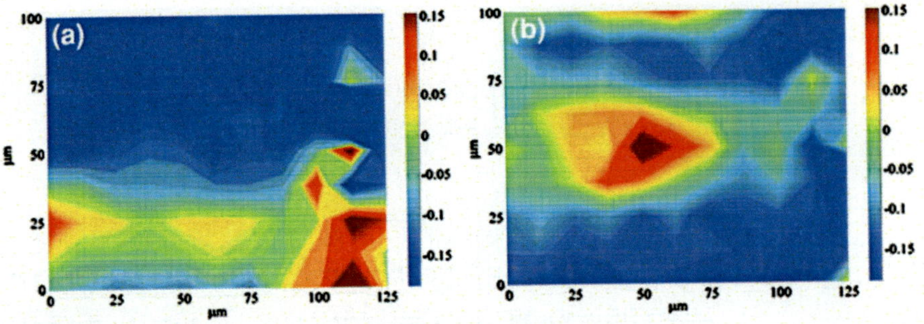

Fig. 6.12 Micro-FTIR line array maps obtained in the sample removed from the *Lady's* mantle, in the area indicated by a black rectangle in Fig. 6.8: contour plot of the distribution of azurite, intensity of the $\upsilon_1 + \upsilon_3$ mode at ca. 2400 cm^{-1} (**a**), and contour plot of the distribution of Prussian blue, intensity of ν(CN) at ca. 2090 cm^{-1} (**b**). The corresponding spectra are not shown

is a synthetic pigment discovered in Berlin around 1710 [43] and, therefore, not original in this context. Prussian blue was not observed in the Raman mapping measurements due to the fact that the Raman ν(CN) stretching band appears at ca. 2150 cm^{-1}, above the range selected for the experiments.

It is possible to differentiate natural lapis lazuli and synthetic ultramarine blue by FTIR. The asymmetric stretching of carbon dioxide, an impurity entrapped in the sodalite structure of the natural pigment, has been used as a marker for lapis lazuli [48]. This band was not detected by micro-FTIR confirming that the ultramarine blue pigment is the synthetic form.

By allowing non-destructive chemical and structural mapping of heterogeneous samples with a micrometer spatial resolution, Raman offers unique capabilities for assessing the spatial distribution of both mineral and organic phases within paint samples. However, due to the complex composition of most paintings, it is not always possible to describe the entire stratigraphy by using a single spectroscopic method. As SEM-EDS, Raman, and FTIR spectroscopies are complementary in nature, their combined use offers the opportunity to describe heterogeneous mixtures in more detail.

6.3 Raman Mapping Using Scanning Mirrors

6.3.1 Experimental Setup

The setup for this mapping approach is based on a combination of mirrors that scan the laser beam along a line or in an area in the object to be studied. The principles of the different modes of operation (step-by-step, averaging, and macro-mapping) are described in detail in this book, in Chap. 1. Two different configurations can be used in which the laser beam is directed either vertically, thru the microscope head, or through a horizontal exit. In the first case, the xy stage is removed and the artwork is placed under the microscope objective. This approach has similar limitations to those of the mapping approach using a motorized xy stage in terms of the size of the work of art that can be fitted under the objective. In the second configuration, the scanning mirrors are placed in a horizontal exit, directing the laser beam to the artwork placed perpendicular to this exit (Fig. 6.13a). The advantage of this approach is that the object to be studied can be larger in size.

In addition to its non-invasive nature, this configuration has the advantages of being fully confocal, not subject to poor coupling efficiency (in contrast to fiber optic probes), and of being compatible with multiple laser excitations, which in the analysis of works of art is indispensable as the material composition is usually heterogeneous. In this experimental setup, the maximum area that can be scanned depends on the objective used, and it is limited to $300 \times 300\,\mu$m when a 10x objective is employed (See Chap. 1).

In one of the modes of operation of this scanning mechanism, called high throughput screening (averaging mode) (See Chap. 1), the scanning mirrors allow to generate a macro-Raman spot (from 1×1 to $300 \times 300\,\mu$m) on the selected area. In this continuous scanning mode, the energy of the laser beam is spread out instead of concentrated in a small spot enabling average Raman measurements on materials (e.g. pigments) that are prone to degrade under the laser beam. It also enables a quick acquisition of an average spectrum on the selected mapping area that can give important preliminary information on the materials present to help set up the optimum experimental conditions (e.g. acquisition time, spectral range) for further step-by-step high-definition area mapping.

This approach has limitations in terms of its axial resolution, especially in cases when the surface of the artwork is not flat, for example when a painted surface has a certain texture, or when the object surface can not be placed totally perpendicular to the microscope head or to the horizontal exit. The axial resolution depends on the laser excitation wavelength and on the numerical aperture of the objective used and it can be easily calculated as shown schematically in Fig. 6.13b. For example, the maximum tolerable angle of displacement (α) of the artwork surface from the optimal perpendicular position when using a 10x objective and 785 nm laser excitation is calculated to be 6.9°.

The Raman maps for the case studies discussed below were obtained with a Horiba Jobin Yvon LabRAM HR800 Raman spectrometer coupled to an Olympus BXFM optical microscope equipped with a scanning mirrors mechanism placed in the horizontal exit. The measurements were done using a 785 nm laser excitation line, 10x or 20x objective lens and a 600 grooves/mm grating, which gave a spectral resolution of 0.83 cm^{-1}/pixel. The power at the samples was set to 6.6 or 33 mW using neutral density filters. A multi-channel, air-cooled CCD detector was used, with integration times between 10 and 25 s. In each case an average spectrum of the selected area was first recorded to set the optimum spectral range and integration time conditions for the subsequent step-by-step mapping. Spectra are presented after baseline correction.

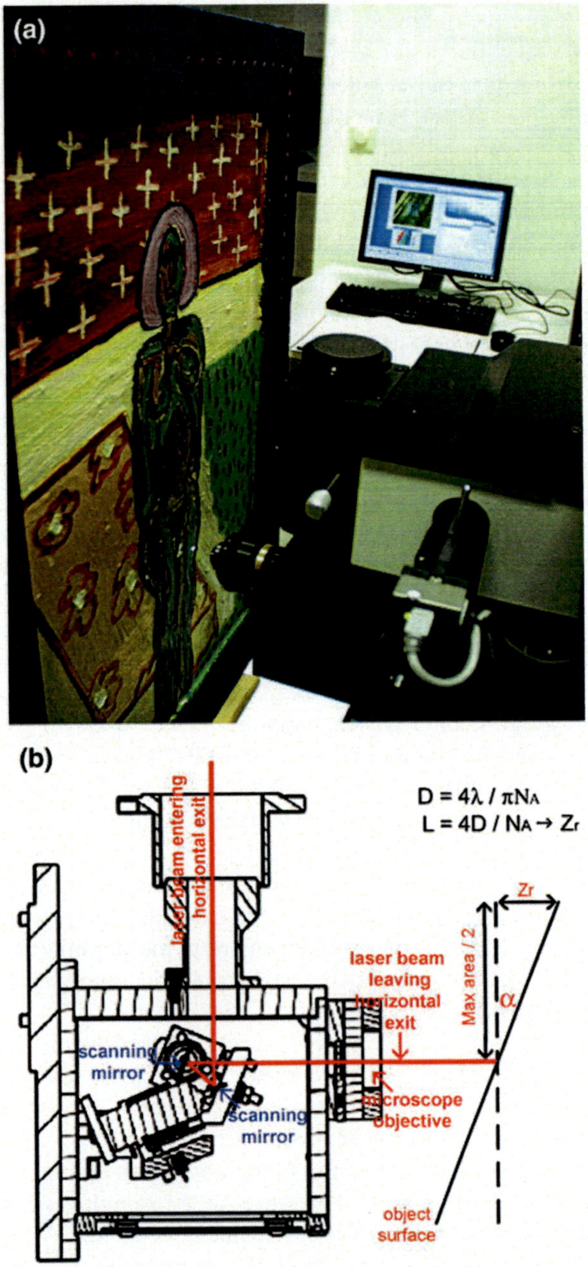

Fig. 6.13 Setup for Raman mapping using scanning mirrors placed in the horizontal exit with the contemporary painting *Gola Muza / Naked Muse* (**a**), and sketch showing the incidence of the laser beam from the horizontal exit onto the artwork surface (**b**). D: lateral resolution, λ: laser beam wavelength, N_A: numerical aperture of the microscope objective, L (Zr): axial resolution

The synthetic organic pigment names are represented according to the Colour Index International classification by their generic names (i.e. PY that stands for pigment yellow, PG for pigment green, etc) and a number [49].

6.3.2 Case studies

The case studies discussed below comprise in situ mapping experiments on the work entitled *Gola Muza / Naked Muse* (2004), by the artist Silvester Plotajs-Sicoe, painted on canvas in an oil technique, and on the work *Untitled* (2001), by the artist Jelka Flis, executed in an acrylic technique on a wood support.

Questions regarding these two contemporary works could be answered by analyzing the surface components, since this was the information required to select suitable solvents for cleaning. Paint layers on modern and contemporary works often comprise synthetic organic pigments and synthetic resins as binding media, such as acrylic, alkyd, and PVA, although traditional oil binders are also used. The paint formulations containing modern materials generally have good mechanical properties and stability except for their tendency to accumulate dirt and air pollutants. However, modern and contemporary works of art often comprise materials that are not compatible, that can show defects due to degradation processes, and that may therefore require sampling in order to better plan a conservation intervention.

6.3.2.1 In Situ Mapping in a Contemporary Oil Painting on Canvas

Details and results for the selected mapping areas on *Gola Muza/Naked Muse* are presented in Figs. 6.14, 6.15, 6.16, 6.17, 6.18, 6.19, where mainly synthetic organic pigments, along with inorganic extenders, were identified. Figure 6.15 shows a Raman spectrum recorded in the averaging mode in the mapping area indicated in Fig. 6.14, along with reference spectra for the synthetic organic pigments identified: PG 7, PY 74, and PY 1. PY 1 and PY 74 belong to the monoazo pigment subgroup and PG 7 is a chlorinated Cu (II) phthalocyanine [49, 50]. The relevant modes of synthetic organic pigments commonly found in works of art have been previously reported and discussed by several authors [14–16, 19]. Bands observed for PY1, at ca. 1668, 1619, 1485, 1388, 1307, 1254, 1190, 1135, 998, 951, 784, 511, and 390 cm^{-1} are weak or very weak, as the proportion of this pigment is probably low compared to those of PG 7 and PY 74.

Three inorganic white pigments were also identified in this area: calcite, barium white ($BaSO_4$), and titanium white (TiO_2). Calcite and barium white have characteristic bands at ca. 1086 and 987 cm^{-1}, respectively. Titanium white was found to be present in its rutile form, with bands at 610, 447, 236, and 143 cm^{-1}, and also in its anatase form with a strong feature at ca. 142 cm^{-1} [21].

The spectral range for the step-by-step mapping experiments was set between 50 and 1800 cm^{-1}. Even though weak Raman bands due to the aromatic C-H stretches

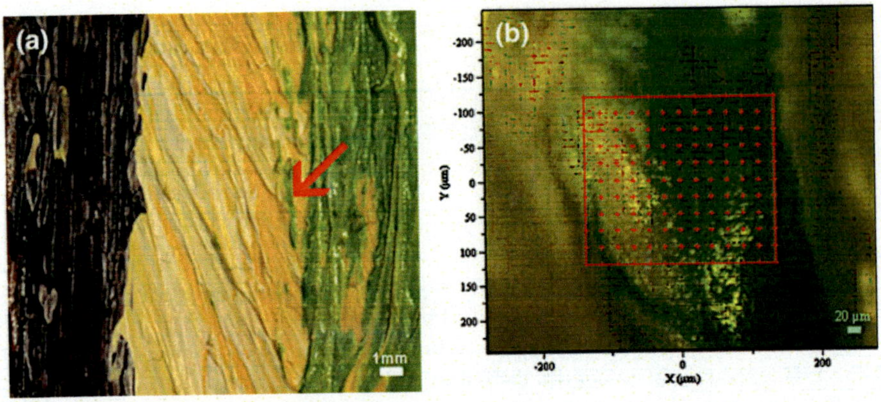

Fig. 6.14 Details of one of the areas selected for Raman mapping on *Gola Muza / Naked Muse*. In Figure **a** the red arrow indicates the area selected at the border of green and yellow paints. Figure **b** corresponds to the same area photographed using an 10x objective lens, with a mapping grid (271.3 × 239.9 µm) superimposed. At each point the Raman spectrum was recorded using a 25 s integration time, and a 6.6 mW power

are expected in the range 3100-3000 cm^{-1} for synthetic organic pigments, the analysis of the features in the functionality range below 1800 cm^{-1} is generally sufficient to identify and differentiate these pigments [11, 14–16, 19].

The spatial distribution of the pigments obtained by step-by-step mapping on the area shown in Fig. 6.14 is presented in Fig. 6.16, where it is demonstrated that the green paint is composed of PG 7, PY 1, and barium white, while the yellow paint contains PY 74 and calcite.

The map distributions of the anatase and rutile forms of TiO_2 perfectly correlate, indicating that both were present in the commercial white paint used by the artist. In addition, the distribution of titanium white does not match the green nor the yellow paints' distributions alone, suggesting that the white paint was not added to any of these two paints before their application, and that, instead, it was applied as such, without mixing.

Figure 6.17 displays the second area selected for mapping in *Gola Muza/Naked Muse*, at the border of the violet and the turquoise paint passages. The average spectrum recorded in this area shows the presence of PB15:3, PG 7, PV 23, and barium white (Fig. 6.18). Bands at ca. 1528, 1451, 1340, 1306, 1191, 1142, 1108, 951, 745, 681, 590, 482, 289, and 259 cm^{-1} in this spectrum are consistent with those observed for a PB15:3 reference sample (Ostacolor, Fig. 6.18) and with data previously published for PB15:3 [19]. A weak band at ca. 701 cm^{-1} in the spectrum shown in Fig. 6.18, and not observed in the reference samples for any of the pigments identified in the area, could not be assigned with certainty.

The distributions of the pigments identified in this second area on *Gola Muza / Naked Muse* are shown in Fig. 6.19. In the violet paint, PV 23, belonging to the dioxazine subgroup [49, 50], and barium white were identified while it was observed that the turquoise color was achieved by mixing PG 7 and PB15:3.

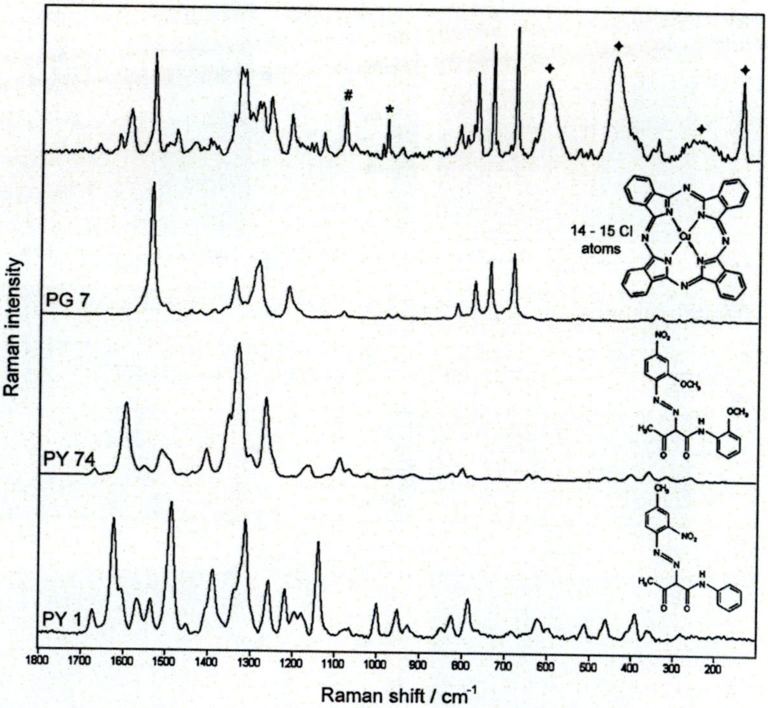

Fig. 6.15 Average Raman spectrum for the area shown in Fig. 6.14 (*top spectrum*), showing the inorganic pigments identified: # stands for calcite, * for barium white, and ✦ for both the anatase and rutile forms of titanium white; and reference spectra and formulas for the synthetic organic pigments observed in the same area: PG 7, PY 74, and PY 1, manufactured by Clariant, Ciba, and SunChemical, respectively

PB15:3, as PG 7, belongs to the Cu (II) phthalocyanine group. Members of the PB15 group are known to occur in no less than five crystal modifications: α, β, γ, δ, ε, though the most important forms from the commercial point of view are the α and β. The α-form (PB15) is not stable and tends to revert to the β-form. PB15:3 and PB 15:4 are the β-form, and PB15:1 and PB15:2 are stabilized α-forms, achieved by partial chlorination to prevent the reversion to the β-form. Depending on the synthetic route, other members of the PB 15 family may also contain minor amounts of chlorine [50]. Tackley et al. reported that differences among the Raman spectra acquired with 514 nm excitation on samples of the α, β, γ, and ε forms of PB 15 were not sufficient to distinguish them and that the pigments degraded when subjected to 785 nm excitation, though the laser powers used are not indicated by these authors [51]. On a later work, different spectra acquired with 785 nm excitation were reported for α, β, and ε forms by Scherrer et al [19]. Due to the sensitivity of the PB 15 members to degradation, and even though the data presented was acquired at low powers, the identification of the β form in *Gola Muza/Naked Muse* should be considered tentative. XRD, both in microsamples and in a non-invasive

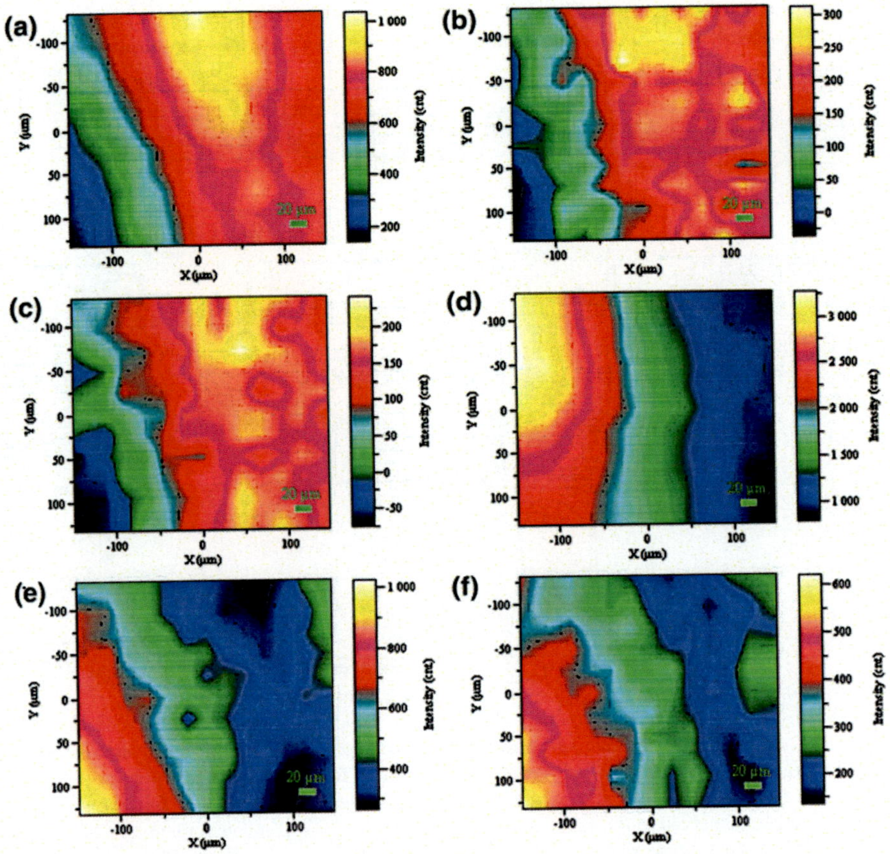

Fig. 6.16 Raman maps of the different pigments identified in the area shown in Fig. 6.14: PG 7 (**a**), PY 1 (**b**), barium white (**c**), rutile (**d**), PY 74 (**e**), and calcite (**f**), respectively. The distribution of rutile observed in this area matched that of anatase

open architecture configuration, has proved to be a powerful tool to unambigously differentiate synthetic organic pigments polymorphs [52], but the dimensions of this painting did not allow to fit it in the platform of an open architecture diffractometer for non-invasive analysis.

6.3.2.2 In situ Mapping in a Contemporary Acrylic Painting on a Wood Support

The painting *Untitled* by Jelka Flis shows a superposition of color layers, with a semi-transparent yellow layer at the very top that makes the underpaint layers partially visible to the naked eye. A detail of this painting that includes the area selected for mapping at the border of the yellow and orange paint passages is shown in Fig. 6.20. In Fig. 6.21, the average area spectrum shows the presence of two synthetic organic

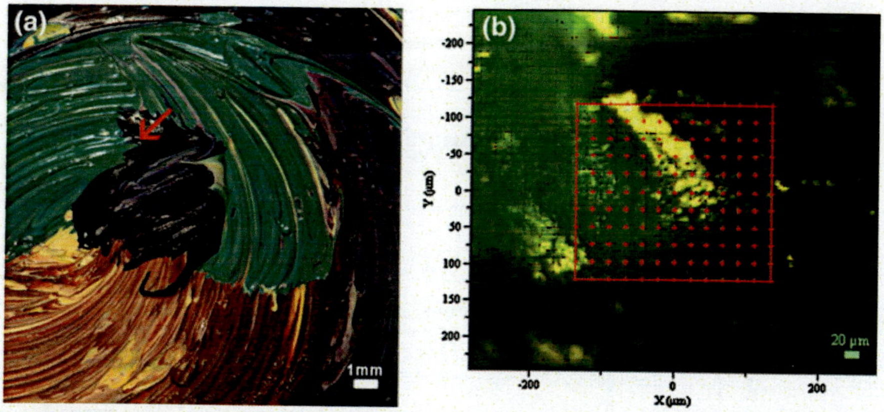

Fig. 6.17 Details of the second area selected for Raman mapping on *Gola Muza / Naked Muse*. In Figure **a** the red arrow indicates the area selected at the border of the turquoise and violet paints. Figure **b** shows the same area photographed using an 10x objective lens, with a grid (271.3 × 239.9 μm) superimposed. At each point the Raman spectrum was recorded using a 25 s integration time, and a 6.6 mW power

pigments PO 34 and PY 3, that belong to the disazopyrazolone and monoazo pigment subgroups, respectively [49, 50]. PO 34 shows modes at 1595, 1541, 1476, 1419, 1290, 1272, 1241, 1156, 1046, 914, 667, 537, 392, and 366 cm^{-1}, while bands for PY 3 appear at ca. 1672, 1611, 1565, 1545, 1494, 1443, 1387, 1336, 1307, 1191, 1138, 1101, 1034, 955, 914, 743, 647, 412, and 392 cm^{-1}.

Inorganic white pigments were also identified in this painting. Titanium white in its rutile form is present with bands at ca. 610, 447, and 230 cm^{-1}, while the medium intensity band at ca. 142 cm^{-1} in the spectrum shown in Fig. 6.21 indicates that the anatase form is also present. The band at ca. 988 cm^{-1} in this spectrum can be assigned to the SO_4^{2-} symmetric stretch of barium white.

As in the case of the first area discussed for *Gola Muza / Naked Muse*, the distributions of the anatase and rutile forms perfectly correlate indicating that both are present in the white paint used by the artist. These inorganic pigments were most likely added to the yellow paint, as can be noticed by comparing the distribution of PY 3, titanium white, and barium white (Figs. 6.22b–d). However, the distribution of PO 34 and PY 3 (Figs. 6.22a,b) in the selected mapping area does not entirely match the orange or the yellow paint locations. A possible explanation for these results is that the axial resolution when using an 20x objective and a 785 nm laser excitation line is ca. 25 μm and that the yellow layer on top is likely to be thinner than this, in addition to be unevenly distributed, allowing the PO 34 pigment in underpaint layer to be detected.

The wood support of this painting was found to be far more stable than the canvas support of *Gola Muza/Naked Muse*. When a canvas support is loose a secondary support to prevent vibrations from interfering with the Raman measurements is necessary in the case of long Raman mapping acquisitions.

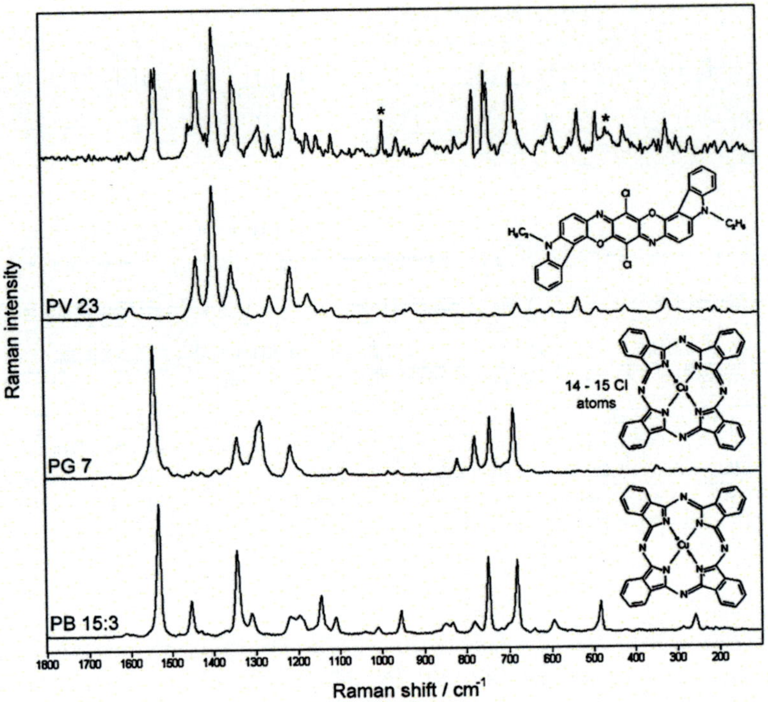

Fig. 6.18 Average Raman spectrum recorded in the area shown in Fig. 6.17 (top spectrum), with the characteristic peak for barium white indicated by ∗, along with the reference spectra and formulas for the synthetic organic pigments identified in the same area: PV 23, PG 7, both manufactured by Clariant, and PB 15:3, supplied by Ostacolor

6.4 Conclusions

In the cases of the two contemporary paintings presented, the Raman mapping approach using scanning mirrors allowed to identify the pigment mixtures used by the artists non-invasively. This not only gave insight into their painting techniques but also provided information that is crucial to determine what solvents are appropriate for cleaning the paintings' surfaces. In situ non-invasive mapping is also possible using a motorized xy stage when the art objects are small enough to fit on the microscope platform, as in the case of the Medieval Persian ceramic fragments discussed above.

However, non-invasive analyses not always give sufficient information, for example when a work of art has been repainted or reworked in the past and the original materials lie below restoration layers, as in the case of the polychrome wooden sculpture discussed, in cases where conservation- restoration treatments involve more than surface cleaning (e.g. consolidation), or when the stratigraphy and the composition of every layer in the object is necessary to elucidate the technological process involved

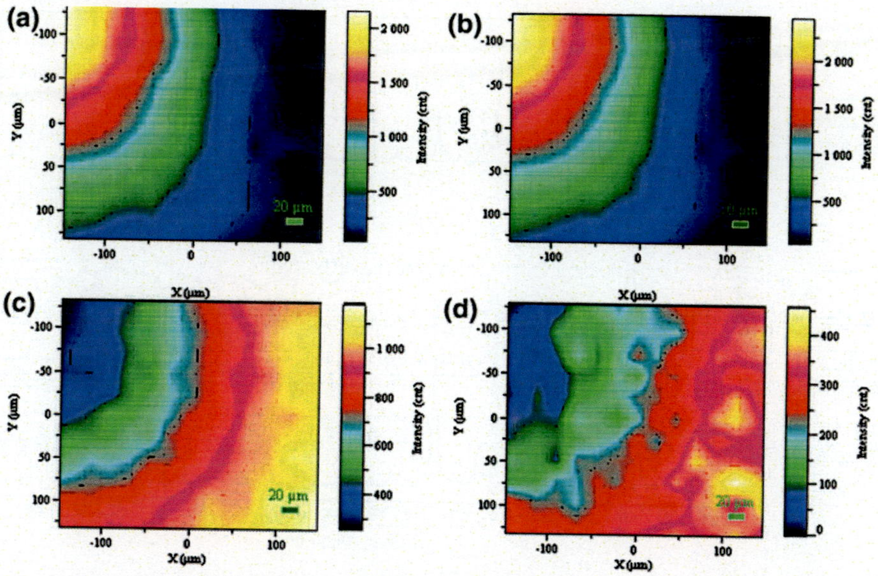

Fig. 6.19 Raman maps of the different pigments identified in the area shown in Fig. 6.17: PB 15:3 (**a**), PG 7 (**b**), PV 23 (**c**), and barium white (**d**), respectively

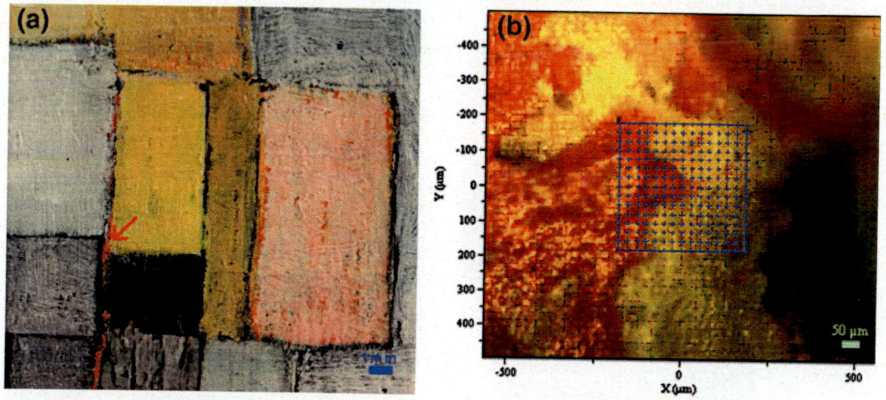

Fig. 6.20 Details of the Raman mapping area on the painting *Untitled*. In Figure **a** the red arrow indicates the area selected at the border of orange and yellow paints. Figure **b** shows the same area photographed using an 20x objective lens, with a grid (360.9 × 360.5 μm) superimposed. At each point the Raman spectrum was recorded using a 10 s integration time, and a 33 mW power

in its manufacture. Therefore, sampling and a detailed identification of the materials in all layers is required. In these cases, the mapping approach using a motorized xy stage is very powerful, particularly when used in combination with other microanalytical techniques.

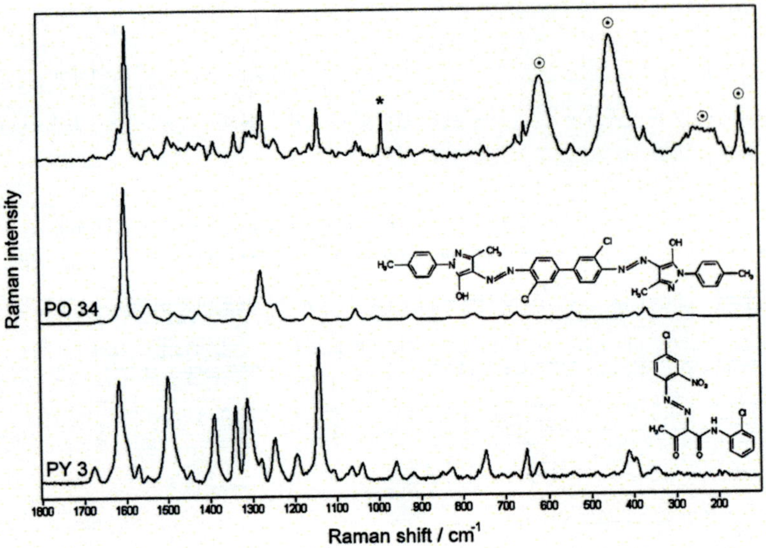

Fig. 6.21 Average spectrum recorded in the area shown in Fig. 6.20 (*top spectrum*) indicating the inorganic pigments identified: ∗ stands for barium white, and ⊙ for titanium white, present in both its anatase and rutile forms; along with the reference spectra and formulas for the synthetic organic pigments observed in the same area: PO 34 and PY 3, both manufactured by SunChemical

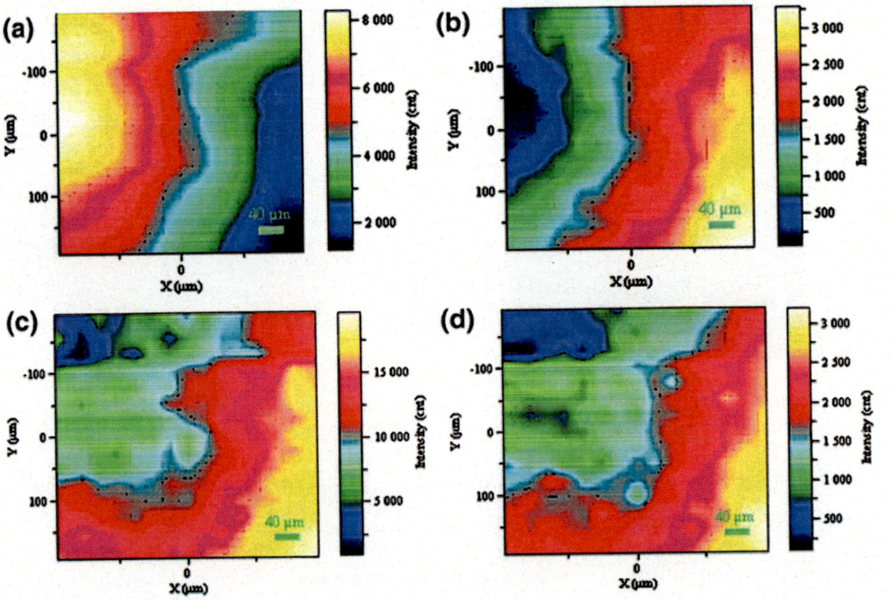

Fig. 6.22 Raman maps of the different pigments identified in the area shown in Fig. 6.20: PO 34 (**a**), PY 3 (**b**), rutile (**c**), and barium white (**d**), respectively. The distribution of rutile observed in this area matched that of anatase

Acknowledgments The authors would like to thank Studio Černe, Ljubljana, for lending the contemporary works of art analyzed for the present study; Črtomir Tavzes, from the Institute of the Protection of Cultural Heritage of Slovenia, for help with the experimental setup of the Raman mapping utilizing the horizontal exit; and especially Arnaud Zoubir, from Horiba Jobin Yvon, for the fruitful collaboration in the development of the technology behind the scanning mirrors system. CM thanks Francesca Rosi and Valentina Manuali (from the Centro SMAArt, Università di Perugia, Italy) for help in the acquisition of mapping data using the xy stage, Luca Nodari (Dipartimento di Chimica, Università di Padova, Italy) for the loan of the cross-section from the Hotel Hungaria tile and the Islamic Museum of Tehran for the loan of the Iranian lustred fragments.

References

1. J. Raman Spectrosc. **35**, 605–818 (2004)
2. J. Raman Spectrosc. **37**, 961–1237 (2006)
3. J. Raman Spectrosc. **39**, 961–1124 (2008)
4. J. Raman Spectrosc. **41**, 1389–1561 (2010)
5. N. Stavisky, S.A. Centeno, in *The Prato Haggadah. Companion Volume to the Facsimile Edition*, ed. by N.M. Steinberger. The Prato Haggadah: An Investigation into the Artists' and Scribes' Materials and Techniques, and Conservation Treatment, (The Library of the Jewish Theological Seminary and Patrimonio Ediciones, New York, 2007)
6. B. Edelstein, S.A. Centeno, M.T. Wypyski, in *The Object in Context: Crossing Conservation Boundaries*, ed. by D. Saunders, J.H. Townsend, S. Woodcock. Illuminating a Complex History: A Study of the Materials and Techniques of the Tombs of Urgell at the Cloisters, (Contributions to the Munich Congress, International Institute for Conservation of Historic and Artistic Works, London, 2006)
7. F. Rosi, C. Miliani, I. Borgia, B. Brunetti, A. Sgamellotti, Identification of nineteenth blue and green pigments by in situ x-ray fluorescence and micro-Raman spectroscopy. J. Raman Spectrosc. **35**, 610–615 (2004)
8. S.A. Centeno, D. Mahon, M.T. Wypyski, Examination of a spanish medieval processional crucifix substantially reworked in the 20th century. J. Raman Spectrosc. **35**, 774–780 (2004)
9. H.G.M. Edwards, P. Wyeth, in *Raman Spectroscopy in Archaeology and Art History. RSC Analytical Spectroscopy Monographs*, ed. by H.G.M. Edwards, J.M. Chalmers. Case Study: Ancient Textile Fibers, (The Royal Society of Chemistry, Cambridge, 2005)
10. G.D. Smith, R.J.H. Clark, Raman spectroscopy in archaeological science. J. Archaeol. Sci. **31**, 1137–1160 (2004)
11. S.A. Centeno, V. Lladó Buisan, P. Ropret, Raman study of synthetic organic pigments and dyes in early lithographic inks (1890–1920). J. Raman Spectrosc. **37**, 1111–1118 (2006)
12. C. Ricci, I. Borgia, B.G. Brunetti, C. Miliani, A. Sgamellotti, C. Seccaroni, P. Passalacqua, The Perugino's palette: integration of an extended in situ XRF study by Raman spectroscopy. J. Raman Spectrosc. **35**, 616–621 (2004)
13. S.A. Centeno, T. Meller, N. Kennedy, M. Wypyski, The daguerreotype surface as a SERS substrate: characterization of image deterioration in plates from the nineteenth-century studio of Southworth and Hawes. J. Raman Spectrosc. **39**, 914–921 (2008)
14. P. Ropret, S.A. Centeno, P. Bukovec, Raman identification of yellow synthetic organic pigments in modern and contemporary paintings: reference spectra and case studies. Spectrochim. Acta. Part A **69**, 486–497 (2008)
15. F. Schulte, K-W. Brzezinka, K. Lutzenberger, H. Stege, U. Panne, Raman spectroscopy of synthetic organic pigments used in 20th century works of art. J Raman Spectrosc. **39**, 1455–1463 (2008)
16. P. Vandenabeele, L. Moens, H.G.M. Edwards, R. Dams, Raman spectroscopic database of azo pigments and application to modern art studies. J Raman Spectrosc. **31**, 509–517 (2000)

17. M. Leona, J. Stenger, E. Ferloni, Application of surface-enhanced Raman scattering techniques to the ultrasensitive identification of natural dyes in works of art. J Raman Spectrosc. **37**, 981–992 (2006)
18. A.V. Whitney, R.P. Van Duyne, F. Casadio, An innovative surface-enhanced Raman spectroscopy (SERS) method for the identification of six historical red lakes and dyestuffs. J Raman Spectrosc. **37**, 993–1002 (2006)
19. N.C. Scherrer, S. Zumbuehl, F. Delavy, A. Fritsch, R. Kuehnen, Synthetic organic pigments of the 20th and 21st century relevant to artist's paints: Raman spectra reference collection. Spectrochim. Acta Part A **73**, 505–524 (2009)
20. I.M. Bell, R.J.H. Clark, P.J. Gibbs, Raman spectroscopic library of natural and synthetic pigments (pre∼1850 AD). Spectrochim. Acta Part A **53**, 2159–2179 (1997)
21. L. Burgio, R.J.H. Clark, Library of FT-Raman spectra of pigments, minerals, pigment media and varnishes, and supplement to existing library of Raman spectra of pigments with visible excitation. Spectrochim. Acta Part A **57**, 1491–1521 (2001)
22. P. Vandenabeele, B. Wehling, L. Moens, H. Edwards, M. De Reu, G. Van Hooydonk, Analysis with micro-Raman spectroscopy of natural organic binding media and varnishes used in art. Anal. Chim. Acta **407**, 261–274 (2000)
23. M. Bouchard, D.C. Smith, Catalogue of 45 reference Raman spectra of minerals concerning research in art history or archaeology, especially on corroded metals and coloured glass. Spectrochim. Acta Part A **59**, 2247–2266 (2003)
24. C. Fischer, I. Kakoulli, Multispectral and hyperspectral imaging technologies in conservation: current research and potential applications. Rev. Conserv. **7**, 3–16 (2006)
25. S. Aze, J-M. Vallet, A. Baronnet, O. Grauby, The fading of red lead pigment in wall paintings: tracking the physico-chemical transformations by means of complementary micro-analysis techniques. Eur. J. Miner. **18**, 835–843 (2006)
26. D. Lau, C. Villis, S. Furman, M. Livett, Multispectral and hyperspectral image analysis of elemental and micro-Raman maps of cross-sections from a 16th century painting. Anal. Chim. Acta **610**, 15–24 (2008)
27. D. Neff, L. Bellot-Gurlet, P. Dillmann, S. Reguer, L. Legrand, Raman imaging of ancient rust scales on archaeological iron artefacts for long-term atmospheric corrosion mechanisms study. J. Raman Spectrosc. **37**, 1228–1237 (2006)
28. L.I. McCann, K. Trentelman, T. Possley, B. Golding, Corrosion of ancient chinese bronze money trees studied by Raman microscopy. J. Raman Spectrosc. **30**, 121–132 (1999)
29. S. Murcia-Mascarós, C. Domingo, S. Sánchez-Cortés, J.V. García-Ramos, C. Roldán, A. Muñoz Ruiz, Composition and Distribution of Crusts in Weathered Stained Glass Windows Through Mapping and Confocal Micro-Raman Spectroscopy. in *Art'05-8th International Conference on Non Destructive Investigations and Microanalysis for the Diagnostics and Conservation of the Cultural and Environmental Heritage*, Leece, Italy, 2005
30. P. Ropret, C. Miliani, S.A. Centeno, C. Tavzes, F. Rosi F, Advances in Raman mapping of works of art. J Raman Spectrosc. **41**, 1172–1177 (2010)
31. I.N.M. Wainwright, J.M. Taylor, R.D. Harley, in *Artists' Pigments: A Handbook of Their History and Characteristics*, ed. by R.L. Feller. Lead Antimonate Yellow, vol. 1 (Cambridge University Press, Cambridge, 1986)
32. C. Sandalinas, S. Ruiz-Moreno, A. Lopez-Gil, J. Miralles, Experimental confirmation by Raman spectroscopy of a Pb-Sn-Sb triple oxide yellow pigment in sixteenth-century Italian pottery. J. Raman Spectrosc. **37**, 1146–1153 (2006)
33. S. Ruiz-Moreno, R. Pérez-Pueyo, A. Gabaldón, M.J. Soneira, C. Sandalinas, Raman laser fibre optic strategy for non-destructive pigment analysis. Identification of a new yellow pigment (Pb, Sn, Sb) from the Italian XVII century painting. J. Cult. Herit. **4**, 309–313 (2003)
34. D. Hradil, T. Grygar, J. Hradilová, P. Bezdicka, V. Grunwaldová, I. Fogaš, C. Miliani, Micro-analytical identification of Pb-Sb-Sn yellow pigment in historical European paintings and its differentiation from lead tin and Naples yellows. J. Cult. Herit. **8**, 377–386 (2007)
35. F. Rosi, V. Manuali, C. Miliani, B.G. Brunetti, A. Sgamellotti, T. Grygar, D. Hradil, Raman scattering features of lead pyroantimonate compounds. Part I: XRD and Raman characterization of $Pb_2Sb_2O_7$ doped with tin and zinc. J. Raman Spectrosc. **40**, 107–111 (2009)

36. G. Cruciani, M. Dondi, M. Ardit, T. Stoyanova Lyubenova, J.B. Carda, F. Matteucci, A. L. Costa, Malayaite ceramic pigments: A combined optical spectroscopy and neutron/X-ray diffraction study. Mater. Res. Bull. **44**, 1778–1785 (2009)
37. A.M. Heyns, P.M. Harden, Evidence for the existence of Cr(IV) in chromium-doped malayaite Cr^{4+}:$CaSnOSiO_4$: a resonance Raman Study. J. Phys. Chem. Solids **60**, 277–284 (1999)
38. J. Pérez-Arantegui, J. Molera, A. Larrea, T. Pradell, M. Vendrell-Saz, I. Borgia, B.G. Brunetti, F. Cariati, P. Fermo, M. Mellini, A. Sgamellotti, C. Viti, Luster pottery from the thirteenth century to the sixteenth century: a nanostructured thin metallic film. J. Am. Ceram. Soc. **84**, 442–446 (2001)
39. I. Borgia, B.G. Brunetti, A. Giulivi, A. Sgamellotti, F. Shokouhi, P. Oliaiy, J. Rahighi, M. Lamehi-Rachti, M. Mellini, C. Viti, Characterisation of decorations on Iranian (10th-13th century) lustreware. Appl. Phys. A **79**, 257–261 (2004)
40. S. Padovani, D. Puzzovio, C. Sava, P. Mazzoldi, I. Borgia, A. Sgamellotti, B.G. Brunetti, L. Cartechini, F. D'Acapito, C. Maurizio, F. Shokouhi, P. Oliaiy, J. Rahighi, M. Lamehi-Rachti, M. Pantos, XAFS study of copper and silver nanoparticles in glazes of medieval middle-east lustreware (10th-13th century). Appl. Phys. A **83**, 521–528 (2006)
41. I. Borgia, B.G. Brunetti, I. Mariani, A. Sgamellotti, F. Cariati, P. Fermo, M. Mellini, C. Viti, G. Padeletti, Heterogeneous distribution of metal nanocrystals in glazes of historical pottery. Appl. Surf. Sci. **185**, 206–216 (2002)
42. Ph.Colomban, H.D. Schreiber, Raman signature modification induced by copper nanoparticles in silicate glass. J. Raman Spectrosc. **36**, 884–890 (2005)
43. N. Eastaugh, V. Walsh, T. Chaplin, R. Siddall, *Pigment Compendium. A Dictionary of Historical Pigments* (Elsevier Butterworth-Heinemann, Amsterdam, 2004)
44. D. Mahon, S.A. Centeno, M.T. Wypyski, X.F. Salomon, A. Bayer, *Technical study of three allegorical paintings by Paolo Veronese: the choice between virtue and vice, wisdom and strength and mars and venue united by love.* (Metropolitan Museum Studies in Art, Science and Technology), **1**, 83–108 (2010)
45. G. Chiavari, D. Fabbri, S. Prati, A. Zoppi, Identification of indigo dyes in painting layers by pyrolysis methylation and silylation. A case study: The dinner of Emmaus by G. Preti. Chromatographia **61**, 403–408 (2005)
46. G. Shan, L. Xu, G. Wang, Y. Liu, Enhanced Raman Scattering of ZnO Quantum Dots on Silver Colloids. J. Phys. Chem. C **111**, 3290–3293 (2007)
47. C. Miliani, F. Rosi, I. Borgia, P. Benedetti, B.G. Brunetti, A. Sgamellotti, Fiber-optic mid-FTIR reflectance spectroscopy: a suitable technique for in-situ studies of mural paintings. Appl. Spectrosc. **61**, 293–299 (2007)
48. C. Miliani, C. Daveri, B. Brunetti, A. Sgamellotti, CO2 entrapment in natural ultramarine blue. Chem. Phys. Lett. **466**, 148–151 (2008)
49. Colour Index International, 4th ed. Online. The Society of Dyers and Colourists and the American Association of Textile Chemists and Colorists. http://www.colour-index.org. Accessed June 2009
50. W. Herbst, K. Hunger, *Industrial Organic Pigments. Production, Properties, Applications*, 3rd edn. (Wiley, Weinheim, 2004)
51. D.R. Tackley, G. Dent, W.E. Smith, Phthalocyanines: structure and vibrations. Phys. Chem. Chem. Phys. **3**, 1419–1426 (2001)
52. L. Brostoff, S.A. Centeno, P. Ropret, P. Bythrow, F. Pottier, Combined XRD and Raman identification of synthetic organic pigments in works of art: from powder samples to artists' paints. Anal. Chem. **81**, 6096–6106 (2009)

Chapter 7
Raman Microscopy: A Versatile Approach to Bio-Imaging

J. J. McGarvey and J. Renwick Beattie

Abstract Raman microscopy has become established as a key probe technique in biology and biomedicine. In combination with imaging and mapping it has been employed in the investigation of a diverse array of problems ranging from ex vivo and in vivo single cell studies to elucidation of the often complex, interacting structures which constitute human and animal tissues. This chapter emphasises the unique attributes of Raman microscopy as a bioimaging technique, including its non-invasive, spectral multiplexing ability, allied with high spatial resolution and underpinned by a range of multivariate data processing methods. A number of illustrative examples have been selected for discussion from the fields of molecular biology, ophthalmology, respiratory medicine as well as some non-medical examples. Recent advances and pointers to future activity in the uses of Raman microscopy as a structurally and functionally informative bioimaging technique are briefly considered.

7.1 Introduction and Scope

The appearance of the paper by Delhaye and Dhamelincourt in 1975 entitled 'Raman microprobe and microscope with laser excitation' [1] heralded a new chapter in the development of Raman scattering as a probe for the physical and chemical characterization of materials by adding spatial resolution to the already well-recognised attribute of molecular specificity. The rapid technical advances which continue to take place in Raman spectroscopy and microscopy are paralleled by the increasing

J. J. McGarvey (✉)
School of Chemistry and Chemical Engineering, Queen's University, Belfast, UK
e-mail: j.mcgarvey@qub.ac.uk

J. Renwick Beattie
Centre for Vision and Vascular Sciences, School of Medicine, Dentistry and Biomedical Sciences, Queen's University of Belfast, Belfast, BT7 1NN, UK
e-mail: r.beattie@qub.ac.uk

A. Zoubir (ed.), *Raman Imaging*, Springer Series in Optical Sciences 168,
DOI: 10.1007/978-3-642-28252-2_7, © Springer-Verlag Berlin Heidelberg 2012

number of applications of the technique, especially in areas allied to biology and biomedicine [2]. On the technical side, the advances are evident in regard to spectral acquisition, in the form of both compactness of optical layouts and detectors, the Raman signal acquisition geometry itself, involving either conventional back- or obtuse- angle scattering or, more recently, in certain circumstances, transmission Raman [3]. The large volume of information acquired in a mapping experiment requires suitable tools for identifying and extracting only the useful parameters (there can be in excess of 50 detectable constituent signals characterizing one biological tissue) leading to the widespread use of chemometrics for Raman-based bioanalysis and there has also been significant development in the approaches to data processing [2, 4–10].

The increasing popularity of Raman spectral mapping and imaging (terms discussed in Sect. 7.1.1 below) in the biomedical and biochemical fields is primarily due to the low Raman scattering cross-section of water (many biological materials contain >75 % water) coupled with the high level of information that is available on both the chemical and physical composition of the sample. Because the spatial resolution of Raman microscopy is diffraction limited, it is ideally suited to biological imaging, as most biological cells are on the scale of microns.

Biological tissues do present a number of common challenges for Raman imaging, one of the most important being the widespread incidence of fluorophores in nature. Because the fluorescence is so much more efficient than Raman scattering, it commonly swamps the weaker Raman signal. In biological materials this is particularly crucial as many common biochemicals are not particularly strong Raman scatterers. The magnitudes of scattering cross-sections can have implications for technical matters such as signal acquisition times and detection sensitivity, as will be considered later. A large number of strategies have been employed to reduce the impact of fluorescence including use of deep UV [11] and NIR [12] laser excitation, Kerr gating [13], subtracted shifted Raman spectroscopy [14] and shifted excitation Raman difference spectroscopy [15] as well as a large array of signal processing methods applied post acquisition. Some recent approaches which address what has been regarded from a practical standpoint almost as the 'mutual exclusion' between Raman scattering and fluorescence will be briefly considered in the final section of this chapter.

7.1.1 Terminology

While mapping and imaging techniques are fully dealt with in Chap. 1, it is convenient here to briefly summarise a few salient points for the specific applications to be discussed later. Although an operational distinction between mapping and imaging has been pointed out [5], in common parlance the terms are sometimes used rather more loosely and even interchangeably. Mapping involves recording of a spectrum at a point with coordinates (x, y) on the sample, followed by repeated measurements at further defined positions until the sample area of interest has been probed. An important point is that this mapping procedure provides chemical information for each mapped point across the entire area of sample investigated [5, 16]. Raman imaging techniques broadly subdivide into direct and serial (i.e., raster) approaches, involving in the former case simultaneous imaging of the entire sample area or a selected segment of sample at a specific wavenumber, while in the latter the final Raman image is reconstructed from a (generally large) number of signals recorded by raster scanning at defined, small steps (e.g. 1–10 micron increments) and at a selected wavenumber [17]. One significant drawback of Raman microscopy as a mapping technique is the length of time needed to complete even a spatially limited scan [18]. This, however, is a situation which is changing rapidly as a result of advances in the technology.

A range of signal analysis methods has been used to reduce the multiplex Raman information to a more manageable number of variables that also make more intuitive sense. Linear Combination Analysis (LCA), Principal Component Analysis (PCA), Cluster Analysis (CA), Discriminant Analysis (DA), Partial Least Squares (PLS) and a number of variants of each have all been applied to reduce the spectral information, but do so from different approaches that make their applicability dependent on the information desired. LCA requires extensive prior knowledge of the systems being studied as all the major Raman contributors must be recorded for reference and the range of biochemicals in a tissue can be considerable. A further possible confounding issue may arise from physical interactions between the constituents in a native matrix, which can considerably alter the Raman signal so these also need to be modelled. If too few or too many reference spectra are included in the model, there is a very high risk of under or overfitting the data, respectively. However, in situations where the biochemical composition is limited and well characterised and the target-matrix interactions are easily modelled, then LCA is a very powerful method and very readily interpretable. PCA, DA and CA are all multivariate methods of data reduction, with spectral signals identified after the analysis is complete. CA aims to assign a spectral signal to a specific group membership based on similarity. These groups are defined by the mixture of biochemical constituents present and in imaging experiments these groups often naturally correspond to domains, regions, morphological features, tissue layers or cell types. Thus, CA is a very powerful tool for defining these parameters, making it ideal for histochemical analysis. In contrast, PCA splits the spectra according to differences and reduces the spectral signals into individual components. Consequently, PCA is a powerful tool to map the distribution of individual biochemicals. A combination of CA to define areas within a sample and PCA to track specific biochemical constituents may sometimes offer a very useful strategy. PLS works by regressing the Raman signal against a reference parameter such that new signals can then be used to predict that parameter on new samples. First a calibration is created by preparing a set of standards and recording their Raman signal. PLS calculates a transformation that will convert the Raman signal to the predicted value, such that Raman signals within an imaging experiment can be used to create a map of the reference parameter.

The term 'constituent' will be used in this chapter when referring to individual biochemical entities in order to avoid confusion with the statistically meaningful term 'component'. 'Component' will refer exclusively to principal components calculated by PCA, which balance the relative contribution of two 'constituents'.

7.1.2 *Scope of the Chapter*

Attention in this chapter will be directed mainly to biological or biomedical systems, with particular reference to the topic of bioimaging, for which Raman microscopy offers a number of advantages in addition to high spatial resolution. Although the number of applications of Raman microscopy continues to burgeon, no attempt will be made here at comprehensive coverage. Instead, a limited number of topics will be selected to illustrate some of the significant recent trends and to highlight a few of the important issues, both in regard to areas of application and the most efficient and informative methods of spectral data processing.

Table 7.1 below presents a selection of applications of Raman imaging in the biomedical field, and illustrates the range and versatility of Raman microscopy as an imaging and mapping technique. Not all of the topics listed here are discussed in this chapter, but appropriate references are given which provide further information.

7.2 Single Cell and Subcellular Studies

The central feature of Raman scattering and the one, perhaps more than any other, which is responsible for its increasingly prominent role as a biomedical probe is the structural information it provides at the molecular level. This feature, along with the further important attribute that no labelling of biomolecules is required comes very much to the fore in the context of studies at the cellular and subcellular levels, in which imaging and mapping methods are employed [18, 45]. Several variations on the procedures are in use.

One approach for single cell studies is to record a complete Raman spectrum at each spatial position of the laser beam, an approach referred to as the multiplex mode [18]. Subsequently, Raman images of the single cell for any vibrational band of interest can then be constructed as a function of position. An example [45] is shown in Fig. 7.1.

This example shows Raman imaging and specific spectral data for a freshly prepared neutrophil on a CaF_2 slide. The actual image shown here is constructed in the second stage of data processing, involving hierarchical cluster analysis (HCA) on the primary 3D data matrix . Each colour in the image corresponds to a distinct cluster and from the corresponding average Raman spectra, diagnostic vibrational features can be discerned. In this instance the data led to the conclusion [45] that fresh neutrophils are largely devoid of lipid bodies (LBs), one of the many points of fine detail and biomolecular insight that can be gleaned from Raman microscopy imaging, contributing ultimately to understanding of much wider issues, as in this

Table 7.1 Selected applications of Raman imaging with biological sciences

Topic	Technique	Comments	References
Breast tissue	SORS[1], Raman microscopy	Characterising normal duct tissue, benign and malignant growths	[19–25]
Subcellular studies	Raman microscopy	Induced stress on cells	[18]
Single cell imaging	Hybrid Raman fluorescence	Novel use of quantum dots	[26]
Cell and tissue death	Raman microscopy	Nucleic acid distribution in apoptotic cells	[27, 28]
Live cancer cells	Raman microscopy		[29]
Atherosclerosis	Raman microscopy, Ultrasound	Deposition of triglycerides, cholesterols and calcium salts.	[23, 24, 30–34]
Bladder	Raman microscopy	Characterising normal, obstructed and cancerous tissue	[35, 36]
Malaria	Raman and FTIR microscopy	Characterising changes in hemoglobin	[37]
Cell colonies	Raman microscopy	Biofilms, nutrient uptake	[38, 39]
Nutrient distribution in food	Raman microscopy	Antioxidants, fatty acids	[40–43]
Lung	Raman spectroscopy	Cancer, blood and antioxidants	[18, 43, 44]

[1]SORS = Spatially Offset Raman Spectroscopy

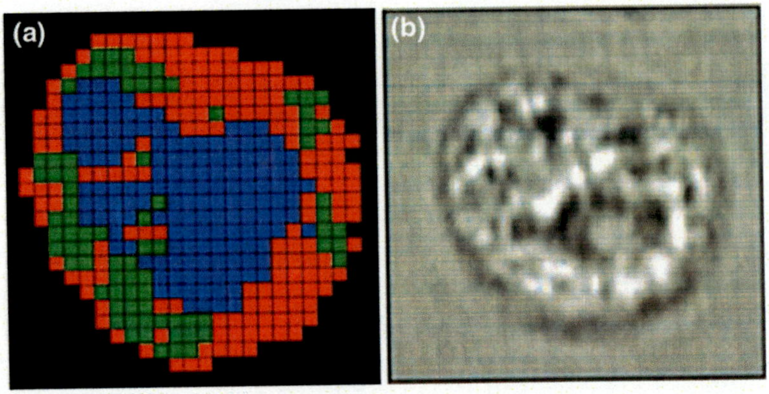

Fig. 7.1 Confocal Raman image of a quiescent neutrophil. **a** HCA image constructed from Raman imaging data of a single neutrophil. Each colour represents a different cluster. **b** Corresponding white-light transmission image. Adapted from [45]

particular example, where the findings may help to clarify the possible role of LBs in innate immune response.

A further example of the tremendous advantages which can accrue to key questions in areas such as cellular biology from multivariate analysis of Raman microscopy data sets is displayed in a recent study in which Raman microscopy has been used to investigate the subcellular structure of cytoplasm [46]. Here, a raster technique was employed on fixed cells to collect series of individual Raman spectra, from which high resolution pseudocolour Raman maps were assembled. The level of spectral specificity available via use of multivariate analysis (HCA) is exemplified in Fig. 7.2, which shows the results of cluster analysis carried out for the spectral range 1200–1800 cm^{-1}. Distinct nucleoli (blue areas) are obvious within the nucleus. The choice of vibrational frequency region for image construction is important. Thus, the equivalent image constructed for the range 2800–3020 cm^{-1}, the C-H vibrational stretch region, failed to reveal any contrast between nucleus and nucleoli.

If studies are conducted under non-resonant Raman excitation, use of high laser powers may be required, >50 mW CW, depending on other factors such as the Raman scattering cross-section of the species under investigation [47] and the detection sensitivity [18]. For peripheral blood lymphocyte cells fixed in formaldehyde, for instance, CW laser powers up to 70 mW appeared to cause no changes in the Raman bands, even for exposure times up to 100 s. Fixed lung fibroblast cells have been imaged in subcellular detail by means of non-resonant Raman microscopy, enabling individual cells to be visualized, and in particular to investigate changes in the cell morphology induced by the fixation conditions [18].

Of course, in working with living cells at the molecular level additional challenges [48] can arise, compared to working with fixed cells, such as changes in molecular composition during a cell cycle. Additionally, the excitation laser energy becomes an important consideration and it may be necessary to minimise laser excitation

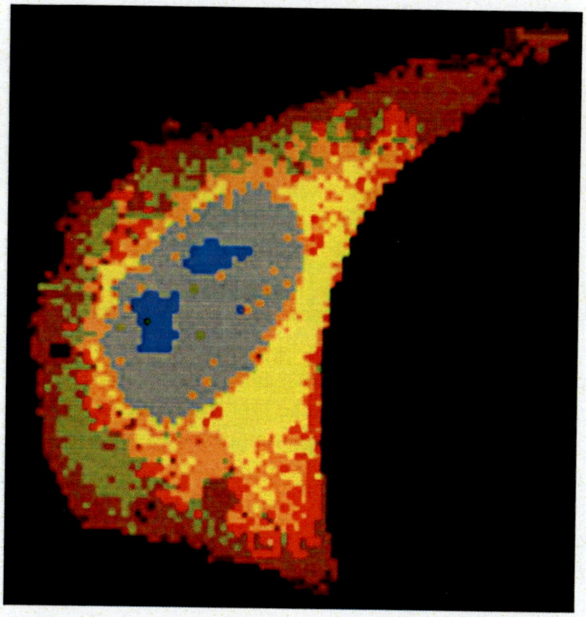

Fig. 7.2 Pseudocolour map derived from HCA of Raman data in the spectral range (1200–1800) cm^{-1} (adapted from [46])

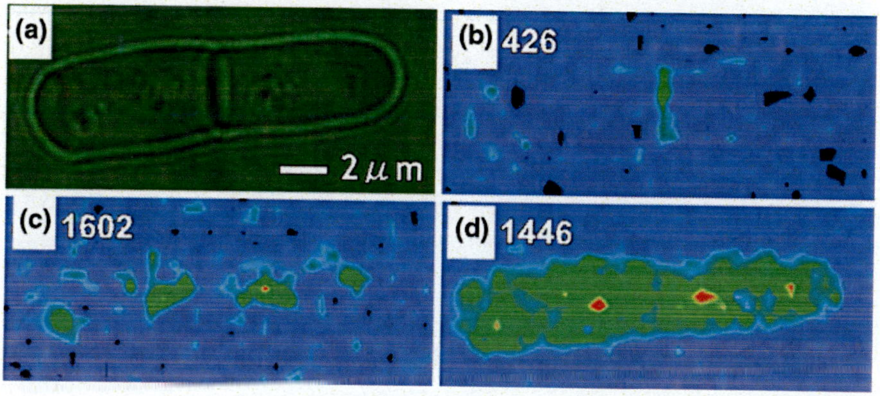

Fig. 7.3 Raman mapping of a living G1/S phase *S. pombe cell*: **a** microscopic optical image, Raman intensity maps using **b** 426 cm^{-1} **c** 1602 cm^{-1}, **d** 1446 cm^{-1}, adapted from [48]

power levels and data acquisition times to avoid species damage. Nevertheless, long acquisition times may still be required in order to achieve a sufficiently high signal-to-noise ratio so as to assemble as much information as possible about the chemical composition of the sample [48]. In the data displayed in Fig. 7.3 showing Raman maps recorded for living yeast cells for several vibrational bands, exposure times up to 300 s were used [48].The maps were constructed from the recorded Raman intensities of individual bands in the spectra at each scanned point across an entire yeast cell. Of particular note is the map for a band at $1602\,cm^{-1}$. This feature has been described as a 'Raman signature of life in a living yeast cell', though no specific assignment of a molecular species responsible for it has so far been possible. More recent studies, however, have established lifetime information for the species [49]. This example does provide a nice illustration of the level of chemical detail and spatial distribution which can be achieved by Raman imaging at the cellular level. In the case of the $1602\,cm^{-1}$ band it is possible to conclude that, whatever the precise biochemical nature of the species responsible, it is localised within mitochondria [48].

Raman microscopy has also been used to investigate normal and malignant prostatic cell lines, in vitro [50]. Using PCA, the distribution of the primary biochemical constituents, DNA and proteins, was mapped, enabling differences in composition and distribution for each cell line to be highlighted. The work is a pointer to the potential of Raman microscopy for discriminating and predicting the group membership of benign and malignant prostatic cell lines.

The choice of laser excitation wavelength itself may also be a consideration, governed partly by whether studies are to be carried out under resonant or non-resonant Raman excitation conditions. With use of a Raman excitation wavelength coincident with an absorption band of the species under study, considerably enhanced (resonance Raman, RR) scattering may be accompanied by other advantages including reduced signal acquisition times at lower excitation laser power levels, as well of course as chromophoric selectivity.

As an example, the subcellular distribution of haeme-containing enzymes such as the NADPH oxidase subunit cyt b_{558} in neutrophils was investigated [51, 52] by recording Raman images in single neutrophilic granulocytes, using 413.1 nm as the excitation wavelength, in resonance with the haeme Soret band region. In this instance RR spectroscopy effectively enabled the intracellular enzyme activity of NADPH oxidase activity to be probed, with the prospect [51] that such investigations might lead ultimately to a better understanding of the function and activity of this and related enzymes in other cells and tissues.

7.3 Ex Vivo Analysis of Tissue

Ex vivo analysis of biological material allows more time to acquire high volumes of information taken from tissues that have been 'frozen' in time. Because of the weak nature of the Raman scattering process there has been an understandable emphasis on the analysis of ex vivo tissue. The topics covered range from medical diagnostics, through biochemistry to plant science and nutrition. Because of the ability of Raman methods (with appropriate chemometrics) to report on a vast number of molecular constituents, it has proved an effective tool to address a wide range of biomedical questions. These questions range from basic anatomy (by defining structures based on composition) through biochemistry (studying the interaction of the various molecular species) and pathology (studying the progression of a disease) to detection of

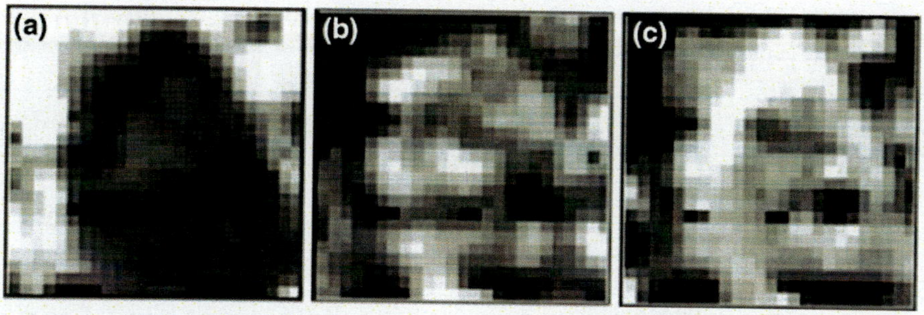

Fig. 7.4 Raman/HCA images of **a** collagen, **b** cell cytoplasm and **c** cell nucleus in normal breast duct. Adapted from [24]

pathologic events (e.g. cancer, necrosis, etc.) in histochemical mapping of a wide range of tissues. Raman microscopy was able to directly and simultaneously detect disparate biochemical constituents that would otherwise require a battery of different analyses.

7.3.1 Breast Tissue

Raman microscopy has been used to probe the histochemistry of breast duct epithelia, demonstrating the potential role of Raman methods in the understanding of disease processes and in diagnostics. The relative contribution of a number of biochemical constituents to the Raman signal across sections of breast duct was determined by the use of LCA [19]. Normal physiological constituents such as collagen, fat and β-carotene were included along with the pathogenic markers calcium hydroxyapatite and calcium oxalate. By employing linear combinations of pure reference compounds, it is possible to achieve a high level of biochemical interpretability, but this approach can suffer where the constituents of interest are affected by the physical interactions with its matrix in situ, as will be highlighted below. Because the pathogenic markers are crystallised salts that do not interact with the surrounding matrix, the approach worked well in this study. A different statistical approach using HCA to determine the different signals present in the dataset without prior knowledge was applied in an investigation by Raman microscopy to probe the distribution of biochemical constituents in breast tissue [22, 24] (Fig. 7.4). The signals identified for the clusters (which are not signals of individual constituents, but of combinations of constituents that occur together) were then identified post hoc using reference signals. In effect, this means that any matrix interactions are inherently incorporated into the analysis. By mapping the histochemical distribution within a section it is possible to more precisely differentiate anatomical features that may otherwise require a range of stained sections while simultaneously assessing the biochemical status. Further work has been carried out on characterising breast tissue, with some

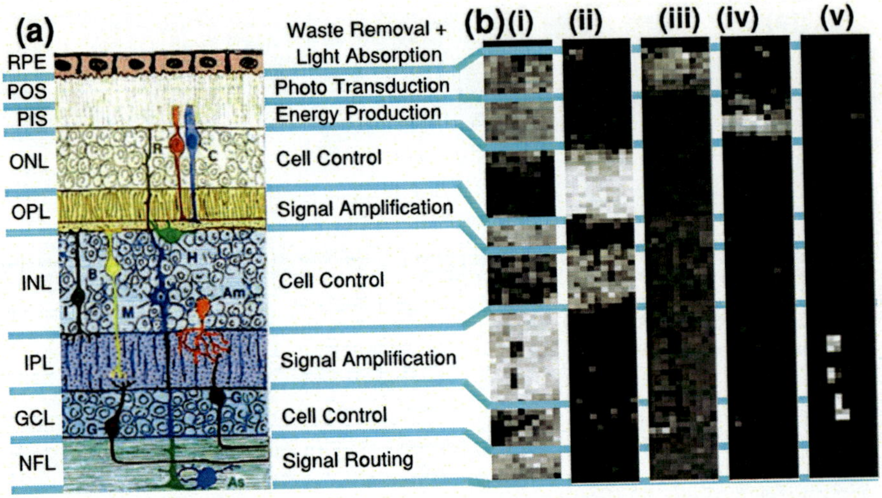

Fig. 7.5 Schematic diagram of the retina. RPE Retinal Pigmented Epithelium; POS/PIS Photoreceptor Outer /Inner Segments, O/INL Outer/Inner Nuclear Layer, O/IPL Outer/Inner Plexiform-Layer, GCL Ganglion Cell Layer, NFL Nerve Fibre Layer **b** Raman images of i protein ii DNA iii fatty acid lipids iv cytochrome c v heme. Schematic adapted from http://webvision.med.utah.edu/

noteable developments including the assessment of the lesions margins in order to precisely define the volume of material that needs to be excised [20, 21] and also the development of in vivo location of lesions using SORS [3, 25]. By collecting the Raman signal laterally offset from the incident radiation, it is possible to correlate the spatial offset of the signals with depth, thereby providing simultaneous biochemical and spatial information critical in a surgical situation.

7.3.2 Eye

Conventional biochemical stains allow only a limited number of biochemical constituents to be mapped at once. Raman has the ability to simultaneously map the distribution of many constituents, often as many as 50. The usefulness of Raman microscopy in this regard has been well illustrated in recent mapping studies of the layered structure of porcine retina, which revealed the anatomical structure of this particular biological tissue [53, 54]. The retina is the photosensitive tissue at the back of the eye that is responsible for the transduction and amplification of visual stimuli. This tissue is delineated into very clear layers that correspond to particular regions within cells and each layer performs a dedicated function (see Fig. 7.5).

The region of transduction, i.e. the photoreceptor outer segments, shows a high concentration of highly unsaturated fatty acid, while the inner layers show a lower content of more saturated fatty acid. The photoreceptor discs, responsible for

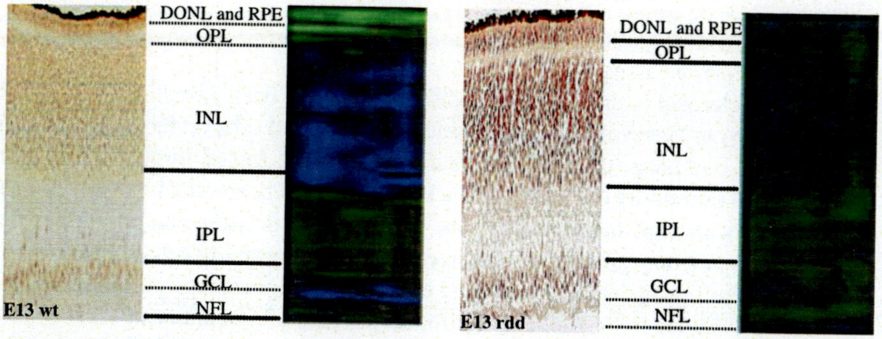

Fig. 7.6 Raman maps from embryonic chick day (*E*) 13 wild type (*wt*) and retinal dysplasia and degeneration (*rdd*) retina with optical histological image. *Green* = fatty acid-based lipid, *blue* = nuclear protein/DNA. Retinal layers have been annotated. Adapted from [55]

absorbing the light photons and converting them to an electrical signal, are composed primarily of dodecahexenoic acid, a fatty acid with six unsaturated bonds per chain. The inner layer fat is more similar to that of intramuscular fat with a high proportion of 16- and 18- carbon saturated and mono-unsaturated fatty acids. The presence of a DNA signal allows mapping of nuclei distribution. Most of the nuclei are restricted to the Outer and Inner Nuclear Layers, with a few isolated nuclei from Muller cells dotted around the ganglion cell layer. The Raman map readily demonstrates the much higher density of nuclei in the outer layer, and the association of fatty acid with the nuclei in the inner layer, in contrast to the almost lipid-free outer nuclear layer. The Raman studies also identified a number of different protein populations, all of which have been combined to produce the protein map shown in Fig. 7.5. In addition to major macronutrients, Raman also has the ability to provide useful information on the distribution of more minor constituents such as the two distinct porphyrins, Cytochrome c and haeme. The former, which is the primary enzyme used in energy production by mitochondria, is shown to be localised to the Photoreceptor Inner Segment, the region responsible for providing the energy needs of the photoreceptor outer segments (POS). In contrast, the signal revealing the presence of haeme, the oxygen transport molecule of red blood cells, is found to be confined to a vertical strip in the inner plexiform layer that matches a blood vessel. As with the breast tissue work discussed above, the foundation of a detailed Raman model of normal tissue acts as a valuable starting point for assessing biochemical and histological alterations that occur in retina. Indeed, Raman microscopy has recently been used to compare the biochemical status of normal chicken retina with that of diseased retina [55]. In this case the disease was *retinitis pigmentosa*, a progressive ocular disease in which the photoreceptors in the peripheral retina are degraded leading to tunnel vision. Raman was able to demonstrate that, even early in the onset of retinitis pigmentosa, the photoreceptor outer layers lose fatty acids, but fats in the inner layers are relatively unaffected, while the nuclear layers show some decrease in DNA density (Fig. 7.6).

7.3.3 Artery

Atherosclerosis is another major disease that has been investigated using Raman mapping, and it involves the deposition of lipids and salts on the wall of arteries. As material builds up, the blood pressure rises as it is forced through an ever smaller gap until eventually the artery is blocked, leading to heart attack or stroke depending on the location of the blockage. The manner in which a lesion progresses depends more on its composition than its size. So it is important to obtain chemical as well as morphological data in an imaging experiment. In that respect, Raman microscopy has been shown to be an ideal tool for identifying and quantifying a range of chemicals that are deposited within arteries during atherosclerosis [56]. The list includes cholesterol and its esters, triglycerides, phospholipids and various calcium salts. Linear combination analysis was used, in conjunction with a series of standard reference compounds selected to match the range anticipated within the hardening arteries. Seven compounds were found to accurately model the sample spectra. It also proved possible to use the Raman signal to determine the composition of the different types of plaque that occur [57]. Intimal fibroplasias showed mainly protein and fatty acid spectra, atheromatous plaque showed mainly protein and cholesterols, while calcified atheromatous plaque showed mainly calcium salts. This initial investigation laid the groundwork for chemical mapping [32] using the reference spectra to map the distribution of DNA, cytoplasm, phospholipids, cholesterol ester and triglycerides in single human colonic carcinoma (HT29) cells.

7.3.4 Lung

Another organ that has received attention in the literature is the lung, the site of oxygen transfer between the environment and bloodstream. Due to its function it is exposed to aerosol pollutants and high oxidative stress and presents a warm moist environment for airborne bacteria to thrive. Researchers have used Raman mapping to probe a range of tissues and diseases within the lung.

In initial exploratory studies, an 'expansive' approach was adopted to map the lung, by employing a larger than normal step size of 100 μm so as to map a large area of the lung (ca. 35 mm^2) [58, 59]. This enabled rapid identification of potentially more interesting areas that could then be mapped in higher resolution. It was found that the dominant variation in the mapping experiments originated from haemoglobin in the blood. It is known that haeme is still resonantly enhanced at 785 nm [53], while the alveolar region the authors were studying contains a very high density of capillaries in order to capture the incoming oxygen as efficiently as possible. Signals of mucus and collagen could be detected too, but the haeme signal was overwhelming. The differing concentrations of blood still proved useful in pinpointing areas of more interest, namely congenital cystic adenomatoids (CCAM), which could then be mapped at higher (10 μm) resolution. It proved possible to compare the biochemical

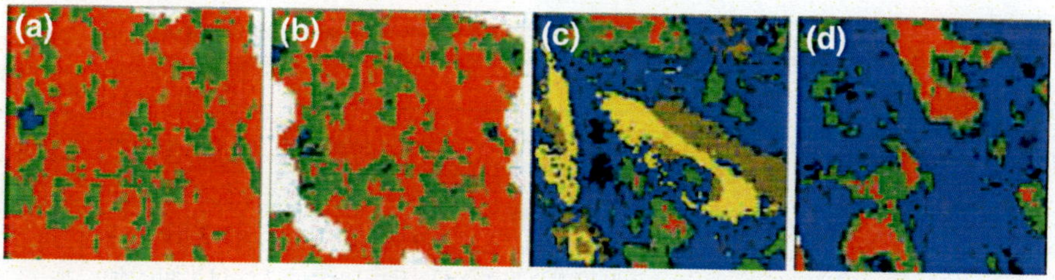

Fig. 7.7 Raman/Cluster Analysis maps of two normal (**a**+**b**) and two CCAM (**c**+**d**) lungs. *Red* = haemoglobin, *Blue* = cellular material, *Green* is intermediate combination of *red* and *blue* signals, *Yellow* = collagen, *grey/brown* = phospholipid. Adapted from [58]

composition of normal and CCAM tissue, which in turn enabled high contrast definition of several morphological features (Fig. 7.7). The level of haeme was found to be highest in normal tissue, and lowest in CCAM, but this variation was not specific to these two tissue types. Lipids and glycogen were more specifically elevated in CCAM, but displayed very low variation in normal tissue.

The airways that connect the alveoli to oxygen of the outside world are a series of branching tubes comprised of the central trachea (windpipe) which branches into two bronchi that further subdivide into the bronchioles that terminate in little air sacs called alveoli. Raman microscopy was employed to define the structures of the bronchus by identifying the unique blend of biochemical constituents that typify different anatomical structures [44]. By carrying out PCA and then applying the K-means CA, the spectral data were first summarised according to variation in the biochemical constituents and then according to the relative contribution of those constituents to the spectra at each point in the map. Use of CA allows biochemically related regions to be defined, and because anatomical features usually have precise biochemical jobs, anatomical features can be identified quite readily. It proved possible to map the distribution of mucus, epithelium, stroma, glands/ducts and cartilage based on the unique composition of each of these distinct tissues. Investigation of mucus revealed Raman signal due to carotenoids, with a high proportion of fatty acid also present and a small contribution from a protein high in phenylalanine and tyrosine. The combination of various constituents including DNA, lipids, collagen, actin, haemes and chondroitin allows the Raman to distinguish between epithelial tissue, stroma, glands and cartilage.

Raman microscopy can also provide useful information on the biochemical processes occurring within biological tissue. Vitamin E is an important family of antioxidants which share the features of a chromanyl ring and a branched acyl ('phytyl') chain, collectively known as tocopherols [43]. Raman spectroscopy is ideally suited to this, as the signals for each tocopherol and metabolite give diagnostic features [43]. The site at which oxygen is absorbed into the body (consequently a high oxygen tension site) is the alveoli, small inflatable sacs of cells that bunch at the end of the bronchioles (Fig. 7.8a). Figure 7.8b and c compares a histological H+E stain with the equivalent low resolution (3 μm) Raman map of protein (green), prophyrin (red) and tocopherol (blue). The Raman map reveals

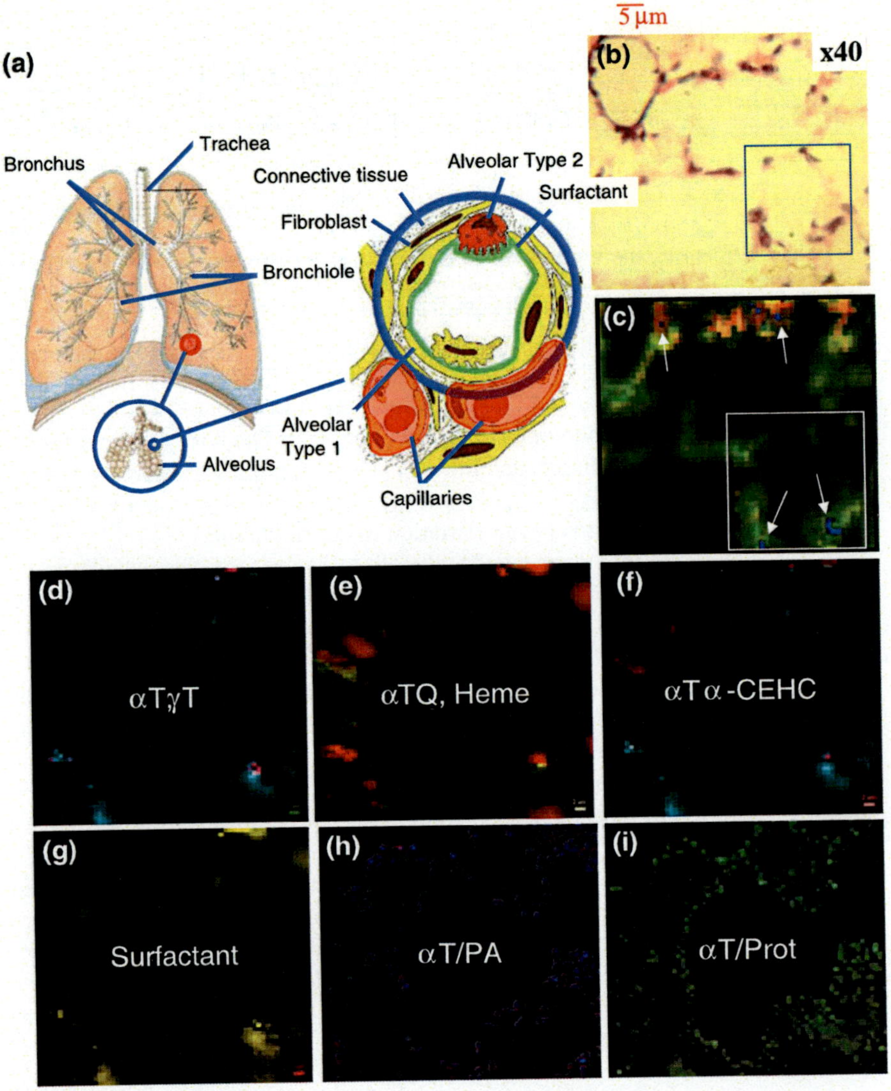

Fig. 7.8 **a** Schematic diagram of an alveolus in the lung **b** histological image of a region previously scanned by Raman microscopy **c** Low resolution (3 μm) map of the region shown in (**b**). High resolution (1 μm) pseudocolour Raman images of **d** $\alpha + \gamma$-tocopherol **e** α-tocopherol quinone and haeme **f** α-tocopherol and α-carboxylethyl hydroxychroman **g** lung surfactant **h** $\alpha-$ tocopherol/lipid ratio **f** α-tocopherol/protein ratio. Adapted from [43]

two main regions where the tocopherol was localised: in the walls of the blood vessels (white arrows along the top of Fig. 7.8c) and in the walls of an alveolus (white arrows in the bottom of Fig. 7.8c). Figure 7.8d shows that the two tocopherol homologues colocalise, with γ present in approx one-tenth the concentration of α.

Their range also overlaps considerably with the range of the phospholipid surfactant (Fig. 7.8g), but is very distinct from that of the oxidation products (Fig. 7.8e, f). The direct ($2e^-$) oxidation product, α-tocopherol quinone, is more concentrated than the metabolic ($1e^-$) oxidation product, α-carboxyethyl hydroxychroman and both are found in close association with porphyrin molecules. These data suggest that the porphyrin groups, which are typical of redox enzymes, are involved in metabolising the quinone by removing its phytyl chain through a further controlled $1e^-$) oxidation step. The data used to produce these maps were not normalised in advance of the PCA, with the consequence that the maps relate to the total amount of the various compounds within each sampling pixel. These maps show that the type II cells accumulate the majority of the vitamin E homologues within the centre of the cell, but this is not the site of greatest oxygen tension. In this case looking at data ratioed against another constituent gives a completely different angle on the data. When the α-tocopherol concentration is normalised to either protein or lipid, a very different appearance becomes evident. The α-tocopherol versus lipid map also shows that although the largest absolute quantity of α-tocopherol is in the centre of the cell, there is an elevated relative concentration along the lining of the alveolus—the very site of highest oxygen tension. The α-tocopherol versus protein map gives a similar view of the lining, but in the centre of cells the tocopherol reading is elevated due to a protein being precluded from the lipophilic area.

7.4 Non-Medical Bioimaging

7.4.1 Plants

Antioxidants have also been imaged in plants by Raman microscopy, as exemplified by a mapping study of the distribution of carotenoids [40]. Carrot roots were sliced and the Raman image was recorded and analysed for carotenoids and polyacetylene content. Because both these molecular families contain large conjugated double bond regions, they give strongly resonance-enhanced signals at a wide range of excitation wavelengths and even off resonance the signal is strong due to the high polarisability of the π-bonding. Because of the intensity of these constituents, the authors were able to employ straightforward band intensities at selected wavelengths to track the concentrations of the various constituents in Fig. 7.9. This study clearly showed that the two classes of compound had unique distributions, with the carotenoids present throughout the root, but elevated in the centre of the phloem parenchyma while the polyacetylenes were highly localised to the pericyclic parenchyma (outer edge of the root) and inner edge of the phloem parenchyma. The polyacetylenes are involved in disease resistance, which explains their elevation in the vascular bundle at the outer edge of the root. The other vascular bundle is located at the inner edge of the phloem parenchyma, suggesting an involvement of these channels in the transport and accumulation of the polyacetylenes. While the poly-

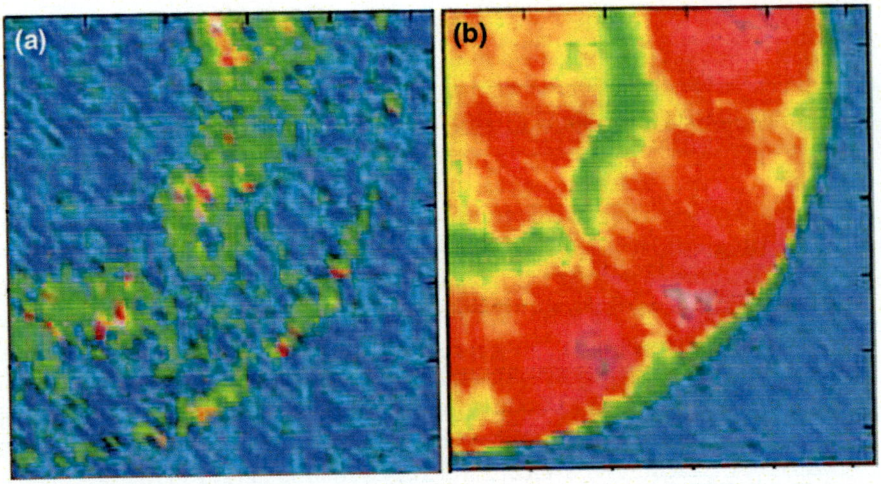

Fig. 7.9 Raman maps of **a** polyacetylenes and **b** carotenes in carrot root (transverse section). Adapted from [40]

acetylene provides some resistance to storage diseases, they are also believed to be responsible for a bitter off-taste that can occur in carrots and also provide some anti-cancer protection making this an important molecule from a food production point of view.

In addition to antioxidants and antifungal agents, plants contain a wide range of metabolites that are responsible for flavour and smell, making them commercially important molecules. The ability of Raman spectroscopy to map the distribution of a series of plant metabolites has also been demonstrated [41] , including anethole and fenchone in fennel fruit, spiroethers in chamomile and curcumin in curcumin root.

7.4.2 Animal

The ability of Raman spectroscopy to predict the fatty acid composition of foods has also been demonstrated. Reliable estimates of the content of over 15 major fatty acids in adipose tissue proved possible [60, 61]. In addition, clear differences between adipose tissue from various domesticated food animals were discernible, allowing very high rates of discrimination even between biochemically similar animals [62]. It was also possible to use Raman spectroscopy to map the distribution of fatty acids in the adipose tissues (Fig. 7.10), which have a very different physical structure. Beef, lamb and chicken adipose consist of irregular-shaped packets of adipose tissue packed together, rather than forming the smooth layers seen in the pig [63]. This prevents use of freeze fracture followed by GC, and consequently the biochemical distribution of fatty acids had remained untested. Raman spectroscopy was able to

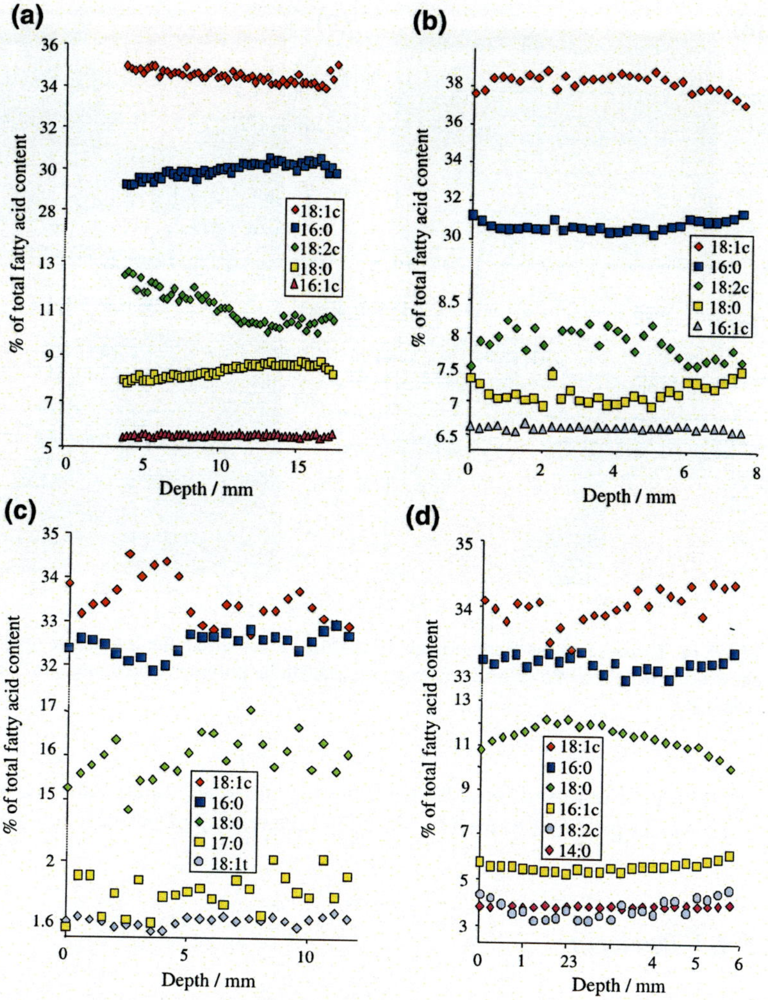

Fig. 7.10 Depth profiles of Raman predicted fatty acids composition in adipose tissue of **a** pork, **b** chicken, **c** beef and **d** lamb. Fatty acids indicated are oleic (18:1c), palmitic (16:0), linoleic (18:2c), stearic (18:0), palmitoleic (16:1c), margaric (17:0), elaidic (18:1t) and myristic (14:0) acid

demonstrate that the fatty acid composition not only varied from the outer edge through to the inner edge of the tissue, but also within the individual packets of adipose tissue. Lamb and Chicken adipose showed opposite trends with unsaturation equal on both surfaces, but reaching a peak in the middle. Beef adipose showed much less consistent trends, with large fluctuations across the depth of the tissue. Comparing the trend with the physical distribution of adipose packets shows that these are in fact localised trends within each packet. At the boundary of each packet

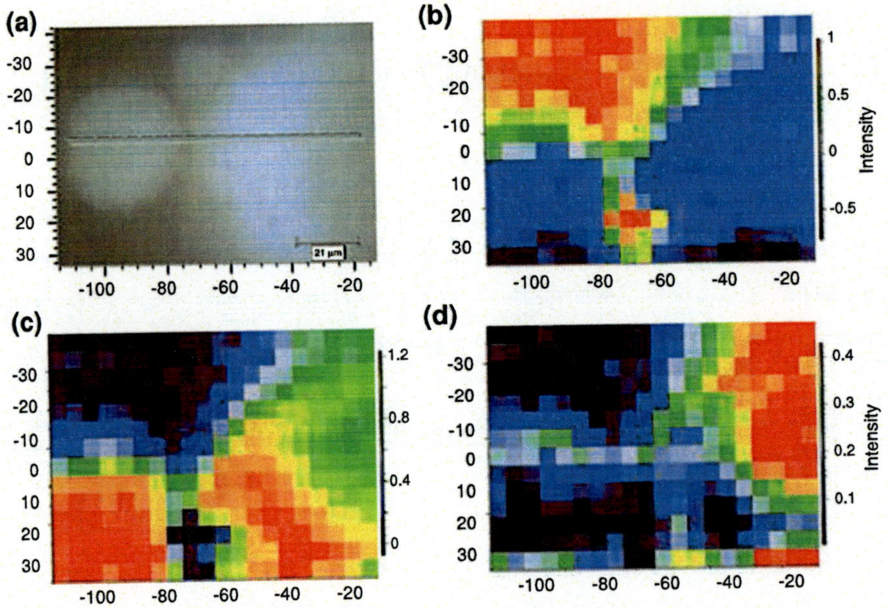

Fig. 7.11 Cross-sectional Raman maps of biofilms taken along the line in the optical image in **a** showing distribution of **b** cellular material, **c** extracellular polymers and **d** medium (water). Adapted from [38]

the fatty acid composition was significantly more saturated than the interior. This trend was largely determined by the relative amounts of palmitic and oleic acids.

7.4.3 Bacteria

Raman has also been employed in mapping the uptake of nutrient media in live bacterial biofilms over both space and time. The development of biofilms of *Pseudomonas aeruginosa* in a minimal nutrient medium has been studied using Raman microscopy [39]. The group found that the cell colonies excreted an extra cellular polymer to enhance cell adhesion in the presence of a flow, and this accumulated over time (Fig. 7.11). The group was also able to distinguish clumps of cells that became liquid, as part of the colonies seed dispersal process. It has been demonstrated that Raman could map the accumulation of carotenoids across a bacterial colony (*Halobacterium salinarium*) [38] and this uptake correlated with the rate of growth. The study demonstrated that the carotenoids were localised within concentric rings across the colony and that the formation of these rings was not related to the diurnal cycle of light and dark. The pattern of rings fitted a model of bacterial colony formation in which

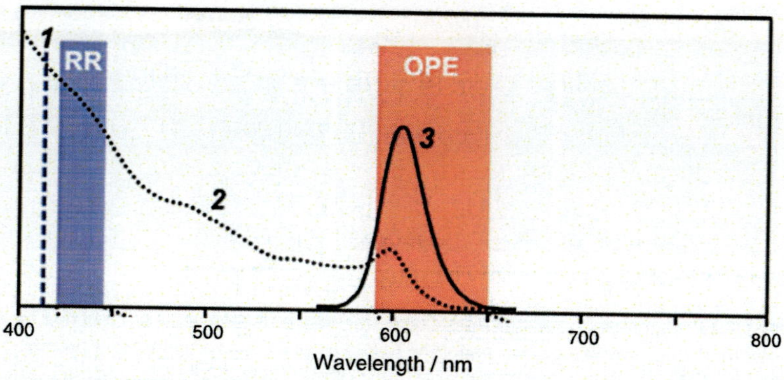

Fig. 7.12 Well-separated spectral windows for detection of resonance Raman (RR) and OPE signals, both on Stokes side of excitation wavelength, 413.1 nm, positioned at 1. Curves 2 and 3 show the QD absorption and emission spectra. Adapted from [26]

periods of consolidation and migration alternate that are solely triggered by local cell density and not by external biological or chemical factors.

7.5 Recent Advances

The applications of Raman microscopic imaging as a uniquely informative biological and biomedical probe continue to grow, frequently paralleled by considerable ingenuity in both the manner of application and in the choice of other techniques or procedures with which it is coupled. An interesting example [26] is the use of semiconductor quantum dots (QD) to overcome the traditionally regarded 'impasse' between Raman scattering and fluorescence[1] in which a hybrid Raman fluorescence technique was employed for single cell imaging. The basic principle underlying the approach, of spectral separation of fluorescence emission and Raman scattering, had already been demonstrated [64] by combining two-photon excitation fluorescence with non-resonant Raman scattering, the former emission occurring on the anti-Stokes side of the laser excitation wavelength but the Raman scattering being observed on the Stokes side. In contrast, in the more recently proposed hybrid Raman/one-photon excitation (OPE) QD fluorescence combination, both the emission and the Raman scattering occur on the Stokes side, but advantage is taken of the large Stokes shifts displayed by semiconductor QD nanocrystals, so that the spectral detection windows for resonance Raman scattering and fluorescence are well separated, as shown in Fig. 7.12:

Experiments were carried out by incubation of cells with QD nanocrystals. Neutrophilic granulocytes were chosen for study in view of considerable previous work

[1] though there are fortunate exceptions [45] where, even with near-UV or visible excitation of biological samples, interference in the Raman scattering due to fluorescence does not arise.

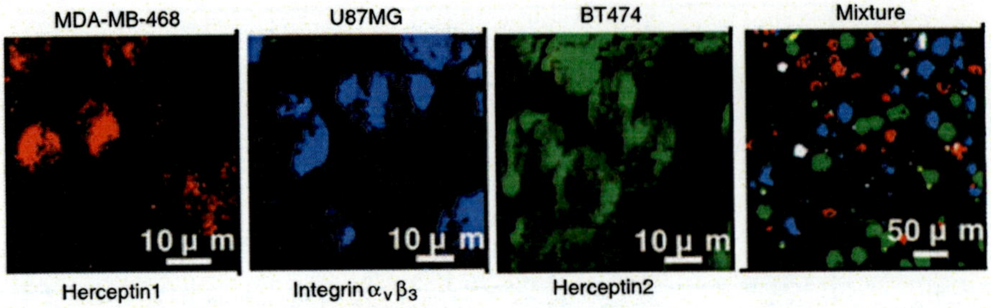

Fig. 7.13 Raman maps of cell lines expressing different proteins, each labelled by an isotopically modified single-walled carbon nanotube linked to an antibody against that protein. MDA-MB-468 and BT474 are different breast cancer cell lines while U87MG is a glioblastoma. Adapted from [29]

on these systems [47, 52]. An elegant feature of the strategy is that use of QDs enables them to be targeted to specific proteins. Moreover, it also proved possible to conduct the strategy on live cells, incubated with the QDs.

Aside from the rapidly increasing use of SERS/SERRS techniques for Raman bioimaging studies (see several articles in [65]), an additional instance of the use of nanomaterials in Raman-based bioimaging is displayed in a report on the application of single-walled carbon nanotubes (SWNTs) as Raman 'tags' [29], illustrated in Fig. 7.13. A novel feature is the judicious use of isotopic combinations (C12, C13 or C12/C13 mixtures) to shift the peak position of the Raman carbon G-band, thus enabling *multicolour* Raman imaging. Live cells (human breast cancer Her2+, among others) were incubated with isotopic mixtures of SWNTs conjugated with appropriate antibody ligands. High resolution, multicolour Raman microimages displaying clearly differentiated Raman labels with the ability to distinguish different cell types in a given sample could be generated in 1 h or less, using 0.5 s/pixel collection times.

Very recently, stimulated Raman scattering has been introduced [66] as a vibrational imaging technique with a sensitivity considered to be much superior to conventional spontaneous Raman microscopy, in addition to being background free. It has been successfully applied to several biomedical topics, including differential mapping of saturated lipid and fatty acid distributions in live cells and other areas as diverse as the imaging of brain and skin tissues.

In recent Raman scattering studies of canine bone, Raman has been combined with computed tomography (CT) in order to enhance the quantitative recovery of the Raman signals [67]. The important feature of CT is that it enabled imaging at the micron resolution level, with the Raman part of the hybrid providing specific molecular signatures. Interestingly, both reflectance and transmission geometries were used to record the Raman scattering. In either geometry, the use of the CT images enabled recovery of the characteristic Raman signals for dog tibia with contrast ratios up to 145:1, vastly superior to what would be possible from Raman scattering alone, where in the dense samples being studied high background scatter effectively limits the spatial resolution to the millimetre range.

7.6 Concluding Remarks

The range of topics across the fields of biomedicine and biology which have been discussed in this chapter are adequate testament to the extent to which Raman microscopy has been a nucleation point for the development of innovative probing and bioimaging techniques centred on Raman scattering. It seems evident that the unique attributes of the Raman phenomenon, especially biochemical specificity, non-invasive character, multiplexing ability and a growing realisation of its potential to link structural elucidation with enhanced understanding of biological function and role will ensure a central position for Raman microscopy as a bioimaging technique in the future.

Acknowledgments The authors wish to gratefully acknowledge the many contributions from collaborators and colleagues to the QUB-based studies included in this chapter. Vital financial support has been provided over several years from the UK Biotechnology and Biological Sciences Research Council, the Medical Research Council, Research and Development Office of NI Health Service and the Leverhulme Trust.

References

1. M. Delhaye, P. Dhamelincourt, Raman microprobe and microscope with laser excitation. J. Raman Spectrosc. **3**, 33–43 (1975)
2. E.B. Hanlon, R. Manoharan, T.W. Koo, K.E. Shafer, J.T. Motz, M. Fitzmaurice, J.R. Kramer, I. Itzkan, R.R. Dasari, M.S. Feld, Prospects for in vivo Raman spectroscopy. Phys. Med. Biol. **45**, R1–R59 (2000)
3. N. Stone, P. Matousek, Advanced transmission Raman spectroscopy: A promising tool for breast disease diagnosis. Cancer Res. **68**, 4424–4430 (2008)
4. J.R. Beattie, J.V. Glenn, M.E. Boulton, A.W. Stitt, J.J. McGarvey, Effect of signal intensity normalization on the multivariate analysis of spectral data in complex 'real-world' datasets. J. Raman Spectrosc. **40**, 429–435 (2009)
5. D. Clark, S. Sasic, Chemical images: technical approaches and issues. Cytom. Part A. **69A**, 815–824 (2006)
6. C. Gendrin, Y. Roggo, C. Collet, Pharmaceutical applications of vibrational chemical imaging and chemometrics: A review. J. Pharm. Biomed. Anal. **48**, 533–553 (2008)
7. A. Jirasek, G. Schulze, M.M.L. Yu, W. Blades, R.F.B. Turner, Accuracy and precision of manual baseline determination. Appl. Spectrosc. **58**, 1488–1499 (2004)
8. B. Lavine, J. Workman, Chemometrics. Anal. Chem. **80**, 4519–4531 (2008)
9. C.A. Lieber, A. Mahadevan-Jansen, Automated method for subtraction of fluorescence from biological Raman spectra. Appl. Spectrosc. **57**, 1363–1367 (2003)
10. G. Schulze, A. Jirasek, M.M.L. Yu, A. Lim, R.F.B. Turner, M.W. Blades, Investigation of selected baseline removal techniques as candidates for automated implementation. Appl. Spectrosc. **59**, 545–574 (2005)
11. S. Bykov, I. Lednev, A. Ianoul, A. Mikhonin, C. Munro, S.A. Asher, Steady-state and transient ultraviolet resonance Raman spectrometer for the 193–270 nm spectral region. Appl. Spectrosc. **59**, 1541–1552 (2005)
12. Z. Huang, A. McWilliams, H. Lui, S. Lam, H. Zeng, Optical diagnosis of lung cancer using near-infrared Raman spectroscopy. In proceeding of SPIE: The International Society for Optical Engineering, 2004

13. P. Matousek, M. Towrie, A. Stanley, A.W. Parker, Efficient rejection of fluorescence from Raman spectra using picosecond Kerr gating. Appl. Spectrosc. **53**, 1485–1489 (1999)
14. S.E.J. Bell, E.S.O. Bourguignon, A. Dennis, Analysis of luminescent samples using subtracted shifted Raman spectroscopy. Analyst **123**, 1729–1734 (1998)
15. N.J. Cherepy, A.P. Shreve, L.J. Moore, S. Franzen, S.G. Boxer, R.A. Mathies, Near-infrared resonance Raman spectroscopy of the special pair and the accessory bacteriochlorophylls in photosynthetic reaction centers. J. Phys. Chem. **98**, 6023–6029 (1994)
16. J. Barbillat, Raman Imaging, in *Raman Microscopy: Developments and Applications*, ed. by G. Turrell, J. Corset, (Academic Press, London, 1996), pp. 175–196
17. S. Schlucker, M.D. Schaeberle, S.W. Huffman, I.W. Levin, Raman microspectroscopy: a comparison of point, line, and wide-field imaging methodologies. Anal. Chem. **75**, 4312–4318 (2003)
18. C. Krafft, T. Knetschke, R.H.W. Funk, R. Salzer, Studies on stress-induced changes at the subcellular level by Raman microspectroscopic mapping. Anal. Chem. **78**, 4424–4429 (2006)
19. A.S. Haka, K.E. Shafer-Peltier, M. Fitzmaurice, J. Crowe, R.R. Dasari, M.S. Feld, Identifying microcalcifications in benign and malignant breast lesions by probing differences in their chemical composition using Raman spectroscopy. Cancer Res. **62**, 5375–5380 (2002)
20. A.S. Haka, K.E. Shafer-Peltier, M. Fitzmaurice, J. Crowe, R.R. Dasari, M.S. Feld, Diagnosing breast cancer by using Raman spectroscopy. Proc. Natl. Acad. Sci. U. S. A. **102**, 12371–12376 (2005)
21. A.S. Haka, Z. Volynskaya, J.A. Gardecki, J. Nazemi, J. Lyons, D. Hicks, M. Fitzmaurice, R.R. Dasari, J.P. Crowe, M.S. Feld, In vivo margin assessment during partial mastectomy breast surgery using Raman spectroscopy. Cancer Res. **66**, 3317–3322 (2006)
22. J. Kneipp, T. Bakker Schut, M. Kliffen, M. Menke-Pluijmers, G.J. Puppels, Characterization of breast duct epithelia: a Raman spectroscopic study. Vib. Spectrosc. **32**, 67–74 (2003)
23. K.E. Shafer-Peltier, A.S Haka, M. Fitzmaurice, J. Crowe, J. Myles, R.R. Dasari, M.S. Feld, Chemical basis for breast cancer diagnosis using Raman spectroscopy. Lasers Surg. Med. **3**(Suppl 14), 2 (2002)
24. K.E. Shafer-Peltier, A.S. Haka, M. Fitzmaurice, J. Crowe, J. Myles, R.R. Dasari, M.S. Feld, Raman microspectroscopic model of human breast tissue: implications for breast cancer diagnosis in vivo. J. Raman Spectrosc. **33**, 552–563 (2002)
25. N. Stone, R. Baker, K. Rogers, A.W. Parker, P. Matousek, Subsurface probing of calcifications with spatially offset Raman spectroscopy (SORS): future possibilities for the diagnosis of breast cancer. Analyst. **132**, 899–905 (2007)
26. H.J. van Manen, C. Otto, Hybrid confocal Raman fluorescence microscopy on single cells using semiconductor quantum dots. Nano Lett. **7**, 1631–1636 (2007)
27. N. Amharref, A. Bejebbar, S. Dukie, L. Venteo, L. Schneider, M. Pluot, M. Manfait, Discriminating healthy from tumor and necrosis tissue in rat brain tissue samples by Raman spectral imaging. Biochimica Et Biophys. Acta-Biomembr. **1768**, 2605–2615 (2007)
28. S. Koljenovic, L.P. Choo-Smith, T.C.B. Schut, J.M. Kros, H.J. van den Berge, G.J. Puppels, Discriminating vital tumor from necrotic tissue in human glioblastoma tissue samples by Raman spectroscopy. Lab. Invest. **82**, 1265–1277 (2002)
29. Z. Liu, X. Li, S.M. Tabakman, K. Jiang, S. Fan, H. Dai, Multiplexed multicolor Raman imaging of Live cells with isotopically modified single walled carbon nanotubes. J. Am. Chem. Soc. **130**, 13540–13541 (2008)
30. H.P. Buschman, G. Deinum, J.T. Motz, M. Fitzmaurice, J.R. Kramer, A. van der Laarse, A.V. Bruschke, M.S. Feld, Raman microspectroscopy of human coronary atherosclerosis: Biochemical assessment of cellular and extracellular morphologic structures in situ. Cardiovasc. Pathol. **10**, 69–82 (2001)
31. H.P. Buschman, J.T. Motz, G. Deinum, T.J. Romer, M. Fitzmaurice, J.R. Kramer, A. van der Laarse, A.V. Bruschke, M.S. Feld, Diagnosis of human coronary atherosclerosis by morphology-based Raman spectroscopy. Cardiovasc. Pathol. **10**, 59–68 (2001)

32. K.E. Shafer-Peltier, A.S. Haka, J.T. Motz, M. Fitzmaurice, R.R. Dasari, M.S. Feld, Model-based biological Raman spectral imaging. J. Cell. Biochem. Suppl 39, 125–137 (2002)
33. S.W.E. van de Poll, T.J. Romer, O.L. Volger, D.J.M. Delsing, T.C.B. Schut, H.M.G. Princen, L.M. Havekes, J.W. Jukema, A. van der Laarse, G.J. Puppels, Raman spectroscopic evaluation of the effects of diet and lipid-lowering therapy on atherosclerotic plaque development in mice. Arter. Thromb. Vasc. Biol. **21**, 1630–1635 (2001)
34. H.W. Wang, T.T. Le, J.X. Cheng, Label-free imaging of arterial cells and extracellular matrix using a multimodal CARS microscope. Opt. Commun. **281**, 1813–1822 (2008)
35. B. De Jong, T. Bakker Schut, J. Coppens, K. Wolffenbuttel, D. Kok, G.J. Puppels, Raman spectroscopic detection of changes in molecular composition of bladder muscle tissue caused by outlet obstruction. Vib. Spectrosc. **32**, 57–65 (2003)
36. B. De Jong, T. Bakker Schut, K. Wolffenbuttel, J. Nijman, D. Kok, G.J. Puppels, Identification of bladder wall layers by Raman spectroscopy. J. Urol. **168**, 1771–1778 (2002)
37. B.R. Wood, S.J. Langford, B.M. Cooke, F.K. Glenister, J. Lim, D. McNaughton, Raman imaging of hemozoin within the food vacuole of plasmodium falciparum trophozoites. Febs Lett. **554**, 247–252 (2003)
38. J.R. Goodwin, L.M. Hafner, P.M. Fredericks, Raman spectroscopic study of the heterogeneity of microcolonies of a pigmented bacterium. J. Raman Spectrosc. **37**, 932–936 (2006)
39. C. Sandt, T. Smith-Palmer, J. Pink, L. Brennan, D. Pink, Confocal Raman microspectroscopy as a tool for studying the chemical heterogeneities of biofilms in situ. J. Appl. Microbiol. **103**, 1808–1820 (2007)
40. M. Baranska, H. Schulz, Spatial tissue distribution of polyacetylenes in carrot root. Analyst **130**, 855–859 (2005)
41. M. Baranska, H. Schulz, P. Rosch, M.A. Strehle, J. Popp, Identification of secondary metabolites in medicinal and spice plants by NIR-FT-Raman microspectroscopic mapping. Analyst **129**, 926–930 (2004)
42. J.R. Beattie, S.J. Bell, A.M. Fearon, B.W. Moss, Variation in fatty acid profile of porcine adipose tissue measured by Raman spectroscopy, in *International Workshop on Muscle Spectroscopy*, Matforsk, Oslo, 27–28 Sept 2007
43. J.R. Beattie, C. Maguire, S. Gilchrist, L.J. Barrett, C.E. Cross, F. Possmayer, M. Ennis, J.S. Elborn, W.J. Curry, J.J. McGarvey, B.C. Schock, The use of Raman microscopy to determine and localize vitamin E in biological samples. Faseb J. **21**, 766–776 (2007)
44. S. Koljenovic, T.C.B. Schut, J.P. van Meerbeeck, A. Maat, S.A. Burgers, P.E. Zondervan, J.M. Kros, G.J. Puppels, Raman microspectroscopic mapping studies of human bronchial tissue. J. Biomed. Opt. **9**, 1187–1197 (2004)
45. H.J. van Manen, Y.M. Kraan, D. Roos, C. Otto, Single-cell Raman and fluorescence microscopy reveal the association of lipid bodies with phagosomes in leukocytes. Proc. Natl. Acad. Sci. U. S. A. **102**, 10159–10164 (2005)
46. C. Matthaus, T. Chernenko, J.A. Newmark, C.M. Warner, M. Diem, Label-free detection of mitochondrial distribution in cells by nonresonant Raman microspectroscopy. Biophys. J. **93**, 668–673 (2007)
47. N. Uzunbajakava, A. Lenferink, Y. Kraan, E. Volokhina, G. Vrensen, J. Greve, C. Otto, Nonresonant confocal Raman imaging of DNA and protein distribution in apoptotic cells. Biophys. J. **84**, 3968–3981 (2003)
48. Y.S. Huang, T. Karashima, M. Yamamoto, H. Hamaguchi, Molecular-level investigation of the structure, transformation, and bioactivity of single living fission yeast cells by time- and space-resolved Raman spectroscopy. Biochem. **44**, 10009–10019 (2005)
49. Y.S. Huang, T. Nakatsuka, H. Hamaguchi, Behaviors of the "Raman spectroscopic signature of life" in single living fission yeast cells under different nutrient, stress, and atmospheric conditions. Appl. Spectrosc. **61**, 1290–1294 (2007)
50. A. Taleb, J. Diamond, J.J. McGarvey, J.R. Beattie, C. Toland, P.W. Hamilton, Raman microscopy for the chemometric analysis of tumor cells. J. Phys. Chem. B. **110**, 19625–19631 (2006)

51. H.J. van Manen, Y.M. Kraan, D. Roos, C. Otto, Intracellular chemical imaging of heme-containing enzymes involved in innate immunity using resonance Raman microscopy. J. Phys. Chem. B. **108**, 18762–18771 (2004)
52. H.J. van Manen, N. Uzunbajakava, R. van Bruggen, D. Roos, C. Otto, Resonance Raman imaging of the NADPH oxidase subunit cytochrome b(558) in single neutrophilic granulocytes. J. Am. Chem. Soc. **125**, 12112–12113 (2003)
53. J.R. Beattie, S. Brockbank, J.J. McGarvey, W.J. Curry, Effect of excitation wavelength on the Raman spectroscopy of the porcine photoreceptor layer from the area centralis. Mol. Vis. **11**, 825–832 (2005)
54. J.R. Beattie, S. Brockbank, J.J. McGarvey, W.J. Curry, Raman microscopy of porcine inner retinal layers from the area centralis. Mol. Vis. **13**, 1106–1113 (2007)
55. S. Finnegan, Proteomic and raman analysis of the developing wild type and retinal dysplasia and degeneration chick retina. Ph.D. in Faculty of Medicine, Health and Life Sciences. Queen's University Belfast, 2008
56. J.F. Brennan, T.J. Romer, R.S. Lees, A.M. Tercyak, J.R. Kramer, M.S. Feld, Determination of human coronary artery composition by Raman spectroscopy. Circulation **96**, 99–105 (1997)
57. T.J. Romer, J.F. Brennan, M. Fitzmaurice, M.L. Feldstein, G. Deinum, J.L. Myles, J.R. Kramer, R.S. Lees, M.S. Feld, Histopathology of human coronary atherosclerosis by quantifying its chemical composition with Raman spectroscopy. Circulation **97**, 878–885 (1998)
58. C. Krafft, D. Codrich, G. Pelizzo, V. Sergo, Raman and FTIR imaging of lung tissue: Methodology for control samples. Vib. Spectrosc. **46**, 141–149 (2008)
59. C. Krafft, D. Codrich, G. Pelizzo, V. Sergo, Raman mapping and FTIR imaging of lung tissue: congenital cystic adenomatoid malformation. Analyst **133**, 361–371 (2008)
60. J.R. Beattie, S.E.J. Bell, C. Borgaard, A. Fearon, B.W. Moss, Prediction of adipose tissue composition using Raman spectroscopy: Average properties and individual fatty acids. Lipids **41**, 287–294 (2006)
61. J.R. Beattie, S.E.J. Bell, C. Borgaard, A.M. Fearon, B.W. Moss, Multivariate prediction of clarified butter composition using Raman spectroscopy. Lipids **39**, 897–906 (2004)
62. J.R. Beattie, S.E.J. Bell, C. Borggaard, A.M. Fearon, B.W. Moss, Classification of adipose tissue species using raman spectroscopy. Lipids **42**, 679–685 (2007)
63. W.W. Christie, D. McEwan Jenkinson, J.H. Moore, Variation in lipid composition through the skin and subcutaneous adipose tissue of pigs. J. Sci. Food Agric. **23**, 1125–1129 (1972)
64. N. Uzunbajakava, C. Otto, Combined Raman and continuous-wave-excited two-photon fluorescence cell imaging. Opt. Lett. **28**, 2073–2075 (2003)
65. D. Graham, R. Goodacre, Chemical and bioanalytical applications of surface enhanced Raman scattering spectroscopy. Chem. Soc. Rev. **37**, 883–884 (2008)
66. C.W. Freudiger, W. Min, B.G. Saar, S. Lu, G.R. Holtom, C. He, J.C. Tsai, J.X. Kang, X.S. Xie, Label-free biomedical imaging with high sensitivity by stimulated Raman scattering microscopy. Science **322**, 1857 (2008)
67. S. Srinivasan, M. Schulmerich, J.H. Cole, K.A. Dooley, J.M. Kreider, B.W. Pogue, M.D. Morris, S.A. Goldstein, Image-guided Raman spectroscopic recovery of canine cortical bone contrast in situ. Opt. Express **16**, 12190–12200 (2008)

Chapter 8
Mapping Chemical and Structural Composition of Pharmaceutical and Biological Samples by Raman, Surface-Enhanced Raman and Fluorescence Spectral Imaging

Igor Chourpa, Simone Cohen-Jonathan and Pierre Dubois

Abstract Raman spectroscopy is an analytical technique recognised for its structural and conformational specificity. The efficient discrimination of molecular species by Raman is particularly potent for multidimensional microscopic imaging of complex biological environment, as demonstrated in the present book. The commonly admitted problem of Raman, low sensitivity, can often be circumvented due to high output instruments and via approaches like RRS (resonance Raman scattering), SERS (surface-enhanced Raman scattering), TERS (tip-enhanced Raman scattering) or CARS (coherent anti-Stokes Raman scattering). In contrast to the latter, RRS and SERS are realizable with less sophisticated set-up based on common Raman systems. Although more invasive than RRS, SERS provides better sensitivity and quenching of fluorescence. SERRS (surface-enhanced resonance Raman scattering) spectroscopy can be used in coupling with fluorescence and competes in selectivity and sensitivity with spectrofluorimetry. In the chapter below, we use recent applications made in our group to illustrate the use of Raman and SERRS spectral imaging for characterization of biological samples (animal subcutaneous tissue, human cancer cells) and pharmaceutical samples (microparticles for drug delivery, fibres for wound dressing). After a brief description of experimental details on spectral imaging, the chapter will focus on results concerning (i) biocompatible pharmaceutical materials made of alginates and (ii) anticancer drugs in pharmaceutical forms and in biological systems.

8.1 Introduction

Raman spectroscopy is an analytical technique recognised for its structural and conformational specificity. The efficient discrimination of molecular species by Raman is particularly potent for multidimensional microscopic imaging of complex biological

I. Chourpa (✉) · S. Cohen-Jonathan · P. Dubois
EA 6295, Nanomédicaments et Nanosondes,
Université François Rabelais de Tours, 37200 Tours, France
e-mail: chourpa@univ-tours.fr

A. Zoubir (ed.), *Raman Imaging*, Springer Series in Optical Sciences 168,
DOI: 10.1007/978-3-642-28252-2_8,

environment, as demonstrated in the present book. The commonly admitted problem of Raman, low sensitivity, can often be circumvented due to high output instruments and via approaches such as resonance Raman scattering (RRS), surface-enhanced Raman scattering (SERS), tip-enhanced Raman scattering (TERS) or coherent anti-Stokes Raman scattering (CARS). In contrast to the latter, RRS and SERS are realizable with less sophisticated setup based on common Raman systems. Although more invasive than RRS, SERS provides better sensitivity and quenching of fluorescence. SERRS spectroscopy can be used in coupling with fluorescence and competes in selectivity and sensitivity with spectrofluorimetry. In this chapter, we use recent applications made in our group to illustrate the use of Raman and SERRS spectral imaging for characterization of biological samples (animal subcutaneous tissue, human cancer cells) and pharmaceutical samples (microparticles for drug delivery, fibres for wound dressing). After a brief description of experimental details on spectral imaging, this chapter will focus on results concerning (i) biocompatible pharmaceutical materials made of alginates and (ii) anticancer drugs in pharmaceutical forms and in biological systems.

8.2 Experimental

8.2.1 Instrumental Setup

All the results presented in this chapter were obtained with a commercially available dispersive Raman confocal microspectrometer (LabRam, Horiba Jobin Yvon, Villeneuve d'Ascq, France). Thus, the applications we describe here are accessible to a wide research community. The experimental setup is schematised in Fig. 8.1. The key feature of our instrument is that it operates at either Raman- or fluorescence-compatible spectral resolution that can be chosen by rapid switching between the two gratings: 1,800 or 300 grids/mm. We equipped our spectrometer with three lasers to have excitation wavelengths from visible to near-infrared: 457.9, 488 and 514.5 nm with an air-cooled argon laser (35-LAP-431-230, Melles Griot, Voisins Le Bretonneaux, France), 632.8 nm with the built-in air-cooled He–Ne laser and 785 nm with a diode laser. Laser power on the samples never exceeded 8 mW (used to study fibers and animal tissue) and was most often attenuated with a set of optical density filters (OD from 0.3 to 4), down to less than 100 μW when applied to study cancer cells. The Olympus BX40 microscope of the microspectrometer is equipped with four objectives ($10\times, 50\times, 50\times$ LWD and $100\times$) which are employed to focus laser light onto the sample and to collect the backscattered Raman signal from the small volume in the focal plane of the objective hereafter called focal volume. The dimensions of the focal volume are set by means of an adjustable confocal hole (spatial filter). Prior to pass to spectrometer and to an air-cooled CCD detector, the light collected from the sample is also spectrally filtered by a Notch filter which rejects the excitation wavelength. The spectral window for Raman and/or fluorescence spectra detection varied from experiment to experiment. The spectra were collected at least over the Raman fingerprint region (200–1,800 cm^{-1}), while the high frequency region (up to 5,000 cm^{-1}) was recorded only for certain applications.

White light microscopic images of samples were captured with a colour CCD camera (WV-CP454E, Panasonic, La Plaine Saint Denis, France). To map the surface or the optical section of the sample, selected area of typically 25 × 25 μm (for cells) to 50 × 50 μm (for tissues and fibres) was scanned in a point-by-point mode by displacement of the motorised table with intervals of 0.8–2 μm (*X*, *Y*) and 2–5 μm (*Z*). The acquisition time was from 0.02 to 20 s per fluorescence/SERS and Raman spectrum, respectively. Multispectral matrix of 900 (30 × 30)–2,500 (50 × 50) spectra were recorded and treated by means of Labspec software (Horiba Jobin Yvon, Villeneuve d'Ascq, France).

8.2.2 *Brief Description of Principle of Spectral Imaging*

The first step of spectral imaging consists in acquisition of a matrix file where multiple Raman spectra are assigned to a defined number of sample points within a scanned region. The scan can cover either isolated points or neighbouring points distributed along one or more spatial axes. Pretreatment step of a multispectral matrix may include baseline correction, elimination of detector-related artefacts and extraction of a spectral region of interest (for instance, of the Amide I region related to protein secondary structure).

The following steps of spectral analysis are critically important since they are intended to generate one or more parametric images from spectral features characteristic of the sample composition. The successfully chosen spectral features discriminate the sample regions and provide resolved and morphologically relevant images. The algorithms of generating parametric images and their graphical presentation are determinant with respect to qualitative and quantitative interpretation of the Raman multispectral maps.

Basic treatment options available in most of the spectral mapping software consist in generating images corresponding to an average intensity (or integrated area) in one or more spectral regions (determined with cursors). The use of the intensity ratio of different Raman bands allows to compensate the variable overall spectral intensity through the sample. For instance, most often, Raman images of tissues are generated using the intensity of a selected Raman band normalised on the intensity of the band situated at ca. 1,450 cm^{-1} (CH_2 scissor mode influenced by environment but considered as nonspecific, i.e. present for any molecule containing CH_x groups, including proteins and lipids) [1].

More advanced algorithms of generating images can be developed based on curve-fitting options, enabling to fit the experimental spectra either with theoretical profiles (Lorentzian for Raman and Gaussian for fluorescence bands) or with experimentally determined reference spectra. The reference spectra can be established from independent measurements with reference samples (well-defined composition, environment) and/or extracted from characteristic regions of the same or similar map (region-average spectra, spectra representative of clusters or of principal components revealed by statistical analysis). In this case, the images correspond to fitting coefficients (scores) assimilated to the semi-quantitative determination of molecular composition/structure. These advanced options require some additional expertise in experimental modelling and in statistical methods, but are really interesting since permit to develop more specific approaches of molecular imaging of so complex samples as biological ones.

8.3 Results

8.3.1 I. Biocompatible Materials Made of Alginates

8.3.1.1 (a) Alginate Fibres as Material for Wound Dressing

Alginates are biocompatible hydrophilic polysaccharides, usually extracted from brown algae. Alginates are found as stabilising compounds in the extracellular polymeric substances of certain biofilms. An alginatic acid is a linear copolymeric chain consisting of α–*L* guluronic (G) and β–*D* mannuronic (M) acids organised in three kinds of sequence blocks, G–G, M–M (Fig. 8.1) and M–G (not shown). Bacterial alginates are partially acetylated on the O-2 and O-3 positions of the mannuronate units [2] and their guluronate/mannuronate ratio is usually lower than in algal alginates.

The ability of alginates to form highly viscous solutions or gels by complexing with divalent metal ions as Ca($^{++}$) has lead to their extensive use in biotechnology and biomedicine: to immobilize or encapsulate champagne yeast, enzymes, living cells and drugs [3]. If to extrude aqueous sodium alginate solution into a calcium chloride spinning bath [4] one can obtain calcium alginate fibres used to make wound dressings with remarkable moist healing properties.

The M/G ratio and the sequential arrangement of the two uronic acids are naturally critical for physico-chemical properties of alginates and therefore for their use [5]. We have demonstrated [6] that the M- and G-rich alginates can be rapidly characterised from their characteristic Raman spectra, namely from the intensity ratio of the bands at $811/959\,\mathrm{cm}^{-1}$ (skeletal stretching ν(C–C) and deformational δ(C–C–H) and δ(C–O) modes) (Fig. 8.1). Raman characterization is rapid and non-destructive compared to the chemical techniques traditionally used NMR alternative methods of the M/G determination [7]. Later study by Salomonsen et al. [8] compared Raman with two complementary vibrational spectroscopic techniques, IR and NIR, as methods for a rapid determination of the mannuronic acid/guluronic acid ratio (M/G ratio) of commercial alginate samples. They concluded that the chemometrics model based on the Raman spectra was the simplest and therefore probably also more

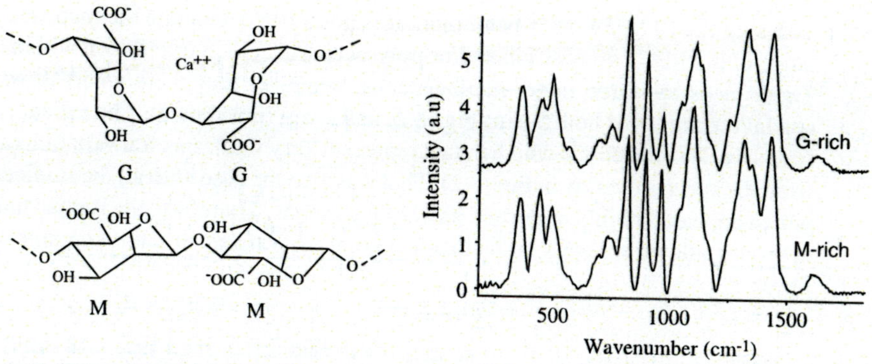

Fig. 8.1 *Left* Structural formulae of G–G and M–M blocks of alginate polysaccharides. *Right* Characteristic Raman spectra of calcium salts of G-rich and M-rich alginates

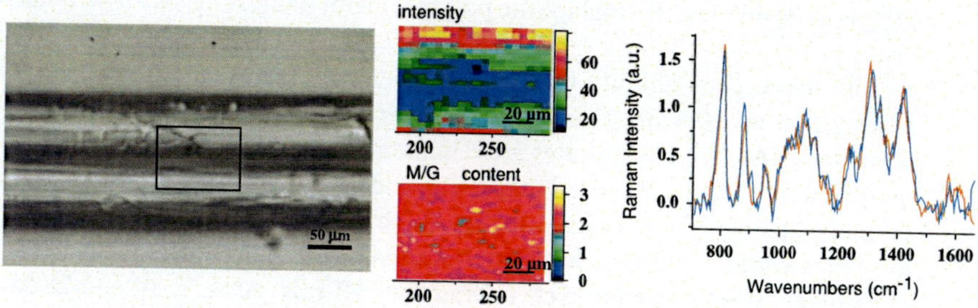

Fig. 8.2 *Left* Video capture of a fibre made of calcium salt of a G-rich alginate; the frame indicates the scanned region (*XY* optical section situated at the half-depth of the fibre). *Middle* Spectral maps of the average Raman intensity (bands between 750 and 1,650 cm^{-1}) and M/G content (ratio of the bands at 811/959 cm^{-1}). *Right* Spectra from the *blue* and *yellow* points of the "M/G content" image

robust to predicting new samples. An error was comparable to that of the solution-state ^{1}H NMR reference method which requires the destructive and time-consuming hydrolysis procedure. Otherwise one may apply a more sophisticated approach based on the CP-MAS NMR (C-13 cross-polarisation magic angle spinning nuclear magnetic resonance) spectrometry coupled to multivariate curve resolution (MCR) [9]. In contrast to the NMR, the vibrational spectroscopic techniques have a high industrial potential for at- or online quality control as well as screening of large numbers of samples in industry and research as illustrated with the results below.

We used the confocal Raman spectral imaging with red laser excitation (632.8 nm, 8 mW, 5 s/spectrum) to study the structural organisation of the fibres of calcium alginate (G-rich, Fig. 8.2) [6]. The changes noted in the relative intensity of Raman bands, namely the skeletal stretching modes cited above, when the fibres were oriented parallel or perpendicular to laser polarisation [6], indicated anisotropy of Raman effect that signifies that the alginate chains are highly oriented parallel to the fibre axis.

Using the 811/959 cm^{-1} band ratio, it is possible to generate the maps representative of the M/G composition and/or polysaccharide chain orientation within fibres. These maps appeared rather homogeneous, wherever they were recorded with the spatial resolution of both 5 μm (Fig. 8.2) and 1 μm (not shown). The enhanced contrast reveals some points which in fact represent very weak spectral variation (as seen from the spectra overlay in Fig. 8.2). That was also the case with all the studied fibres which appeared highly organised and homogeneous. Therefore, the Raman mapping approach allows to assess both chemical composition and structural organisation of fibres.

8.3.1.2 (b) Bioresorption of Natural and Amphiphilic Alginates

Chemical substitution of uronic residues of alginates with Cn hydrophobic chain results in novel class of amphiphilic polymers with interesting self-assembling properties [10].

The tissue compatibility of newly synthetized amphiphilic alginates (AA, C_{12} substitution on 10% of uronic residues) was examined in comparison with native alginates (NA). These substrates were implanted in rats, by intracutaneous injection on the back, a site widely used in earlier toxicity and biocompatibility studies. Subdermal implants of the alginate matrix (both NA and AA) did not exhibit toxicity in vivo and did not induce an acute inflammatory response in the host. Both type of implants were clinically well tolerated by all rats. At 4, 7, 14 and 28 days post implantation, the implants with surrounding tissue were retrieved and processed both histologically and by confocal Raman microspectroscopy [11].

Within short-term implantation (4–7 days), the clinical observations were similar regardless of site or type of polymer implant used. Histopathology revealed well-circumscribed implants with pseudoencapsulation and mixed inflammatory cells predominating at the periphery, as could be expected with implantation of any synthetic substance. Some foreign-body giant cells did accumulate around the polymer, suggesting its degradation is facilitated by solubilisation, hydrolysis as well as by cell activity. Long-term subdermal implants of both polymers allowed marked infiltration of vascular and connective tissue, suggesting the free flow of fluids and nutrients in the implants. Over 14 days of implantation, there were substantial differences between the implanted AA and NA. The long-term bioresorption proceeded more rapidly with AA implants than with NA ones. This was concluded from comparing the volume lost and biocolonization of implants, both being particularly pronounced with AA.

The molecular composition of the implants as well as of the surrounding tissues has been further analysed using confocal Raman spectral imaging (Fig. 8.3) with red excitation (632.8 nm, 8 mW, 10 s/spectrum). The Raman spectra revealed that both types of the implants were largely enriched in plasma proteins, including the very short-term implantation. The Raman patterns from the tissue/implant interface were in agreement with results of the studies demonstrating electrostatic interactions of alginates with serum proteins [12]. Plasma protein adsorption to the implant surface may be the first reaction between the implant and the surrounding tissue. On the

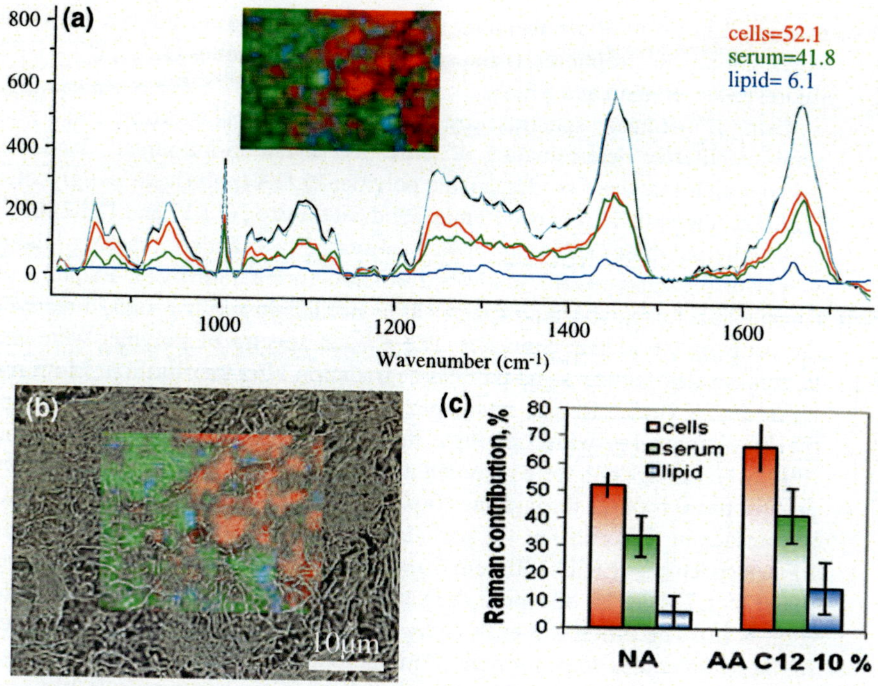

Fig. 8.3 Raman image of a rat tissue (cryo-section, dye-free) in the region at the interface implant of NA. **a** Example of the spectrum deconvolution into three reference spectra representative of cells (*red*), serum proteins (*green*) and lipids (*blue*) established from the characteristic regions of tissue maps. The inset image shows the merge of their distribution over the scanned region. **b** Merge of the Raman image with *white* light image of the tissue section. **c** Semi-quantitative results averaged from ten Raman images (5 for AA and 5 for NA)

other hand, the AA implants exhibited an additional lipid component, increasing with longer implantation time. It appeared in correlation with the increased lipophilicity of the AA molecules. The increased colonisation of the AA implants by cells is also indicating the good biocompatibility of these materials. These results may form the basis for further research into the real merits of the amphiphilic alginates, namely, in tissue-engineering applications.

8.3.2 Anticancer Drugs in Pharmaceutical Forms and in Biological Systems

8.3.2.1 (a) Anticancer Drug in Microparticles for Drug Delivery

Nowadays, an extensive pharmaceutical research concerns development of novel systems for targeted drug delivery using micro- and nanoparticles. In our laboratory, the chemical composition of micro- and nanoparticular vectors of anticancer drugs is studied using multiple methods including separative techniques (chromatography, capillary electrophoresis) and spectroscopic techniques (AAS, UV–visible, IR, fluorescence, Raman and SERS).

Using FT-Raman spectroscopy, we developed an approach that allows a semi-quantitative determination of anticancer agent (5-iodouracil) loaded within microparticles made of biodegradable polymer PLGA (poly(lactic-co-glycolic acid) [13]. The intensity ratio of two Raman bands of the drug (1,617 and 1,682 cm^{-1}) over one of the polymer (1,774 cm^{-1}) was calibrated to calculate the drug loading (%, w/w) with the standard error as low as 2%. The determination by FT-Raman is fast and is applicable to analyse microspheres after sterilisation with γ-rays, directly though the walls of the plastic container. The Raman spectra of polymer were analysed to confirm that there was no particle degradation after sterilisation. More recently, by means of visible Raman microspectrometer, we performed Raman mapping of the drug repartition within a single microsphere loaded with ~10% of the drug (Fig. 8.4). The spectra were recorded using the 632.8 nm laser excitation (8 mW, 5 s per spectrum) focused through the 100x objective to an optical section ~5 μm below the surface in the centre of the particle. The maps were generated by fitting each Raman spectrum (1,550–1,850 cm^{-1}) with two model spectra: that of pure polymer (band at 1,774 cm^{-1}) and pure drug in powder (bands at 1,617 and 1,682 cm^{-1}) (Fig. 8.4d). The fitting was enabled by the fact that no spectral changes were found for these molecules when formulated in microparticle. The polymer map (Fig. 8.4b) is less intense inside thus indicating that the particle has a capsule-like structure. The drug map (Fig. 8.4c) reveals its non-homogeneous repartition in the form of microcrystals is located within the capsule. This kind of information, i.e. concerning both size and location of microcrystals of API, is useful for pharmaceutical research and development, namely for optimising formulation protocols to obtain more or less rapid drug delivery. Furthermore, the Raman microspectroscopy can be used to follow in situ PLGA microsphere degradation inside macrophages, as demonstrated by Van Apeldoorn et al. [14].

8.3.2.2 (b) Anticancer Drug Within Living Cancer Cell

Numerous anticancer drugs possess a conjugated aromatic chromophore moiety that enables their detection by fluorescence and Raman spectroscopy. In view of the high cytotoxicity, the study of these active molecules within living cancer cells implies to deal with micromolar and sub-micromolar concentrations: the typical lethal concentrations (LC_{50}) of the chemotherapy agents are often as low as few nM. For spectral imaging of intracellular anticancer drugs at micromolar and sub-micromolar concentrations, fluorescence and SERS have to be used instead of conventional Raman (reviewed in [15]). In most of our experiments, silver nanoparticles were incubated into cells already treated with the drug (typically 0.5–2 μM), to minimise their possible influence on the natural intracellular repartition of the drug. However, only few colloids can be internalised without dramatic perturbation of the cell integrity. Furthermore, the intracellular distribution of aggregated colloids able to provide SERS

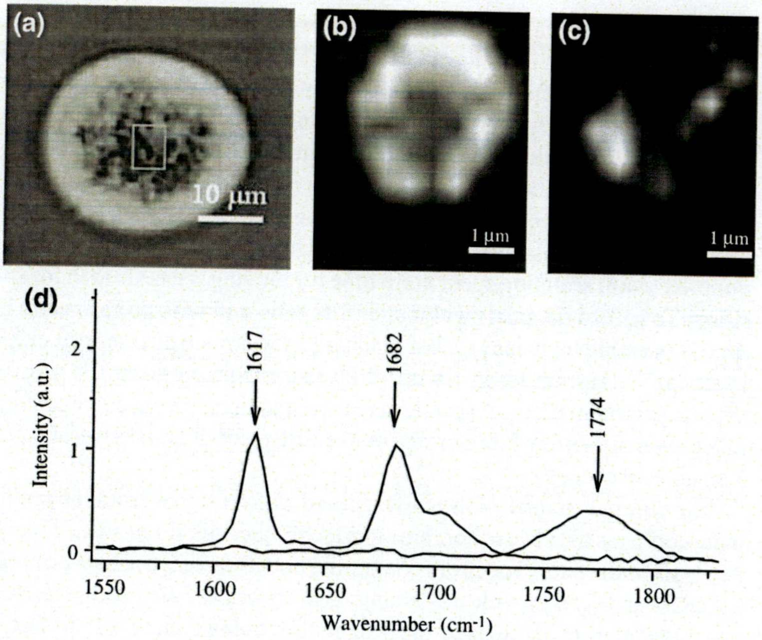

Fig. 8.4 Repartition of anticancer drug 5-iodouracil loaded within a microparticle of biodegradable polymer PLGA. **a** Video capture of the particle (negative image contrast) and the scanned region. **b** and **c** are the spectral maps of the polymer (band at 1,774 cm^{-1}) and of the drug (bands at 1,617 and 1,682 cm^{-1}), respectively. The drug/polymer bands intensity ratio reached 2,6 in the brightest zones of the map in **c**

is random and they can enhance Raman signal only from the molecular species adsorbed on their surface. As a result, with this kind of nano-substrates the SERS map cannot provide as continuous and morphologically objective image of the cellular compartments as enabled by use of fluorescence. To join the homogeneous mapping with structurally specific vibrational spectra, our suggestion is to combine both SERS and fluorescence techniques on the same map [15]. This method is illustrated below using anticancer drug mitoxantrone (MTX) incubated within MCF7 cancer cells (human breast cancer).

In the study presented here, 10^6 cells were treated for 1 h with 0.5 μM MTX solution in RPMI medium, then washed with PBS buffer pH 7.4 and incubated with silver colloids for 5 min. Mild centrifugation (100 g for 2 min) has been applied to stimulate mechanically the nanoparticle internalisation by cells. Spectral map acquisition was performed with a long distance (50×) objective, by focusing a 632.8 nm laser beam through the thin glass cover of the microscopy chamber (Harvard Instruments). The chamber was thermostabilised to 37°C to respect the cell physiology. High signal level allows to record MTX fluorescence and SERRS spectra of exploitable S/N ratio using laser power at the sample estimated to be below 50 μW and integration time as short as 0.05 s per spectrum. These conditions respect cell integrity, as checked by the fact that it was possible to record at least 2 reproducible fluorescence maps from each cell (data not shown).

As can be seen from the results shown below, the main subcellular features of interest can be clearly seen with the scanning step of 0.7–1 μm, i.e. close to the limit of the lateral resolution [16]. Even higher quality of the chemical images potentially useful to map subcellular organelles can be attained with the use of higher numerical aperture objective coupled with a scanning step 5–10 times smaller than the laser spot size, as demonstrated recently [17]. Unfortunately, our instrument working in point-by-point scanning spent more time for moving the table than for spectral acquisition. To record the maps with better S/N ratio and with more spectra in reasonable time is possible by means of fast scanning modes such as line scan or SWIFT™ and DuoScan™ (Horiba Jobin Yvon). With these technologies, it is possible to reduce the measurement time per spectrum to 5 ms and to collect about 50,000 spectral maps in less than 10 min, while preserving the full spectral and confocal advantage of the Raman microscope.

The complex nature of the SERRS (and SERS) signal from intracellular environment confirms the need to combine this technique with fluorescence spectral imaging. Analysis of the entire spectrum containing both fluorescence background profiles and SERRS bands can provide the complementary information on the microenvironment and molecular interactions of the drug within living cancer cell. In fact, the drug fluorescence band position and shape depict the local microenvironment in different cellular compartments and the position and ratio of SERS bands indicate the drug moiety(ies) involved in the interaction in this compartment.

Under a resonance excitation (632.8 nm for red-absorbing MTX), it is possible to detect simultaneously fluorescence and SERRS of the drug (Fig. 8.5). This is the case if the microvolume contains both SERRS-active drug molecules adsorbed on nanoparticle surface (where the fluorescence is quenched) and fluorescence-active drug molecules bound to cellular components comprised in the volume near and/or around nanoparticles. Depending on the presence of nanoparticles and on the possibility of the drug adsorption on their surface, the ratio of the SERRS/fluorescence intensity varies from zero to dominant SERRS signal.

Spectral imaging of fluorescence emission of anticancer drugs like mitoxantrone (Fig. 8.5) demonstrates the ability of this technique to monitor the spatio-temporal evolution of active molecule in treated cells. According to its main action (inhibition of nuclear enzyme topoisomerase), MTX is preferentially stained in nucleus where the drug is bound to DNA by intercalation mode. The drug intercalated between DNA bases undergoes $\pi - \pi^*$ interactions that leads to a bathochromic shift of its fluorescence emission spectrum and the decrease of its main SERS band at ca. 1,300 cm^{-1} [18]. Indeed, the red-shifted fluorescence of MTX can be detected from the intracellular regions corresponding to nucleus. We have previously assigned [19] four model fluorescence spectra (Fig. 8.5d) characteristic of MTX in nucleus, in cellular membrane, in cytosolic complex as well as the spectrum of an oxidative metabolite of the drug bound to endoplasmic reticulum (ER). The main emission maxima in these spectra are situated, respectively, at 700, 683, 696 and 654 nm. This set of model spectra allows to fit each intracellular fluorescence spectrum with error below 2%. The curve-fitting results (deconvolution coefficients expressed in % of spectral contribution to the total spectrum) result in four fluorescence-related maps that can be merged (Fig. 8.5c) for comparison with the white light image of the cell (Fig. 8.5a). The current spectral resolution of the spectrometer of 2 nm allows to resolve signal responses from multiple fluorescence species located within different compartments of the same cell. Being based on weak spectral changes such as shape modification and maxima shift of few nm, most of these maps would not be accessible with bandpass fluorescence microscopy where the main discriminating criteria is intensity. The most remarkable is the fact that MTX fluorescence from membrane is seen on the spectral map, although it is about 50 times weaker than that from nucleus. This is enabled by the fact that the shape and position of blueshifted fluorescence is recognised by spectral analysis before the intensity scale is applied.

In addition to subcellular structures seen from fluorescence, the spectral map in Fig. 8.5c shows localised spots corresponding to SERRS signal of MTX originating from micron-scale and sub-micron-scale aggregates of silver nanoparticles adhered to the internal side of the membrane (point 1) and to the ER region in the cytosol of the cell (point 2). The corresponding spectra and their deconvolution are shown in Figs. 8.5e and f, respectively. The fitting results illustrate how SERRS and fluorescence data can correlate in describing a given intracellular environment: ongoing from point 1 to point 2, both fluorescence and SERRS components characteristic to membrane decrease and those characteristic of cytosolic complex of MTX increase. The point 2 is dominated by the fluorescence of the cytosolic complex of MTX (53.9%) but has also an increased fluorescence contribution from the metabolite (21.9%). Among the challenges of coupling fluorescence with SERRS there is a possibility to identify the possible drug metabolites within cells. However, the origin of the subcellular spectra made the assignment of the SERRS model components more difficult than it was the case with fluorescence models. The SERRS spectra in the intracellular regions are variable from cell to cell, from aggregate to aggregate within the same cell and from time to time on the same aggregate. This is not surprising, since in addition to the overall microscopic situation, the SERRS signal depends on the mechanisms of adsorption–desorption of chromophores taking place in vicinity of the metal surface (discussed in [15]). To resolve this kind of problem one needs to cumulate a larger bank of experimental data over many cells and to sort the spectra by means of multivariate statistical approaches [20].

The main disadvantage of the conventional, electrostatically stabilised NP colloids, is that they aggregate and cannot be distributed homogeneously in cellular compartments. This limitation can be circumvented with use of plasmonic NPs coated with neutral polymers such as poly-ethylene glycol (PEG) that stabilise the colloids due to sterical hindrance. We have recently demonstrated that aggregation-free Ag–PEG NPs enable SERS detection of MTX due to the drug diffusion through the hydrated polymeric layer [21]. In addition, the SERS spectra of the drug on the Ag–PEG NPs were affected by addition of Cl-ions, thus confirming that ions and small chromophores can be studied with help of such substrates. Furthermore, the polymer can be used to carry biological ligands to target specific cells/intracellulaer compartments.

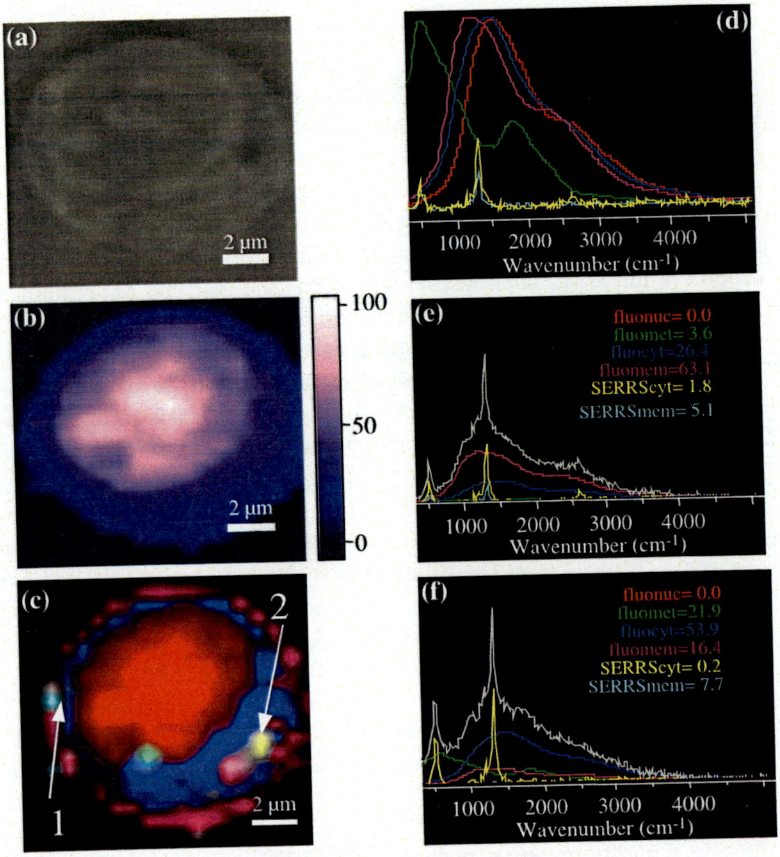

Fig. 8.5 Coupling fluorescence and SERRS on the same spectral map of a cancer cell (MCF7 line) treated with anticancer drug mitoxantrone (MTX). **a** Video capture of the cell in white light. **b** Spectral intensity (400–5,000 cm^{-1}) map normalised from 0 to 100%. **c**. Merged maps of spectra shown in window D and encoded with the same colours: *red*—fluorescence of MTX in nucleus; *blue*—fluorescence of MTX in cytosol; *magenta*—fluorescence of MTX in membrane; *green*—fluorescence of oxidative metabolite of MTX; *yellow*—SERRS of MTX in cytosol; *cyan*—SERRS of MTX in membrane. **d** Model spectra used to generate the maps in window C. Windows *e* and *f* illustrate deconvolution of the spectra from the points 1 and 2 of the map shown in window C

8.3.3 Conclusion and Perspectives

The applications cited in this chapter illustrate the structure-specific information on pharmaceutical and biological samples brought by Raman and SERS spectral imaging and demonstrate the complementarities of this method with other spectroscopic, microscopic and biochemical techniques. The perspectives of Raman bio-imaging are dependent on instrumental progress, chemometrical advances and the ability to integrate recent biological data. The potent of SERS imaging is in strong relation with enhancement substrates which may be named "optical contrast agents" similar to MRI contrast agents. The new generation of hybrid, organic–inorganic nanoparticles are being created for SERS imaging and are revolutioning the applicability and the reliability of this technique.

Acknowledgments The authors acknowledge Région Centre (France), Ligue Nationale Contre le Cancer (France), ARC (France), IFR-135 (Tours, France) and Laboratoire Brothier (Fontevraud l'Abbaye, France) for the financial support.

References

1. J.R. Beattie et al., Mol. Vis. **11**, 825–832 (2005)
2. T. Windhues, W. Borchard, Carbohydr. Polym. **52**, 47–52 (2003)
3. M. Rajaaonarivony, C. Vauthier, G. Courraze, F. Puisieux, P. Couvreur, J. Pharm. Sci. **82**, 912–917 (1993)
4. J.B. Speakman, N.H. Chamberlain, J. Soc. Dye. Colour. **60**, 264–274 (1944)
5. K.I. Draget, G. Skjäk-Braek, O. Smidsrød, Carbohydr. Polym. **25**, 31–38 (1994)
6. I. Chourpa, P. Carpentier, P. Maingault, P. Dubois, Proc. SPIE **3608**, 48–54 (1999)
7. G. Annison, N.W.H. Cheetham, I. Couperwhite, J. Chromatogr. **264**, 137–143 (1983)
8. T. Salomonsen, H.M. Jensen, D. Stenbaek, S.B. Engelsen, Carbohydr. Polym. **72**, 730–739, 2008
9. T. Salomonsen, H.M. Jensen, F.H. Larsen, S. Steuernagel, S.B. Engelsen, Carbohydr. Res. **344**(15), 2014–2022 (2009)
10. J.S. Yang, Y.J. Xie, W. He, Carbohydr. Polym. **84**, 33–39 (2011)
11. I. Chourpa, P. Carpentier, P. Maingault, F. Fetissoff, P. Dubois, Proc. SPIE **3918**, 166–173 (2000)
12. Y. Zhao, F. Li, M.T. Carvajal et al., J. Colloid Interface Sci. **332**, 345–353 (2009)
13. A. Geze, F. Boury, J.P. Benoit, I. Chourpa, P. Dubois, Analyst **124**, 37–42 (1999)
14. A.A. Van Apeldoorn, H. Van Manen, J.M. Bezemer, J.D. de Bruijn, C.A. Van Blitterswijk, C. Otto, J. Am. Chem. Soc. **126**(41), 13226–13227 (2004)
15. I. Chourpa, F.H. Lei, P. Dubois, M. Manfait, G.D. Sockalingum, Chem. Soc. Rev. **37**, 993–1000 (2008)
16. S. Sharonov, I. Nabiev, I. Chourpa, A. Feofanov, P. Valisa, M. Manfait, J. Raman Spectrosc. **25**, 699–707 (1994)
17. E. Lee, F. Adar, A. Whitley,The Application Notebook, Spectroscopy 21 (2006)
18. I. Nabiev, A. Baranov, I. Chourpa, A. Beljebbar, G.D. Sockalingum, M. Manfait, J. Phys. Chem. **99**, 1608–1613 (1995)
19. S. Vibet, K. Mahéo, J. Goré, P. Dubois, P. Bougnoux, I. Chourpa, Drug Metab. Dispos. **35**, 822–828 (2007)
20. S.J. Clarke, R.E. Littleford, W.E. Smith R. Goodacre, Analyst, **130**, 1019–1026 (2005)
21. A. Shkilnyy, M. Soucé, P. Dubois, F. Warmont, M.L. Saboungi, I. Chourpa, Analyst **134**, 1868–1872 (2009)

Chapter 9
Raman Imaging of Grain Orientation, Strain, Crystallinity and Doping Levels in Solar Silicon

M. Becker, G. Sarau, H. P. Strunk and S. Christiansen

9.1 Introduction

Micro-Raman spectroscopy is a method that has been used since the 1960s in solid state physics to investigate orientation of crystals, mechanical stresses, phase changes as well as dopant concentrations in semiconductor materials [35, 37, 71]. Raman spectroscopy measures these effects in crystals via an inelastic interaction of laser light with lattice vibrations. The method probes the material of interest nondestructively without requiring any complex sample preparations. Another beneficial aspect of Micro-Raman spectroscopy is that, compared to many other methods, no expensive vacuum equipment is needed for the measurements. Micro-Raman spectroscopy allows for a lateral resolution of the order of at least $\sim$200 nm $-$ 1.5 μm when focusing the incident light beam through the objective on the sample surface. In addition, Raman signals can be mapped, providing images of stress/strain-, orientation-, doping level distributions can be obtained [38].

In this contribution we explain how to use Micro-Raman spectroscopy to map and display some very important parameters for solar silicon, such as the spatial

M. Becker (✉)
Experimental Department 2, Max-Planck-Institut für Mikrostrukturphysik, Halle, Germany
e-mail: mbecker@mpi-halle.mpg.de

G. Sarau
Leuchs, Micro- and Nanostructuring, Max-Planck-Institut für die Physik des Lichts, Erlangen, Germany

H. P. Strunk
Institut für Werkstoffwissenschaften, Lehrstuhl Mikrocharakterisierung, Universität Erlangen-Nürnberg, Erlangen, Germany

S. Christiansen
Leuchs, Micro- and Nanostructuring, Max-Planck-Institut für die Physik des Lichts, Erlangen, Germany

S.Christiansen
Institut für Photonische Technologien, Jena, Germany

A. Zoubir (ed.), *Raman Imaging*, Springer Series in Optical Sciences 168,
DOI: 10.1007/978-3-642-28252-2_9, © Springer-Verlag Berlin Heidelberg 2012

distribution of stress/strain, crystal grain orientation, dopant level distribution, and crystallinity of the sample. We will present several qualitative maps of these parameters and describe in detail how to extract quantitative information from the respective images.

Though the theoretical descriptions and calculation procedures to obtain quantitative information are sometimes complicated and lengthy, often all necessary parameters can be obtained from one measurement cycle and once the calculation procedures are implemented, the calculations can be carried out rather rapidly. Our contribution starts with the description of the procedure to determine the crystal orientation of arbitrarily oriented silicon grains in polycrystalline solar cell wafers. Once the crystal orientation is determined, a detailed analysis of the stress/strain states within the different grains becomes feasible. A detailed description of the physical background concerning the stress/strain state measurements in polycrystalline materials as well as some illustrative examples of stress/strain measurements will be presented in the second part of this chapter.

Highly doped regions in mono or polycrystalline silicon solar cells, such as the emitter or the BSF region, are essential for the proper and effective operation of solar cells. A highly n-doped layer (emitter) at the upper surface of a silicon solar cell is necessary to produce a pn-junction and a highly p-doped layer BSF of a few μm thickness is necessary to prevent the light generated free carriers to recombine at the back surface of the solar cell. In the fourth part of this chapter we introduce a Raman-based method to quantitatively determine the free carrier concentration within these regions and to map their spatial distribution.

Finally, in the last part of our contribution, we shortly describe how information about the crystal quality, defect densities and the crystalline/amorphous ratio of the sample can be directly obtained from an analysis of the modified shape of the respective Raman peaks.

9.1.1 Imaging of Crystal Orientations

9.1.1.1 Qualitative Imaging

The intensity of the 3-fold degenerate F_{2g}—first-order Si Raman band depends on its Raman tensors, the scattering geometry during Raman measurements, e.g the polarization direction of the incident and backscattered light, and on the respective crystal orientation (with reference to a fixed reference coordinate system). The Raman tensors R'_j [65] for the three (degenerate) Raman peaks are given below and correspond to a Si crystal whose axes are exactly aligned along the axes of the reference coordinate system (e.g. $x = x' = (100)$, $y = y' = (010)$, $z = z' = (001)$ see Fig. 9.1 for the descriptions):

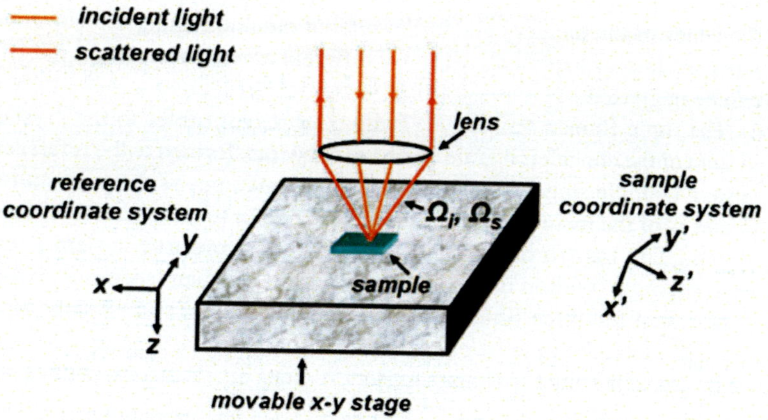

Fig. 9.1 Schematic drawing of a Micro-Raman spectroscopy setup. The incident laser beam is focused by a lens on the sample surface and the scattered light is collected by the same lens. Ω_i and Ω_s are the angles for the cones of the incident and scattered light underneath the objective lens. The polarization directions of the incident and backscattered beams are characterized by the unit vectors $\vec{e}_i$ and $\vec{e}_s$ respectively. The reference coordinate system remains fixed whereas all physical properties of the sample (e.g. Raman tensors, etc.) refer to the sample (*crystal*) coordinate system

$$R'_{\mathbf{x}} = \begin{pmatrix} 0 & 0 & 0 \\ 0 & 0 & d \\ 0 & d & 0 \end{pmatrix} R'_{\mathbf{y}} = \begin{pmatrix} 0 & 0 & d \\ 0 & 0 & 0 \\ d & 0 & 0 \end{pmatrix} R'_{\mathbf{z}} = \begin{pmatrix} 0 & d & 0 \\ d & 0 & 0 \\ 0 & 0 & 0 \end{pmatrix}, \tag{9.1}$$

where d represents a material constant that is only important in case a comparison of the Si Raman peak intensities with the Raman peak intensities of other materials becomes necessary. Usually the material parameter d is set to 1 to make the calculations easier. In the above described case, the totally measured Raman peak intensity is given by the simple approximation [73, 101]:

$$I(\vec{e}_i, \vec{e}_s) \approx I_0 \cdot \sum_{j=1}^{3} \left| \vec{e}_i \cdot R'_j \cdot \vec{e}_s \right|^2 \tag{9.2}$$

$\vec{e}_i$, $\vec{e}_s$ are the polarization vectors of the incident and scattered light, respectively. The scaling parameter I_0 contains all fixed experimental parameters (laser intensity, laser wavelength, measurement time, etc.). The index j discriminates the three phonon polarization directions x, y, z and R'_j are the corresponding Raman tensors as given in Eq. (9.1). The approximation given above represents the Raman intensity—scattering geometry relationship rather well when an objective with a small numerical aperture (e.g. a 10x objective) is used to focus the laser spot on the sample surface (see. Fig. 9.1).

Deviations from Eq. (9.2) occur when objectives with rather large numerical apertures are used. In this case the strict polarization settings for the incident and backscattered beam are somewhat "diluted" by the continuous polarization directions within the cones of the incident and backscattered radiation under the objective. However, when evaluating a qualitative map of different crystal orientations these inaccuracies can be neglected.

The three Raman tensors R'_j (with $j = x, y, z$) refer to the crystal coordinate system of the considered grain and have to be transformed to the reference coordinate system (sample stage) for arbitrary grain orientations, as the polarization directions $\vec{e}_i$ and $\vec{e}_s$ of the incident and scattered beams refer to the global reference coordinate system. The transformation into the reference coordinate system is accomplished by applying a rotation matrix $T(\alpha, \beta, \gamma)$ to the Raman tensors R'_j, where α, β and γ represent the three Euler angles. The explicit expression for the rotation matrix $T(\alpha, \beta, \gamma)$ can be found e.g. in [10].

Inserting the rotated Raman tensors R_j into Eq. (9.2), one obtains:

$$I(\vec{e}_i, \vec{e}_s, \alpha, \beta, \gamma) \approx I_0 \cdot \sum_{j=1}^{3} \left| \vec{e}_i \cdot \left(T^{-1}(\alpha, \beta, \gamma) \cdot R'_j \cdot T(\alpha, \beta, \gamma) \right) \cdot \vec{e}_s \right|^2. \quad (9.3)$$

A qualitative map of different crystal orientations within a certain mapping region can now easily be achieved by fixed settings of the polarization directions of the incident and backscattered beams. The intensities of the Raman peaks then depend only on the components of the rotation matrix $T(\alpha, \beta, \gamma)$ and therefore only on the specific crystal orientation. An example of such a qualitative orientation map is shown in Fig. 9.2. The left image displays an optical micrograph of a polycrystalline silicon wafer that is used for solar cells. After mechanical and chemical polishing and subsequent Secco etching [91], crystal defects such as e.g. dislocations, random- and twin boundaries become visible. The grain sizes (areas) in this case vary from a few μm^2 to many hundred μm^2. The right image represents the intensities of the first-order silicon Raman peak. The intensities vary according to the relation given in Eq. (9.3), where in this case the incident and backscattered polarization directions where both set to the y-position with respect to the reference coordinate system, or in Porto notation [30] $z(yy)\bar{z}$, as the direction of the incident beam is along the z-axis of the reference coordinate system and the backscattered signal is collected along the negative z-axis.

The image of the Raman intensity mapping shown in Fig. 9.2 (right image) reproduces very well the different grain orientations already visible due to the Secco-etched grain boundaries in the light optical micrograph (left image). But it also reveals grain boundaries that are not visible in the optical micrograph (white ellipse) due to an insufficient etching. Therefore, crystal orientation imaging performed by Micro-Raman spectroscopy provides more detailed and reliable informations about the crystal orientation distribution then a simple etching procedure.

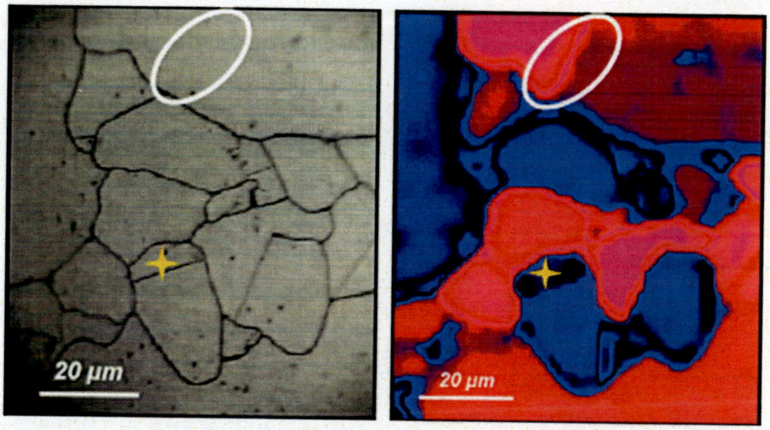

Fig. 9.2 *Left image* light optical micrograph of a polycrystalline silicon solar cell wafer. The grain boundaries are made visible by Secco-etching. The right image shows a qualitative mapping of the Raman intensities acquired within the same region. *Blue/black colour* coding corresponds to high Raman intensities and *red/purple* to low intensities. The mapping shows that the Raman intensities can drastically change with different grain orientations, where the intensities follow the relation given in Eq. (9.3). The grain boundary within the *white* ellipse is only visible in the Raman intensity mapping and not in the light optical micrograph since it might have been etched not properly. This observation shows that a simple qualitative Raman intensity map can reveal details about crystal orientations that would not be observable after a simple defect etching

9.1.1.2 Quantitative Determination of Crystal Orientations

A rather accurate quantitative determination of crystal orientation can be achieved when an objective with a small numerical aperture is used for the measurements of the Raman intensities. Equation (9.3) can be used as a basis for orientation measurements by Micro-Raman spectroscopy. The polarization direction $\vec{e}_i$ of the incident laser beam is usually adjusted by a rotatable $\lambda/2$-plate (see Fig. 9.3 for a detailed schematic drawing of the used Raman spectrometer setup and the paths of the incident and backscattered light).

Unfortunately, the influence of the mirrors and the notch-filter on the polarization direction of the incident beam after passing the $\lambda/2$-plate results in a dependence of the intensity of the incident laser beam on its polarization direction [51] and a difference between the effective polarization direction ϕ on the sample surface and the original polarization direction $\tilde{\phi} = 2\theta$ that is adjusted with the $\lambda/2$-plate [51] (θ is the rotation angle of the $\lambda/2$-plate). These slight deviations can be accounted for by introducing a correction factor k, e.g. the polarization vector of the incident beam is then elliptically dependent on the effective polarization direction ϕ at the sample surface and can be written as:

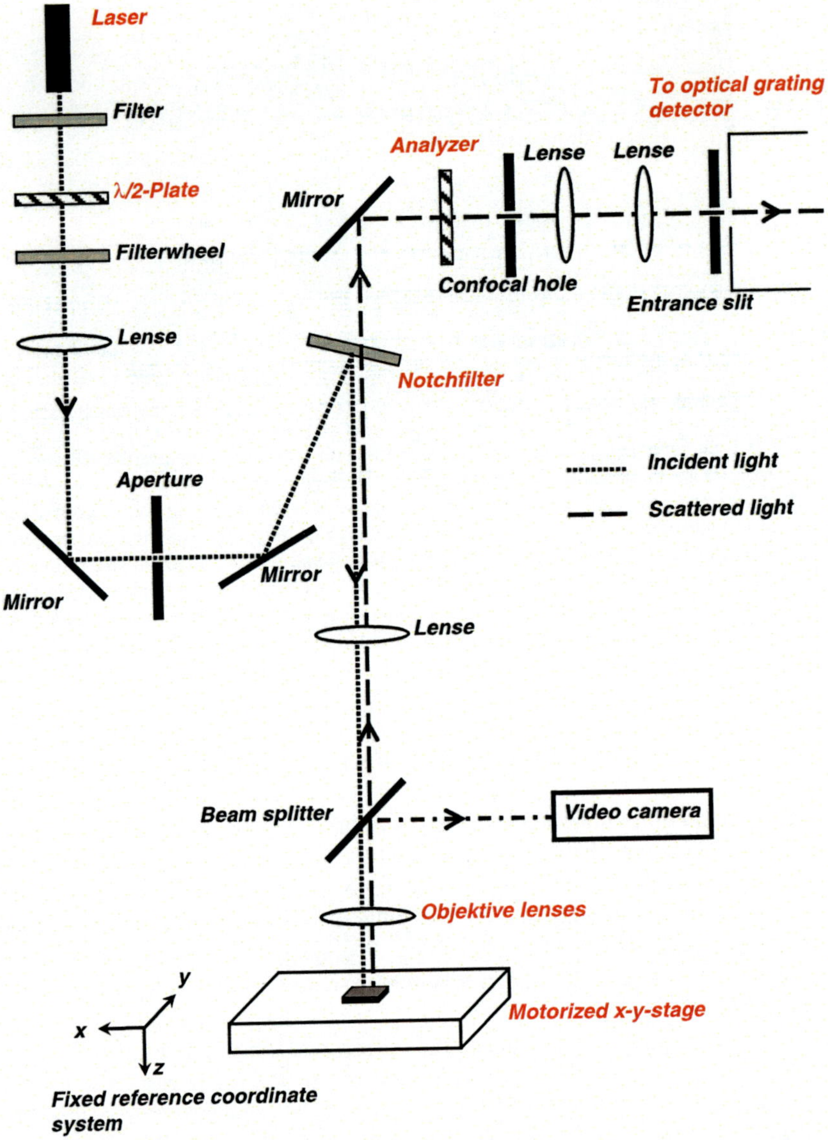

Fig. 9.3 Detailed schematic drawing of the Raman spectrometer setup (Jobin Yvon LabRam HR 800). With the rotatable $\lambda/2$-plate, the polarization direction of the incident beam can be adjusted continuously. With the analyzer in the backscattered path, the polarization direction of the backscattered signal is set. In contrast to the $\lambda/2$-plate, the analyzer can only be set to the H (horizontal)—or V (vertical)- polarization direction with respect to the reference coordinate sytem (in case of a continuously adjustable $\lambda/2$-plate for the incident polarization direction, an additional continuously adjustable analyzer is redundant in for polarization dependent Raman measurements)

$$\vec{e}_i = \begin{pmatrix} \cos\phi \\ k \cdot \sin\phi \\ 0 \end{pmatrix} \quad \text{with} \quad 0.9 \le k \le 1 \quad \text{(usually)}. \tag{9.4}$$

As the optical grating of the spectrometer also affects the polarization dependence of the backscattered beam, the effective polarization vector $\vec{e}_s$ of the scattered beam has to be modified according to:

$$\vec{e}_s = \begin{pmatrix} m \cdot \cos\eta \\ \sin\eta \\ 0 \end{pmatrix} \quad (m \approx 0.5), \tag{9.5}$$

where η describing the analyzer position ($\eta = 0°$ for analyzer in *H*-position, $\eta = 90°$ for analyzer in *V*-position) . The correction factor m accounts for the optical grating to modify the polarization direction of the scattered beam. The influence of the optical grating could also be eliminated by putting a $\lambda/4$-plate (scrambler) into the path of the scattered light. As we will see, the influence of the grating is of no importance for the determination of the crystallographic orientations. Details concerning the correction factor m will therefore not be discussed here.

After inserting Eqs. (9.4) and (9.5) into Eq. (9.3), the intensities of the orientation dependent Raman peaks can be expressed through the following simple matrix equation:

$$I(\phi, \eta, \alpha, \beta, \gamma) \approx I_0 \cdot \begin{pmatrix} \cos^2\phi \\ k \cdot \cos\phi \sin\phi \\ k^2 \cdot \sin^2\phi \end{pmatrix}^T \cdot \begin{pmatrix} f_{11} & f_{12} & f_{13} \\ f_{12} & f_{22} & f_{23} \\ f_{13} & f_{23} & f_{33} \end{pmatrix} \cdot \begin{pmatrix} m^2 \cdot \cos^2\eta \\ m \cdot \cos\eta \sin\eta \\ \sin^2\eta \end{pmatrix}, \tag{9.6}$$

where the matrix functions $f_{\mathrm{ij}} = f_{\mathrm{ij}}(\alpha, \beta, \gamma)$ are only dependent on the three Euler angles α, β, γ. The explicit expressions for the functions f_{ij} are given in [10]. As there are only two possible analyzer positions (*H* and *V*), there are only two ϕ dependent functions that can be obtained from Eq. (9.6):

$$\begin{aligned} I_H(\phi) &= I_0 \left(f_{11} \cos^2\phi + k f_{12} \cos\phi \sin\phi + k^2 f_{13} \sin^2\phi \right) m^2 \\ I_V(\phi) &= I_0 \left(f_{13} \cos^2\phi + k f_{23} \cos\phi \sin\phi + k^2 f_{33} \sin^2\phi \right). \end{aligned} \tag{9.7}$$

It is therefore reasonable to fit the experimental determined Raman peak intensity curves with the following two fit functions:

$$\begin{aligned} F_H(\phi) &= U_1 \cos^2\phi + U_2 \cos\phi \sin\phi + U_3 \sin^2\phi \\ F_V(\phi) &= V_1 \cos^2\phi + V_2 \cos\phi \sin\phi + V_3 \sin^2\phi, \end{aligned} \tag{9.8}$$

where the factors U_1, V_1, etc. are the fitting parameters. By taking the four ratios of the fitting parameters $u_1 = U_1/U_3$, $u_2 = U_2/U_3$, $v_1 = V_1/V_3$, $v_2 = V_2/V_3$ and comparing it to the corresponding ratios of theoretical factors contained in Eq. (9.7), one obtains a system of equations that is used to numerically determine the three

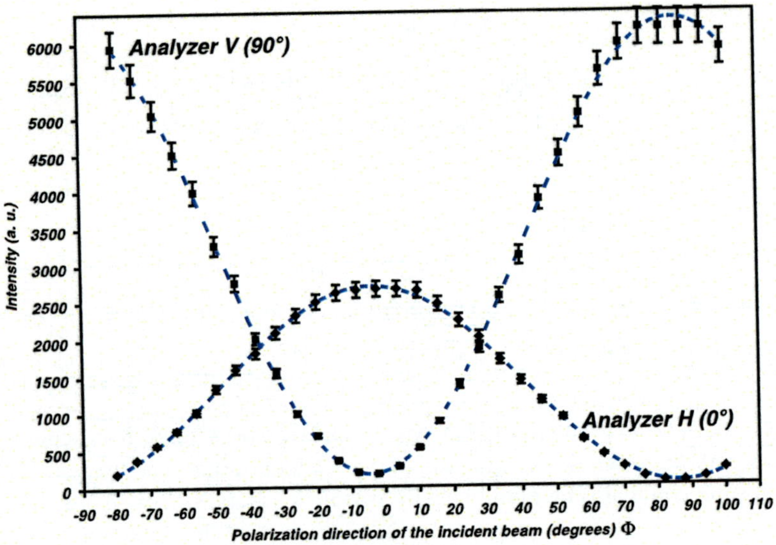

Fig. 9.4 Variations of the Raman peak intensities with the polarization direction ϕ of the incident beam obtained from the grain marked with the *yellow asterisk* in Fig. 9.2. The two curves belong to the two different analyzer positions (H and V). The *black points* represent the measurement data, whereas the *blue dotted curves* correspond to the respective best fit functions $F_H(\phi)$ and $F_V(\phi)$ from which the three Euler angles α, β, γ and the rotation matrix $T(\alpha, \beta, \gamma)$ can be determined. The error bars account for the variations of the incident laser intensity with time. The maximum Raman peak intensities for the H- and V- analyzer positions are different mainly due to the influence of the optical grating used for the spectral analysis (correction factor m). As explained in the text, this general difference in Raman intensities do not affect the determination of the crystal orientations

Euler angles α, β, γ from the Raman intensity measurements:

$$\begin{aligned}(k^2u_1)f_{13} - f_{11} &= 0\\ (ku_2)f_{13} - f_{12} &= 0\\ (k^2v_1)f_{33} - f_{13} &= 0\\ (kv_2)f_{33} - f_{23} &= 0,\end{aligned} \tag{9.9}$$

where the Euler angles are contained in the functions $f_{\text{ij}} = f_{\text{ij}}(\alpha, \beta, \gamma)$ (the correction factor m does not occur in the equation system (9.9) and therefore the influence of the grating on the polarization direction of the backscattered Raman signal is eliminated). In addition, as the experimental scaling factor I_0 cancels in the procedure described above, it is unnecessary to spend time on calibration measurements to determine I_0.

Equation (9.9) represents an overdetermined system, as it contains four independent equations to calculate the three Euler angles α, β, γ. Therefore an exact solution for the equation system (9.9) is not defined. Instead, the set of Euler angles has to be used that yields the minimal norm of the error vector that corresponds to the equation system (9.9). Once the Euler angles are determined, the whole information about the respective crystal orientation is known as the rotation matrix $T(\alpha, \beta, \gamma)$ can easily be re-calculated from the Euler angles. As the rotation matrix $T(\alpha, \beta, \gamma)$ can also be written in terms of the mutual angles between the axes of the crystal and the axes of the reference coordinate system as:

$$T = \begin{pmatrix} \cos(\sphericalangle xx') & \cos(\sphericalangle xy') & \cos(\sphericalangle xz') \\ \cos(\sphericalangle yx') & \cos(\sphericalangle yy') & \cos(\sphericalangle yz') \\ \cos(\sphericalangle zx') & \cos(\sphericalangle zy') & \cos(\sphericalangle zz') \end{pmatrix}. \tag{9.10}$$

The position of the respective crystal axes system with respect to the reference coordinate system can often be estimated at a glance. An experimental example measurement is shown in Fig. 9.4, where the Raman peak intensity data are obtained from the grain marked in Fig. 9.2 with a yellow asterisk. The measurement points are represented by the black squares and the blue dotted lines represent the best fit functions $F_x(\phi)$ and $F_y(\phi)$ from which the three Euler angles α, β, γ and the rotation matrix $T(\alpha, \beta, \gamma)$ and finally the values for $\cos(\sphericalangle xx')$, etc. are determined.

From the experimental curves shown in Fig. 9.4 we finally find the rotation matrix that contains the full information of the crystal orientation of the grain marked with the yellow asterisk in Fig. 9.2:

$$T_{\text{Asterisk}} = \begin{pmatrix} 0.681 & -0.731 & 0.044 \\ 0.728 & 0.682 & 0.069 \\ -0.080 & -0.015 & 0.997 \end{pmatrix}$$

which shows that the angles $\sphericalangle xx'$, $\sphericalangle yy'$, and $\sphericalangle zz'$ are ~47°, ~47°, and ~5°, respectively e.g. a crystal that has its z'-axes almost exactly aligned along the z-axes of the reference coordinate system and where the x' and y' axis are rotated at an angle of ~47° around the z-axes of the reference coordinate system.

9.1.1.3 Comparison with Other Orientation- Measurement and Imaging Methods

There are a few other competing methods to determine crystal orientations, each with certain advantages and drawbacks compared to the orientation measurements performed by Micro-Raman spectroscopy. These are: Etching techniques [60, 43, 13], methods based on electron diffraction, such as transmission electron microscope (TEM) [99] or electron backscattering diffraction (EBSD) [58] and X-ray diffraction [77]. Etching techniques use selective etchants whose etching rates depend on crystallographic orientation. The orientations of the remaining facets can then be determined by light reflection techniques. This technique needs no complicated equipment but is intrinsically destructive. In addition, this method lacks a really good lateral resolution (a few tens of μm) and the determination of orientations is rather inaccurate. Much higher accuracy and better lateral resolution (few tens of nm) are achieved with electron diffraction (TEM, EBSD). However, prior to the measurements, the samples have to be destroyed because a thin foil has to be prepared for the TEM analysis or the sample surface needs special preparation when EBSD is used. Another drawback of these techniques is the expensive vacuum equipment. Less complex equipment is needed in the case of orientation determination by X-ray diffraction. Though the crystal orientations can be determined very accurately and the samples need not to be destroyed prior to the X-ray measurements, the lateral resolution lies only in the range of a few tens of μm, when standard instrumentation is used. Orientation measurements with Micro-Raman spectroscopy do not need special sample preparation, as long as the surfaces are not too rough, which would lead to optical artifacts. Other important advantages of Raman spectroscopy are that the measurements can be performed under ambient conditions and within the range of a few hundred nm to a few μm (depending on the wavelength and the objective used), the lateral resolution is rather good. However, there are also a few drawbacks to the Raman-based orientation measurement method. One lies in the measurement principle, which presumes that the material under investigation possesses at least one Raman active vibrational mode e.g. an optical phonon mode in case of crystals. This is usually not the case for most metals as they only possess acoustic modes due to their simple crystal structure. These modes can be detected through Brillouin scattering [86], e.g. inelastic light scattering of acoustic modes. However, Brillouin scattering experiments need more sensitive and intricate spectral detection techniques than what is usually used for Raman scattering experiments. Another drawback of orientation measurements with Raman spectroscopy is that they need several Raman intensity acquisitions for different positions of the $\lambda/2$-plate and the analyzer. As long as these optical devices cannot be adjusted automatically, their positions have to be adjusted by hand. In this case, orientation measurements become rather time consuming. The accuracy of the Raman-based orientation measurements lies at $\sim \pm 2°$ for the determination of the Euler angles. Though this error value is sufficient for many other measurements that need the information about the crystal orientation (e.g. successive stress/strain measurements), the accuracy is far from the accuracy which can be achieved with X-ray diffraction. However, the great advantage of Raman-based orientation measurements is that subsequent Raman measurements that need the information about the crystal orientation (stress/strain analysis) can be performed in the same reference coordinate system. This would not be possible, if another method to determine the crystal orientation is used.

9.1.2 *Imaging of Stress/Strain States*

9.1.2.1 General Theoretical Description

Mechanical stresses in crystals cause distortions of the crystal structure as well as a change in the average atomic distances. Changes in the crystal structure cause changes of the Raman tensors which finally result in modified Raman selection rules

[3]. The modified average atomic distances cause a shift of the phonon eigenfrequencies. This effect can only be described if one assumes that the components of the force-constant matrix $\boldsymbol{K}$ depends on the average atomic distances and therefore on the stress/strain states within the specific crystal (i.e. deviations from the harmonic approximation have to be made) [70]. However, to calculate the strain-induced shift of the phonon eigenfrequencies, a quasi-harmonic approach is used [42]. That is, the components of the force constant matrix $\boldsymbol{K}$ are assumed to depend on the average atomic distances but stay independent of the atomic displacements during a lattice oscillation. The strain/stress-induced changes of the force constant matrix $\boldsymbol{K}$ are then simply described by a perturbation matrix $\Delta \boldsymbol{K}$ [42, 20, 2].

Besides strain/stress in crystals, alloying crystals with foreign atoms leads also to a change in phonon eigenfrequencies. Analog to the perturbation matrix $\Delta \boldsymbol{K}$, a second perturbation matrix $\Delta \boldsymbol{M}$ is introduced and added to the mass matrix $\boldsymbol{M}$ to describe the influence of the different masses of foreign atoms on the phonon frequencies. For further details concerning the mass effects, see e.g. [70, 22, 69]. The general eigenvalue problem to determine the phonon frequencies in the presence of strain/stress and isotopes or foreign atoms with different masses can then be written as [70]:

$$\left[(K + \Delta K) - \omega_r^2(\vec{q}) \cdot (M + \Delta M)\right] \cdot \vec{a}_r = 0, \qquad (9.11)$$

where $\omega_r(\vec{q})$ describes the eigenfrequency of a specific phonon mode r with wavevector $\vec{q}$. The corresponding eigenvector $\vec{a}_r$ describes the relative atomic displacements of atoms in an elementary cell of the crystal during a lattice oscillation r. In the following, $\Delta M = 0$ is set as only the shifts of phonon frequencies due to strain/stress will be discussed here.

For the force constant matrix $K^{(\boldsymbol{\varepsilon}')} = K + \Delta K$, usually a linear approach in the six strain tensor components $\varepsilon'_{\mathrm{kl}}$ is used when stress/strain is present in the crystal [42](prime quantities correspond to the crystal coordinate system).

The components of the force constant matrix $K^{(\boldsymbol{\varepsilon}')}$ can then be written as:

$$K_{\mathrm{ij}}^{(\varepsilon')} = K_{\mathrm{ij}}^{0} + \sum_{k,l} (\partial K_{\mathrm{ij}}/\partial \varepsilon'_{\mathrm{kl}}) \cdot \varepsilon'_{\mathrm{kl}} = K_{\mathrm{ij}}^{0} + \sum_{k,l} K_{\mathrm{ijkl}}^{(\varepsilon')} \cdot \varepsilon'_{\mathrm{kl}} \qquad (9.12)$$

with $K_{\mathrm{ij}}^{0} = \omega_0^2 \cdot \delta_{\mathrm{ij}}$, where ω_0 is the phonon frequency of a certain mode without stress/strain.

The strain tensor $\boldsymbol{\varepsilon}'$ is a symmetric 3×3-matrix with six different components [61]:

$$\boldsymbol{\varepsilon}' = (\varepsilon'_{\mathrm{kl}})(k, l = x, y, z). \qquad (9.13)$$

The components $K_{\mathrm{ijkl}}^{(\varepsilon')}$ in Eq. (9.12) form a fourth rank tensor and are called phonon deformation potentials (PDP's). The structure of this tensor and the number of its independent components are determined by the specific crystal symmetry. In case of crystals with cubic symmetry (e.g. silicon), the phonon deformation tensor contains only three independent components p, q, and r [4, 2, 20]. PDP's are material constants and can be determined either by theoretical calculations [42] or measured experimentally [26, 5]. Solving the eigenvalue Eq. (9.11) with Eq. (9.12) and the three PDP's *p, q,* and *r* for the three optical phonon modes of crystals with diamond structure, leads finally to the following secular equation (for a detailed derivation see e.g. [42] and [70]) :

$$\begin{vmatrix} p \cdot \varepsilon'_{xx} + q \cdot (\varepsilon'_{yy} + \varepsilon'_{zz}) - \Delta\omega_r^2 & r \cdot \varepsilon'_{xy} & r \cdot \varepsilon'_{xz} \\ r \cdot \varepsilon'_{xy} & p \cdot \varepsilon'_{yy} + q \cdot (\varepsilon'_{xx} + \varepsilon'_{zz}) - \Delta\omega_r^2 & r \cdot \varepsilon'_{yz} \\ r \cdot \varepsilon'_{xz} & r \cdot \varepsilon'_{yz} & p \cdot \varepsilon'_{zz} + q \cdot (\varepsilon'_{xx} + \varepsilon'_{yy}) - \Delta\omega_r^2 \end{vmatrix} = 0, \tag{9.14}$$

where $\Delta\omega_r^2 = (\omega_r^2 - \omega_{r0}^2)$ $(r = 1, 2, 3)$ are the differences of the squared eigenfrequencies of the three phonon modes in the strained and unstrained state of the silicon crystal. Using the approximation

$$\lambda_r = \left(\omega_r^2 - \omega_{r0}^2\right) \approx 2 \cdot \omega_{r0} \cdot \Delta\omega_r \Rightarrow \Delta\omega_r \approx \frac{\lambda_r}{2 \cdot \omega_{r0}} \tag{9.15}$$

the shifts in phonon frequencies $\Delta\omega_r$ can be easily determined from the eigenvalues λ_r.

Using the secular Eq. (9.14) it is a rather simple task to determine the shifts of the phonon frequencies when the stress/strain state within the crystal is known. But for experimental applications it is more important to determine the stress/strain states in crystals from the strain-induced shifts of the phonon frequencies. This is more difficult, as in the ideal case only three values of the frequency shifts are experimentally accessible at maximum. From these three experimental values it is in principal not possible to determine all six strain or stress tensor components. In practice, some simplifying assumptions are made, concerning the strain/stress states which lead to a reduction of the number of tensor components that have to be calculated.

The simplest Raman stress measurement scenario in case of silicon is a uniaxial stress state (only one nonzero stress tensor component) in a silicon wafer when the Raman signal is backscattered from the [001] surface. In this case Eq. (9.14) leads to the rule of thumb, that a measured Raman peak shift of $\sim$1 cm^{-1} (wavenumbers) corresponds to a uniaxial stress of $\sim$500 MPa. (see e.g., [34] for the numerical values of the PDP's and all other material parameters). Though it has limited applicability in practice, this memorable relation can be used to estimate the stresses in more complex cases at a glance, e.g. general stress states in arbitrarily oriented Si crystals. In these cases, the average value of the shifts of the three optical phonons and the sum of all the determined stress components should approximately be related according to the above given rule of thumb.

9.1.2.2 Quantitative Strain/Stress Analysis in Polycrystalline Silicon Wafers

The determination of several stress tensor components in crystals with diamond structure has already been shown [16, 17, 84, 74, 64]. But the stress/strain analysis has been carried out only for monocrystalline samples with known orientation. Stress analysis in these cases is rather straightforward. Similar accurate stress tensor determination have not been shown for polycrystalline materials with diamond structure e.g. polycrystalline silicon wafers for solar cells or thin film transistors (TFT's) [18] due to the fact that grain orientation variations in these materials make stress measurements increasingly difficult. In this section it will be demonstrated how several stress tensor components using Micro-Raman spectroscopy can be measured in polycrystalline silicon within grains of arbitrary orientations. Examples of practical interest in the field of polycrystalline photovoltaic silicon material will be presented.

Assumptions

Equation (9.14) serves as the basis for determining the mechanical stress components also in multicrystalline silicon wafers. The strain-tensor components, which occur in Eq. (9.14), refer to the specific crystal coordinate system (prime quantities). Also all other physical properties of the crystal, such as the phonon polarization directions and the phonon frequency shifts, refer naturally to the respective crystal coordinate systems. In multicrystalline wafers, a large number of crystallites with many different coordinate systems are present. Therefore, it is useful to refer the stress and strain states in the different crystallites to reference strain and stress tensors $\boldsymbol{\varepsilon}$ (strain) and $\boldsymbol{\sigma}$ (stress) that are defined in a fixed reference coordinate system (unprimed quantities).

At the laser wavelengths (488, 514, and 633 nm) used for these measurements, the light penetrates only a few hundred nm to a few μm into silicon and only the stress/strain states close to the crystallite surface can be detected. The stress state directly at the crystallite surface is necessarily a planar stress state.

Therefore, the reference stress tensor $\boldsymbol{\sigma}$ is chosen as follows:

$$\upsilon = \begin{pmatrix} \sigma_{xx} & \tau_{xy} & 0 \\ \tau_{xy} & \sigma_{yy} & 0 \\ 0 & 0 & \Delta_z \end{pmatrix}. \tag{9.16}$$

The additional component $\Delta_z(<\sigma_{xx}, \sigma_{yy})$ serves as a correction component that accounts for the slight deviations from the ideal planar stress state due to the finite penetration depth of the laser light.

Numerical Determination of Stress Components

The components of the reference stress tensor given in Eq. (9.16) can be determined from the experimentally three measured phonon frequency shifts $\Delta\omega_1$, $\Delta\omega_2$, and $\Delta\omega_3$ and Eq. (9.14) according to the following procedure.

The reference stress tensor (9.16) has to be transformed into the respective crystal coordinate system by use of the rotation matrix $T(\alpha, \beta, \gamma)$ (known from the orientation measurements). This gives the corresponding stress tensor $\boldsymbol{\sigma}'$:

$$\sigma' = \begin{pmatrix} \sigma'_{xx} & \tau'_{xy} & \tau'_{xz} \\ \tau'_{xy} & \sigma'_{yy} & \tau'_{yz} \\ \tau'_{xz} & \tau'_{yz} & \sigma'_{zz} \end{pmatrix} = T^{-1}(\alpha, \beta, \gamma) \cdot \sigma \cdot T(\alpha, \beta, \gamma). \tag{9.17}$$

The strain tensor components ε'_{kl} in Eq. (9.14) then have to be expressed in terms of the stress tensor components σ'_{kl} given in Eq. (9.17), through the inverse Hook's law [48, 56] :

$$\boldsymbol{\varepsilon}' = \underline{\underline{\mathbf{S}}} \cdot \boldsymbol{\sigma}', \tag{9.18}$$

where $\underline{\underline{\mathbf{S}}}$ designates the elastic compliance tensor. As the strain and stress tensors are matrices, the compliance tensor is consequently a fourth rank tensor. However, as the strain and stress tensors are symmetric, Eq. (9.18) can also be written in vector notation, where the compliance tensor then becomes a 6×6 matrix. In case of cubic crystal structures (Diamond, Si), the structure of the compliance tensor becomes rather simple and the number of independent components reduces to three. The values for these material constants for silicon (S_{11}, S_{12} and S_{44}) can be found in the literature [98].

The secular Eq. (9.14) can now be expressed in terms of the stress tensor components given in Eq. (9.18) by use of Eqs. (9.16) and (9.17). This finally yields:

$$\left| A \cdot \sigma_{xx} + B \cdot \tau_{xy} + C \cdot \sigma_{yy} + D \cdot \Delta_z - \lambda_r \cdot E \right| = 0, \tag{9.19}$$

where the components of the four symmetric 3×3 matrices $\boldsymbol{A}$, $\boldsymbol{B}$, $\boldsymbol{C}$, and $\boldsymbol{D}$ are simple but lengthy expressions that depend on the elastic compliances S_{11}, S_{12}, S_{44}, on the PDP's *p,q,r*, and on the components of the rotation matrix $T(\alpha, \beta, \gamma)$. Therefore, the components of these four matrices are constants for a specific crystallite. $\boldsymbol{E}$ is the unit matrix.

The secular Eq. (9.19) leads to the characteristic polynomial:

$$P(\lambda_r) = \lambda_r^3 + a \cdot \lambda_r^2 + b \cdot \lambda_r + c, \tag{9.20}$$

where the coefficients *a,b,c* are only functions of the stress components defined in (9.16). The roots of the polynomial correspond to the measured phonon frequency shifts $\lambda_r \approx 2\omega_0 \cdot \Delta\omega_r$. Comparing coefficients (rule of Vieta) finally leads to the equation system that is used to determine the stress components (9.16) numerically from the phonon frequency shifts:

$$\begin{aligned} a &= -2\omega_0 \cdot (\Delta\omega_1 + \Delta\omega_2 + \Delta\omega_3) \\ b &= (2\omega_0)^2 \cdot (\Delta\omega_1\Delta\omega_2 + \Delta\omega_1\Delta\omega_3 + \Delta\omega_2\Delta\omega_3) \\ c &= -(2\omega_0)^3 \cdot (\Delta\omega_1\Delta\omega_2\Delta\omega_3). \end{aligned} \tag{9.21}$$

In practice, one uses the following way to calculate the stress components from the measured phonon frequency shifts: at first, an ideal plane stress state is assumed, i.e. $\Delta_z = 0$ is set. If then the numerical error of the system (9.21) is too large, Δ_z is used as a correction parameter and is iteratively adjusted to $\Delta_z \neq 0$ (in a reasonable range) until the numerical error corresponds to the minimal detectable peak shift resolution of $\sim$0.05 cm^{-1} for all the three phonon frequency shifts $\Delta\omega_1$, $\Delta\omega_2$, and $\Delta\omega_3$.

Experimental Procedure to Determine the Phonon Frequency Shifts

According to Eq. (9.21) all three phonon frequency shifts of the three optical phonons of silicon have to be measured experimentally to determine the components of the stress tensor. However, these three optical modes are always 3-fold degenerate in case of stress-free silicon. In case of stresses in the GPa range, the degeneracy is lifted and the three frequency shifts can be measured separately [6, 84] (not in case of hydrostatic stress, as then the three phonon frequency shifts are the same [2]). In polycrystalline silicon wafers the expected stress values are usually below 1 GPa and mechanical stresses of this magnitude cause phonon frequency shifts that are smaller than the experimentally measured half width of a silicon Raman band. In this case, the lifting of degeneracy is not clearly visible and an accurate direct determination of the three frequency shifts $\Delta\omega_1$, $\Delta\omega_2$, and $\Delta\omega_3$ is not possible. However, by using three special polarization settings during the Raman measurements, it becomes possible to excite each of the three phonon modes nearly separately. These three polarization settings can be found when the full information about the respective crystal orientation is known (from the orientation measurements). Mapping the regions of interest three times with these three special polarization settings yields then the three phonon frequency shifts $\Delta\omega_1$, $\Delta\omega_2$, and $\Delta\omega_3$ nearly separately. Details can be found in e.g. [10]. The three values for $\Delta\omega_1$, $\Delta\omega_2$ and $\Delta\omega_3$ can then be read off and with the procedure described above, the stress tensor components can be calculated, even though the lifting of degeneracy is not visible in the experimental spectra.

Additional Influences on the Phonon Frequency Shifts

Temperature

The phonon frequencies, the half widths of the Raman bands as well as the intensity ratio I_{AS}/I_S of the Anti-Stokes and Stokes Raman bands are temperature dependent.

A rise in sample temperature leads to a decrease in phonon frequencies, an increase in half widths as well as an increase in the I_{AS}/I_S ratio [8, 19, 95, 85]. In the case of silicon, the phonon frequencies are rather sensitive to changes of the sample temperature, i.e. a change of ~4°C in sample temperature causes a shift of phonon frequencies of ~0.1 cm^{-1} [36]. That is, slight temperature changes lead to frequency shifts that are already above the experimental detection limit (≈0.05 cm^{-1}) for the phonon frequency shifts. Thus, great care has to be taken during mechanical stress analysis with Raman spectroscopy that no significant temperature gradients arise within the sample regions of interest during the measurements. Temperature gradients during Raman maps within the sample might arise due to a change in the laser spot size and/or a change in the laser power. Both might lead to different energy densities and therefore to different temperatures in between different measurement points. On the other hand, structural inhomogeneity within the sample (interfaces, agglomerates of defects, etc.) and on the sample surface (differences in oxide layer thickness, contaminations, etc.) can cause inhomogeneous heat generation and transport and therefore also temperature gradients in between different measurement points. The average temperature within a measurement point can be roughly determined by the I_{AS}/I_S—ratio [8]. It is clear that the average temperature within a measurement point becomes higher when objectives with high numerical apertures are used (as the laser spot becomes smaller). For example, in case of a HeNe-laser with a power of ~ 10 mW on the sample and a 100x objective (N.A. 0.95), a stationary state temperature within a measurement point of ~45° C can be estimated from the I_{AS}/I_S—ratio. In the experiments shown here, noticeable changes in temperature within a mapping region could only be detected in the vicinity of sample edges, where temperature gradients of ~1°C – 2°C are measured. In the vicinity of e.g. grain boundaries or other crystal defects, changes in sample temperature that would affect the stress-induced phonon frequency shifts were not detected.

Drift of the Spectrometer Calibration

Depending on the size and resolution of a Raman map, the needed measurement time can easily exceed several hours. During such a long acquisition time, the wavenumber position might drift due to the drift of the optical grating. This grating drift becomes noticeable in a constant drift of Raman peaks with time into one direction on the wavenumber axis. Usual drift rates lie in the range of ~0.02–0.04 cm^{-1}/hour. The grating drift-induced Raman peak shifts can be easily eliminated, when a plasma line of the probe laser is used as a wavenumber reference. Therefore, Raman measurements without an interference filter (that only lets the main excitation wavelength of the laser pass) should be performed. The plasma lines then also reach the detector and can be observed in the spectra. As the plasma lines have physically fixed wavelengths [78], a drift of plasma lines with time exactly resembles the grating drift. The phonon frequency shifts that are solely produced by the sample (stress or other influences) are then obtained by subtracting the respective shift of a plasma line from the particular measured Raman peak. The experimental procedure is depicted in Fig. 9.5. The Raman peaks as well as the plasma lines are fitted with a mixed Gaussian/Lorentzian

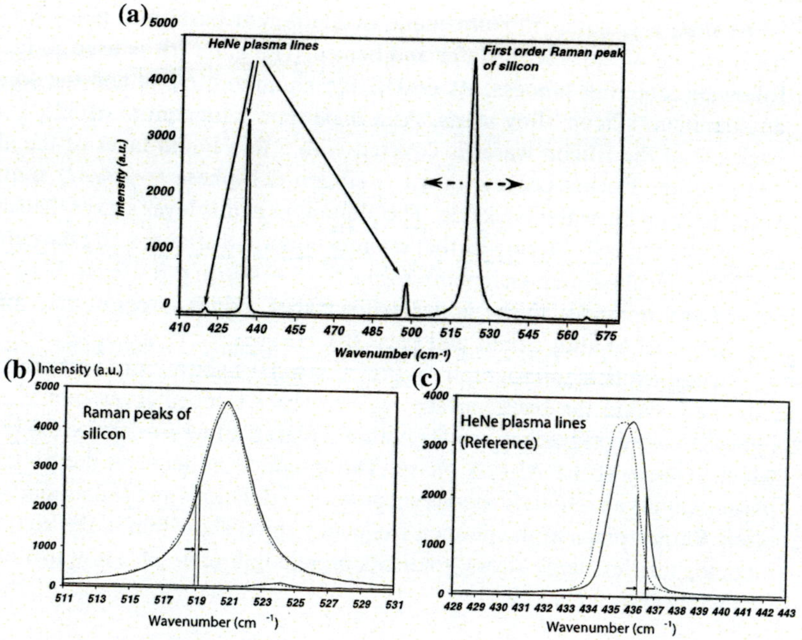

Fig. 9.5 **a** Raman spectrum of Silicon with the additional HeNe plasma lines. **b** Raman peak shifts of two Si Raman peaks and **c** the corresponding shifts of the plasma lines due to a spectrometer drift. The compensation of the drift of grating with the plasma lines even leads to a change in sign of the Raman peak shifts in **b**

fit function [36]. From the fitting data, the peak shifts can be determined with an accuracy of $\sim 0.05\,\mathrm{cm}^{-1}$. When the spectrometer drift-compensated Raman peak shifts are then plotted versus the mapping coordinates, a corrected peak shift map is obtained from which the mechanical stress states can then be detected accuractely.

9.1.2.3 Applications

Mechanical Stresses at the Backside of Silicon Solar Cells

Silicon solar cells are usually provided with a laminar, highly p-doped region ($\sim 10^{18} - 10^{19}$) at the backside [46, 68, 45]. This leads to a space charge region in between the standard p-doped (10^{15})–(10^{16}) base of a silicon solar cell and the highly p-doped region at the backside BSF. The light generated electrons within the base are then kept away from the back surface of the cell, which avoids charge carrier recombination at the free surface. This finally leads to a relative increase of the solar cell efficiency to ~5–10 % [44]. As a dopant for this highly doped layer, usually Aluminum is used. The fabrication of the BSF is usually carried according to the following process steps [53–55]:

At first, a granular mixture made of aluminum paste and glass frit is put on the backside of the silicon wafer and heated. The glass frit is used to facilitate the following sintering process. At 660°C the aluminum melts and the formation of an aluminum-silicon alloy starts. At a maximum temperature of 825°C the entire backside of the silicon wafer is covered with a thin liquid layer of the aluminum-silicon alloy. During cooling down, a silicon-rich phase ($\sim 1\,\%Al$) starts to grow epitaxially on the wafer backside. The aluminum in this layer serves then as a dopant to generate the BSF. During further cooling down and after passing the eutectic point at 577°C the aluminum-rich phase forms a $\sim 5 - 10\mu$m thick layer. Underneath this layer (with reference to the upper cell surface) resides a region that consists of a mixture of Al-Si alloy grains and glass frit.

The light optical micrographs in Figs. 9.6 and 9.7 show cross-sections of a silicon solar cell within the back surface region. The Al-doped layer (not visible in the light optical micrographs) resides within a region a few μm above the bright layer, which represents the Al-rich phase. The interface in between the Si- and Al-rich phase exhibits an inhomogeneous shape with undulations and spikes that occur rather often. Raman maps of the phonon frequency shifts show that in the vicinity of such inhomogeneities, large stress gradients on a length scale of a few μm occur rather frequently (Fig. 9.6 and 9.7):

$$\Delta\sigma = \begin{pmatrix} -224 \pm 22 & -3 \pm 1 & 0 \\ -3 \pm 1 & -197 \pm 19 & 0 \\ 0 & 0 & -138 \pm 10 \end{pmatrix} \text{MPa}$$

$$\Delta\sigma = \begin{pmatrix} -46 \pm 9 & -11 \pm 3 & 0 \\ -11 \pm 3 & -65 \pm 12 & 0 \\ 0 & 0 & -28 \pm 5 \end{pmatrix} \text{MPa}$$

These large stress gradients are introduced during the epitaxial growth process due to the different thermal expansion coefficients of the Si-rich and the Al-rich phase. Large stresses within the BSF region turn out to be a severe problem in view of the mechanical stability of the readily processed silicon solar cells. As the firing of the BSF is one of the last process steps for a Si solar cell, this last step introduces sources for cracks and therefore breakage of the solar cell if additional external mechanical stress is applied due to bending/deformation of the cell. As breakage of a readily processed silicon solar cell is a much greater loss then the breakage of an unprocessed silicon wafer, inhomogeneities within the BSF region that are prone to large mechanical stress values should be avoided by better processing steps such as e.g. a smoother firing of the BSF by the solar cell producers.

Stress Fields at Micro Cracks in Polycrystalline Silicon Wafers

Often cracks with lengths in the range of a few ten μm (micro cracks) occur in polycrystalline silicon at the wafer edges. These micro cracks are one of the main

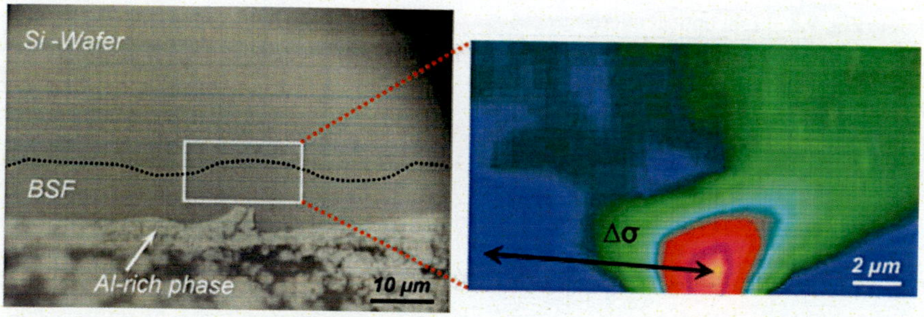

Fig. 9.6 *Left image* cross-section of a silicon solar cell close to the back surface field (BSF) region (light optical micrograph). The highly Al-doped region resides up to ~10 μm above (*black dotted line*) the brighter Al-rich phase. The *right* image shows the mapping of the frequencies of one phonon mode shortly above the spike within the Al-rich phase (*white rectangle*). With reference to the position at the *left* tip of the *black arrow* (which is assumed to be nearly stress free), the region above the spike is under compressive stress.

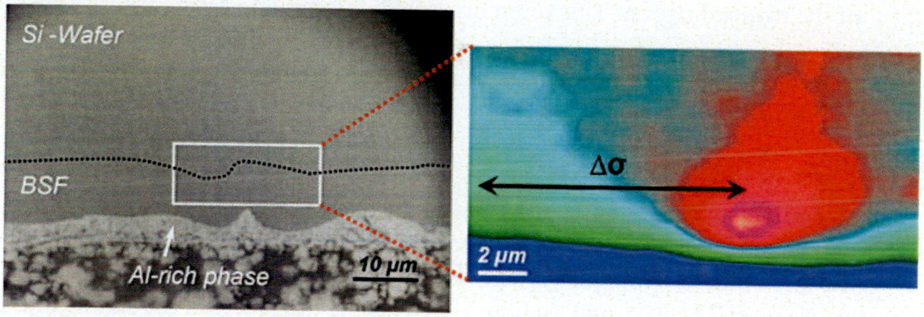

Fig. 9.7 Analog to Fig. 9.6. The triangular shaped spike within the Al-rich phase underneath the BSF region induces a fan-shaped stress field (compression) into the silicon wafer. Again, the stress state is referred to a reference position which is assumed to be nearly stress-free. However, the induced compressive stress is lower than in the example shown in Fig. 9.6.

reasons for wafer breakage during the solar cell processing steps [59]. Large stress gradients on a length scale of a few μm occur at the tips of the micro cracks. These μm-sized stress fields can be measured with Micro-Raman spectroscopy and from the phonon frequency shifts the mechanical stress components can be calculated. However, specific problems regarding fracture mechanics will be not discussed here, though some characteristic material parameters concerning fracture mechanics can also be determined by Micro-Raman measurements [81, 80, 82, 83, 79]. This experimental example mainly serves to show how accurate mechanical stress fields can be measured and mapped on the μm scale by Micro-Raman spectroscopy.

Figure 9.8 displays the edge region of a polycrystalline silicon wafer where a micro crack of $\sim 100 \mu$m length resides. The region close to the crack tip where the stress mapping is performed is marked with a white rectangle.

After determining the crystal orientation, the region marked with the white rectangle in Fig. 9.8 is mapped with the three special polarization settings where each of the

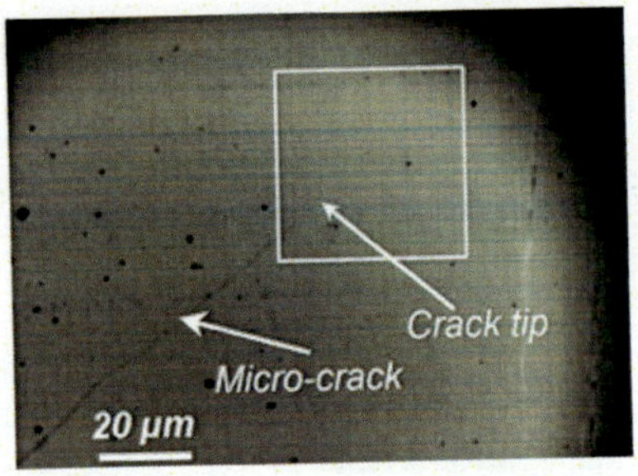

Fig. 9.8 Light optical micrograph of the surface of a polycrystalline silicon wafer where a micro crack resides. The *white rectangle* (42 × 45μm) marks the region where the phonon frequency maps are carried out. For the measurements 40 × 40 measurement points are used

three optical phonon modes of silicon is respectively excited separately. Figure 9.9 shows the maps of the three phonon frequency shifts $\Delta\omega_1$, $\Delta\omega_2$, and $\Delta\omega_3$ determined in the way described above. Again, the stress state that is measured in the vicinity of the crack tip is referred to a measurement position at the left upper corner of the mapped region. This reference position is assumed to be nearly stress-free.

$$\Delta\sigma = \begin{pmatrix} 57 \pm 12 & -7 \pm 3 & 0 \\ -7 \pm 3 & 54 \pm 11 & 0 \\ 0 & 0 & 19 \pm 4 \end{pmatrix} \text{MPa.}$$

According to theory [10], the stresses in the vicinity of the crack tip are tensile and the amounts of the stress components lie in the range of what is expected from the three measured frequency shifts at the indicated position: $\Delta\omega_1 = -0.20\,\text{cm}^{-1}$, $\Delta\omega_2 = -0.27\,\text{cm}^{-1}$, and $\Delta\omega_3 = -0.29\,\text{cm}^{-1}$. Considering the course of the crack (~45° angle with the axis of the reference coordinate system) and the nearly symmetrical distribution of the phonon frequency shifts, it is reasonable that the two stress components in the x- and y- direction (reference coordinate system) $\Delta\sigma_{xx}$ and $\Delta\sigma_{yy}$ have nearly the same value.

Stress States at Grain Boundaries in Polycrystalline Silicon Solar Cell Material and the Relation to the Grain Boundary Microstructure and Electrical Activity

Micro-Raman spectroscopy, EBSD, and electron beam-induced current (EBIC) [47] techniques can be combined to investigate the relation between mechanical stress at grain boundaries (GBs) in polycrystalline silicon wafers and the microstructure as well as the electrical activity of grain boundaries. The EBSD technique is used during these investigations to determine the character of the large amount of grain boundaries present in polycrystalline silicon wafers. In principle, grain boundary characteristics can also be determined with Micro-Raman spectroscopy, since orientation measurements are possible (see Section 9.1.1). However, grain boundary analysis with EBSD works in this case faster as it is an already established and automated measurement method that is able to handle regions with a large amount of

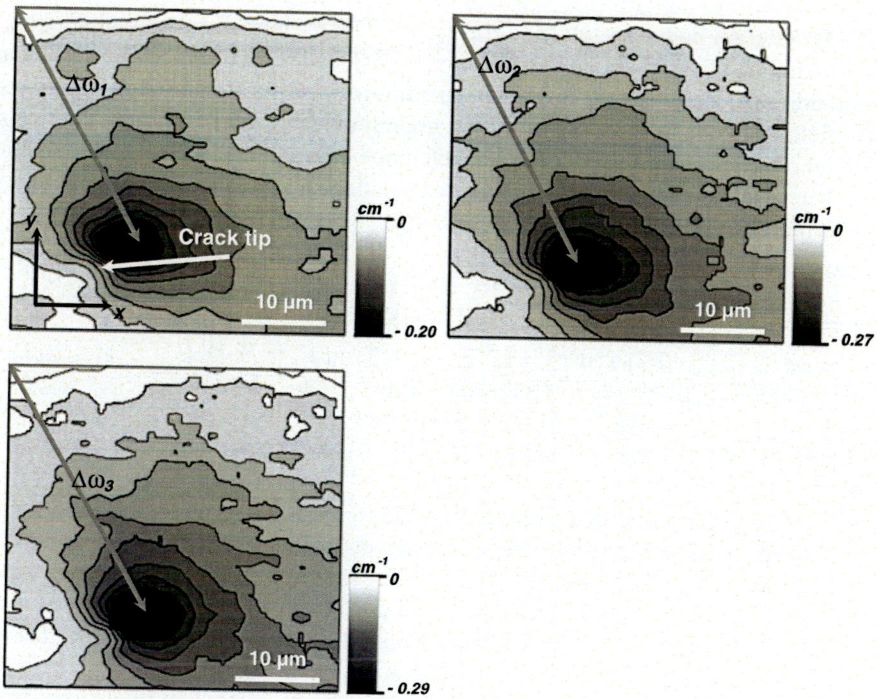

Fig. 9.9 Three maps of the phonon frequency shifts $\Delta\omega_1$, $\Delta\omega_2$, and $\Delta\omega_3$ close to the tip of a micro crack shown in Fig. 9.8 (*white rectangle*). The three mappings correspond to the three polarization settings where each of the three phonon modes are respectively excited separately. The calculated stress state at the tip of the micro crack is referred to the stress state at the *upper left corner*, which is assumed to be stress-free. The ranges of the depicted *gray* scales are different, as they correspond to the maximum amount of frequency shifts of the respective phonon mode. By using the procedure described above, the following stress components can be determined at the indicated position (lower tip of the *gray arrow*) a few μm away from the tip

crystallites. EBIC is a standard method to investigate the electrical activity of grain boundaries. EBIC and EBSD are both methods that can be operated in a scanning electron microscope (SEM), a fact that makes it appealing to combine both methods for grain boundary and polycrystalline silicon wafer analysis. In the following experimental results, the misorientation between adjacent grains is described in terms of the Σ- value which gives the fraction of atoms in the grain boundary plane that coincide with both adjacent lattices. Figure 9.10a shows a SEM image of a sample region within a polycrystalline silicon wafer which includes the Raman map shown in Fig. 9.10b. The marker, produced by a focused ion beam (FIB), allows exact lateral correlation between the different measurement techniques.

A kink in the Σ27a grain boundary trajectory is denoted with *I* and a filled blue circle, present in all images of Fig. 9.10, indicates an inclusion (probably SiC). The Raman map reveals very localized and symmetric stress fields, including both compressive (red region in the map) and tensile (blue) stresses, in the vicinity of the

$\Sigma 27$a grain boundary, close to the kink *I*. The reasons for a stress field with such a kind of shape might be the presence of dislocations within the grain boundary or additional inclusions. A detailed polarization dependent Raman analysis of the stress fields on either side of the indicated grain boundary gives the following two stress tensors (referring to the indicated reference coordinate system) which again have been evaluated relative to the indicated positions away from the grain boundary and which are assumed to be nearly stress-free.

$$\Delta\sigma_1 = \begin{pmatrix} -40 \pm 10 & -14 \pm 1 & 0 \\ -14 \pm 1 & -38 \pm 10 & 0 \\ 0 & 0 & -25 \pm 10 \end{pmatrix} \text{MPa}$$

$$\Delta\sigma_2 = \begin{pmatrix} 33 \pm 10 & -7 \pm 1 & 0 \\ -7 \pm 1 & 31 \pm 10 & 0 \\ 0 & 0 & 34 \pm 10 \end{pmatrix} \text{MPa.}$$

Although the stress fields become visible in the Raman map (Fig. 9.10), the determined stress components are rather small and close to the resolution limit of ~20 MPa. The EBIC images (Fig. 9.10d, e) show a strong contrast (i.e. a reduced carrier lifetime) close to the kink *I* both at 77 K and 300 K and a contrast variation with temperature along the $\Sigma 27$a grain boundary. The region enclosed by the rectangle corresponds to the stress map shown in Fig. 9.10b. By comparing these images it becomes clear that the stressed area close to the kink *I* and the stress free area above it exhibit strong but similar EBIC contrasts and therefore similar electrical activity. This example shows that mechanical stresses of this magnitude are probably not able to change the recombination activity of polycrystalline silicon due to a stress-induced change in band structure. However, as the largest EBIC contrast is visible at the position of the kink *I*, where the dipole like stress field resides, most probably a cluster of edge dislocations or inclusions are responsible for the visible stress field.

General Remarks Concerning the Measurement Error

The errors that correspond to the stress tensor components result in general from the limited precision in measuring the Raman peak shifts of $\sim\pm 0.05\,\text{cm}^{-1}$, from the uncertainty of the determination of the crystal orientations and from the errors of the material parameters that are used for the calculations. For the error estimation of the presented stress tensor components, the measurement data are put into the numerical solving procedure with their largest possible deviations. A closed, analytical error calculation is in this case not possible due to the large and complex equation systems.

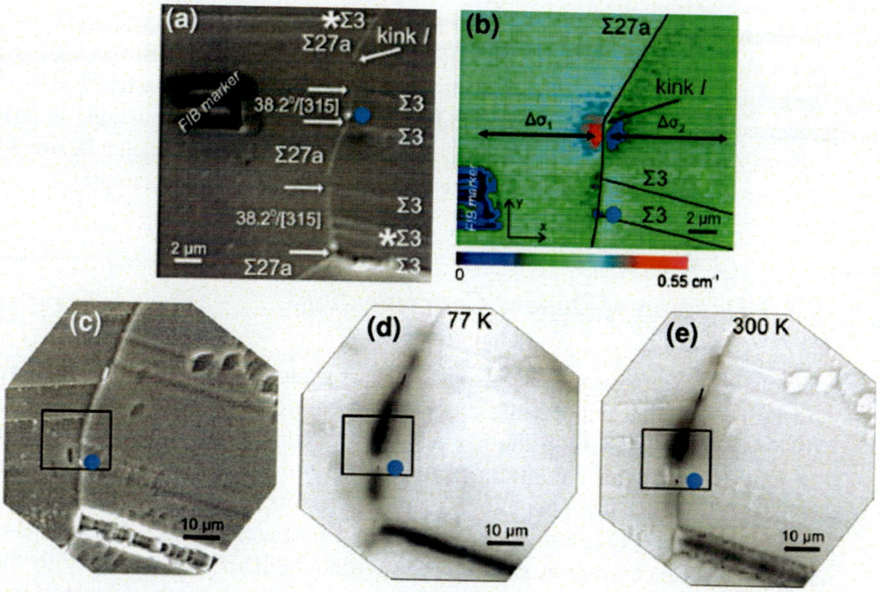

Fig. 9.10 **a** SEM image of the sample's region including the Raman phonon frequency map shown in **b**. The compressive (*red color*) and tensile (*blue color*) stress fields around the Σ27a grain boundary is probably produced by a cluster of edge dislocations or additional inclusions located at the kink position I. **c** SEM image and the corresponding EBIC image taken at **d** 77 K and **e** 300 K where the area enclosed by the rectangle represents the region of the Raman mapping displayed in **b**. The filled *blue circle* visible in all images denotes the position of an inclusion

9.1.2.4 Comparison with Other Stress/Strain Measurement Methods

Besides Micro-Raman spectroscopy, there are a few other methods that are frequently used to determine mechanical stress/strain in solids. But again, each of these methods also has its specific drawbacks concerning practical applicability, lateral resolution, or the destructive nature of the measurement technique.

In the case of stress/strain analysis in crystals by standard X-ray diffraction, all the stress/strain tensor components can be detected with some effort [89]. But the method is rather time-consuming, the lateral resolution of the stress analysis is small and the obtained data are stress values that are averaged over a rather large sample volume. Much better lateral resolution can be obtained with the brilliant X-ray radiation of a synchrotron [89], but the measurements then become even more complex.

With stress-induced birefringence, i.e. the analysis of stress-induced optical anisotropies in crystals [49], the mechanical stress/strain distributions within a large sample region (a few cm^2) can be detected rather fast. However, the lateral resolution of stress-induced birefringence is much too low to analyze stresses on the μm-scale and in addition, only the differences of absolute values (i.e. without sign) of the principle stress components can be determined.

Mechanical stress analysis on the nm-scale can be performed with the convergent beam electron diffraction (CBED) technique in a TEM [100]. But as CBED needs the preparation of thin samples for the TEM, this technique is destructive and needs rather complex instrumentation. Additionally, as the investigated TEM samples are rather thin, artifacts such as thin foil relaxation [29] lead to a change in stress states with respect to the stresses that were present in the unprepared samples.

9.1.3 Imaging of Dopant Concentrations

Highly doped regions in silicon solar cells are important for the proper and effective operation of solar cells. While the *n*-doped emitter region is produced by dopant (phosphor) diffusion from the gas phase, the highly *p*-doped BSF is generated by alloying silicon with aluminum from an *Al*-paste at the back surface of a silicon solar cell [46, 68, 45]. This alloying process results in a few μm thick silicon layer where roughly $\sim 1/1{,}000$ atoms consist of aluminum [53–55]. Usually, the exact free-carrier concentrations and gradients within the BSF's are measured by the electrochemical capacitance–voltage (ECV) method [1, 14]. A method which measures the free carrier concentrations rather accurately but which lacks good lateral resolution as the measurement results are integrated and therefore averaged, over an area of a few mm^2 at the back surface of the respective solar cell.

Raman specific Fano resonances, combined with small angle beveling techniques can be used to determine the concentration of free holes within the BSF regions and to map the free hole concentration and gradients on an effective length scale of a few 10's of *nm*. In this chapter, the basic features of Raman specific Fano resonances will be discussed shortly and measurement results, which show that the free carrier concentrations within the BSF regions can be qualitatively and quantitatively mapped and analyzed by Micro-Raman spectroscopy, will be presented.

9.1.3.1 Theory

As discussed in the original articles of Fano [40, 41, 39], a resonant interaction of a discrete energy level with a continuum of energy levels in a given system leads to a characteristic Fano-type peak asymmetry which can be usually observed during spectroscopic experiments of such kind of quantum mechanically coupled systems (see Fig. 9.11).

Cerdeira and other authors [23, 24, 7, 96, 92], explain the occurrence of the Fano-type peak asymmetry of the F_{2g} Raman band of highly *p*-doped silicon with a resonant interaction of the discrete optical phonon (phonon Raman scattering) state with the continuum of energy levels, corresponding to the hole transitions from filled to empty valence states (electronic Raman scattering [33, 62]) due to electron-phonon coupling.

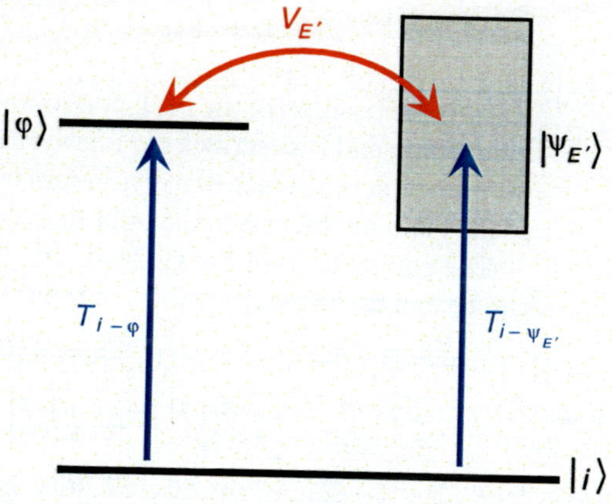

Fig. 9.11 Interaction of a discrete state $|\varphi\rangle$ with a continuum of states $|\psi_{E'}\rangle$ that leads to Fano resonances. From an initial state $|i\rangle$ the system can be excited into the continuum of states and into the discrete state by the arbitrary transition operators $T_{I-\varphi}$ and $T_{I-\psi_{E'}}$. If there is an interaction $V_{E'}$ between the discrete state and the continuum, the transition probabilities interfere, leading to Fano-type peak asymmetries which are observable in spectroscopic experiments

According to Fano's fundamental theory, if a system exhibits an energetic continuum of states $|\psi_{E'}\rangle$ and a discrete state $|\varphi\rangle$, the corresponding elements of the Hamiltonian matrix can be written as:

$$\begin{aligned} \langle\varphi|\,H\,|\varphi\rangle &= E_\varphi = \hbar\Omega \\ \langle\psi_{E'}|\,H\,|\varphi\rangle &= V_{E'} \\ \langle\psi_{E''}|\,H\,|\psi_{E'}\rangle &= E\cdot\delta(E''-E'). \end{aligned} \tag{9.22}$$

The parameter $V_{E'}$ represents the coupling strength of the discrete state with the continuum of states, H is the Hamilton operator, and Ω specifies the frequency that corresponds to the respective discrete state. The new eigenvector $|\Psi_E\rangle$ of the coupled system, which now corresponds to the new composite state given by Eq. (9.22) becomes then a linear combination of the discrete eigenstate and the continuum of states:

$$|\Psi_E\rangle = a_E\,|\varphi\rangle + \int b_{E'}\,|\psi_{E'}\rangle\,\mathrm{d}E'. \tag{9.23}$$

The coefficients $a_E$and b_E are functions of the energy. If there is an arbitrary initial state $|i\rangle$ and a respective dipole transition operator T (see Fig 9.11), the transition probability to the unperturbed continuum $|\psi_{E'}\rangle$becomes $|\langle\psi_{E'}|\,T\,|i\rangle|^2$, and the transition probability to the new composite state $|\Psi_E\rangle$ is defined as $|\langle\Psi_E|\,T\,|i\rangle|^2$. The ratio of the transition probabilities to the new composite state and to the unperturbed continuum gives then the general shape of the measured spectral peaks that are now distorted by the occurring Fano resonances [40, 41, 39]:

$$\frac{|\langle\Psi_E|\,T\,|i\rangle|^2}{|\langle\psi_{E'}|\,T\,|i\rangle|^2} = \frac{(q+\varepsilon)^2}{1+\varepsilon^2}, \tag{9.24}$$

where ε indicates the renormalized energy or, in the spectroscopic experiments, the renormalized wavenumber parameter:

$$\varepsilon = \frac{2 \cdot (\Omega - \Omega_{\text{Max}})}{\Gamma} \qquad (9.25)$$

Ω_{Max} is the position of the peak maximum, which is also slightly shifted by the Fano interaction, and Γ describes the linewidth of the peak, where $\Gamma = \Gamma_0 + \Delta\Gamma$ with Γ_0 the linewidth in absence of a continuum of states and $\Delta\Gamma$ the symmetric contribution of the Fano interaction to the total linewidth of the peak.

The parameter q in Eq. (9.24) is frequently called the symmetry parameter, and can be expressed as:

$$q = \frac{\langle \Phi | T | i \rangle}{\pi \cdot \langle \varphi | H | \psi_E \rangle \cdot \langle \psi_E | T | i \rangle} = \frac{\langle \Phi | T | i \rangle}{\pi \cdot V^*_{E'} \cdot \langle \psi_E | T | i. \rangle} \qquad (9.26)$$

The state $|\Phi\rangle$ defines the discrete state $|\varphi\rangle$ modified by the additional continuum of states. P means the "principal part of ":

$$|\Phi\rangle = |\varphi\rangle + P \int V_{E'} \, |\psi_{E'}\rangle / (E - E') \mathrm{d}E'. \qquad (9.27)$$

The symmetry parameter q determines the characteristic asymmetric shape of a spectral peak influenced by Fano resonances. The spectral line becomes symmetric if $q \to \infty$, whereas if $q \to 0$, the more asymmetric the peak becomes. Especially for highly doped silicon, $1/q$ becomes roughly proportional to the free carrier concentration [24, 7, 67]. $1/q$ is positive in case of highly p-doped silicon and negative in case of highly n-doped silicon, whereas the sensitivity of experimentally measurable Fano resonances is much higher in case of p-doped then in case of n-doped silicon [67, 57, 25, 75, 26].

9.1.3.2 Experimental Examples

Small Angle Beveling and Nomarski Differential Interference Contrast Micrographs

For the following examples of Raman-based dopant analysis, readily processed, standard polycrystalline silicon solar cells, provided with a BSF were cut into pieces (area of a few cm^2). These pieces were put on wedges with opening angles α ranging from ~5°–10°. The samples residing on the wedges were then embedded in a cylindrical shaped basin with resin. After drying, the embedded samples were mechanically and chemically polished. Finally, all samples were Secco [91] etched to make the crystal defects (random, twin boundaries, dislocations, etc.) as well as the BSF region visible. By the small angle beveling process, the length scales of the free carrier concentration profiles in the solar cell samples are artificially increased in one direction by a stretching factor $s(\alpha)$, which solely depends on the opening angle α. For the stretching factor $s(\alpha)$, the simple relation holds (see Fig. 9.12):

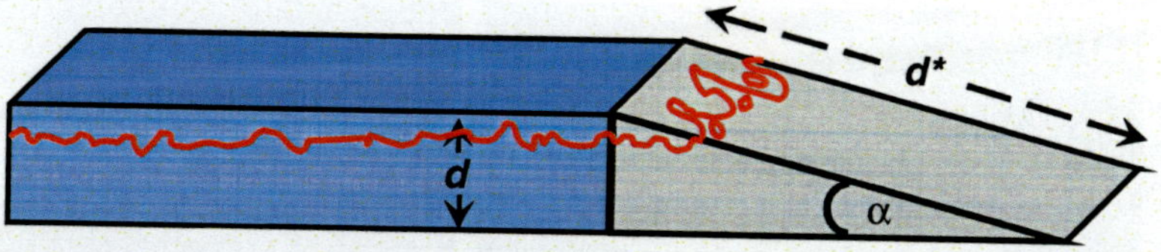

Fig. 9.12 Schematic drawing of a piece of a standard polycrystalline silicon solar cell containing a BSF (*above red line*) after small angle beveling. Due to the small angle beveling process, the original thickness d of the cell is increased to an apparent thickness d* according to the relation $d^* = s(\alpha) \cdot d$, where $s(\alpha)$ is the stretching factor given in Eq. (9.28). Small angle beveling reveals details of the BSF shape and microstructure that are usually not visible

$$d^* = \frac{1}{\sin(\alpha)} \cdot d = s(\alpha) \cdot d, \qquad (9.28)$$

where d^* is the increased but oblique thickness of the solar cell after the small angle beveling process and d indicates the original thickness.

The small angle beveling process reveals much more details of the BSF shape and its adjacent microstructure then a standard polishing process of the cross section of a silicon solar cell would reveal. On the other hand, small angle beveling introduces some irritating artefacts, such as e.g. island formation within an originally wrinkled but continuous interface (see Fig. 9.13).

The highly p-doped BSF region exhibits a different Secco edging rate compared to the rest of the crystal due to the additional Al-atoms in the Si-crystal lattice. Therefore, a sharp step of a few 100 nm in height (depending on the etching time) arises in between the highly p-doped BSF region and the usually weakly ($\sim 10^{15}\,\text{cm}^{-3} - 2 \times 10^{16}\,\text{cm}^{-3}$) boron doped bulk wafer region after Secco etching. This step, as well as the trenches and pitches produced by the grain boundaries and dislocations, become visible in a Nomarski differential interference contrast (NDIC) [72] microscope. Figure 9.13a–c show some NDIC-images of the BSF—bulk wafer interface region, where the different samples where beveled by using different wedge angles. The resulting stretching factors $s(\alpha)$ are given in brackets above the scale bars and the stretched directions are indicated by the dotted white arrow.

Evaluation of the Raman Data

For the Raman spectroscopic investigations, regions of interest within the small angle beveled samples were chosen and mapping areas were defined. Within the mapping regions, Raman spectra of the 3-fold degenerate first-order optical phonon (F_{2g}) mode of silicon were acquired every $1 - 3\,\mu$m in both x- and y-directions. All Raman measurements shown here were performed with the 633 nm emission of a HeNe-laser, as the peak asymmetry due to Fano resonances becomes more pronounced when longer excitation wavelengths are used [24].

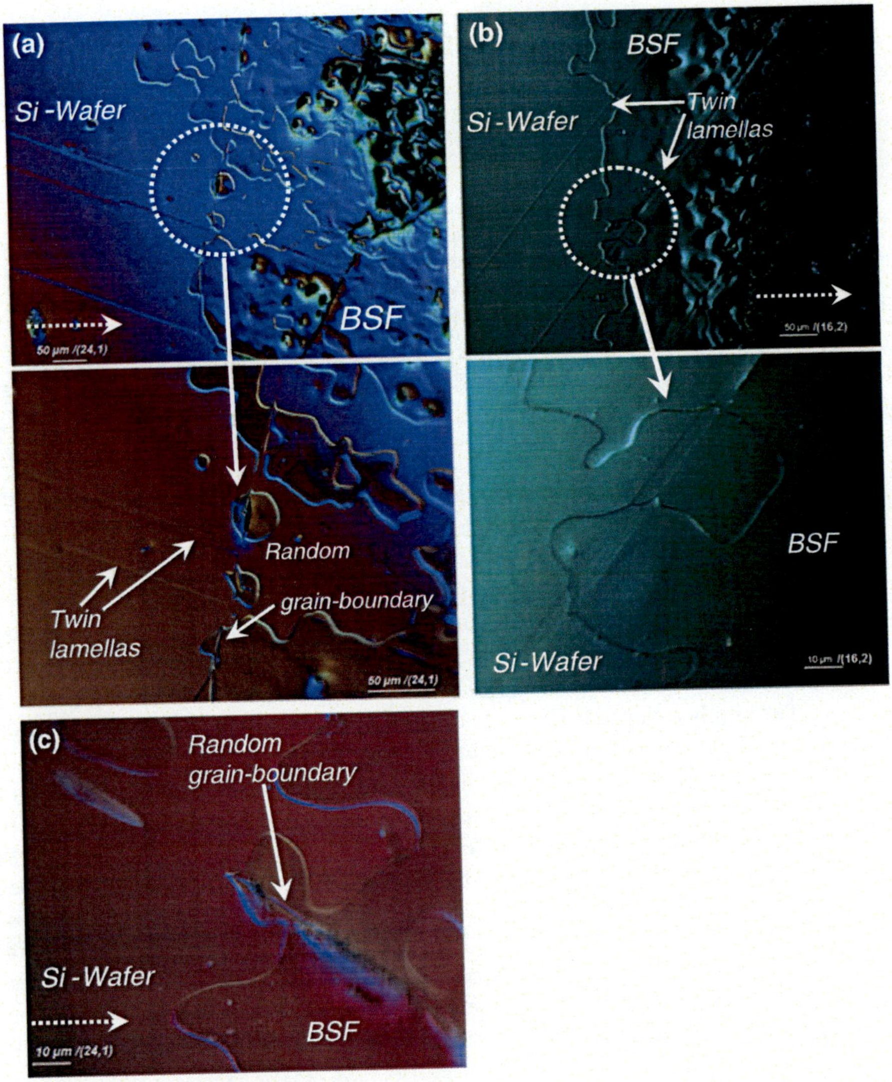

Fig. 9.13 Images of the BSF-bulk wafer interface region of standard polycrystalline silicon solar cells. The images were taken with a Nomarski differential interference contrast (NDIC) microscope which allows the visualization of the height difference that occurs after Secco etching in between the BSF- and the weakly *p*-doped bulk wafer region. The stretching factors $s(\alpha)$ are given in brackets above the scale bars and the stretched directions are indicated by the white dotted arrows

To obtain quantitative information about the free carrier concentrations from Raman maps, the intensity $I(\Omega, q, \Gamma, \Omega_{\text{Max}})$ of the first-order *Si* Raman peak should be fit with a function resulting from Eqs. (9.24) and (9.25):

$$I(\Omega, q, \Gamma, \Omega_{\mathrm{Max}}) = I_0 \cdot \frac{[q + 2(\Omega - \Omega_{\mathrm{Max}})/\Gamma]^2}{1 + [2(\Omega - \Omega_{\mathrm{Max}})/\Gamma]^2}, \tag{9.29}$$

where in the experiments, Ω simply represents the wavenumber axis (in cm^{-1}) and I_0 is just a scaling factor. If the exact values of the physically important parameters q, Ω_{Max}, and Γ are needed, all Raman spectra of interest should be fitted with Eq. (9.8). However, as Raman mappings usually consist of many thousand measurement points, fitting all the corresponding spectra with iterative routines would take unnecessarily long time. To determine the free carrier concentrations in highly doped regions with sufficient accuracy, mainly the values of the symmetry parameter q are of importance. The symmetry parameter q can be easily determined and with sufficient accuracy from the experimental data by the following procedure:

The integrated intensity of the Raman peak $I(\Omega, q, \Gamma, \Omega_{\mathrm{Max}})$ from the maximum position Ω_{Max}up to a larger wavenumber value away from the maximum, where the $I(\Omega, q, \Gamma, \Omega_{\mathrm{Max}})$ curve becomes close to zero (in the experiments, $100\,\mathrm{cm}^{-1}$ are sufficient) has to be calculated.

The right intensity integral I_{R}, which contains the asymmetric flank due to Fano resonances in case of high p-doping, can be evaluated according to:

$$I_{\mathrm{R}} = \int_{\Omega_{\mathrm{Max}}}^{\Omega_{\mathrm{Max}}+n\Gamma} I(\Omega, q, \Gamma, \Omega_{\mathrm{Max}})\mathrm{d}\Omega, \tag{9.30}$$

where the upper integration limit is defined in multiples n of the peak width Γ. In our experiments, Γ lies usually in the range of $\sim 3.1\,\mathrm{cm}^{-1}$ to $4.4\,\mathrm{cm}^{-1}$ (depending on the doping level) i.e. we define an average Γ- value of $3.8\,\mathrm{cm}^{-1}$. To obtain a sufficient upper limit of $\sim 100\,\mathrm{cm}^{-1}$ away from the maximum, n then has to be set to $n \sim 27$. Analog to Eq. (9.30), the left integrated intensity I_{L}can be calculated.

With the total integrated intensity $I_{\mathrm{Tot}} = I_{\mathrm{R}} + I_{\mathrm{L}}$ the Ratio $R(q, n)$ is defined as:

$$R(q, n) = \frac{I_{\mathrm{R}}}{I_{\mathrm{Tot}}} = \frac{I_{\mathrm{R}}}{I_{\mathrm{L}} + I_{\mathrm{R}}} \tag{9.31}$$

which gives the ratio of the intensity of the asymmetric peak portion to the total Raman peak intensity. The lengthy analytical expression for $R(q, n)$ can be found in [11].

The advantage of the procedure described above is that the ratio function $R(q, n)$ mainly depends on q but only changes slightly with a varying n. The dependencies on all other parameters are cancelled.

The maps shown in this chapter plot the $R(q, n)$ values that are determined by the procedure given above. In general, the yellow/red color coding represents high $R(q, n)$-values and small q-values (high p-doping levels) and the blue/black color coding displays the opposite conditions.

Calibration Measurements

It is possible to determine the free carrier concentration levels in silicon from the q-parameter directly from theoretical calculations. But it has been shown in literature [26] that the theoretically calculated free carrier concentrations substantially deviate from the experimentally determined ones (almost one order of magnitude). Therefore, it is preferable to use data obtained from the ECV method to calibrate the Raman data. The q-parameters and the free carrier concentration levels that are obtained from an analysis of the Raman peaks are calibrated in the following way: the free hole concentration profiles within the BSF region of a standard mc-silicon solar cell were first measured with the ECV method. Then a piece of the same solar cell was small angle beveled and secco etched in the way described above. The BSF region was then analyzed by Raman spectroscopy. Depending on the solar cell, the standard free hole concentration level within the bulk wafer region lies usually in the range of $\sim 5 \times 10^{15}\text{cm}^{-3} - 2 \times 10^{16}\,\text{cm}^{-3}$. According to the literature [67, 24] the standard p-doping levels of the respective bulk wafers still produce some slight Fano resonances when the 633 nm excitation wavelength is used, but the Raman peaks become almost symmetric in practice and the values for the parameter q take values larger than 150. Therefore, the q-value of $q = 150$, obtained within the standard p-doped region, is used as another calibration value and the corresponding free hole concentration is defined as the threshold concentration for the onset of slight Fano resonances in practice (using the 633 nm excitation wavelength). It is known, that the reciprocal symmetry parameter $1/q$ is roughly proportional to the free hole concentration in silicon [24, 7, 67]. Therefore, the free hole concentrations that are obtained from the ECV-measurements can be plotted versus $1/q$ (obtained from the Raman measurements) which results in a roughly linear correlation between the free hole concentration and the $1/q$ values. This plot is shown in Fig. 9.14. The obtained (roughly) linear curve is used for the free hole concentration measurements and maps with Micro-Raman spectroscopy.

The left image in Fig. 9.15 shows an enlarged optical micrograph of the spike-containing BSF feature already shown in Fig. 9.13c. The right image represents the Raman map of the Fano asymmetry (ratio function $R(q, n)$) produced by the high p-doping within the BSF region. At the six positions, marked with an asterisk, the symmetry parameter q as well as the free hole concentration were determined by the more accurate peak fitting routine using Eq. (9.29) (see Table 9.1) rather than by the less accurate inversion of the experimentally determined ratio function $R(q, n)$. The highest free hole concentration levels are measured at positions (position 4, 5, 6) in the vicinity of a random grain boundary (dotted line). These regions occur much darker in the light optical micrograph (red arrow), compared to the bright bulk wafer region and the slightly darker BSF region. It is assumed, that the somewhat higher etching rate at the random grain boundaries is responsible for the dark contrast: close to the random grain boundaries, the silicon becomes rather thin due to the higher etching rate. The dark Al-rich phase underneath the Si-rich phase (BSF-region) then becomes better visible and already influences the free carrier concentration measurements due to the additional Al-atoms. The q-value and the corresponding free hole concentration

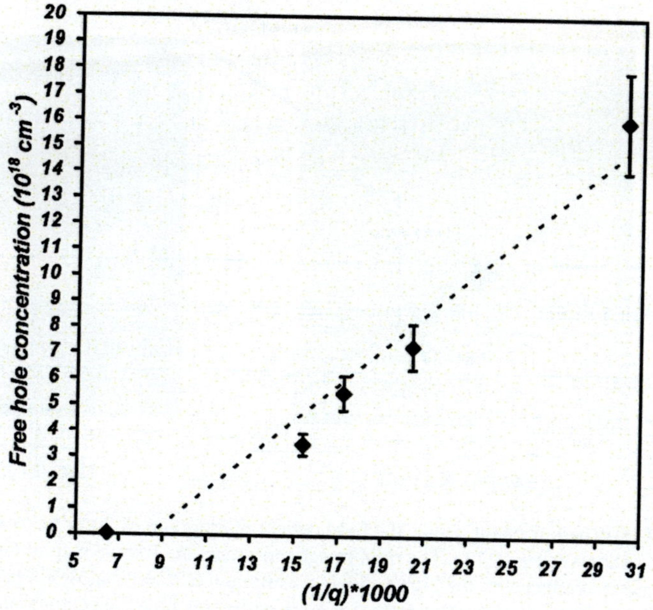

Fig. 9.14 Free hole concentrations within the BSF region determined by the ECVmethod plotted versus the reciprocal Fano symmetry parameter $1/q$. The roughly linear correlation is indicated by the dotted calibration line. This line is used to determine the free hole concentrations within the BSF regions from the following Raman maps. The error bars account for the inaccuracy of the ECV measurements, for the variations of the p-dopant concentrations with the measurement locations within a single solar cell and for the errors made by determining the q-values from the experimental Raman spectra by the numerical fitting routines

level at position 3 probably gives a more accurate value of the free hole concentration contained in the major part of the BSF region. The fact, that the small spike BSF—bulk wafer interface close to the random grain boundary can be resolved by the Raman map, shows how good the lateral resolution of the combined small angle beveling—Raman measurements of the highly p-doped regions are. In Fig 9.16, the baselines of the two first-order Raman peaks of silicon, measured within the bulk wafer (dotted grey) and the BSF region (continuous black) within another region (not shown here) of the same sample are displayed. High p-doping (black curve) leads not only to a broadening of the higher wavenumber flank of the Raman peak (q positive but small), but also to a larger Γ (peak width) value, as well as an also slightly broader lower wavenumber flank of the silicon Raman peak. From the two integrated half-peak intensities I_L and I_R (see dotted vertical lines in the diagram), the ratio function $R(q, n)$ (Eq. (9.12)), and the q-parameter is determined. The black curve in Fig. 9.16 shows one of the maximal Fano-induced peak asymmetry of all the Raman measurements presented here. The corresponding free hole concentration level is $> 2 \times 10^{19}\,cm^{-3}$.

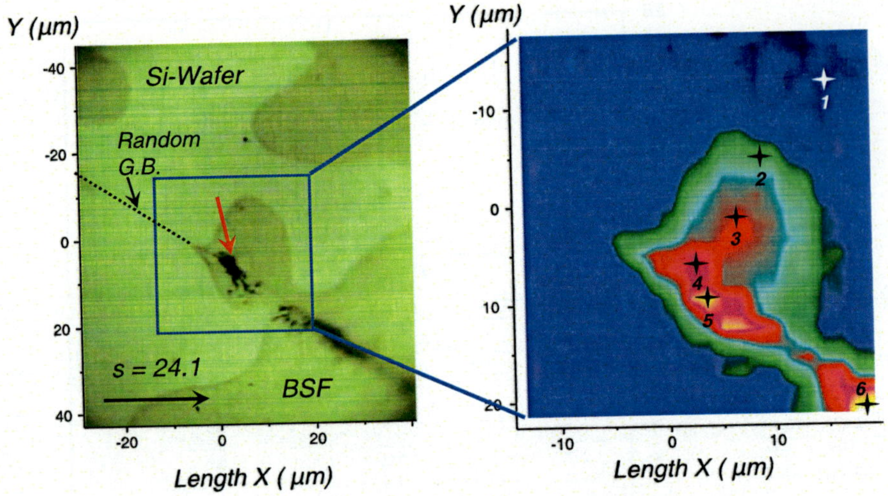

Fig. 9.15 The left image shows the light optical micrograph of the spike-containing BSF feature already shown in Fig. 9.13c. The right image represents the Raman map of the Fano-induced peak asymmetry produced by the high p-doping within the BSF region. At six positions, marked with an asterisk, the symmetry parameter q as well as the free hole concentrations were determined by the more accurate peak fitting routine using Eq. (9.29) (see Table 9.1)

Table 9.1 Symmetry parameters q and the resulting reciprocal values $1/q$ as well as the free hole concentrations determined by Raman measurements. The data correspond to the six positions, marked with an asterisk in the Raman map shown in Fig. 9.15

Position	Symmetry parameter q	$1/q$*1,000	Free hole concentration (10^{18} cm^{-3})
1	>150	<6.67	<0.025
2	65	15.38	4.7
3	54	18.52	6.8
4	51	19.61	7.6
5	39	25.64	11.7
6	29	34.48	17.7

The transition from the standard low doping level within the bulk wafer (dark blue regions) to the highly p-doped regions BSF usually takes place within less then ~4 μm in the Raman mappings. If we take into account, that this transition elongation into the small angle beveled direction has to be divided by the stretching factor $s(\alpha)$, the effective transition length would lie in the range of ~250–700 nm. However, it can be assumed that the dopant atom concentrations at the bulk wafer—BSF interface follows the theoretical step functional shape and that the measured gradient of the free hole concentration results from 1) the blurring effect of the focused (and sometimes defocused) laser spot and 2) the elongation of the free hole diffusion of a few hundreds of nm [9] into the weakly doped bulk wafer region.

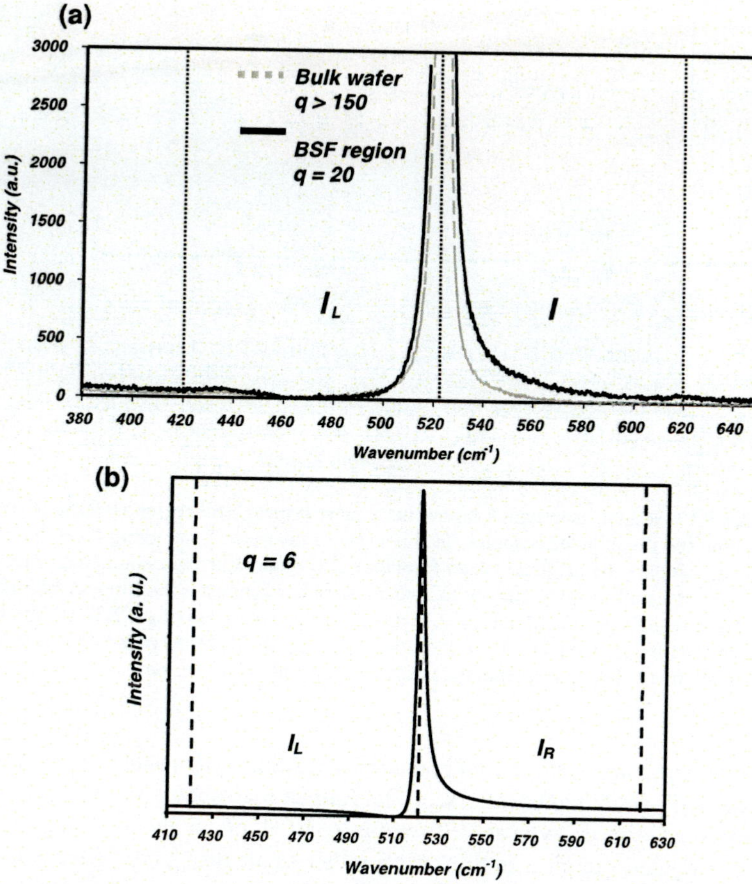

Fig. 9.16 Baselines of the two first-order Raman peaks of silicon measured within the bulk wafer (dashed gray curve) and the BSF region (continuous black curve). The spectra were acquired from a different mapping region (not shown here) within the same sample. The two peaks are normalized and therefore the maximum intensities are the same (not visible in the plot). High p-doping leads to a broadening of the higher wavenumber-flank of the Raman peak (q positive but smaller) as well as to a larger Γ value, which manifests in an also slightly broader lower wavenumber-flank of the Raman peak. b) Calculated Fano Raman peak for a hypothetical q-value of $q = 6$ (very high p-doping). The antiresonance near the lower wavenumber-flank of the peak becomes clearly visible. With the two integrated peak intensities I_L and I_R up to the dotted vertical lines in the diagrams we define the ratio function $R(q, n) = I_R/(I_R + I_L)$ from which the symmetry parameter q can also be determined rather fast but with an error of $\sim \pm 2$

The BSF region shown in the light optical micrographs and Raman map in Fig. 9.15 exhibits a rather exceptional rough and wrinkled shape. One reason for the unusual shape is that in this case a large stretching factor $s(\alpha)$ was used, resulting in small angle beveling artifacts such as island formation and rather distorted BSF shapes. Another reason is that the region shown in Fig. 9.15 contains a more complex

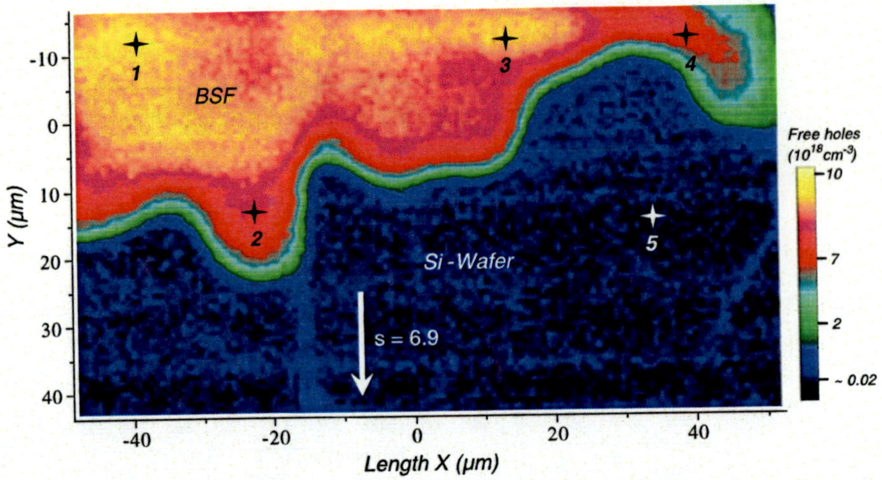

Fig. 9.17 Raman map which represents a more typical BSF region (~95 % of all BSF regions) within the investigated samples. In this case a rather small stretching factor ($s(\alpha) = 6.9$) was used, avoiding the artificial island formation. No random or twin boundaries were present in the analyzed sample region, leading to a much more homogeneous and smooth BSF shape. As in the sample regions presented in Figs. 9.15 and 9.17, the elongation of the BSF to bulk wafer free hole concentration gradient is $\sim 3\mu m$, resulting from the blurring effect of the finite size of the laser spot and the free hole diffusion into the weakly doped bulk wafer region

microstructure, e.g. random boundaries, leading to a pronounced spike formation at the BSF—bulk wafer interface. The Raman map shown in Fig. 9.17 represents the shape of a more typical BSF region (~95 % of all the BSF regions) within the investigated samples. In this case a rather small stretching factor ($s(\alpha) = 6.9$) was used, avoiding the artificial island formation. Also no random or twin boundaries are present in the analyzed sample region, leading to a more homogenous and smooth BSF shape. As in the sample region presented in Fig. 9.15, the elongation of the BSF to bulk wafer free hole concentration gradient is $\sim 3\mu$m, resulting from the blurring effect due to the finite size of the laser spot and the free hole diffusion into the weakly doped bulk wafer region. The variations of the q parameter in this case are rather small (43–51), corresponding to small free hole concentration variations of $7.6 \times 10^{18}\,cm^{-3} - 10.1 \times 10^{18}\,cm^{-3}$. Again, there are 5 measurement points (marked with an asterisk) that were analyzed in more detail. The obtained data are presented in Table 9.2.

9.1.3.3 Comparison with Other Dopant Measurement Methods

Besides Raman spectroscopy and the ECV method, there are two other dopant concentration measurement techniques that are worth mentioning as they are frequently used. The standard method when B-doped silicon samples are analyzed is secondary-

Table 9.2 Symmetry parameters q and the resulting reciprocal values 1/q as well as the free hole concentrations determined by Raman measurements. The data correspond to the five positions, marked with an asterisk in the Raman map shown in Fig. 9.17

Position	Symmetry parameter q	$1/q$*1,000	Free hole concentration (10^{18} cm^{-3})
1	43	23.26	10.1
2	49	20.41	8.1
3	46	21.74	9.0
4	51	19.61	7.6
5	>150	<6.67	<0.025

ion-mass-spectroscopy (SIMS) [21, 50, 12]. SIMS is a technique that measures directly the concentrations of dopant atoms and not the free carrier concentrations. With SIMS it is possible to perform maps of dopant atom concentrations with ∼ μm resolution in 2 or even 3 dimensions and in addition, with SIMS it is possible to distinguish between the different dopant species. However, SIMS needs high vacuum equipment which makes the method complex and expensive. Another drawback of SIMS is that quantitative measurement results are often difficult to obtain, especially if silicon samples that are highly doped with other dopants than boron (e.g. aluminum) are investigated [90]. A method that measures lateral free carrier concentration gradients instead of absolute values of dopant concentrations is the lateral photoscanning method (LPS) [66]. But the lateral resolution of the LPS method is limited to ∼ 30 μm and therefore not suited to analyze and map free carrier concentration variations within the few μm thick BSF and other highly doped regions (e.g. the emitter) within a silicon solar cell.

9.1.4 Raman Imaging of Defect Densities and Crystallinity

9.1.4.1 General Remarks

As shown in the previous sections, Raman scattering measurements can provide detailed information about semiconductor materials. Especially in case of solar silicon, the crystal orientations, the distribution of residual stresses, and free carrier concentrations can be determined rather fast and accurate. In addition to the intensity, position, and symmetry of the first-order Raman modes, their linewidths mainly relate to the presence of extended crystal defects, such as dislocations, low angle grain boundaries, cracks, nanocrystalline/amorphous regions, and their combinations [87, 15]. Such defects act as anharmonic perturbations leading to reduced phonon lifetimes that manifest themselves as a broadening of the first-order Raman peaks described by the full-width at half maximum Γ (FWHM) values. In contrast, in a perfect crystal, the phonon lifetimes are theoretically infinite in the harmonic approximation (that neglects third- and higher order derivatives of the crystalline potential) with narrow delta function-like linewidths of the optical phonon Raman bands [97].

The anharmonic lifetimes of phonons defined as $\sim 1/\Gamma$ are usually evaluated using first principles calculations including both kinematic effects; i.e. the decay of phonons into vibrations of lower frequency and dynamic effects, i.e. the magnitude of the tensor of the third derivatives of the crystal potential with respect to atomic displacements that describes the instability of one-phonon states. The calculated Γ values of the Raman-active optical phonons were found to agree well with those determined experimentally in the case of single crystalline semiconductors, (defect free) such as diamond, Si, Ge, GaAs, GaP and InP [31, 32]. In the case of single crystalline Si at room temperature, a FWHM value of $\sim 3.6\ \text{cm}^{-1}$ is measured [93].

Broadening of the Si Raman peak occurs in industrial relevant materials such as polycrystalline silicon for solar cells due to anharmonic effects produced not only by defects, but also by the intrinsic stress fields of defects, the presence of residual stresses and impurities as well as by the grain size. On the one hand, even relatively low stress values are known to induce Raman frequency shifts either to higher wavenumbers implying tensile stresses or to lower wavenumbers in case of compressive stresses. However, they can also broaden the Raman peak when the level of internal or applied stresses approach the stress-induced splitting limit of the peak (lifting of degeneracy) or when there are large stress gradients within the probed volume [34]. On the other hand, impurities can either produce new Raman peaks through their own localized vibrational modes or alter the Raman spectrum of the host material through the change in mass and bond length (atomic effects) as well as through the resonant interaction of free carriers (donors or acceptors) with the lattice vibrations (electronic effects). Characteristic for the electron-phonon coupling in highly doped silicon is the Fano-type Raman peak asymmetry (see Section 9.1.3). The Raman linewidths were also found to broaden in the case of relatively small grains (in the nm range) due to the phonon confinement effect, that is, the frequency distribution of the scattered light comes from a broader interval in k-space around the Γ-point (zone centre) in the Brillouin-zone since the $\Delta k = 0$ selection rule is broadened to higher values due to the space-momentum uncertainty principle [15, 94]. Because the results presented in the following were obtained on large grained solar silicon, the broadening of the FWHM due to small grain sizes does not occur.

9.1.4.2 Examples

One example of the comprehensive insight into microstructural and mechanical properties of silicon provided by the analysis of the first-order Raman spectra is demonstrated in Fig. 9.18. Here typical FWHM, stress, free carrier concentration, and intensity distribution maps are shown for the seed layer of a thin film solar cell. They were obtained by fitting the Raman spectra with a Gauss-Lorentzian fit function or with a Fano-type fit function for the doping evaluation (Section 9.1.3). These data are complemented by EBSD measurements at the same position shown in Fig. 9.19b–

d. The Raman measurements were performed using polarization settings for incident and scattered light (see Section 9.1.1). The stress map was corrected for the peak shift contributions produced by the Fano effect. These results were obtained on a sample consisting of a 110 nm thick boron doped (2.1×10^{19} cm^{-3}) silicon seed layer on a borosilicate glass (BSG) substrate that was crystallized with a diode laser array. The seed layer which serves as the solar cell emitter in a superstate cell concept is then epitaxially thickened (to $>1\mu m$) by solid phase epitaxy (SPE) to form a complete thin film solar cell. The laser crystallization process was optimized to avoid cracking of the substrate and the seed layer. However, the preparation conditions involving inhomogeneous temperature gradients during melting and recrystallization of the amorphous silicon seed layer lead to thermal stresses as well as to sources of intrinsic local stress concentrations including both compressive and tensile stresses in the seed layer. These localized stresses are attributed to particular defect configurations/structures formed to relax thermal stresses as long as the plastic deformation is allowed by temperature, as well as to residual (remaining) stresses that were not relieved by defect formation and are then present at room temperature. The presence of defects is confirmed by TEM and EBSD studies displayed in Fig. 9.19. The difference in the coefficients of thermal expansion (CTE) between the crystalline silicon film and the BSG substrate was predicted to result in a uniformly distributed extrinsic tensile stress of ~3 MPa in the seed layer when cooling from 560°C (annealing point of BSG) to room temperature [87]. This value is below the stress resolution limit of 12 MPa of our Raman spectrometer. Therefore, for the investigated samples, the measured shift of the Si Raman peak toward lower wavenumbers with respect to the stress-free reference of 520 cm^{-1} is attributed to the presence of defects and the CTE mismatch between the silicon film and the surface silicon oxide film (see TEM image in Fig. 9.19a). More information about the laser crystallization process and its impact on the stress states and microstructure is given elsewhere [87, 28].

It is well-known that the broadening of the FWHM of Raman spectra is mainly determined by high defect densities. Indeed, larger full width at half maximum values of the Si peak shown in Fig. 9.18a correlate to the existence of low angle grain boundaries as measured by EBSD and indicated by arrows in Fig. 9.19b–d. These types of grain boundaries are known to consist of dislocation networks/arrays [27]. This is further supported by the continuously changing crystal orientation within the studied grain as indicated by the gradual changing color code in the intergranular misorientation gradient map by EBSD in Fig. 9.19c that was suggested to be associated with geometrically necessary dislocations forming low angle grain boundaries [76].

As mentioned earlier in this section, there are also other factors that may modify the Raman linewidths such as the intrinsic stress fields of defects, residual stresses, and impurities. By comparing the FWHM and residual stress maps displayed in Fig. 9.18a–b, respectively, it can be seen that the two patterns are different exhibiting quite similar FWHM values in regions of different stress levels (different Raman peak positions) along the two defect lines marked by dashed lines indicating virtually no influence of stresses on the FWHM. The fact that the dislocation networks/arrays are only partly accompanied by stresses can be explained by the local rearrangement in configurations in which the superimposed long-range intrinsic stress fields of the dislocations cancel partially or totally [52, 88]. In addition, the measured stress variations within the probed volume (~1μm diameter for the 633 nm excitation wavelength) are found not to broaden the FWHM. Moreover, the measured stresses are significantly smaller than those close to the splitting of the Raman Si peak that would eventually lead to a stress-induced widening of the peak [7]. Concerning the influence of impurities (doping) on the FWHM, there is nearly no correlation between them as seen by comparing the free hole concentration map obtained from the symmetry parameter q (defined in Section 9.1.3) with the FWHM map displayed in Fig. 9.18d and a, respectively. For the quantitative doping evaluation, the free carrier concentration vs. q calibration curve shown in Fig. 9.14 was used. The free hole concentrations were evaluated to be lower than the nominal doping presumably due to incomplete activation of dopants.

To conclude, the FWHM distribution shows the high defective regions, while the stress distribution reflects either residual stresses or particular low density defect configurations/structures that do not necessarily result in a marked broadening of the FWHM. Doping was found not to influence the FWHM in the studied samples.

9.1.5 Concluding Remarks

Micro-Raman spectroscopy has established as a valuable analysis tool in semiconductor science and technology as well as microelectronics within the last decade. Micro-Raman spectroscopy has become a standard instrument in research and development departments of most semiconductor and microelectronics industries. There, Raman spectroscopy is used to analyze mechanical stress, composition, dopant distributions, and crystallinity within semiconductor devices or for failure analysis on the μm-scale. However, in photovoltaics science and technology, Raman spectroscopy has only emerged within the last few years from a niche method to a routinely used analysis tool. One reason for the delayed establishment of Raman spectroscopy in this field might be that the usually more complex microstructure structure of photovoltaic materials made it difficult to simply transfer the measurement and evaluation procedures used in microelectronics to photovoltaic devices. Another reason is that until the last few years there was no real need for solar cell material analysis with Raman spectroscopy. Due to the rather thick and therefore mechanically stable silicon wafers, questions concerning the mechanical stress states and wafer breakage played only a minor role. Due to the already well-established and automated dopant analysis methods (e.g. ECV measurements) there was also no need to develop a new doping analysis tool based on Raman spectroscopy. However, things changed drastically, as the solar cell producers were recently forced to reduce the thickness of the bulk silicon wafers or to use thin film solar cell concepts in order to reduce the overall solar cell production costs and to finally offer photovoltaic energy at competitive market prizes. However, as thinner wafers and thin silicon films (a few μm thick) are prone, to a greater extent, to breakage due to internal mechanical stress fields, crack formation and propagation, mechanical stress/strain analysis on the μm scale

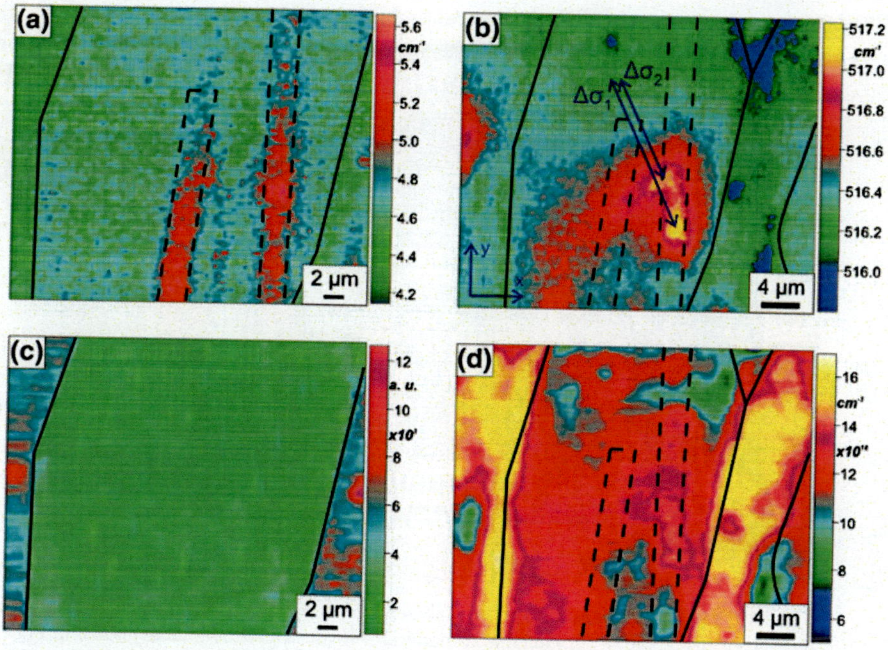

Fig. 9.18 Raman maps obtained from the first-order Raman peak of silicon: (a) the full width at half maximum (FWHM) map. The dashed lines delimitate regions of lower crystallinity corresponding to low angle grain boundaries marked by arrows in the EBSD maps of Fig. 9.19. (b) The peak position map shows the stress distribution attributed to the intrinsic stress fields of defects and thermal stresses. Two lateral stress gradients ($\Delta\sigma_1 = 227$, $\Delta\sigma_2 = 197 \pm 12$ MPa) are indicated with respect to the reference coordinate system. (c) The peak intensity map relates to the crystallographic orientation of the grains. The solid lines in all maps represent grain boundaries inferred from intensity maps. (d) The free hole concentration map (10^{18} cm^{-3}) calculated from the symmetry parameter q. By comparing these maps, it can be seen that the FWHM map does not correlate with both stress and doping maps suggesting that there is virtually no influence of them on FWHM

has become increasingly important. As Micro-Raman spectroscopy is one of the few methods that enables quantitative and nondestructive mechanical analysis with μm spatial resolution and offers quantitative analysis of dopant concentrations along the way, this method has become particularly important in solar materials characterization within the last few years. Though Raman measurements can be performed rather fast and easily in practice, the interpretation of the measurement results can become increasingly tricky and puzzling. Therefore, the main purpose of this contribution was to equip the potential applicant of Raman spectroscopy in the field of photovoltaics with the necessary physical background and knowledge of spectroscopy of solids to use the presented methods effectively or to improve and further develop the described evaluation procedures.

However, besides mono, poly, and multicrystalline silicon wafers, there are many other solar cell material systems where Raman spectroscopy might also serve as an important analysis tool in future, but that cannot be discussed in this contribution.

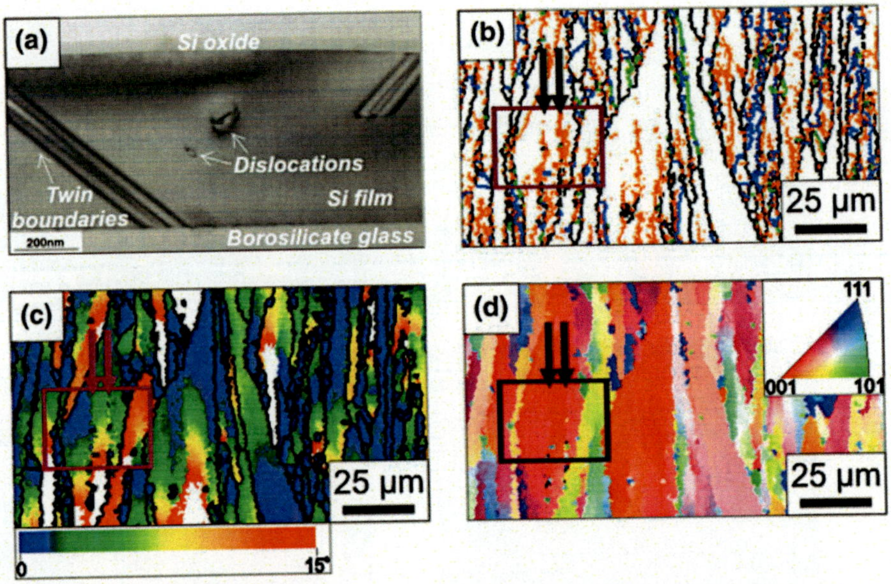

Fig. 9.19 (a) TEM image of a laser crystallized silicon seed layer. The defects are inhomogeneous distributed inside the Si film leading to nonuniform stress distributions as shown in Fig. 9.18 (b). (b) Grain boundary map: (*black lines*) high angle GBs > 15° , (*orange lines*) low angle GBs < 15° , (*blue lines*) Σ3 GBs, (*green lines*) Σ9 GBs. The area enclosed by the rectangle in all EBSD maps represents the Raman mapped area displayed in Fig. 9.18 (b) and (d), while the arrows indicate the low angle GBs marked by dashed lines in Fig. 9.18. (c) Intergranular misorientation gradient map. (d) Inverse pole figure map

e.g. in amorphous or microcrystalline laser crystallized thin film solar cells. There, the crystalline/amorphous ratio as well as the average grain size could be determined and imaged with Micro-Raman spectroscopy. Also in chalcopyrite based solar cells Raman spectroscopy could serve to analyse the composition, material structure or the influences of mechanical stresses.

References

1. T. Ambridge, M.M. Faktor, J. Appl. Electrochem. **5**, 319 (1975)
2. E. Anastassakis, E. Burstein, J. Phys. Chem. Solids **32**, 563 (1971)
3. E. Anastassakis, A. Pinczuck, E. Burstein, F.H. Pollack, M. Cardona, Solid State Commun. **8**, 133 (1970)
4. E. Anastassakis, E. Burstein, J. Phys. Chem. Solids **32**, 313 (1971)
5. E. Anastassakis, A. Cantarero, M. Cardona, Phys. Rev. B **41**, 7529 (1990)
6. E. Anastassakis, J. Appl. Phys. **86**, 249 (1999)
7. M. Balkanski, K.P. Jain, R. Beserman, M. Jouanne, Phys. Rev. B **12**, 4328 (1975)
8. M. Balkanski, R.F. Wallis, E. Haro, Phys. Rev. B **28**, 1928 (1983)
9. P.A. Basore, D.A. Clugston, G. Thorson, A. Smith, D. Rover, B. Hansen, G. Anderson, *Solar Cells Simulation Software*, (University of New South Wales, Syeney, 2003)

10. M. Becker, H. Scheel, S. Christiansen, H.P. Strunk, J. Appl. Phys. **101**, 0635135 (2007)
11. M. Becker, U. Gösele, A. Hofmann, S. Christiansen, J. Appl. Phys. **106**, 074515 (2009)
12. A. Benninghoven, Z. Phys. **230**, 403 (1970)
13. K.A. Bezjian, H.I. Smith, J.M. Carter, M.W. Geis, J. Electrochem. Soc. **129**, 1848 (1982)
14. P. Blood, Semicond. Sci. Technol. **1**, 7 (1986)
15. X.-Z. Bo, N. Yao, S.R. Shieh, T.S. Duffy, J.C. Sturm, J. Appl. Phys. **91**, 2910 (2002)
16. E. Bonera, M. Fanciulli, D.N. Batchelder, Appl. Phys. Lett. **81**, 3377 (2002)
17. E. Bonera, M. Fanciulli, D.N. Batchelder, J. Appl. Phys. **94**, 2729 (2003)
18. S.D. Brotherton, J.R. Ayres, M.J. Edwards, C.A. Fisher, C. Glaister, J.P. Gowers, D.J. McCulloch, M. Trainor, Thin Solid Films **337**, 188 (1999)
19. H.H. Burke, I.P. Herman, Phys. Rev. B **48**, 15016 (1993)
20. H. Callen, Am. J. Phys. **36**, 735 (1968)
21. R. Castaign, B. Jouffrey, G. Slodzian, Compt. Rend. **251**, 1010 (1960)
22. F. Cerdeira, M. Cardona, Phys. Rev. B **5**, 1440 (1972)
23. F. Cerdeira, T.A. Fjeldly, M. Cardona, Solid State Comm. **13**, 325
24. F. Cerdeira, T.A. Fjeldly, M. Cardona, Phys. Rev. B **8**, 4734 (1973)
25. M. Chandrasekhar, M. Cardona, O.E. Kane, Phys. Rev. B **16**, 3579 (1977)
26. M. Chandrasekhar, J.B. Renucci, M. Cardona, Phys. Rev. B **17**, 1623 (1978)
27. J. Chen, T. Sekiguchi, Jpn. J. Appl. Phys. **46**, 6489 (2007)
28. S. Christiansen, P. Lengsfeld, J. Krinke, M. Nerding, J. Appl. Phys. **89**, 5348 (2001)
29. L. Clément, R. Pantel, L.F.T. Kwakman, J.L. Rouvière, Appl. Phys. Lett. **85**, 651 (2004)
30. T.C. Damen, S.P.S. Porto, B. Tell, Phys. Rev. **142**, 570
31. A. Debernardi, S. Baroni, E. Molinari, Phys. Rev. Lett. **75**, 1819 (1995)
32. A. Debernardi, Phys. Rev. B **57**, 12847 (1998)
33. T.P. Devereaux, R. Hackl, Rev. Mod. Phys. **79**, 175 (2007)
34. I. De Wolf, Semicond. Sci. Technol. **11**, 139 (1996)
35. I. De Wolf, H.E. Maes, S.K. Jones, J. Appl. Phys. **79**, 7148 (1996)
36. I. De Wolf, J. Jimenez, J. Landesman, C. Frigeri, P. Braun, E. Da Silva, E. Calvet, *New Developments in Optical Spectroscopy for the Characterization of Semiconductor Devices (Nostradamus)*, (European Commission, Belgium, 1998). ISBN 92-828-5011-0
37. I. De Wolf, J. Raman Spectrosc. **30**, 877 (1999)
38. B. Dietrich, K.F. Dombrowski, J. Raman Spectrosc. **30**, 893 (1999)
39. U. Fano, G. Pupillu, A. Zannoni, C.W. Clark, J. Res. Natl. Inst. Stand. Technol. **110**, 583 (2005). (English Translation of: U. Fano, Nuovo Cimento **12**, 154 (1935))
40. U. Fano, Phys. Rev. **124**, 1866 (1961)
41. U. Fano, Phys. Rev. **137**, 1364 (1965)
42. S. Ganesan, A.A. Maradudin, J. Oitmaa, Ann. Phys. **56**, 556 (1970)
43. M.W. Geis, H.I. Smith, B.Y. Tsaur, J.C.C. Fan, E.W. Maby, D.A. Antoniadis, Appl. Phys. Lett. **40**, 158 (1982)
44. R. Gereth, H. Fischer, E. Link, S. Mattes, W. Pschunder, Energy Convers. **12**, 103 (1972)
45. M.P. Godlewski, C.R. Baraona, H.W. Brandhorst, *Proceedings of the 10th IEEE Photovoltaic Specialists Conference*, (Palo Alto, California, 40, 1969)
46. M.A. Green, *Silicon Solar Cells: Advanced Principles & Practice*, (University of New South Wales, Centre for Photovoltaic Devices and Systems, 1995)
47. J.I. Hanoka, R.O. Bell, Ann. Rev. Mat. Sci. **11**, 353 (1981)
48. K.H. Hellwege, *Einführung in die Festkörperphysik* (Springer, Berlin, 1988). (in german)
49. M. Herms, J. Cryst. Growth **210**, 172 (2000)
50. R.K. Herzog, H. Liebl H., J. Appl. Phys. **34**, 2893 (1963)
51. J.B. Hopkins, L.A. Farrow, J. Appl. Phys. **59**, 1103 (1986)
52. D. Hull, D.J. Bacon, *Introduction to Dislocations*, Chapters 4 and 9, (Butterworth-Heinemann, Oxford, 2001)
53. F. Huster, in *Proceedings of the 20th European Photovoltaic Solar Energy Conference and Exhibition*, (Barcelona, Spain, 2005), p. 635

54. F. Huster, in *Proceedings of the 20th European Photovoltaic Solar Energy Conference and Exhibition*, (Barcelona, Spain, 2005), p. 1462
55. F. Huster, in *Proceedings of the 20th European Photovoltaic Solar Energy Conference and Exhibition*, (Barcelona, Spain, 2005), p. 1466
56. T. Ito, H. Azuma, S. Noda, Jpn. J. Appl. Phys. **33**, 171 (1994)
57. M. Jouanne, R. Beserman, I. Ipatova, A. Subashiev, Solid State Comm. **16**, 1047 (1975)
58. D.C. Joy, D.E. Newbury, D.L. Davidson, J. Appl. Phys. **53**, R81 (1982)
59. D. Karg, K. Roy, W. Koch, in *Proceedings of the 19th European Photovoltaic Solar Energy Conference*, (Paris, France, 2004), p. 923
60. D.L. Kendall, Ann. Rev. Mater. Sci. **9**, 373 (1979)
61. C. Kittel, *Introduction to Solid State Physics*, (Wiley, New York, 1996)
62. M.V. Klein, in *Light Scattering in Solids*, vol. 8, ed. by M. Cardona, (Springer, Berlin, 1975), p. 147
63. P. Lengsfeld, N.H. Nickel, Ch. Genzel, W. Fuhs, J. Appl. Phys. **91**, 9128 (2002)
64. G.H. Loechelt, N.G. Cave, J. Menendez, J. Appl. Phys. **86**, 616 (1999)
65. R. Loudon, Adv. Phys. **13**, 423 (1964)
66. H.J. Schulze, A. Lüdge, H. Riemann, J. Electrochem. Soc. **143**, 4105 (1998)
67. V. Magidson, R. Beserman, Phys. Rev. B **66**, 195206 (2002)
68. J. Mandelkorn, J.H. Lamneck, J. Appl. Phys. **44**, 4785 (1973)
69. M. Melendez-Lira, J. Menendez, W. Windl, O.F. Sankey, G.S. Spencer, S. Sego, R.B. Culbertson, A.E. Bair, T.L. Alford Phys, Rev. B **54**, 12866 (1997)
70. J. Menendez, W.H. Weber, R. Merlin, *Raman Scattering in Materials Science*, (Springer, Berlin, 2000)
71. J. Michler, Y. Kaenel, J. Stiegler, E. Blank, J. Appl. Phys. **83**, 187 (1998)
72. D.C. Miller, G.A. Rozgony, *Handbook on Semiconductors 3*, ed. by S.P. Keller, 217 (1980)
73. K. Mizoguchi, S. Nakashima, J. Appl. Phys. **65**, 2583 (1989)
74. S. Narayanan, S.R. Kalidindi, L.S. Schadler, J. Appl. Phys. **82**, 2595 (1997)
75. N.H. Nickel, P. Lengsfeld, I. Sieber, Phys. Rev. B **61**, 15558 (2000)
76. Ch. Niederberger, J. Michler, A. Jacot, Acta Mater. **56**, 4002–4011 (2008)
77. T. Nishioka, Y. Shinoda, Y. Ohmachi, Appl. Phys. Lett. **43**, 92 (1983)
78. M.-W. Pan, R.E. Benner, L.M. Smith, in *Handbook of Vibrational Spectroscopy*, vol. 1 (Wiley, Chichester 2002), p. 490
79. G. Pezzotti, H. Ichimaru, L.P. Ferroni, K. Hirao, O. Sbaizero, J. Am. Ceram. Soc. **84**, 1785 (2001)
80. G. Pezzotti, Y. Okamato, T. Nishida, M. Sakai, Acta Mater. **44**, 899 (1996)
81. S. Tochino, G. Pezzotti, J. Raman Spectrosc. **33**, 709 (2002)
82. G. Pezzotti, J. Raman Spectrosc. **30**, 867 (1999)
83. G. Pezzotti, N. Muraki, N. Maeda, K. Satou, Y. Nishida, J. Am. Ceram. Soc. **82**, 1249 (1999)
84. P. Puech, S. Pinel, R.G. Jasinevicius, P.S. Pizani, J. Appl. Phys. **88**, 4582 (2000)
85. J. Raptis, E. Liarokap, E. Anastassakis, Appl. Phys. Lett. **44**, 125 (1984)
86. J.R. Sandercock, Phys. Rev. Lett **28**, 237 (1972)
87. G. Sarau, M. Becker, A. Bochmann, A.Gawlik, G. Andrä, S.Christiansen, Proceedings of the 24th European Photovoltaic Solar Energy Conference, Hamburg, p 2494 (2009)
88. G. Sarau, M. Becker, S. Christiansen, M. Holla, W. Seifert, in *Proceedings of the 24th European Photovoltaic Solar Energy Conference*, (Hamburg, Germany, 2009), p. 969
89. M. Schreck, H. Roll, J. Michler, E. Blank, B. Stritzker, J. Appl. Phys. **88**, (2000)
90. D.K. Schroder, in *Semiconductor Material and Device Characterization*, 2nd edn. (Wiley, New York, 1998), pp. 683–689
91. F. Secco D' Aragona, J. Electrochem. Soc. **119**, 948 (1972)
92. A.W. Simonian, A.B. Sproul, Z. Shi, E. Gauja, Phys. Rev. B **52**, 5672 (1995)
93. P.A. Temple, C.E. Hathaway, Phys. Rev. B **7**, 3685 (1973)
94. Z. Iqbal, S. Veprek, A.P. Webb, P. Capezzuto, Solid State Commun. **37**, 993 (1981)
95. P. Verma, S.C. Abbi, K.P. Jain, Phys. Rev. B **51**, 16660 (1995)
96. G.D. Watkins, W.B. Fowler, Phys. Rev. B **16**, 4524 (1977)

97. W.H. Weber, R. Merlin (ed.), Chapter 3 *Raman Scattering in Materials Science*, (Berlin, Heidelberg 2000)
98. C. Weißmantel, C.C. Hamann, *Grundlagen der Festkörperphysik*, (Springer, Berlin, 1979). (in german)
99. D.B. Williams, C.B. Carter, *Transmission Electron Microsopy*, (Plenum, New York, 1996)
100. R. Wittmann, C. Parzinger, D. Gerthsen, Ultramicroscopy **70**, 145 (1998)
101. P.Y. Yu, M. Cardona, *Fundamentals of Semiconductors*, (Springer, Berlin, 1996)

Chapter 10
Tip-Enhanced Near-Field Optical Microscopy of Carbon Nanotubes

Carsten Georgi, Miriam Böhmler, Huihong Qian, Lukas Novotny and Achim Hartschuh

Abstract We review recent experimental studies on single-walled carbon nanotubes on substrates using tip-enhanced near-field optical microscopy TENOM). High-resolution optical and topographic imaging with sub 15 nm spatial resolution is shown providing novel insights into the spectroscopic properties of these nanoscale materials. In the case of semiconducting nanotubes the simultaneous observation of Raman scattering and photoluminescence (PL) is possible enabling a direct correlation between vibrational and electronic properties on the nanoscale. So far applications of TENOM were focused on the spectroscopy of localized phonon modes, local band energy renormalizations induced by charge carrier doping, the environmental sensitivity of nanotube PL, and inter-nanotube energy transfer. At the end of this review we discuss remaining limitations and challenges.

10.1 Introduction

Conventional optical microscopes rely on the detection of propagating electromagnetic waves in the far field of the radiation source at distances larger than the wavelength of the emitted light λ. Since propagating waves carry only limited spatial frequencies according to $|k| = 2\pi/\lambda$ detailed information about the nanoscale structure of the emitter is lost. As an example, two radiation sources with identical emission properties can only be distinguished using conventional microscopy if they are separated by more than about half the wavelength of the emitted light,

---

C. Georgi · M. Böhmler · H. Qian · A. Hartschuh (✉)
Department Chemie und Biochemie and CeNS, Ludwig-Maximilians-Universität München, 81377 München, Germany
e-mail: achim.hartschuh@cup.uni-muenchen.de

L. Novotny
University of Rochester, The Institute of Optics, Rochester, Rochester, NY 14627, USA

A. Zoubir (ed.), *Raman Imaging*, Springer Series in Optical Sciences 168,
DOI: 10.1007/978-3-642-28252-2_10,

i. e., $\Delta x \approx \lambda/2$. Concepts extending optical microscopy below this limit down to nanometer length scales are distinguished into far-field and near-field techniques.

Far-field techniques require situations in which the radiation sources within the illuminated sample volume do not emit simultaneously at all times with the same emission properties. This can be achieved either by active control, e.g., through stimulated emission depletion (STED) (see e.g.[1, 2]), or by exploiting the stochastical character of the emission due to photo-physical and photo-chemical effects of the sources such as photo-blinking, -bleaching, or -switching. Schemes following the latter approach are termed photoactivated localization microscopy (PALM) and stochastic optical reconstruction microscopy (STORM) [3, 4]. The concept of STORM and PALM relies on high signal-to-noise ratios that need to be sufficient to discriminate the signal from single sources within the sample. In this case, the problem of spatial resolution is translated into the problem of position accuracy that describes the task of localizing a single emitter [5]. Using this principle, subdiffraction photoluminescence (PL) imaging on a length scale of about 100 nm has recently been demonstrated for single-walled carbon nanotubes (SWCNTs) [6]. Here, PL blinking has been induced by acid treatment of the nanotube causing stochastic charge carrier doping followed by local quenching of the PL along the nanotube. At present, existing far-field schemes are not applicable to Raman scattering.

Near-field techniques collect the evanescent signal in the near-field of the sample, which contains the nanoscale spatial information that is not limited by diffraction. Therefore, a pointed probe is brought very close to the sample surface with the tip-sample distance controlled by a feedback mechanism based on either force or tunneling current detection. Images are acquired by scanning the tip across the sample, while collecting the optical signal at each image pixel. This approach is termed scanning near-field optical microscopy (SNOM) [5, 7–9]. The optical signal can be collected with a subwavelength aperture at the end of the probe, where the spatial resolution is given by the aperture size of typically 50–100 nm (aperture SNOM). Unfortunately, the low throughput of subwavelength apertures limits the applicability of this technique. The other concept is to use apertureless tips [10] which scatter or enhance the optical near-fields, which are then detected in the far-field (apertureless SNOM). Here, the resolution is only limited by the size of the tip apex that can be made smaller than 10 nm. Using the near-field enhancement in the vicinity of a sharp metal tip, not only the resolution but also the detection efficiency is greatly enhanced allowing for nanoscale imaging of weak optical signals such as Raman scattering or fluorescence of single molecules [11, 12]. This technique termed tip-enhanced near-field optical microscopy (TENOM) is subject of this chapter, in which we review the applications of TENOM to study single-walled carbon nanotubes (SWCNTs). SWCNTs are quasi one-dimensional nanostructures with great promise for novel applications in nanotechnology. As outlined above, conventional confocal micro-Raman or PL spectroscopy only provide limited resolution resulting in substantial spatial averaging. We show that nanoscale heterogeneities such as localized defects and structural transitions occurring along the nanotube that remain hidden in conventional microscopy can be resolved using TENOM. In addition, nanotubes form a model system for the development of TENOM providing both strong optical responses (Raman scattering and PL) and the possibility to directly correlate optical and topographical data.

10.2 Tip-Enhanced Near-Field Optical Microscopy

In tip-enhanced near-field optical microscopy, enhanced optical fields in the proximity of a sharp metallic probe are used to locally excite the sample surface and to extract the emitted light. Local signal enhancement not only serves to achieve high spatial resolution but also increases the detection sensitivity enormously. The technique is therefore applicable to weak emitters with low quantum yield and intrinsically weak signals such as Raman scattering. Moreover, the versatility of the technique allows the study of a variety of spectroscopic signals including local time-resolved fluorescence or Raman spectra (see e.g. [13, 14]).

Since Raman scattering and fluorescence are distinct optical processes involving different electronic states of the sample material with different degree of coherence, the enhancement effects for either type of signal and thus the image contrast is expected to vary. The enhanced fields at the tip cause enhanced excitation rates corresponding to an increase of the excitation density provided by the light source. In general, the enhancement of the excitation rate k^{ex} will be identical for Raman scattering and fluorescence if electronic dephasing is not changed. By defining the field enhancement factor f as the ratio between the electric field at the tip $E_{tip}(x)$ and without tip $E_0(x)$, the excitation rate enhancement is expressed as $k_{ex}^{tip}/k_0^{ex} = f^2$. In analogy to a radio which can be used both as receiver and transmitter, the enhanced fields can also promote radiative decay expressed by the rate k^{rad}. Radiative rate enhancement can be discussed in terms of the Purcell effect and is expressed by Fermi's golden rule in the weak coupling limit [15, 16]. In the case of Raman scattering, the total signal depends on the product of the transition rates $k^{ex}(\lambda^{ex}) \cdot k^{rad}(\lambda^{rad})$. As a consequence the total signal enhancement scales with the fourth power of the field enhancement for small differences between the excitation and emission wavelength $\lambda^{ex,rad}$ and assuming that the field enhancement at the tip does not depend sensitively on the wavelength.

$$M_{Raman} = k_{tip}^{ex}/k_0^{ex} \cdot k_{tip}^{rad}/k_0^{rad} \approx (E^{tip}/E^0)4 = f^4 \tag{10.1}$$

The fluorescence intensity, on the other hand, depends on the excitation rate k^{ex} and the quantum yield η denoting the fraction of transitions from excited state to ground state that give rise to an emitted photon. The quantum yield is expressed in terms of the radiative rate k^{rad} and the non-radiative rate k^{nonrad} through $\eta = k^{rad}/(^{krad} + k^{nonrad})$. Accordingly, the fluorescence enhancement due to the presence of the metal tip can be written as

$$M_{Flu} = k_{tip}^{ex}/k_0^{ex} \cdot (\eta_{tip}/(\eta_0 = f^2(\eta_{tip}/\eta_0). \tag{10.2}$$

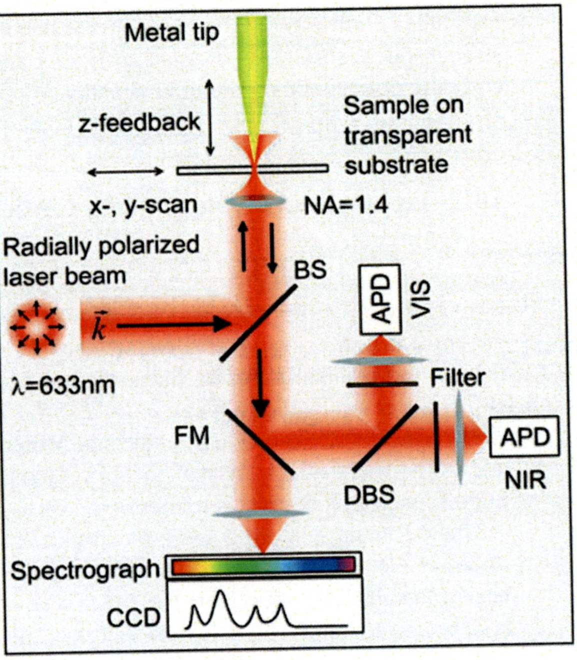

Fig. 10.1 Schematic of the experimental setup employing on-axis illumination of a transparent sample used to observe simultaneous Raman scattering and photoluminescence of carbon nanotubes [13]. A sharp metal tip is positioned in a tightly focused radially polarized laser beam. The optical signal is detected either by two avalanche photodiodes (APDs) for the visible (VIS) and near-infrared (NIR) spectral range or by a combination of a spectrograph and a charge-coupled device (CCD).A dichroic beam splitter (DBS) before the APDs is used to split VIS and NIR radiation

Here, we assume that the system is excited far from saturation. From Eq 10.2 it is clear that TENOM works most efficiently for samples with small fluorescence quantum yield η_0 and fast cycling rates such as, e.g., semiconducting single-walled carbon nanotubes [17, 18]. For highly fluorescent samples such as dye molecules, the quantum yield η0 is already close to unity and cannot be enhanced further.

A TENOM setup consists of a confocal microscope and a focus-centered sharp metal tip. Figure 10.1 illustrates the experimental approach in more detail. The tip sample distance z is typically controlled using a feedback scheme based on either force or tunneling current detection. The configuration in Fig. 10.1 has been optimized for the observation of nanotube PL and Raman scattering and is used by our group. For the simultaneous detection of Raman scattering in the visible (VIS) and PL signals in the near-IR (NIR) spectral range we utilize two independent avalanche photodiodes (APDs) in combination with spectral bandpass filters (APD VIS and APD NIR). In the present example, excitation light is provided by a HeNe laser operating at 632.8 nm. Longitudinal focal field components parallel to the tip axis, that are required to generate efficient local field enhancement [19], are produced by converting the linearly polarized Gaussian laser mode into a radially polarized mode with doughnut-shaped intensity distribution [20]. Although nanotube optical signals are enhanced substantially by the metal tip, maximized collection efficiencies achieved through the use of high numerical aperture (NA) objectives (NA > 1) are paramount for the observation of PL and Raman scattering at low excitation intensities that prevent photochemical bleaching [21].

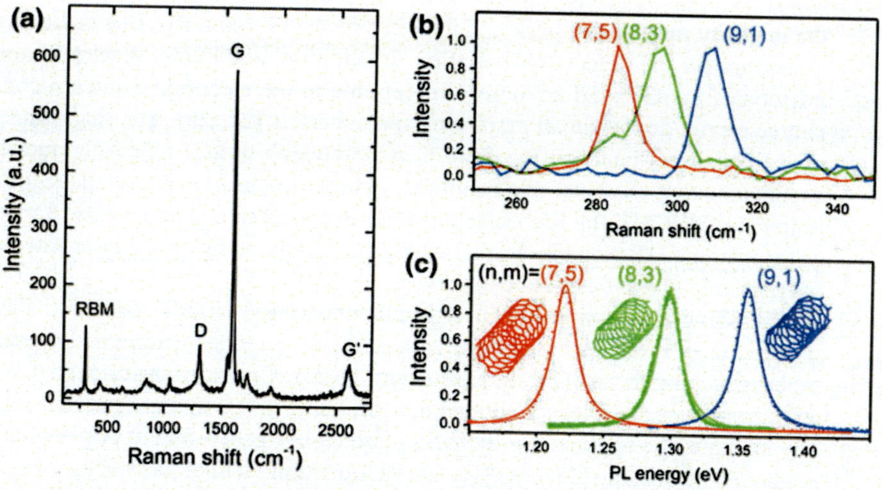

Fig. 10.2 Typical Raman spectrum of an individual SWCNT (**a**), showing the RBM-, D-, G- and G'-band. Depending on the chirality (n, m), the Raman RBM-band and photoluminescence spectra of individual SWCNTs (**b**, **c**) differ and allow for specific structure assignment. The index pair (n, m) defines the nanotube structure based on the unit cell vectors of graphene [24]

10.3 Single-Walled Carbon Nanotubes

This section provides a brief summary of the most relevant physical properties of single-walled carbon nanotubes needed for the discussion of the experimental results in Sect. 10.4. For more detailed information we refer the interested reader to the recent review in [22]. Single-walled carbon nanotubes (SWCNTs) are hollow cylinders consisting of a single sheet of rolled-up graphene. They are typically between 0.5 and 2 nm in diameter, but up to millimeters in length. Because of their unique mechanical and electronic properties, carbon nanotubes are considered for various technological applications, e.g., for biosensors, integrated circuits, nanotube actuators, and optoelectronic devices [22, 23].

The structure of a SWCNT is determined by its circumferential vector $\mathbf{C}_h = n\mathbf{a}_1 + m\mathbf{a}_2$ that is expressed in terms of multiples (n, m) of the unit cell vectors of graphene $\mathbf{a}_1$ and $\mathbf{a}_2$.Tight-binding calculations within a simplified free charge carrier model predict electronic band states and energies depending on the nanotube chirality (n, m). Nanotubes with (n-m) mod 3 = 0 are metallic or semi-metallic. All other nanotubes are semiconducting with a bandgap in the near-infrared to infrared range.

Raman spectroscopy is extremely useful for the structural analysis of SWCNTs, since they show a very distinct Raman spectrum. The most prominent features are the radial breathing mode (RBM), the D-, G-, and G'-band (see Fig. 10.2a) [22, 25, 26]. From the frequency of the RBM, the diameter and the chirality (n, m) of an individual nanotube can be determined (see Fig. 10.2b). The D-band is disorder induced and the intensity ratio of this peak and its overtone the G'-band ($I_D/I_{G'}$) is discussed as a measure for the defect concentration. In addition, amplitude, spectral shape, and position of the G'-band are highly susceptible to the electronic band structure and charge carrier doping in all graphene-type materials [27, 28]. The shape and width of the Raman G-band arising from C=C stretching modes reflect the metallic or semiconducting character of a nanotube. In order to observe the Raman signal of an individual SWCNT, the Raman process has to be resonant with an optical transition of the nanotube (Resonance Raman scattering). Depending on the laser wavelength, only specific chiralities can be observed.

The photoluminescence (PL) of semiconducting SWCNTs occurs in the near-infrared to infrared spectral range with emission energies determined by the nanotube structure (n, m) [24, 29] (see Fig. 10.2c). Optical transitions in SWCNTs arise from excitons (strongly bound electron–hole pairs) that are relevant for virtually all nanotube applications including free carrier generation in photovoltaics and nanoscale biosensing [22, 30, 31]. The PL quantum yield of nanotubes is rather low varying between several percent for high quality material with tight polymer-wrapping to below 10–3 for nanotubes on substrates [22, 32]. Short excited state lifetimes in the picosecond range on the other hand allow for very high cycling rates thereby enabling single nanotube detection [24, 33].

10.4 Examples

In the following sections we review several experimental results achieved by TENOM on SWCNTs highlighting the enormous potential of this microscopic technique.

10.4.1 High-Resolution Optical Imaging of Carbon Nanotubes

First reports on Tip-enhanced Raman scattering (TERS) on SWCNTs nanotubes were published in 2003 [34–36]. In the following several other groups presented TERS results on carbon nanotubes (see e.g. [37–39]). Figure 10.3 shows near-field optical images of the Raman G-band of metallic SWCNTs on glass featuring a spatial resolution of 25 nm. It is apparent that not all nanotubes which can be seen in the topography are also visible in the optical image. Their specific chirality has no optical transition in resonance with the laser wavelength. The Raman signal was collected with an APD after bandpass filtering, only detecting light around 700 nm corresponding to the Raman G-band of SWCNTs (VIS APD in Fig. 10.1). Since the SWCNTs on this sample were metallic, no PL was simultaneously observed with the second APD, collecting light in the NIR region (NIR APD in Fig. 10.1). The simultaneous observation of tip-enhanced Raman scattering and photoluminescence from the same semiconducting nanotube was already presented in [17]. For our setup with laser excitation at 632.8 nm and PL detection sensitivity up to 1,050 nm,

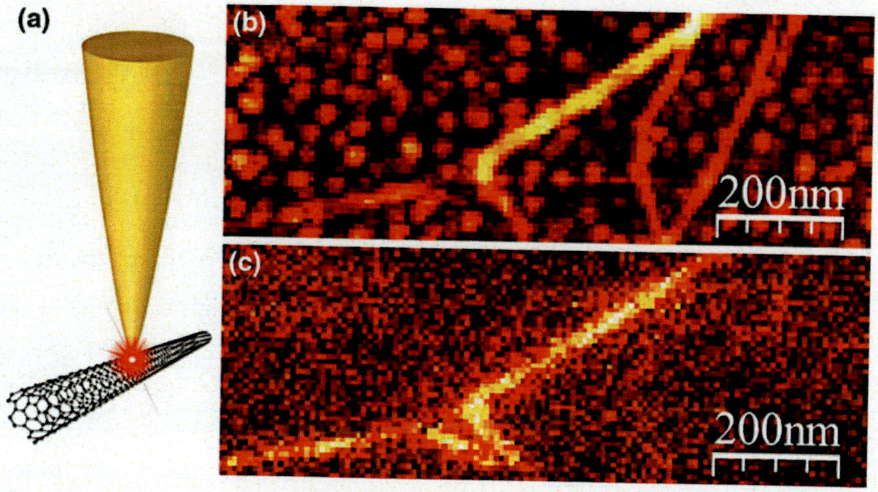

Fig. 10.3 High-resolution near-field optical imaging of carbon nanotubes on glass. **a** Schematic of the experimental configuration. **b**, **c** Simultaneously detected topography and Raman G-band of SWCNTs after laser excitation at 632.8 nm. The optical resolution is about 25 nm, determined by the diameter of the tip apex. Raman scattered light (G-band) is detected at 700 nm using the VIS APD after bandpass filtering (see Fig. 10.1). Only the tubes which are resonantly excited show strong Raman signal, others are not visible in the optical image. Simultaneous photoluminescence on the NIR APD could not be observed, because the deposited nanotubes were metallic

simultaneous Raman and PL can only be observed from the semiconducting (9, 1), (8, 3), and (7, 5) nanotubes showing sufficiently strong resonant Raman scattering. This method, detecting the optical signals with APDs after bandpass filtering, is sufficient to identify and distinguish nanotubes from additional particles on the surface that are visible in the topography. Here, the dots in Fig. 10.3b result from plasma cleaning of the glass substrate. Nevertheless, this measurement yields only limited information about the manifold optical properties of SWCNTs.

Spectroscopic imaging, where a complete spectrum is acquired at every image pixel, gives much more detailed optical information of the sample. Intensity and position of various peaks in the spectrum can be mapped independently and can reveal the structure or environmental influences on SWCNTs with nanoscale spatial resolution. Chirality (n, m)-specific PL imaging of SWCNTs is presented in Fig. 10.4. Energy selective images are established by acquiring spectra at every pixel while raster scanning the sample followed by spectral integration in selected wavelength windows. In Fig. 10.4(a) this window is centered at 900 nm corresponding to the emission energy of DNA-wrapped (6, 4) nanotubes while in (b) (6, 5) nanotubes are observed at 1,000 nm [22]. The insets show the topography and characteristic PL spectra acquired during the scan for the two nanotube chiralities.

The optical resolution apparent in Fig. 10.3b shows that the enhanced field is laterally confined to the size of the metal tip. The confinement in transverse direction can be demonstrated with approach curves, where the optical signal is recorded as

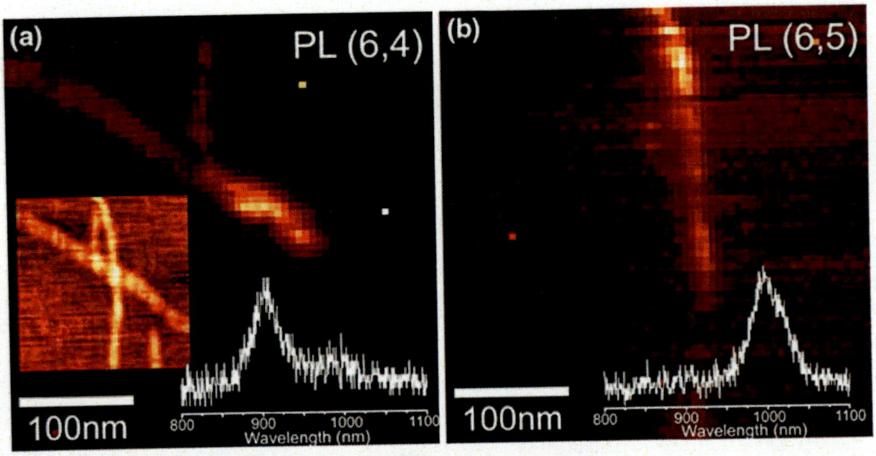

Fig. 10.4 Structural identification of nanotubes using near-field PL spectroscopy. Images (**a**) and (**b**) were formed by detecting PL spectra for every pixel while raster-scanning the sample followed by spectral integration around 900 nm for (**a**) and around 1,000 nm for (**b**) corresponding to PL energies of DNA-wrapped (6, 4) and (6, 5) nanotubes, respectively. Insets: Simultaneously acquired topography (height range 0–3.5 nm) and representative spectra detected during the scan

a function of tip-sample distance. In refs. [34, 40], this is shown for the Raman scattering of nanotubes. Figure 10.5 illustrates the tip-sample distance dependence of the photoluminescence enhancement for a single (6, 4) nanotube. Image (a) shows a 2D PL map formed by measuring PL spectra while first decreasing the tip-sample distance starting at about 20 nm until the tip has approached within a residual distance of about 2 nm followed by retracting the tip. The image clearly shows the strong tip-induced signal increase occurring only for small tip-sample distances below 5 nm. This can also be seen by plotting the observed PL amplitudes in (b).

It is important to note that while the PL intensity enhancement seems to reach a maximum for smallest distances, a decrease of the detected PL due to quenching induced by the metal tip is observed only for very small tip-sample distances of below <2 nm in the case of nanotubes (not shown). In contrast to dye molecules with fluorescence quantum yields already close to unity for which quenching appears to be more significant [42, 43], nanotubes feature only low PL quantum yields (see Sect. 10.3). As a result, locally enhanced electric fields besides increasing the excitation rate can also increase the probability of radiative relaxation. By comparing the tip enhancement of Raman scattering and PL for the same nanotube, PL enhancement appears to be even more efficient supporting the idea of a tip-controlled increase of the quantum yield [17] (see Sect. 10.2).

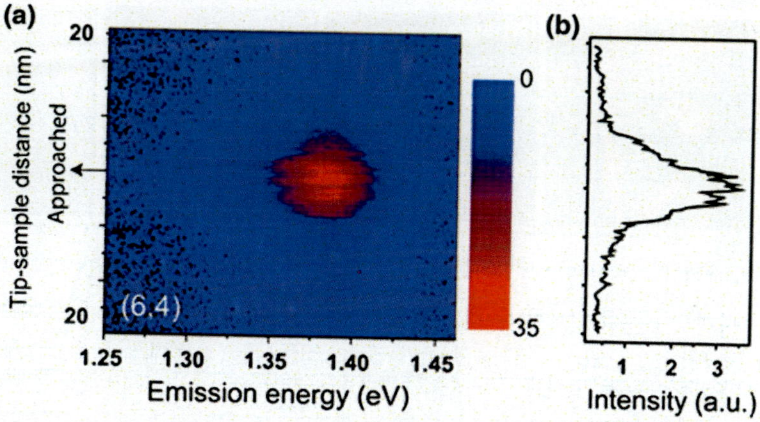

Fig. 10.5 2D photoluminescence maps of a single (6, 4) nanotube (**a**) as a function of tip-sample distance measured during tip approach and retracting. The PL intensity is strongly enhanced at small tip-sample distance and reaches the maximum when the tip is approached marked by the arrow. From [41]

10.4.2 Photoluminescence Quenching and Exciton Mobility

Theoretical calculations and scanning-tunneling experiments indicate that the energy gap in photoluminescent semiconducting SWCNTs vanishes at the ends of the nanotube [44, 45]. Mobile excitons can travel around 100 nm along the nanotube during their lifetime of few tens of picoseconds [21, 46]. Excitons reaching the ends are quenched leading to locally reduced PL intensities toward the ends [47]. Near-field PL images indeed resolved this decrease for several different nanotubes occurring on a length scale of 50–90 nm [48].

Hole doping in the nanotube, induced, e.g., by chemisorption of oxygen, can also lead to local PL quenching. A single hole can effectively quench PL on a length scale of about 100 nm, corresponding to the exciton diffusion length. The Raman G-band signal is not affected, since numerous structural defects and thus substantial damage are required to disturb the Raman process. The data in Fig. 10.6, derived from a spectroscopic image visualizes this phenomenon.

The topography in Fig. 10.6a shows a long and a short SWCNT and some particles, probably excess DNA or some undefined dirt. The spectral information derived from measuring spectra at every image pixel reveals that the long nanotube consists of two nanotubes with different chirality. The lower section emits PL at 975 nm, corresponding to a DNA-wrapped (8, 3)-tube. The upper section with PL at 930 nm and a strong Raman signal is a (9, 1)-tube. These contributions can be clearly seen in the example spectrum (Fig. 10.6e). The respective peak intensities calculated from the spectra are mapped in Fig. 10.6b–d. The short nanotube on the bottom left side is not visible in the optical images, because it is either metallic or emits beyond our detection window above 1,050 nm.

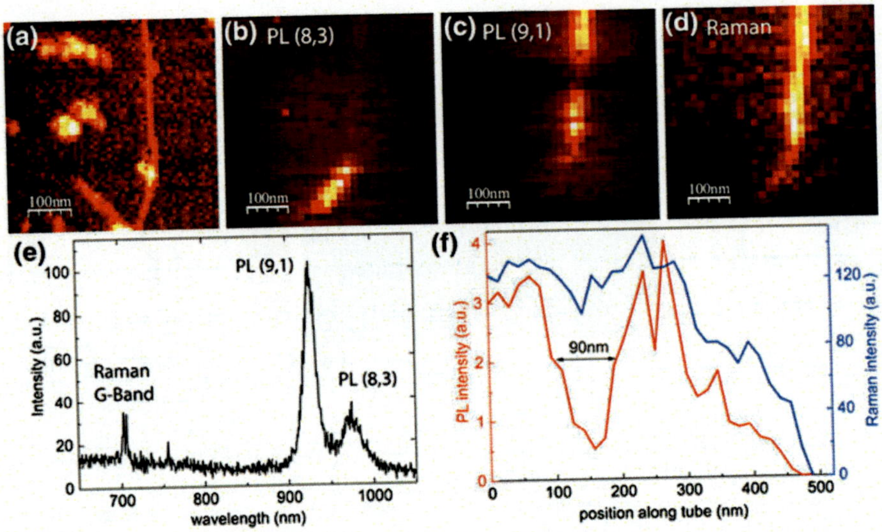

Fig. 10.6 Spectroscopic imaging of DNA-wrapped SWCNTs on mica. Images **(b–d)** were formed by acquiring spectra at every image pixel and subsequent analysis of the specific peaks in the spectra. An example spectrum from the scan **(e)** clearly shows the Raman G-band and the PL peaks from a (8, 3) and a (9, 1) nanotube. **a** Topography. **b** PL intensity from the (8, 3) nanotube at $\lambda = 975$ nm. **c** PL intensity from the (9, 1) nanotube at $\lambda = 930$ nm. In the middle of this tube, the PL is quenched due to a defect. **d** Intensity of the Raman G-band at $\lambda = 700$ nm, much stronger for the resonant (9, 1)-tube. **f** Cross-section of PL and Raman intensity along the (9, 1)-tube. A defect quenches the PL on a length scale of 90 nm while the Raman signal remains unchanged, indicating no substantial damage of the nanotube. After an initial decrease at the end at 300 nm, both signals vanish slowly toward the end indicating small contributions from the far-field background (see Sect. 10.5) that is also visible in the images **(b–d)**

The PL and Raman signal along the (9, 1)-nanotube are plotted together in Fig. 10.6f. From this and also in Fig. 10.6c, d it is apparent that the PL is quenched in the middle of this nanotube, while the Raman signal remains unchanged, indicating that the nanotube is not substantially damaged and the tip enhancement is stable. We can assume a single quenching defect at this location, as, e.g., chemisorbed oxygen or it is possibly related to the small dot on the nanotube visible in the topography. The PL is quenched on a length scale of 90 nm which can be related to the exciton diffusion length, which is therefore directly visualized here with near-field optical imaging.

10.4.3 Localized Phonon Modes

Localized phonon modes in single nanotubes and nanotube bundles were first observed in [34, 36]. A careful analysis in the low frequency range revealed a pronounced spectral variation of RBM-frequencies corresponding to changes in nanotube diameter depending on the growth procedure [49]. In [50] intramolecular junctions within single nanotubes connecting two nanotube sections with different structures (n, m) → (n', m') were studied in detail. TENOM was used to visualize the spatial extension of localized phonon modes and a transition length of about 40–100 nm with enhanced D-band scattering arising from the structural defect forming the junction was determined.

Local pressure-induced modifications of phonon modes were studied in [51]. Upon increasing the tip-nanotube interaction force significant changes in the shape of the G-band were observed indicating the local change of C–C bond length.

Recently, the correlation between highly localized PL emission and the occurrence of charged doping-induced defects affecting the G' band was resolved [27]. Local Raman scattering spectra detected at the position of strong point-like emission were found to exhibit two G' peaks from which the lower energy peak only occurs for n-doping. Together with extensive data obtained by conventional Raman spectroscopy a model was derived in which both phonon and electron bands are renormalized in the vicinity of charged defects.

10.4.4 Environmental Sensitivity of PL and Energy Transfer

Nanotubes consist of surface atoms only. Therefore, the screening of the Coulomb interaction between electron and hole forming the exciton is controlled mainly by the dielectric function of the environment. Emission energy variations on a length scale of few 10 nm were observed for nanotubes deposited from micelle suspensions on substrates due to non-uniform coverage with residual surfactants. Nanotubes wrapped by short pieces of single stranded DNA were studied in [41]. From far-field spectra measured on single nanotubes it was known that DNA-wrapping leads to a redshift of PL emission energies caused presumably by the increased dielectric function provided by the DNA. Near-field PL images on the other hand revealed that this shift is nonuniform with emission energies fluctuating spatially between the energy observed for non-wrapped material and a value twice larger than the redshift determined from far-field data. Since far-field spectra result from spatial averaging over both non-wrapped and wrapped nanotube sections this had to be expected. From conventional microscopy it is well known that the position accuracy can exceed the spatial resolution by more than a factor of 100 given only a single emitter is within the detection area [5]. In [52] this concept was applied to the investigation of distance-dependent energy transfer between two nanotubes of different (n, m) forming a bundle. In the first step, the two nanotubes were resolved spectrally and spatially using tip-enhanced PL microscopy as described in Sect. 10.4 A. These images show an apparent anti-correlation of the respective PL intensities along the bundle, indicating a spatially varying energy transfer between the two nanotubes. In the second step, the exact location of each nanotube was determined from this data with an accuracy of about 5Å, a factor 30 smaller than the spatial resolution of the experiment. These data were

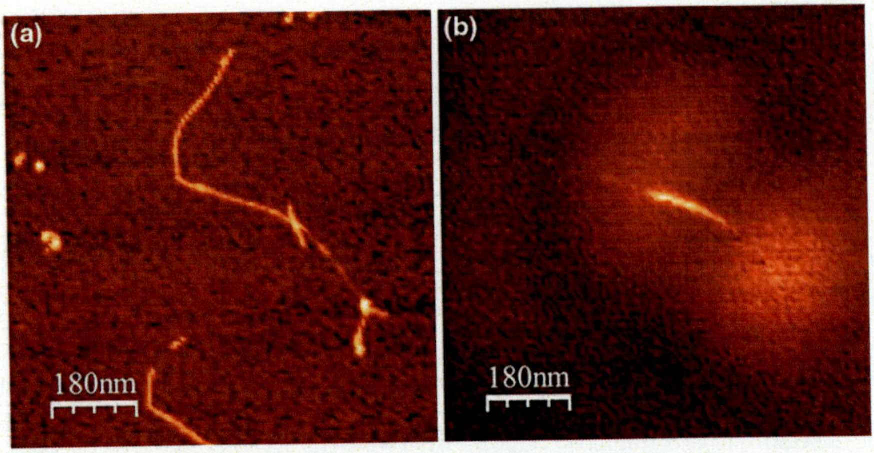

Fig. 10.7 TENOM topography (**a**) and PL image (**b**) of nanotubes. Only a short nanotube section with a length of about 180 nm exhibits PL emission while the remaining sections remain dark. Possible explanations include structural changes (n, m) along the nanotube, a large number of PL quenching defects in the dark sections, and distortions by adherent particles visible in the topography image (**a**). Besides the sharp near-field contrast a characteristic broad double-lobe pattern is observed in(**b**). This pattern corresponds to the confocal far-field signal upon excitation of a dipolar emitter using a radially polarized doughnut mode

then used to resolve the distance between the two nanotubes at each location along the nanotube bundle and correlate this with the local energy transfer efficiency. As a result, efficient energy transfer appears to be limited to few nanometers only and can be described by a Förster mechanism.

10.5 Summary

TENOM opens up unprecedented insights into the properties of carbon nanotubes on the nanoscale. Simultaneous Raman and PL spectroscopy of single SWCNTs becomes possible with sub 15 nm spatial resolution and a series of new results has already been achieved. However, several limitations and experimental challenges remain and need to be addressed. As for any quantum structure that is exposed to oxygen, nanotubes exhibit bleaching due to photochemical reactions [21]. Experiments on the simultaneous detection of Raman scattering and PL at ambient conditions require careful adjustment of the excitation intensity minimizing bleaching while maintaining sufficient Raman scattering signals. Alternatively, experiments could be carried out under UHV conditions [11, 53]. Compared to typical fluorescent dye molecules, however, nanotubes are orders of magnitude more stable enabling long acquisition times of more than 20 min as can be seen for example in the PL image in Fig. 10.7b.

Figure 10.7 illustrates several other issues discussed in the following. Obviously, not all nanotube structures that occur in the topography image (a) are visible in the PL image (b). This will be due to the fact that only thin semiconducting nanotubes exhibit PL within the detection range of the sensitive silicon photodiode used here while thicker nanotubes emit in the IR where suitable detectors are lacking. Other possible explanations for the confinement of near-field PL to the central part of the nanotube (see Fig. 10.7) include changes in (n, m) [50] or a large number of defects that quench the PL in the upper and lower nanotube sections. Based on the present PL data, the final proof for the different explanations cannot be derived. The observation of Raman scattering from single nanotubes on the other hand requires resonant or nearly resonant excitation. In order to completely characterize a SWCNTs sample spectroscopically, substantial additional experimental efforts including tunable laser sources and a series of different photo-detectors would be required.

Besides the sharp tip-enhanced near-field PL, a characteristic broad double-lobe pattern is observed in Fig. 10.7. This pattern corresponds to the confocal far-field signal detected upon excitation of an in-plane dipolar emitter using a radially polarized doughnut mode [54]. In the present example, the emissive length of the nanotube is indeed limited to about 180 nm as can be seen from the near-field contrast. This suggests that a nanotube, to first approximation, could be treated as a chain of point-dipole emitters oriented along the nanotube axis. The strongest tip-enhanced fields, however, are predicted to be polarized perpendicular to the nanotube axis for which weaker absorption probabilities would be expected due to antenna effects [22]. For the quantitative modeling of the image contrast a complete understanding of the tip-nanotube near-field interaction regarding both polarization and the transition from point-like to quasi one-dimensional emitters needs to be developed.

In summary, near-field optical and topographic data, while clearly giving novel and fascinating insights, cannot provide all answers and other complementary techniques providing high, preferentially atomic scale resolution such as high-resolution transmission electron microscopy (HRTEM) are necessary.

Acknowledgments We thank Nicolai Hartmann and Tobias Gokus for sample and experimental support. Financial support by the German Science Foundation DFG through grant HA4405/3-1 and the Nanosystems Initiative Munich (NIM) is gratefully acknowledged.

References

1. M. Dyba, S. Hell, Phys. Rev. Lett. **88**, 163901–163904 (2002)
2. R.J. Kittel, C. Wichmann, T.M. Rasse, W. Fouquet, M. Schmidt, A. Schmid, D.A. Wagh, C. Pawlu, R.R. Kellner, K.I. Willig, S. Hell, E. Buchner, M. Heckmann, S.J. Sigrist, Science **312**, 1051–1054 (2006)
3. E. Betzig, G.H. Patterson, R. Sougrat, O.W. Lindwasser, S. Olenych, J.S. Bonifacino, M.W. Davidson, J. Lippincott-Schwartz, H.F. Hess, Science **313**, 1642–1645 (2006)
4. M.J. Rust, M. Bates, X. Zhuang, Nature Methods **3**, 793–796 (2006)
5. L. Novotny, B. Hecht, *Principles of nano-optics* (Cambridge University Press, Cambridge, 2006)

6. L. Cognet, D.A. Tsyboulski, J.D.R. Rocha, C.D. Doyle, J.M. Tour, R.B. Weisman, Science **316**, 1465–1468 (2008)
7. D.W. Pohl, W. Denk, M. Lanz, Appl. Phys. Lett. **44**, 651–653 (1984)
8. A. Lewis, M. Isaacson, A. Harootunian, A. Muray, Ultramicroscopy **13**, 227–231 (1984)
9. L. Novotny in *Progress in Optics*, vol. 50 ed by E. Wolf, (Elsevier, Amsterdam, 2007), pp. 137–184
10. J. Wessel, J. Opt. Soc. Am. B **2**, 1538–1541 (1985)
11. J. Steidtner, B. Pettinger, Phys. Rev. Lett. **100**, 236101–236104 (2008)
12. H.G. Frey, S. Witt, K. Felderer, R. Guckenberger, Phys. Rev. Lett. **93**, 200801–200804 (2004)
13. A. Hartschuh, Angew. Chem. Int. Ed. **47**, 8178–8198 (2008)
14. S. Kawata, S.V.M. Shalaev (ed.) *Tip enhancement*, (Elsevier, Amsterdam, 2007)
15. E.M. Purcell, Phys. Rev. **69**, 681 (1946)
16. M. Thomas, R. Carminati, J.R. Arias-Gonzalez, J.-J. Greffet, Appl. Phys. Lett. **85**, 3863–3865 (2004)
17. A. Hartschuh, H. Qian, A.J. Meixner, N. Anderson, L. Novotny, Nano Lett. **5**, 2310–2313 (2005)
18. K. Aslan, I. Gryczynski, J. Malicka, E. Matveea, J.R. Lakowicz, C.D. Geddes, Anal. Biotechnol. **16**, 55–62 (2005)
19. L. Novotny, E.J. Sanchez, X.S. Xie, Ultramicroscopy **71**, 21–29 (1998)
20. S. Quabis, R. Dorn, M. Eberler, O. Glöckl, G. Leuchs, Opt. Comm. **179**, 1–7 (2000)
21. C. Georgi, N. Hartmann, T. Gokus, A.A. Green, M.C. Hersam, A. Hartschuh, Chem. Phys. Chem. **9**, 1460–1463 (2008)
22. A. Jorio, M. S. Dresselhaus, G. Dresselhaus (eds.) *Carbon Nanotubes*, vol. 111, (Springer, Berlin Heidelberg New York, 2008)
23. P. Avouris, J. Chen, Materials Today **9**, 46–54 (2006)
24. A. Hartschuh, H.N. Pedrosa, L. Novotny, T.D. Krauss, Science **301**, 1354–1356 (2003)
25. M.S. Dresselhaus, G. Dresselhaus, R. Saito, A. Jorio, Phys. Reports. **409**, 47–99 (2005)
26. S. Reich, C. Thomsen, J. Maultzsch, *Carbon Nanotubes: Basic Concepts and Physical Properties* (Wiley-VCH, Berlin, 2004)
27. I.O. Maciel, N. Anderson, M.A. Pimenta, A. Hartschuh, H. Qian, M. Terrones, H. Terrones, J. Campos-Delgado, A.M. Rao, L. Novotny, A. Jorio, Nature Mat. **7**, 878–883 (2008)
28. A. Das, S. Pisana, S. Piscanec, B. Chakraborty, S.K. Saha, U.V. Waghmare, R. Yiang, H.R. Krishnamurhthy, A.K. Geim, A.C. Ferrari, A.K. Sood, Nature Nanotech. **3**, 210–215 (2008)
29. S.M. Bachilo, M.S. Strano, C. Kittrell, R.H. Hauge, R. Smalley, R.B. Weisman, Science **298**, 2361–2366 (2002)
30. F. Wang, G. Dukovic, L.E. Brus, T.F. Heinz, Science **308**, 838–841 (2005)
31. J. Maultzsch, R. Pomraenke, S. Reich, E. Chang, D. Prezzi, A. Ruini, E. Molinari, M. S. Strano, C. Thomsen, C. Lienau, Phys. Rev. B, **72**, 241402(R)–241405 (R) (2005)
32. S.-Y. Ju, W.P. Kopcha, F. Papadimitrakopoulos, Science **323**, 1319–1323 (2009)
33. T. Gokus, H. Harutyunyan, F. Hennrich, P.T. Araujo, M. Kappes, A. Jorio, A.A. Green, M.C. Hersam, M. Allegrini, A. Hartschuh, Appl. Phys. Lett. **92**, 153116 (2008)
34. A. Hartschuh, E.J. Sanchez, X.S. Xie, L. Novotny, Phys. Rev. Lett. **90**, 095503–095504 (2003)
35. A. Hartschuh, N. Anderson, L. Novotny, J. Microscopy **210**, 234–240 (2003)
36. N. Hayazawa, T. Yano, H. Watanabe, Y. Inouye, S. Kawata, Chem. Phys. Lett. **376**, 174–180 (2003)
37. D. Roy, M.E. Jiang, W. and Welland. Faraday Discussions **132**, 215–225 (2006)
38. S.S. Kharintsev, G.G. Hoffmann, G.de With, J. Loos. Nanotechnology **18**, 315502–315509 (2007)
39. D. Roy, J. Wang, C. Williams, J. Appl. Phys. **105**, 013530 (2009)
40. R.M. Roth, N.C. Panoiu, M.M. Adams, R.M. Osgood Jr, C.C. Neacsu, M.B. Raschke, Optics Express **14**, 2921–2931 (2006)
41. H. Qian, P.T. Araujo, C. Georgi, T. Gokus, N. Hartmann, A.A. Green, A. Jorio, M.C. Hersam, L. Novotny, A. Hartschuh, Nano Lett. **8**, 2706–2711 (2008)

42. S. Kühn, U. Hakanson, L. Rogobete, V. Sandoghdar, Phys. Rev. Lett. **97**, 017402–017405 (2006)
43. P. Anger, P. Bharadwaj, L. Novotny, Phys. Rev. Lett. **96**, 113002–113004 (2006)
44. A. DeVita, J.C. Charlier, X. Blase, R. Car, Appl. Phys. A. **68**, 283–286 (1999)
45. D.L. Carroll, P. Redlich, P.M. Ajayan, J.C. Charlier, X. Blase, A. DeVita, R. Car, Phys. Rev. Lett. **78**, 2811–2814 (1997)
46. L. Cognet, D.T. Tysboulski, J.-D.R. Rocha, C.D. Doyle, J.M. Tour, R.B. Weisman, Science **316**, 1465–1468 (2007)
47. A. Rajan, M.S. Strano, D.A. Heller, T. Hertel, K. Schulten, J. Phys. Chem. B **112**, 6211–6213 (2008)
48. H. Qian, C. Georgi, N. Anderson, A. A. Green, M. C. Hersam, L. Novotny, A. Hartschuh, phys. stat. sol. (b) **245**, 2243–2246 (2008)
49. N. Anderson, A. Hartschuh, L. Novotny, J. Am. Chem. Soc. **127**, 2533–2537 (2005)
50. N. Anderson, A. Hartschuh, L. Novotny, Nano Lett. **7**, 577–582 (2007)
51. T. Yano, Y. Inouye, S. Kawata, Nano Lett. **6**, 1269–1273 (2006)
52. H. Qian, C. Georgi, N. Anderson, A.A. Green, M.C. Hersam, L. Novotny, A. Hartschuh, Nano Lett. **8**(11), 1363–1367 (2008)
53. J. Steidtner, B. Pettinger, Rev. Sci. Instrum. **78**, 103104–103108 (2007)
54. L. Novotny, M.R. Beversluis, K.S. Youngworth, T.G. Brown, Phys. Rev. Lett. **86**, 5251–5254 (2001)

Chapter 11
CARS Microscopy: Implementation of Nonlinear Vibrational Spectroscopy for Far-Field and Near-Field Imaging

Mamoru Hashimoto, Taro Ichimura and Katsumasa Fujita

11.1 Introduction

Raman microscopy has been attracting researchers in biology and medicine due to its capability of detecting molecular vibrations that provide information of molecular species, structures, conditions, and environments. Raman scattering can be obtained by simply illuminating molecules with monochromatic light, and providing molecular vibration frequency as wavelengths of scattered light. This does not require labeling of target molecules, such as chemical or biological staining with fluorophore, which may modify the condition of living biological specimens. Raman spectroscopic approaches have been combined with microscopic imaging techniques for label-free mapping of cellular components (Fig. 11.1) [38, 40, 47, 48, 58], immune response of granulocyte cells [15, 37], and investigation of molecular dynamics in cells under cell division [16, 27, 39].

One of the drawbacks in use of Raman scattering for biological imaging is the extremely low scattering cross-section. A typical Raman scattering cross-section of an organic molecule is about 10^{-30} cm^2, which is considerably smaller than that of single photon absorption (10^{-16} cm^2). Therefore, Raman microscopy requires long

M. Hashimoto
Department of Mechanical Science and Bioengineering
Graduate School of Engineering Science, Osaka University,
1-3 Machikaneyama, Toyonaka, Osaka 560-8531, Japan
e-mail: mamoru@me.es.osaka-u.ac.jp

T. Ichimura
Comprehansive Bioimaging Team, Quantitative Biology Center,
RIKEN, Suita, Osaka 565-0874, Japan
e-mail: tichimura@riken.jp

K. Fujita (✉)
Department of Applied Physics, Graduate school of Engineering,
Osaka University, 2-1 Yamadaoka, Suita, Osaka 565-0871, Japan
e-mail: fujita@ap.eng.osaka-u.ac.jp

A. Zoubir (ed.), *Raman Imaging*, Springer Series in Optical Sciences 168,
DOI: 10.1007/978-3-642-28252-2_11,

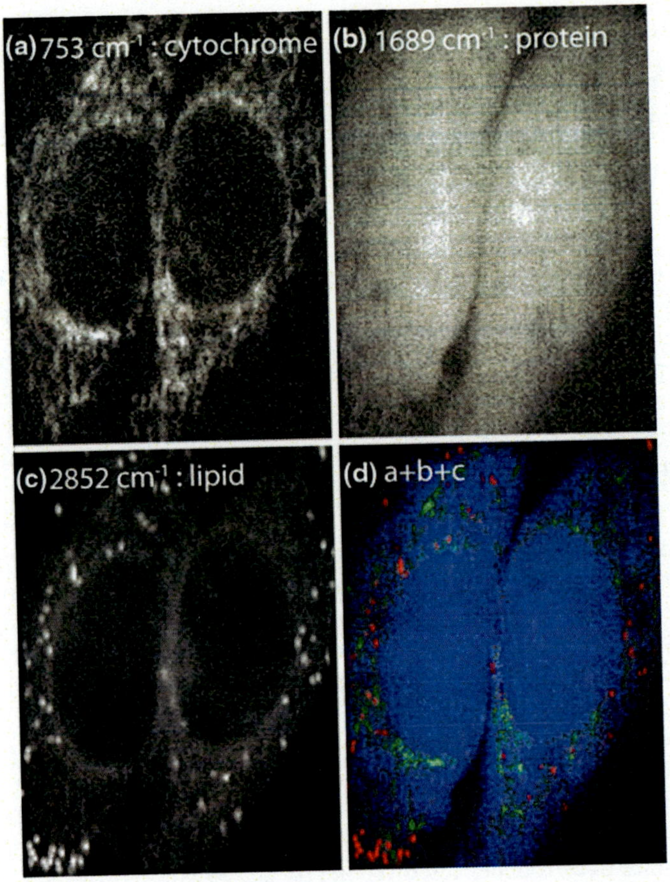

Fig. 11.1 Label-free observation of living HeLa cells by using Raman scattering [16]

exposure time to construct a Raman image, which makes observation of living or moving specimens difficult. Another drawback of Raman scattering is that observable sample is limited to non-fluorescent samples since strong fluorescence emission typically overlaps in the same wavelength range with Raman scattering and makes it difficult to obtain Raman spectra.

To overcome these problems in Raman scattering, [8–10, 18, 64] Coherent Anti-Stokes Raman scattering (CARS) can be used for biological imaging. In CARS microscopy, coherent summation of Raman scattering light increases detectable signal light in the microscope, where image acquisition time can be improved compared to that of using spontaneous Raman scattering. The advantage is significant especially in observation of living sample, where distributions and conditions of molecule change with time to express biological activities. In a typical CARS microscope, near-infrared (NIR) light is used as a light source. Since NIR light shows less scattering compared to visible light, it can penetrate into biological tissues and allow us to observe deep inside of the samples.

Furthermore, CARS allows us to observe even fluorescent samples because light detected in a CARS microscope has a shorter wavelength than that of the incident light. Nonlinear optical process in CARS increases the spatial resolution, which is also beneficial for observation of subcellular structures in a sample. Combination with near-field microscopy further enhances the spatial resolution ~10 nm range.

In this chapter, we introduce the principle of CARS and requirements to induce CARS process in biological molecules. We also describe implementation of CARS into far-field and near-field microscopy that enables vibrational imaging of biological molecules in submicron to nanometer spatial resolutions. The imaging properties and the applications of CARS microscopy are also described.

11.2 Coherent Anti-Stokes Raman Scattering (CARS)

11.2.1 Principles of CARS

CARS is a nonlinear optical phenomenon obtained with the resonant effect to a vibration of molecular bond or crystal lattice, which was first demonstrated by Maker and Terhune in 1965 [36]. Figure 11.2a shows the energy diagram of CARS. When light beams with two different optical frequencies (pump beam: ω_1, stokes beam: ω_2) are incident to a molecule and the difference of the frequencies matches to a frequency of one of vibrational modes of the molecules, the molecule can be excited to vibrational excitation state. In this condition, the pump beam (ω_1) can interact with the molecules to bring about anti-Stokes Raman scattering as shown in Fig. 11.2a and, as a result, CARS light with a frequency of $2\omega_1 - \omega_2$ can be obtained [52, 53].

Detection of CARS light with scanning wavelength of one of the two beams provides CARS spectra since CARS light can be obtained only when the frequency difference of the pump and Stokes beams matches to that of a molecular vibrational mode. Peaks in CARS spectra are positioned similar to that in spontaneous Raman spectra. However, the shape of peaks by CARS is quite different because of interference effect with coherent background light (Fig. 11.3), which is emitted without resonance to molecular vibrations as described below.

CARS spectra can also be obtained by using spectrophotometer when broadband light is used for the Stokes beam. As shown in Fig. 11.2b, with the incidence of the narrow-band pump beam and the broad-band Stokes beam, the molecule can be excited to various vibrational excitation states. As a result, wavelength of CARS light can vary depending on which molecular vibrational excitation state is excited. With repetition of the excitation and scattering processes, measuring CARS light through a spectrometer provides a CARS spectrum without scanning wavelength of incident light. This technique is called multiplex CARS and useful when CARS signals with different Raman shifts have to be obtained simultaneously.

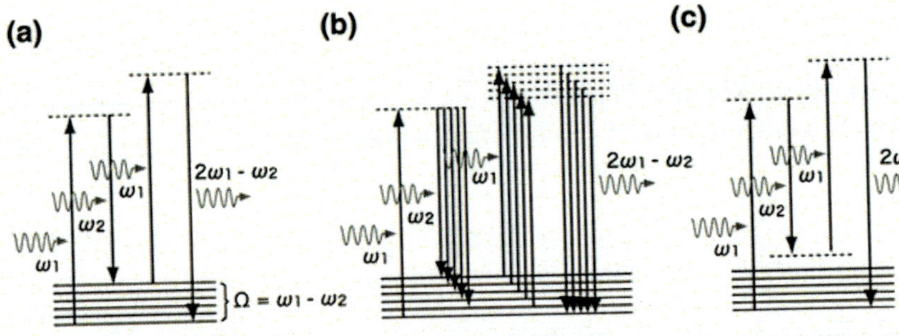

Fig. 11.2 Energy diagrams of **a** CARS, **b** multiplex CARS, **c** four-wave mixing without resonance to molecular vibrations

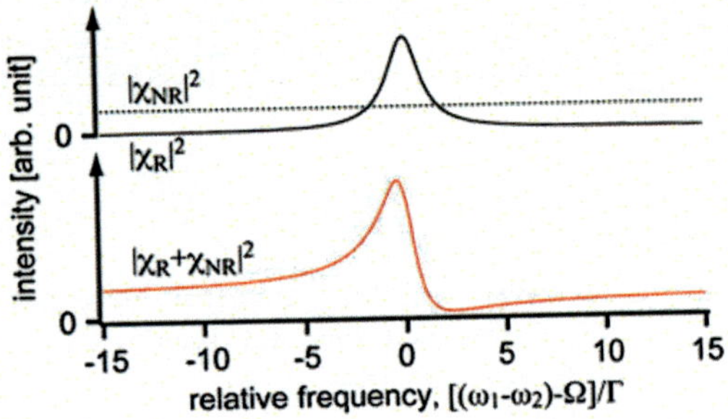

Fig. 11.3 Calculated spectra of $|\chi_R|^2$, $|\chi_{NR}|^2$ and $|\chi_R + \chi_{NR}|^2$. Coherent summation of the resonant and non-resonant components resulting in the asymmetry of the CARS spectrum. In the model calculation, the horizontal axis is represented by relative frequency defined as $[(\omega_1 - \omega_2) - \Omega]/\Gamma$

CARS is a third-order nonlinear effect, and CARS light is given as a result of nonlinear polarization induced by interaction of sample with two different incident light. The nonlinear polarization for CARS ($P^{(3)}_{\text{CARS}}$) can be described as,

$$P^{(3)}_{\text{CARS}}(\omega_3 = 2\omega_1 - \omega_2) \equiv \chi^{(3)} E_1(\omega_1)\, E_2^*(\omega_2)\, E_1(\omega_1) \qquad (11.1)$$

where $\chi^{(3)}$ represents the third-order nonlinear susceptibility, $E_1(\omega_1)$ and $E_2(\omega_2)$ are the electric fields for excitation light. $E_2 * (\omega_2)$ denotes the complex conjugate of $E_2(\omega_2)$. Here, the electric fields are regarded as scalars for simplicity, while the quantities are basically vectors. The nonlinear susceptibility is expressed by vibration-resonant term ($\chi^{(3)}_R$) and the non-resonant term ($\chi^{(3)}_{NR}$).

$$\chi^{(3)} = \chi^{(3)}_R + \chi^{(3)}_{NR} = \frac{A}{\Omega - (\omega_1 - \omega_2) - i\Gamma} + \chi^{(3)}_{NR} \qquad (11.2)$$

The coefficient of the fraction, A, denotes a constant related to the strength of the vibration, Ω denotes one of specific molecular vibrational frequencies of a given sample, and Γ corresponds to the spectral band width of the same vibration mode (damping factor of molecular vibration).

The resonant term ($\chi_R^{(3)}$) produces a strong CARS polarization and subsequent radiation when the frequency difference between the two excitation wavelengths is equal to the vibration frequency (the denominator goes close to zero). The nonresonant term ($\chi_{NR}^{(3)}$) produces light even without resonance to molecular vibration, which is called "non-resonant background", and causes low image contrast and asymmetric shape of Raman peaks due to coherent interference with the resonant CARS signals (Fig. 11.3). Depending on a sample, a strong non-resonant background light is given from the sample and makes it difficult to obtain resonant CARS signal. For observation of those samples, techniques to reduce the non-resonant background by using backward scattering [59], polarization [3], and phase dependence of CARS signal [11, 35], time-resolved CARS [60] have been proposed.

11.2.2 Laser System for CARS Microscopy

11.2.2.1 Laser System

Since multiple photons have to interact with a sample for CARS, photons with a high density are required. In most CARS microscopes, ultrashort pulse lasers are used for both pump and Stokes beams and are focused into a sample with an objective lens. For a typical CARS microscope, laser systems with picosecond pulses are used for both pump and Stokes beams. Laser light with a shorter pulse width is useful to excite nonlinear optical process; however, the pulse is usually composed with a mode locking of a broadband light oscillation and not suitable for high resolution spectroscopy such as Raman spectroscopy. For picosecond pulses, the band width of a mode-locked laser with NIR wavelength is about 0.2 nm, where a spectral resolution of about 3 cm^{-1} can be obtained with 800 nm wavelength of laser light. Figure 11.4 shows the relationship between the spectral resolution (in cm^{-1}) and the efficiency of inducting CARS effect with different pulse widths used for CARS. Depending on the required resolution, one can optimize pulse width to detect CARS signal effectively.

Lasers with NIR wavelengths are typically used for CARS. This is because it is relatively easy to produce pulsed light with various wavelengths in the NIR region. When excited by NIR light, the anti-Stokes scattering is produced in the region of best sensitivity of the silicon detectors, which is another benefit for the use of the NIR excitation. As picosecond laser sources for a CARS microscope, a combination of two mode-locked Ti:Sapphire lasers can be used. Combinations of a pulsed laser with an optical parametric oscillator (OPO) can also be used. Depending on the region of detecting wavenumber, it is important to select a laser system since each laser combination has different emission wavelength. For the laser system with two Ti:Sapphire lasers, it is also important to temporally synchronize the pulses from the different lasers. However, for systems using OPO, pump and Stokes pulses can be synchronized automatically since both pump and Stokes pulses are produced (or one is produced by the other) at the same location and time in the material for OPO.

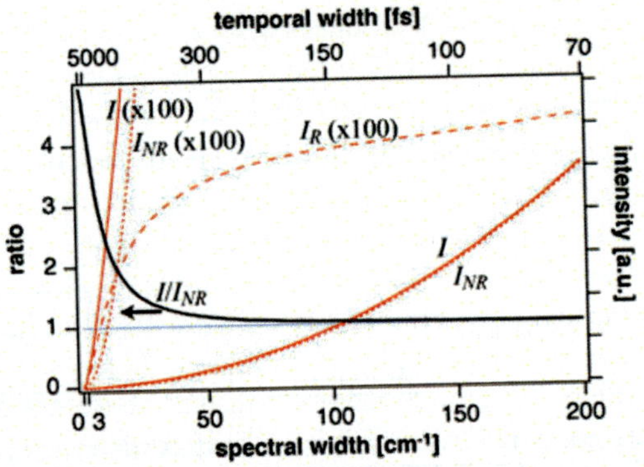

Fig. 11.4 Intensities of the resonant (I_R), non-resonant (I_{NR}), and total signals (I) as a function of the spectral width of pump and Stokes pulses. The ratio of the signal to background (total to non-resonant contribution) is also plotted therein. Both the pulses are assumed to be Fourier-transform limited, and the pulse energies are kept constant. The CARS signals are integrated over the whole spectral range

For multiplex CARS, one laser for the Stokes beam has to be broadband. A femtosecond laser light can be used for Stokes beam. By using a femtosecond pulse with a center wavelength of 800 nm and pulse duration of 100 fs, a range of wavenumber of 100 cm^{-1} can be obtained. For a wider wavenumber range of CARS spectra, white light pulse produced with supercontinuum by, for an example, a photonic crystal fiber can be used [46, 49]. It is also possible to use a portion of broadband laser light as a pump beam and the rest of the band as a Stokes beam [32].

It has also been proposed to obtain CARS signal by using a single broadband pulse, such as white-light continuum light or ultrashort pulse lasers with a pulse width of 20 fs or shorter. By tuning phase of wavelength components in a single pulse so that a certain molecular vibrational mode can be effectively excited by pairs of pump and Stokes wavelength, CARS light representing the vibrational mode can be selectively produced [6].

11.2.2.2 Effect of Jitter Between Two Lasers for CARS

Timing jitter between the excitation beams causes the fluctuation of CARS signal. Therefore, the jitter reduction between the excitation beams is important for CARS microscopy.

A CARS signal excited by two color pulses with a time delay τ is given by

$$I_{\mathrm{CARS}} \propto \int g_1(t)^2\, g_2(t-\tau)\, \mathrm{d}t, \tag{11.3}$$

where $g_1(t)$ and $g_2(t)$ are the temporal shapes of pump and Stokes beams, respectively. The fluctuation of the timing jitter causes the CARS signal fluctuation, and

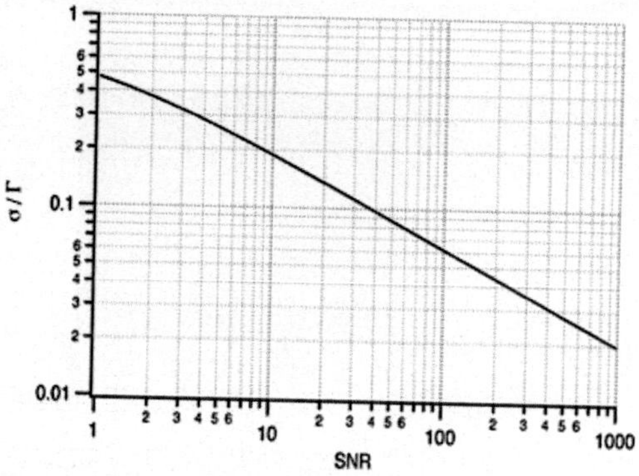

Fig. 11.5 Signal-to-noise ratio of CARS vs jitter (σ) to pulse width (τ) ratio

the signal is represented by a Taylor series expansion:

$$I_{\mathrm{CARS}}(\tau) = I_{\mathrm{CARS}}(\tau)|_{\tau=0} + \left.\frac{\mathrm{d}I_{\mathrm{CARS}}(\tau)}{\mathrm{d}\tau}\right|_{\tau=0} \tau + \frac{1}{2}\left.\frac{\mathrm{d}^2 I_{\mathrm{CARS}}(\tau)}{\mathrm{d}\tau^2}\right|_{\tau=0} \tau^2 + O\left(\tau^3\right). \tag{11.4}$$

The first term of the right side is the CARS intensity without the timing jitter, and other terms indicate the fluctuation of the CARS signal. The second term is negligible because the pulse shapes of the excitation beams are approximated with even functions. The third term contributes to the fluctuation mainly when the pulse durations of excitation beams are larger than timing jitter. If the timing jitter is a main noise source, the signal-to-noise ratio (SNR) of CARS signal is given by

$$\mathrm{SNR} = \frac{I_{\mathrm{CARS}}(\tau)\Big|_{\tau=0} + \frac{1}{2}\left.\frac{\mathrm{d}^2 I_{\mathrm{CARS}}(\tau)}{\mathrm{d}\tau^2}\right|_{\tau=0} E\left(\tau^2\right)}{\left|\frac{1}{2}\frac{\mathrm{d}^2 I_{\mathrm{CARS}}(\tau)}{\mathrm{d}\tau^2}\right|_{\tau=0} \sqrt{V\left(\tau^2\right)}} = \frac{\left(2\Gamma_2^2 + \Gamma_1^2\right) - 8\ln 2 E\left(\tau^2\right)}{8\ln 2\sqrt{V\left(\tau^2\right)}} \tag{11.5}$$

where E(·) and V(·) indicate expectation and variance, and it is assumed that the shapes of ω_1 and ω_2 beams are Gaussian and those pulse durations are Γ_1 and Γ_2, respectively.

When the probability density function of the timing jitter τ is described by the Gaussian distribution with variance σ^2 and mean 0, the expectation and variance of τ^2 are given by σ^2 and $2\sigma^4$, respectively. Hence, the SNR of CARS signal is represented to

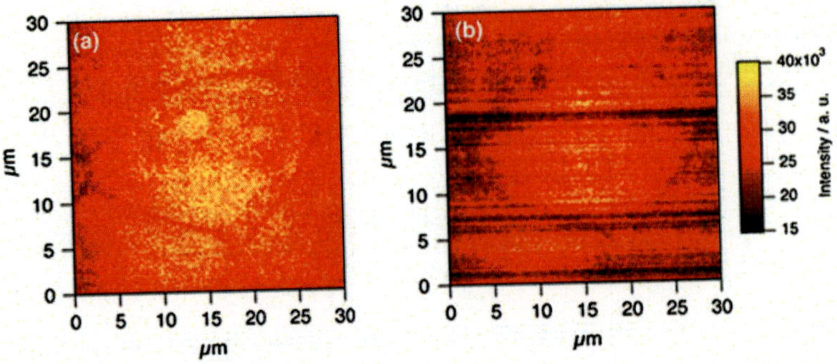

Fig. 11.6 CARS images excited with small jitter (10 fs) **a** and excited with large jitter (≈ 1 ps) **b**

$$\mathrm{SNR} = \frac{3 - 8 \ln 2 \left(\frac{\sigma}{\Gamma}\right)^2}{8\sqrt{2} \ln 2 \left(\frac{\sigma}{\Gamma}\right)^2} \tag{11.6}$$

where the pulse durations of two beams are assumed as $\Gamma_1 = \Gamma_2 = \Gamma$.

Figure 11.5 shows a relationship between the standard deviation of the timing jitter τ to pulse duration ratio (TPR $= \sigma/\Gamma$) and the SNR of CARS. The typical pulse duration of CARS microscopy is 5 ps, which is decided of the typical vibration spectral bandwidth of a few cm^{-1}. Since the Raman bandwidth is typically a few cm^{-1} to several dozen cm^{-1}, a few cm^{-1} bandwidth is required for CARS excitation lasers. Here, we assume to use 5 ps pulse duration lasers, whose spectral bandwidths are about 3 cm^{-1} because of the uncertainty principle or Fourier transform limit. The SNR at the 1 ps timing jitter indicates smaller than 10 from the figure. For SNR higher than 1000, timing jitter smaller than 98 fs is required.

Figure 11.6 shows the effect of the jitter reduction. The jitter between two lasers of Fig. 11.6a was 10 fs and b was about 1 ps. These are far-field CARS images of HeLa cells obtained by single beam scanning with a pair of galvano scanners. The striped noise along the beam scanning was shown in b; however, that noise was rejected in a. The wavenumber difference of excitation lasers was 1440 cm^{-1}. Two picosecond Ti:Sapphire mode-locked lasers synchronized with a PLL technique were used for light source. The jitter between the two lasers was detected with an electric phase detector and a balanced cross-correlator, and the jitter signal was fedback to piezo electric transducer attached on a mirror in the laser cavity to control the repetition rate of laser pulses [42].

11.3 Far-Field CARS Microscopy

Figure 11.7 shows the optical layout of a typical far-field CARS microscope. The excitation beams are combined with a dichroic mirror/filter, and those are tightly focused with an objective lens to increase CARS signal and the spatial resolution

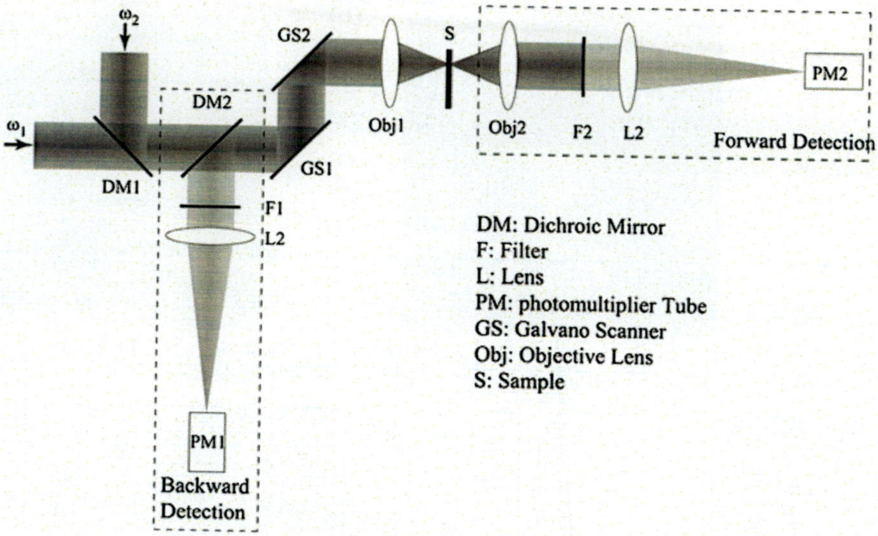

Fig. 11.7 Typical optical layout of CARS microscopy system

by the nonlinear optical effect. In order to obtain an image, the sample (stage) is moved and/or beams are scanned with a pair of galvano mirrors. CARS emission is separated with filters and is detected with a photomultiplier or an avalanche detector.

For the detection of generated CARS, there are two detection methods of forward and backward. The intensity of forward scattering is generally larger than that of backward scattering because of the coherency of CARS emission. Therefore, CARS emission is mainly detected at the forward direction. Although the signal of backward detection is lower than forward detection, backward detection is used for rejection of non-resonant background [59] or observation of thicker tissue inadequate for forward detection, in which case forward CARS emission is either absorbed or scattered backward.

11.3.1 Imaging Property of Far-Field CARS Microscopy

The high spatial resolution of CARS, especially longitudinal resolution, is given by the nonlinearity of CARS. The CARS polarization is proportional to the product of the square intensity of ω_1 beam and the intensity of ω_2 beam (Eq. (11.1)).

Figure 11.8a shows the intensity distribution of the focused beam (NA $=$ 0.9) [61]. The horizontal (r) and the vertical (z) axes indicate lateral and longitudinal directions, respectively. The intensity profiles on focal plane and on optical axis are also shown with solid lines in Fig. 11.8c and d, respectively. The FWHM (full width

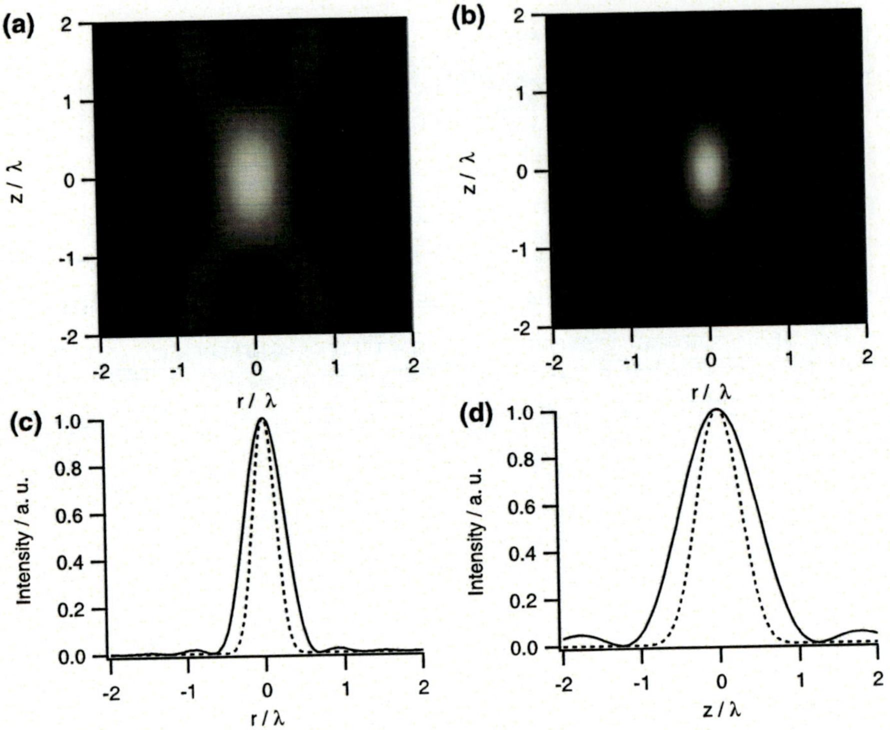

Fig. 11.8 Intensity profiles **a** of focused beam and cubic intensity **b** on r-z plane. Profiles on focal plane **c** and on optical axis. Solid and broken lines are intensity and cubic intensity, respectively

at half maximum) of the lateral direction is about 0.57λ, and that of the longitudinal direction is about 1.09λ.

The nonlinearity of CARS shrinks the volume of excitation region. When the Raman shift is low and the wavelength difference of pump and Stokes beams are negligible, the excitation region of CARS proportional to the cubic intensity. Figure 11.8b shows the cubic intensity distribution and the profiles are shown with broken lines in c and d. The FWHMs of lateral and longitudinal directions are 0.34λ, 0.65λ, respectively. As a result, it is found that the spot size is reduced to about 60%. The resolution enhancement is effective especially for longitudinal direction. The nonlinear effect enables one to observe optically sectioned images [4, 5]. Similar sectioning imaging can be done by the ordinary confocal Raman microscopy, but CARS microscopy enables optical sectioning without a pinhole, which reduces the signal intensity.

Figure 11.9 shows an image of bundles (dia. = 30 nm) of single wall carbon nanotubes (SWCNTs). The sample was dispersed on a coverslip and the backward scattered CARS was detected. The wavenumber difference of incident two beams was 1580 cm^{-1}, which corresponds to the G band of SWCNT [7]. It was found from

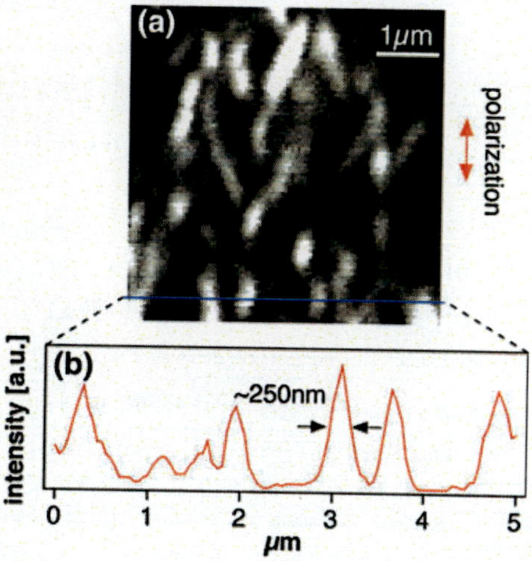

Fig. 11.9 CARS image of single wall carbon nanotube **a** and its line profile **b** The frequency difference of the two lasers were 1580 cm^{-1}

the figure that spatial resolution of the far-field CARS microscopy was about 250 nm. This spatial resolution is about 4 times smaller than the wavelengths of the excitation lasers. The enhancement of spatial resolution is the result of the multiply effect of the nonlinearity of CARS.

The CARS microscopy was applied to living cell imaging. Figure 11.10 shows CARS images of living HeLa cells. By tuning the frequencies of the two excitation lasers, CARS images were obtained at two different molecular vibration modes, (a) 1446 cm^{-1} and (b) 1240 cm^{-1}, which can be assigned to the bending vibration of CH_2 and the Amide-III vibration mode of peptide bonds, respectively. For this imaging, forward scattered CARS was detected. In both the images, lower intensity of CARS was detected at the nuclei than the surrounding cytoplasm. However, nucleoli parts provide strong CARS signals only at 1446 cm^{-1} while they are not apparent at 1240 cm^{-1}. This is the most clear difference between the two images. The difference can be attributed to high density of lipid molecules in the nucleoli.

11.3.1.1 Phase-Matching Condition of CARS Microscopy

The volume of the excitation was shown in Fig. 11.8. However, the emission and detection also have to be considered to estimate the imaging property of CARS imaging, because CARS is a coherent process and the emissions from molecules interfere each other. The constructive interference condition is known as the phase-matching condition, which is important for nonlinear coherent optical phenomena. In the CARS microscopy, the phase-matching condition is relaxed because of tight focusing of the excitation beams [17].

The generated CARS polarization excited by three beams of

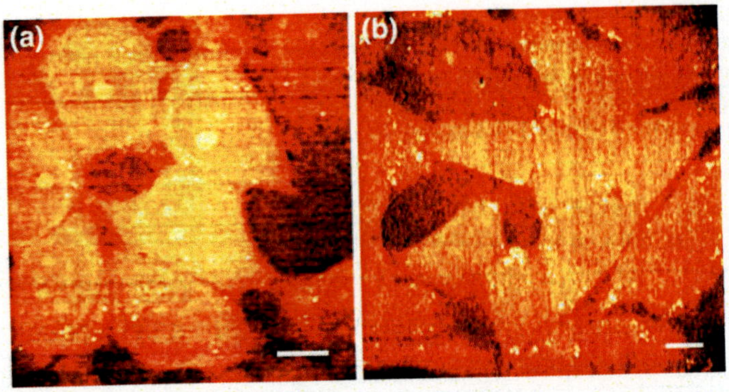

Fig. 11.10 CARS images of living HeLa cells obtained by tuning the frequencies of two excitation lasers for molecular vibration modes at **a** 1446 cm^{-1} and **b** 1240 cm^{-1} that can be assigned to bending vibration of CH_2 and Amide-III vibrational mode of peptide bonds. Scale bars: 10 μm

$$\mathbf{E}_1(\mathbf{r},t) = \mathbf{E}_1 \exp\{i(\omega_1 t - \mathbf{k}_1 \cdot \mathbf{r})\} \tag{11.7}$$

$$\mathbf{E}'_1(\mathbf{r},t) = \mathbf{E}'_1 \exp\left\{i\left(\omega_1 t - \mathbf{k}'_1 \cdot \mathbf{r}\right)\right\} \tag{11.8}$$

$$\mathbf{E}_2(\mathbf{r},t) = \mathbf{E}_2 \exp\{i(\omega_2 t - \mathbf{k}_2 \cdot \mathbf{r})\} \tag{11.9}$$

is given by

$$\begin{aligned}\mathbf{P}_{\mathrm{CARS}} &= \chi^{(3)} : \mathbf{E}_1(\mathbf{r},t)\,\mathbf{E}'_1(\mathbf{r},t)\,\mathbf{E}^*_2(\mathbf{r},t)\\ &= \chi^{(3)} : \mathbf{E}_1\mathbf{E}'_1\mathbf{E}^*_3 \exp[i\{2\omega_1 - \omega_2)t - \mathbf{k}_m \cdot \mathbf{r}\}],\end{aligned} \tag{11.10}$$

where

$$\mathbf{k}_m = \mathbf{k}_1 + \mathbf{k}'_1 - \mathbf{k}_2. \tag{11.11}$$

From the Eq. (11.10), it is found that the generated CARS polarization looks like a wave whose angular frequency is $2\,\omega_1 - \omega_2$ and wavevector is $\mathbf{k}_m$. However, the relationship between the frequency and wavenumber of the propagatable wave is decided by the refractive index dispersion. Therefore,

$$|\mathbf{k}_m| = \frac{n(2\omega_1 - \omega_2)\{2\omega_1 - \omega_2\}}{c_0}, \tag{11.12}$$

where n(ω) and c_0 are the refractive index of material for the light whose angular frequency is ω and the speed of light in vacuum, respectively. This is the phase-matching condition of CARS.

In CARS microscopy, the tightly focused beams are used for the excitation. A tightly focused beam is considered as an ensemble of plane waves with a variety of

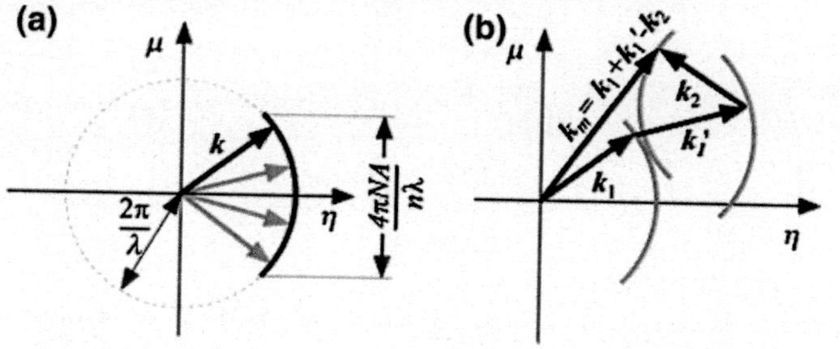

Fig. 11.11 The wavevector distribution of a focused beam **a**, and the relation among $\mathbf{k}_m$, $\mathbf{k}_1$, $\mathbf{k}_1'$, and k_2 **b**

propagation directions. The wavevector of the focused beam distributes on a spherical plane, whose radius is $2\pi/\lambda$, and the direction of the wavevector is limited by the numerical aperture of a focusing objective lens. Figure 11.11a shows this wavevector distribution of the focused beam, where n is the refractive index of a given sample, μ is the axis of the lateral direction in the wavenumber space, and η is the axis parallel to the optical axis.

The vector $\mathbf{k}_m$ of generated CARS polarization represented in Eq. (11.11) can take many kinds of combinations of $\mathbf{k}_1$, $\mathbf{k}_1$', and $\mathbf{k}_2$ as shown in Fig. 11.11b, because each wavevector of $\mathbf{k}_1$, $\mathbf{k}_1$', and $\mathbf{k}_2$ has the distribution shown in Fig. 11.11. As a result, wavevector distribution of excited CARS polarization is widely spread (Fig. 11.12a gray region).

However, the propagatable (detectable) wavevector of CARS radiation is limited by the phase-matching condition. The arc of a circle in Fig. 11.12a ($\mathbf{k}_3$) shows the propagatable wavevector of CARS, and the overlapping between gray region and the arc indicates the satisfaction of phase matching condition. It is found that the phase-matching condition is easily satisfied by focusing the excitation beams.

11.3.1.2 Cutoff Frequency of CARS Imaging

The phase-matching condition described thus far is applicable for homogeneous media. In order to discuss the imaging property, it is necessary to take into account the effect of diffraction.

The wave diffraction from a three-dimensional periodic structure is known as Bragg diffraction. The Bragg diffraction satisfies the following condition,

$$\mathbf{k}_d = \mathbf{K} + \mathbf{k}_i, \tag{11.13}$$

where $\mathbf{k}_i$, $\mathbf{k}_d$, and $\mathbf{K}$ are wavevectors of incident wave, diffracted wave, and an object, respectively. For the CARS, $\mathbf{k}_i$ and $\mathbf{K}$ correspond to wavevectors (= spatial frequencies) of the generated CARS polarization and $\chi^{(3)}$ of the object, respectively.

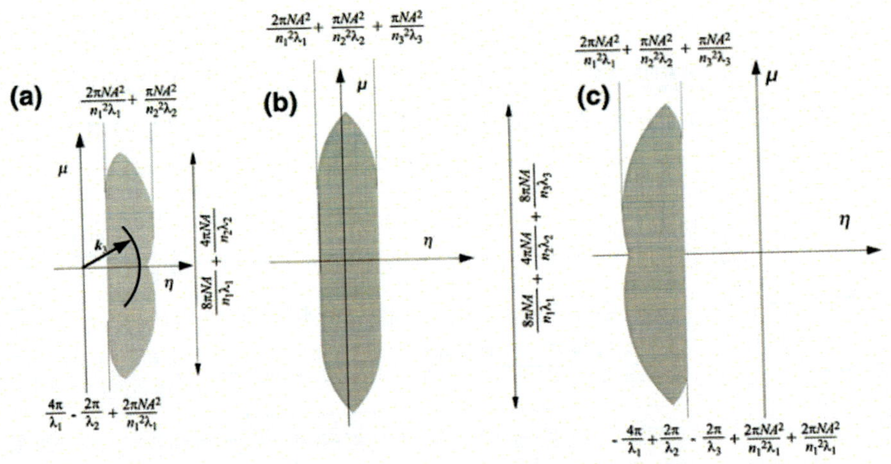

Fig. 11.12 The wavevector distribution of the excited CARS polarization $\mathbf{k}_m (\equiv \mathbf{k}_1 + \mathbf{k}'_1 - \mathbf{k}2)$ (gray region) and the propagatable wavevector of yielded CARS k_3 on the arc of circular **a**, Observable wavevector distribution of object for forward detection **c**, and that for backward detection **c**

As the propagatable (detectable) wavevector of $\mathbf{k}_d$ is restricted as Fig. 11.11 and the wavevector distribution of $\mathbf{k}_i = \mathbf{k}_m$ is given by the gray region in Fig. 11.12a. The available wavevector distribution of **K**, given by Eq. (11.13), in the forward direction and that in the backward (epi) direction is shown in Fig. 11.12b and c, respectively. It was found that **K** in the forward direction is widely distributed from zero to large frequency without any missing region. It indicates that the forward detection of CARS microscopy has three-dimensional resolution.

On the other hand, **K** is distributed in high frequency region for the epi detection. Since the spatial frequency of a homogeneous medium is zero, the CARS signal from the homogeneous medium cannot be detected in the backward direction. However, an object smaller than spot size has a high spatial frequency component, and thus emits CARS signal in the backward direction. This feature is used for non-resonant background suppression from water that homogeneously exists everywhere in biological samples [59].

11.3.2 Implementation of Far-Field CARS Microscopy

The advantage of CARS is the high emission rate compared with spontaneous Raman scattering. It takes about a few minutes to observe an image with the spontaneous Raman microscope; however, real-time imaging is available for CARS microscopy.

Fig. 11.13 Optical layout of the real-time CARS microscopy system using a microlens array scanner

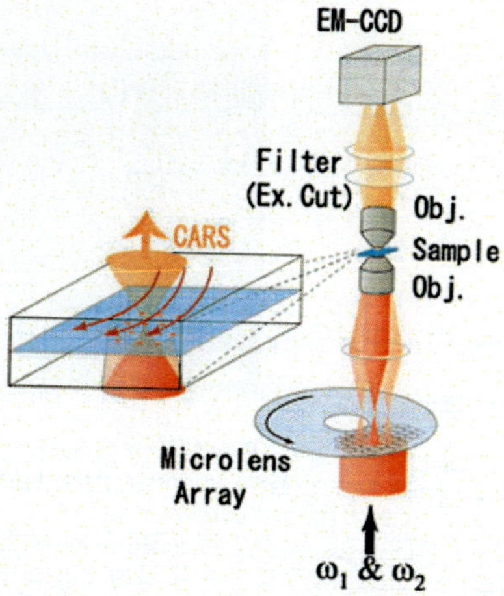

11.3.2.1 Real-Time CARS Imaging

For high-speed imaging, a fast beam scanning system is required. It is known that the fast beam scanning is realized by using a resonant mirror [57], a polygon mirror [54], and a microlens array [28, 34]. Evans et al. developed a real-time CARS microscopy system using a polygon mirror and a galvano mirror pair as a scanner [12].

The authors developed a system using a microlens array for real-time CARS imaging [43]. Figure 11.13 shows the optical layout of the CARS microscopy system using a microlens array scanner for real-time imaging. Temporally and spatially overlapped two laser beams (ω_1 and ω_2 beams) were incident on the microlens array scanner to split them into multiple beamlets. The beamlets were collimated with relay lenses and were focused to multiple spots on the specimen with an objective lens. CARS emission from the multiple focal spots on the specimen were collected by another objective lens and the fundamental signals were cut with optical filters. The CARS signals were observed in parallel with the EM-CCD camera, and a CARS image was obtained by rotating the microlens array disk. The observed molecular vibration was tuned by adjusting the wavelength of the ω_2 light. For the light source, two synchronized Ti: Sapphire lasers described in Sect. 11.2.2.2 were used.

Real-time imaging of Brownian motion of polystyrene beads in water was demonstrated as shown in Fig. 11.14a. The diameter of the beads was 3 μm, the observed Raman shift was 1000 cm^{-1}, and the frame rate was 33 ms/image. In the figure, trajectory of the movement of each bead center is shown. All the beads moved downward owing to slow flow of the water, and random moving by Brownian motion was detected.

Real-time imaging of giant multi-lamellar vesicles (MLV) of dipalmitoylphosphatidylcholinea (DPPC) were demonstrated in Fig. 11.14b. The z-section CARS images were obtained by moving the specimen over 10 μm (200 nm/step) with the PZT stage. The observed molecular vibration was set to 1442 cm^{-1}, which corre-

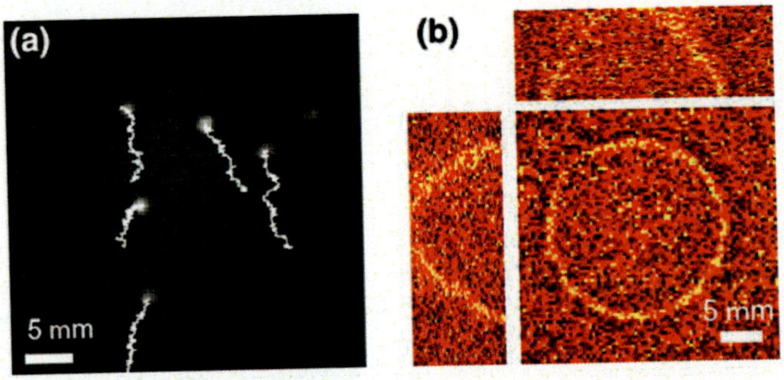

Fig. 11.14 Demonstration of real-time CARS imaging. **a** Brownian motion of polystyrene beads (dia. = 3 μm) in water. The trajectory of each beads were overlapped on an image. The 1000 cm^{-1} Raman band of ring breathing mode of phenyl group were imaged (33 ms/image). **b** Three-dimensional reconstruction of the CARS images of a DPPC MLV. The observed molecular vibration was 1442 cm^{-1} corresponding to CH_2 deformation of the acyl chains. The total image acquisition time of the DPPC MLV was 7 s per 50 slices (100 ms per cross-sectional image). The scale bar represents 5 μm

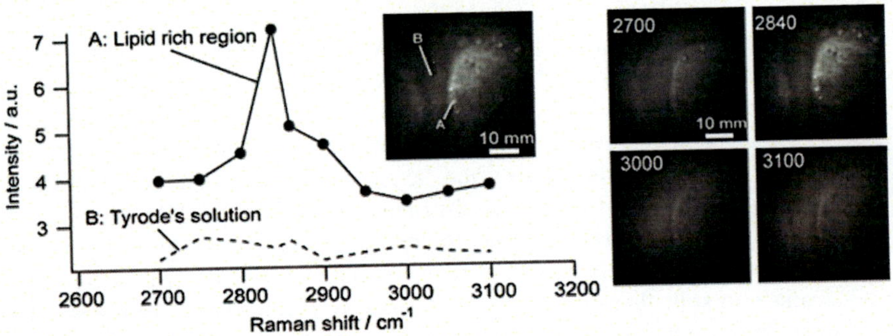

Fig. 11.15 CARS spectra and images of HeLa cell

sponds to CH_2 deformation of the acyl chains. The image acquisition time of each image was 100 ms/image. The three-dimensional CARS image of the DPPC MLV consisted of 50 slices and was obtained within 7 s, and a hollow sphere-shaped DPPC MLV with a diameter of about 20 μm was visualized.

Although the lateral (x-y plane) resolution of the multi-focus imaging is limited by the imaging optics of the detection side, longitudinal (z-axis) resolution is decided by the excitation optics and the third-order nonlinearity of the CARS generation. As a result, the developed multi-focus CARS microscope has a high spatial resolution along the z-axis (~2 μm), and z-section CARS images of the DPPC MLVs were clearly visualized.

Figure 11.15 shows a CARS spectrum and images of a living HeLa cell, where a strong peak around 2840 cm^{-1} due to CH_2 stretching of lipid was clearly seen. By using this CARS peak, a three-dimensional reconstruction image was also

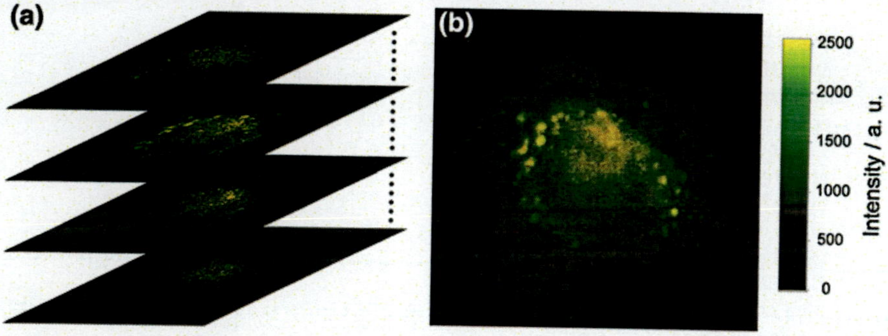

Fig. 11.16 Three-dimensional reconstructed image of HeLa cell visualized by CH_2 stretching vibration of lipid (2840 cm^{-1})

acquired. (Fig. 11.16) The exposure time was 0.2 s for each image, and 85 images were used for reconstruction (total acquisition time was 17 s). The lipid-rich regions in the HeLa cells were clearly detected. The living cells were visualized without any treatment in a short exposure time with the developed microscope.

11.3.2.2 Multimodal System

Since CARS microscopy utilizes pulse lasers for excitation, another nonlinear process could also be excited. Figure 11.17 shows a CARS image and an SHG (second harmonic generation) image of rat cardiomyocyte. Those images were obtained by changing the optical filter in front of the detector. It is also possible to observe them in the same time by use of multiple detectors. Besides, two photon fluorescence is also detectable [45].

11.3.2.3 Multiplex CARS Microscopy

Multiplex CARS microscopy is an excellent method to observe CARS spectra in a wide spectral range [31]. Figure 11.3b shows the energy diagram of the multiplex CARS. In the multiplex CARS microscopy, a pump beam (ω_1) with a narrow spectral width and a Stokes beam (ω_2) with a broad spectral width are used for excitation. The broadband Stokes beam excites a lot of molecular vibrations simultaneously. As a result of the interaction between the excited molecular vibrations and narrowband ω_1 beam, anti-Stokes Raman scattering from the molecular vibrations is generated. The CARS spectrum is obtained by using a spectrometer.

In most of previous works on multiplex CARS, a pair of a femtosecond laser with a broad spectrum and a picosecond laser with a narrow spectrum was used for light source [44]. Recently, photonic crystal fibers are used as wide spectrum light sources. It is known that a photonic crystal fiber generates suppercontinuum with very wide

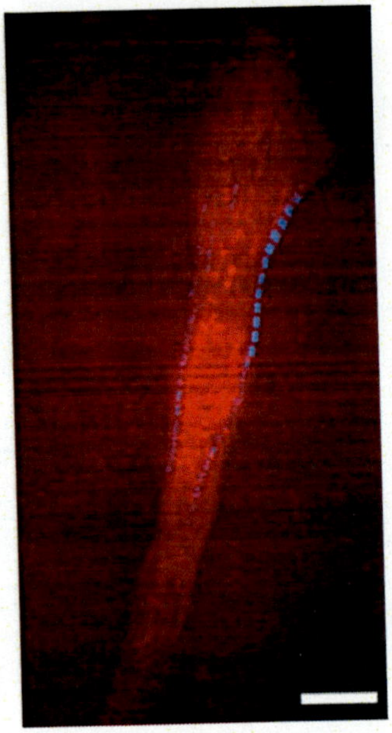

Fig. 11.17 CARS (*red*) and SH (*blue*) image of an unstained living cardiomyocyte. The distribution of CH_2 bending vibration (1446 cm^{-1}) was obtained by CARS, and myosins in myofibril of the cell were detected by SHG. The optical system of CARS microscopy can also be used as other types of nonlinear optical microscopy, such as harmonic generation and multiphoton excitation fluorescence microscopes, allowing microscopic observation of a biological sample with different contrast mechanisms. Scale bar= 10 μm

spectrum [31]. Now, a whole spectral region of fundamental molecular vibrations is covered by the photonic crystal fiber.

The shape of CARS spectrum is affected by the interference with the non-resonant background. As a result, the dispersive-shaped band is observed in the CARS spectrum. It is possible to decompose molecular vibration peaks and non-resonant background component by using techniques of band-shape analysis.

11.3.2.4 Single Laser CARS

A CARS microscopy using a single ultrashort pulse laser with a pulse shaping method has been proposed and demonstrated [6]. An ultrafast pulse with a duration of 10 fs or shorter has a spectral width of 1000 cm^{-1} or more. Such a pulse can excite a lot of molecular vibrations in the same time. Dudovich et al. proposed selective molecular vibration excitation by shaping the ultrafast laser pulse [6]. An ultrashort pulse was converted to a pulse sequence by a pulse shaper, which consist of a grating pair and a spatial phase modulator. The molecular vibration whose period ($2\pi/\omega$) is equal to the integral multiple of the pulse-to-pulse interval is selectively excited. By scanning the pulse interval of the pulse sequence, the spectral information is obtainable.

11.3.3 Application of CARS Microscopy

Biological cells have many functions such as energy generation, migration, cell division, and so on. The CARS microscopy has been applied to the function imaging of cells without staining.

The lipid metabolism of oleic acid (OA) with eicosapentaenoic acid (EPA) [63] was observed. To distinguish the OA from EPA, deuterated OA was used. The frequencies of CD stretching vibrations appear in a "silent" region around 2100 cm^{-1}, where no other peaks appear. The CARS image of the OA was selectively divided from that of the EPA. As a result, CH stretching around 2850 cm^{-1} showed EPA distribution and CD stretching around 2100 cm^{-1} showed OA distribution. The distribution difference of two acids in the cell was studied by CARS microscopy.

Axonal myelin in live spinal tissues was also studied by CARS microscopy [13, 14, 62]. The demyelination of a myelin sheath was visualized by polarization CARS microscopy. The polarization dependency of CH_2 stretching vibration and the information of the ordering degree of intramyelin lipids were visualized.

11.4 Near-Field CARS Microscopy

11.4.1 Resolution Improvement by Tip Enhancement

In recent advances of nanobiology and nanosciences, we often encounter situations where we need to have an observation tool with a nanometric spatial resolution. In order to deal with nanosized samples such as biomaterials, there are two important requirements. One is high spatial resolution of the order of a few nanometers to a few tens of nanometers. Owing to the wave nature of light, the spatial resolution of optical microscopy is limited to about half of the wavelength, which is about a half micron for NIR excitation light. The second important requirement is enhancement of scattered light, because the scattering intensity is often undetectably weak owing to the extremely small scattering volume of the nanosized sample. As the CARS intensity is proportional to the square of the number of molecules (equivalent to the volume of the sample), the CARS intensity rapidly decreases with decrease of the sample size down to nanometer scale. Therefore, it is practically impossible to image many of biomolecules by a far-field CARS microscopy.

Both these requirements can be fulfilled by employing the technique of tip-enhanced Raman scattering (TERS) microscopy [21, 55]. The TERS technique provides a super spatial resolution far beyond the diffraction limits of the probing light, along with an enhanced scattering efficiency, and hence it is suitable for high quality imaging of biomolecules. This technique of TERS was first proposed in 1999 and was later applied to imaging and analysis of several nanomaterials [23, 30]. In this technique, a metal-coated tip with a nanometric apex is brought close to a sample in a focal volume of focused excitation beam. When a metal nanostructure such as a nano-tip is illuminated with light, it generates localized surface plasmon polaritons (LSPPs) at its surface, as shown in Fig. 11.18a. As the LSPPs couple with incident photons resonantly, photons are densely confined in close vicinity of the nano-tip, which results in highly localized and strongly enhanced evanescent field near the tip apex (Fig. 11.18b). The dimension of this enhanced field is comparable to the tip apex. The enhanced field acts as a nano-light source for excitation of CARS polarization, and strong Raman signal is generated from a small volume of the sample in this nano-sized evanescent field. This is referred to as tip-enhanced CARS (TE-CARS). Since Raman signal is selectively enhanced only from the molecules in this local field, TE-CARS provides a high spatial resolution comparable to the size of the tip apex, along with strong enhancement, which ranges from a few thousand times to a million times [29].

Because of the nonlinear nature of CARS, there are several advantages in TECARS spectroscopy and microscopy compared to ordinary TERS. Photons generated by the nonlinear optical effects are spatially confined in a smaller volume than the size of tip enhanced field of fundamental frequencies because the intensity of the nonlinear effects is proportional to the high-order powers (square, cube, etc.) of the excitation light intensity. The spatial distribution of the signal emission becomes narrower than the intensity distribution of the excitation field. This leads to reduction of the effective volume of light-matter interaction. Higher order optical effects give much finer spatial response. In addition to the spatial confinement, because of the nonlinear responses, even a small enhancement of the excitation field could lead to a huge enhancement of the emitted signal. This also allows for effective reduction of relative intensity of the far-field background.

The nonlinear polarization of tip-enhanced CARS (TE-CARS) is expressed by

$$\begin{aligned} P_{\text{CARS}}^{(3)}(\omega_3) &= L(\omega_3)\left\{\chi^{(3)}\left[L(\omega_1)E_1(\omega_1)\right]\left[L(\omega_2)E_2(\omega_2)\right]^*\left[L(\omega_1)E_1(\omega_1)\right]\right\} \\ &= L(\omega_3)L(\omega_1)L(\omega_2)^*L(\omega_1)\left[\chi^{(3)}E_1(\omega_1)E_2^*(\omega_2)E_1(\omega_1)\right], \end{aligned} \tag{11.14}$$

where $(L\,\omega_i)(i = 1, 2, 3)$ is a complex number referred to as "local field factor" which represents changes in amplitude and phase at the wavelengths at a particular position, e.g., at the tip apex. The absolute values of $L(\omega_i)$ are usually greater than unity at surface of metal nanostructures. Field enhancement is effective to each of the fields including the CARS field [33]. On the other hand, tip-enhanced normal Raman scattering can be expressed by

$$\begin{aligned} P_{\text{CARS}}^{(2)} &= L(\omega_2)\left\{\chi^{(1)}\left[L(\omega_1)E_1(\omega_1)\right]\right\} \\ &= L(\omega_2)L(\omega_1)\times\left[\chi^{(1)}E_1(\omega_1)\right], \end{aligned} \tag{11.15}$$

where $\chi^{(1)}$ is a linear susceptibility for normal Raman scattering.

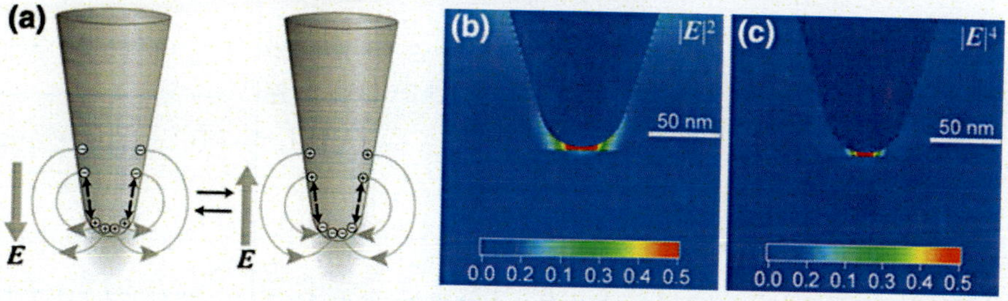

Fig. 11.18 Electric field enhancement due to localized surface plasmon polariton. **a** Temporal variation of electric charge displacement and field lines induced by an external alternating field. **b** Electric field intensity distribution around a conical tip of silver, calculated with the finite-difference time-domain (FDTD) method. **c** Distribution of the square of the electric field intensity shown in **b**. In the calculation **b** and **c**, the tip end is 2 nm away from a surface of a glass substrate. A continuous light ($\lambda = 785$ nm) is introduced with an incident angle of 45° from the glass side. The color scale is normalized by the maximum intensity for each figure

From Eq. (11.14), the net enhancement factor for CARS has higher order dependence on the enhancement factors. Owing to the high-order nonlinearity of the CARS process, one can expect huge enhancement for CARS intensity, which could be much higher than that for normal Raman scattering. The high-order nonlinearity results not only in increase of the signal intensity but also relative reduction of far-field background which is the CARS/Raman signal generated at molecules far from the tip-enhanced spot. The far-field background makes near-field optical images difficult to interpret, and was previously discussed in TERS imaging [22, 41]. In TE-CARS microscopy, the near-field contribution becomes dominant to the far-field contribution, allowing one to interpret the obtained images in a simple way. In terms of spatial resolution, the CARS polarization can be further confined to the very end of the tip in comparison with normal Raman scattering. Under an assumption that a spatial distribution of the local field factor to take a $1/r^3$ form, where r denotes the distance from the tip center [2], spatial distribution of the TE-CARS polarization would have a $1/r^{12}$ dependence, while that of TERS polarization would have a $1/r^6$ dependence. This leads to an essential difference of spatial distribution of TE CARS and TERS.

11.4.2 Imaging System

Figure 11.19 shows a schematic of the experimental setup used in TE-CARS experiments. The modification in the setup for TE-CARS compared to that for far field CARS experiment is the addition of an atomic force microscope (AFM) on a stage of a far-field CARS microscope with epi-detection configuration. The Ti:Sapphire/Ti:Sapphire system (Table 11.1) is employed in the system shown in Fig. 11.19. The AFM controls a metallic tip in the focal area. The metallic tip can

Table 11.1 Comparison of laser systems for CARS microscopy

Laser system	Tunable wavelengths (nm)	Pump wavelength (nm)	Stokes wavelength (nm)	CARS wavelength (nm)	Vibrational frequency (cm^{-1})
Ti:S/Ti:S	700–980/700–980	720	731–980	569–709	210–3680
Ti:S/OPO	700–980/1100–1600(idler)	980	1100–1600	706–884	1110–3950
	700-980/670–1100(signal)	720	731–1011	559–709	209–3998
Nd:YAG/OPO	1064(fixed)/1100–1600(idler)	1064	1100–1600	797–1030	308–3150
	746–1041(signal)/1064(fixed)	746–1041	1064	574–1019	208–4000
OPO/OPO	670–1100(signal)/1100–1600(idler)	976	1100–1600	702–877	1150–4000
		1064	1100–1600	797–1030	308–3150

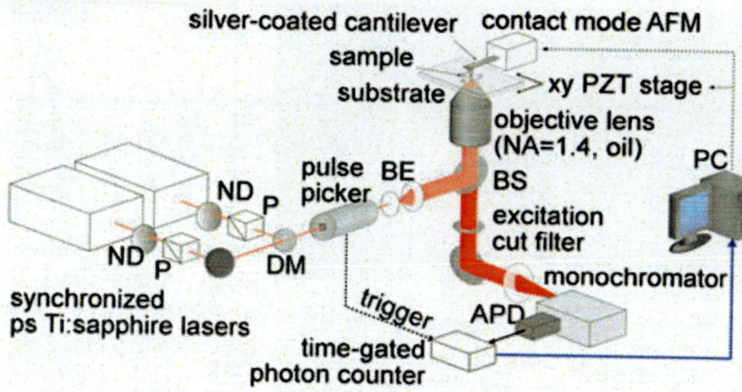

Fig. 11.19 Experimental setup for TE-CARS microscopy

be prepared in several possible ways such as vapor deposition to a silicon cantilever [21], electroless plating on a silicon cantilever [50], electrochemical etching of a metal wire [19], and attaching a metal nanoparticle on an end of a tapered glass fiber [1]. For the following experiments, we prepared silver tips by the vapor deposition process. TE-CARS signals are obtained by bringing the tip close to the sample. A raw TE-CARS signal usually contains a far-field contribution coming from the whole focal area, together with a near-field contribution coming from a small part of the sample adjacent to the tip.

An electro-optically modulated pulse picker controls the repetition rate of the excitation lasers. The backscattered CARS emission enhanced by the probe tip is collected with the objective lens and detected with an avalanche-photodiode (APD)-based photon-counting module through an excitation-cut filter and the monochromator. The pulse signal from the APD is counted by a time-gated photon counter synchronously triggered with the pulse picker, which effectively reduces the dark counts down to almost 0 counts/sec. Scanning the sample stage, while keeping the tip at the focused spot, can acquire two-dimensional TE-CARS images of a specific vibrational mode with a high spatial resolution.

It is well known that LSPPs at a conical tip of metal could be excited more effectively by electric field polarized in z-direction (along the tip axis) than that perpendicular to the tip axis. Therefore, for effective enhancement of local electric field, the tip has to be at a position where the z-polarized component of incident electric field (E_z of $\boldsymbol{E} = (E_x, E_y, E_z)$) is strong [25]. When the excitation laser beams are linearly polarized and focused by a high NA objective lens, there exist two strong peaks of the z-field at the both sides of the center of the focal spot while the z-field is canceled out at the center. Therefore, the tip has to be positioned to one of these peaks to effectively excite the LSPPs and the subsequent CARS signal. Figures 11.20a and b show the intensity distribution of the total field ($|\boldsymbol{E}|^2$) and z-component ($|E_z|^2$), respectively, in the transverse plane involving the geometrical focus corresponding to the interface between the substrate (glass) and the ambient medium (air). The incident field before the lens is set to be x-polarized. Figure 11.20c shows the cross-sections of $|E_z|^2$ along the x axis, and compares the peak positions

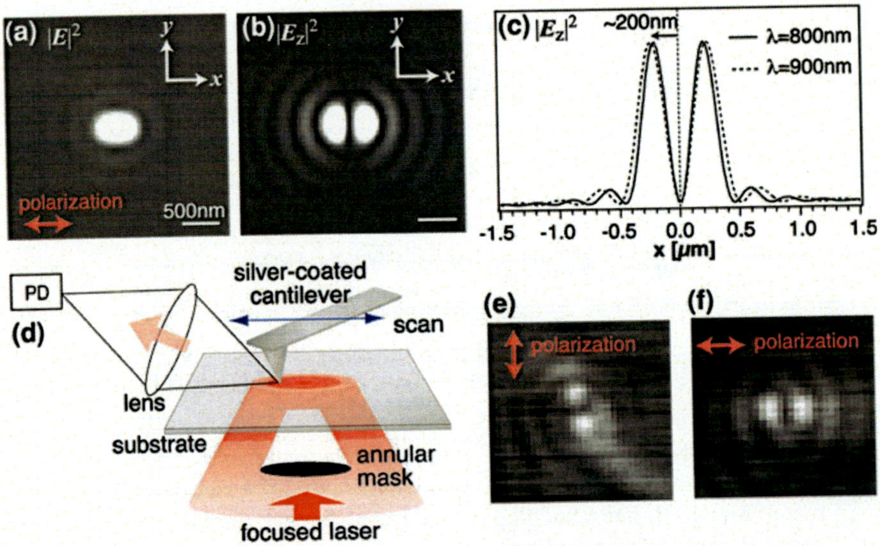

Fig. 11.20 Light intensity distribution in a focal spot and tip positioning. **a** Calculated intensity distribution of electric field, $|\boldsymbol{E}|^2$, focused to an interface between air ($n = 1.0$) and a glass ($n = 1.51$) substrate by an aberration-free lens with 1.4 NA. **b** The intensity distribution of the z-polarized field, $|E_z|^2$, on the interface. The wavelength of the electric field is set to be 800 nm in **a** and **b**. **c** Cross-sections of the line involving the focal point calculated with 800 nm and 900 nm wavelengths. **d** Experimental configuration for monitoring the tip position in the focal region. Scattered light from a silver-coated tip is detected by a photodiode placed at the side. **e-f** Images of the scattered light intensity when the incident light is y-polarized **e** and x-polarized **f**

of the fields with the wavelength of 800 nm and 900 nm. The peaks can be found at ~200 nm and ~225 nm from the center of the focus at the both sides. The difference of the peak positions is small enough with respect to the peak width of ~220 nm. Therefore, as long as the tip is positioned on either peak, the tip can be irradiated with the strong z-polarized components of both the incident fields. By using a radially polarized field instead of the linearly polarized field, one could have the strongest z-field at the center of the focal spot ($x = y = z = 0$) [25, 51].

The tip position with respect to the focused spot can be visualized through observation of light scattered by the tip. The scattered light is monitored by a side optics using an objective lens with long working distance and a photodetector, as shown in Fig. 11.20d. Monitoring the scattered light with scanning the tip within an area involving the focused spot provides us with a two-dimensional image of z-component of electric field. For suppression of the background light, focused evanescent illumination using an annular mask for a high NA (> 1.0) objective lens is employed [20]. Figure 11.20e and f shows the images of the scattered light when the incident light is x- and y- polarized, respectively. It is obvious that the metallic tip strongly scatters the field when the tip is on a peak of the z-polarized component of electric field. The images shown in Fig. 11.20e and f let us know the exact position of the tip and allow for a proper tip positioning.

11.4.3 DNA Imaging

DNA imaging by the TE-CARS microscopy was demonstrated. Two types of samples were prepared; (1) DNA clusters and (2) DNA network. For preparation of DNA clusters, an aqueous solution of single-stranded DNA (poly(dA-dT)) (250 μg/ml) was cast and dried on a coverslip in the room temperature with a fixation time of ~24 hours. The dimensions of the clusters are typically ~20 nm in height and ~100 nm in width. For DNA network, an aqueous solution of double stranded DNA (poly(dA-dT)-poly(dA-dT)) (250 μg/ml) was mixed with $MgCl_2$ (0.5 mM) solution, then the DNA solution was cast on a coverslip and blow-dried after a fixation time of ~2 hours [56]. The DNA chains were self-assembled to form a network structure. The Mg ions electrostatically connect phosphorus atoms of DNA and oxygen atoms of the glass. The DNA network consists of bundles of DNA double-helix filaments aligned parallel on the glass substrate. The height of the bundle structures is ~2.5 nm, which is comparable to the diameter of single DNA double-helix. CARS imaging was performed at 1337 cm^{-1} which coincides with a Raman mode of adenine by tuning the excitation frequencies ω_1 and ω_2 to be 12710 cm^{-1}(λ_1: 786.77 nm) and 11373 cm^{-1}(λ_2 : 879.25 nm), respectively.

Figure 11.21 shows the TE-CARS images of the DNA clusters obtained by our system. Figure 11.21a and b is the TE-CARS image at the on-resonant frequency (1337 cm^{-1}) and the simultaneously acquired topographic AFM image. The DNA clusters of ~100 nm diameter are visualized in Fig. 11.21a. The two DNA clusters with distance of ~160 nm are obviously distinguished by the tip-enhanced CARS imaging. This indicates that the CARS imaging successfully achieved superresolving capability beyond the diffraction limit of light. For comparison, an imaging of the same area was performed at 1278 cm^{-1} where the sample DNA does not have a unique Raman peak. Figure 11.21c shows that the CARS signals mostly vanished at the off-resonant frequency. Figure 11.21a and c verifies that DNA molecules emit vibrationally resonant CARS at the specific frequency. However, there remains some slight signal increase at the clusters at the off-resonant frequency, as seen in Fig. 11.21d, which is the same as Fig. 11.21c but is shown with a different gray scale. This can be caused by both the non-resonant component of the nonlinear susceptibility of DNA (Eq. 11.2) [52, 53] and a topographic artifact [26]. Figure 11.21e shows a CARS image of the DNA network structure at 1337 cm^{-1}. The network structure of the DNA chains was clearly visualized. The full width of half maximum of the finest structure in the image (indicated by yellow arrows) is found to be ~15 nm.

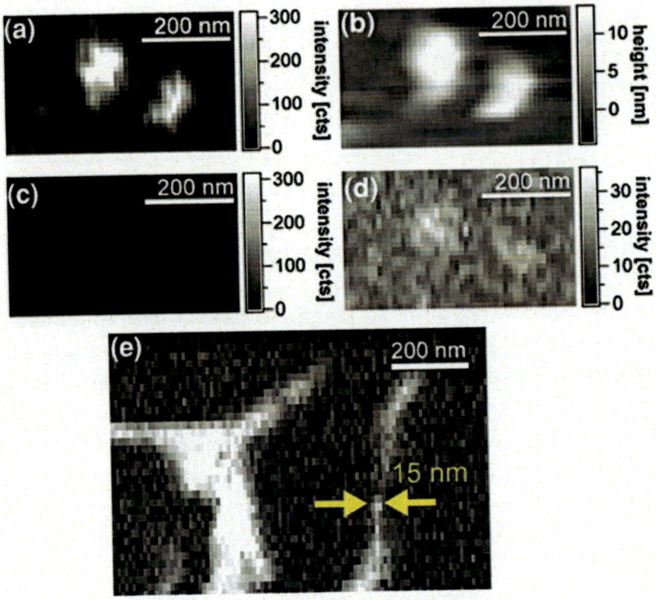

Fig. 11.21 TE-CARS images of DNA molecules. **a** TECARS image at 1337 cm^{-1} of adenine, and **b** the simultaneously obtained topographic image. **c** TE-CARS image at the off-resonant frequency (1278 cm^{-1}). **d** The same image as **c** shown with a different gray scale. **e** TECARS image of the DNA network structure. For all images, the number of photons counted in 100 ms was recorded for one pixel

11.4.4 CNT Imaging

Bundles of SWCNTs of semiconductor-type were used for CARS imaging [24]. The SWCNT bundles were dispersed on a glass substrate by spin-coating. The frequencies of the two beams were set to be $\omega_1 = 12744\,cm^{-1}(\lambda_1 : 784.70\,nm)$ and $\omega_2 = 11163\,cm^{-1}(\lambda_2 : 895.84\,nm)$ such that the frequency difference is 1581 cm^{-1} corresponding to the G-band of SWCNTs [7]. Figure 11.22a and b shows a simultaneously obtained topographic image and a tip-enhanced CARS image of the SWCNT bundles, respectively. Figure 11.22c shows a far-field CARS image obtained without the tip. Figure 11.22d shows the cross-sections of the tip-enhanced and far-field CARS signals of the line indicated by the arrows. By sweeping the frequency of the ω_2 beam from 11521 cm^{-1} (868.00 nm) to 11112 cm^{-1} (899.90 nm) a TECARS spectrum and the corresponding far-field CARS spectrum were obtained, as shown in Fig. 11.22e, where a spontaneous anti-Stokes Raman spectrum is also drawn for comparison. The CARS spectra are in good agreement with the spontaneous Raman spectrum, and the characteristic band shape of the semiconductor type of the SWCNTs are clearly seen in the tip-enhanced CARS spectrum. In the tip-enhanced CARS image in Fig. 11.22b, only several bundles aligned parallel to the polarization of the incident fields (transverse in Fig. 11.5) are selectively visualized, while lots of

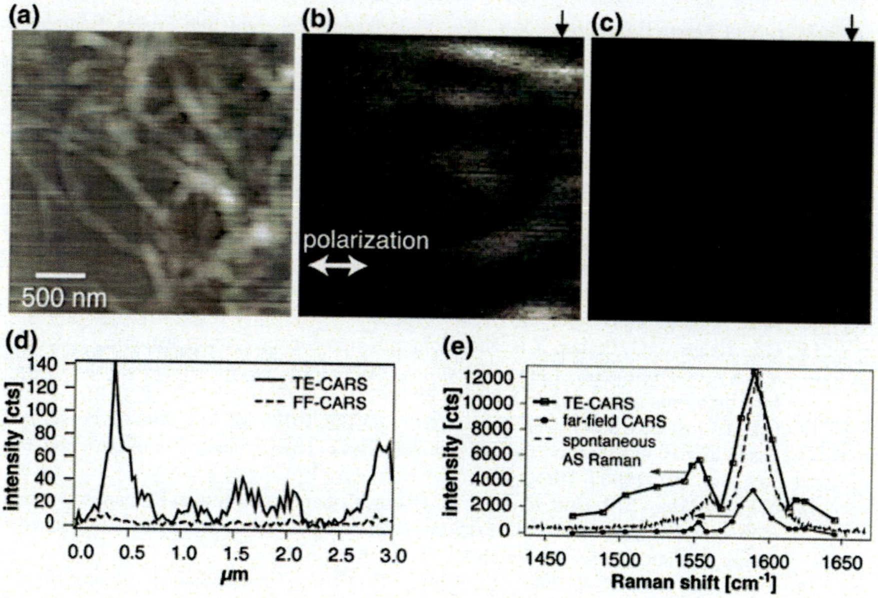

Fig. 11.22 TECARS imaging of SWCNTs. **a** A topographic image of the SWCNTs dispersed on a glass substrate. **b** A TE-CARS image obtained simultaneously with the topographic image in **a**. **c** A far-field CARS image obtained without the tip. **d** The cross-sections of the line indicated by the arrows in **b** and **c**. **e** A TE-CARS spectrum, a far-field CARS spectrum, and spontaneous anti-Stokes (AS) Raman spectrum of a SWCNT bundle. The average powers of the ω_1 beam and ω_2 beam are 12 μW and 6 μW at the 800 kHz repetition rate except for the spectrum measurement in **e** which used the powers of 26 μW and 13 μW

randomly oriented bundles are found in the topographic image in Fig. 11.22a. This selectivity can be explained by a combination of two possible mechanisms following. First, since G-band is a longitudinal mode susceptible to the electric fields parallel to the tubes [7], only some SWCNTs parallel to the polarization of the incident field can be strongly excited. Second, the variety of the SWCNTs which have many different chiralities are contained in the sample, and only some have electronic resonance frequencies matching with either of the excitation fields. Therefore, there can be a large difference in the CARS signal intensity tube by tube. It is verified from all the results shown in Fig. 11.22 that our CARS microscope effectively amplified the CARS signal of the SWCNTs and can obtain vibrational images of SWCNTs.

References

1. P. Anger, P. Bharadwaj, L. Novotny, Enhancement and quenching of single-molecule fluorescence. Phys. Rev. Lett. **96**, 113002 (2006)
2. C.F. Bohren, D.R. Huffman, *Absorption and Scattering of Light by Small Particles* (Wiley, New York, 1983)

3. J.-X. Cheng, L.D. Book, X.S. Xie, Polarization coherent anti-Stokes Raman scattering microscopy. Opt. Lett. **26**, 1341–1343 (2001)
4. W. Denk, J.H. Strickler, W.W. Webb, Science **248**, 73 (1990)
5. W. Denk, K.R. Delaney, A. Gelperin, D. Kleinfeld, B.W. Strowbridge, D.W. Tank, R.J. Yuste, Neurosci. Meth. **54**, 151 (1994)
6. N. Dudovich, D. Oron, Y. Silberberg, Single-pulse coherently controlled nonlinear Raman spectroscopy and microscopy. Nature **418**, 512–514 (2002)
7. G.S. Duesberg, I. Loa, M. Burghard, K. Syassen, S. Roth, Polarized Raman spectroscopy on isolated single-wall carbon nanotubes. Phys. Rev. Lett. **85**, 5436–5439 (2000)
8. M.D. Duncan, J. Reintjes, T.J. Manuccia, Opt. Lett. **7**, 350 (1982)
9. M.D. Duncan, J. Reintjes, T.J. Manuccia, Opt. Eng. **24**, 352 (1984)
10. M.D. Duncan, Opt. Commun. **50**, 307 (1985)
11. C.L. Evans, E.O. Potma, X.S. Xie, Coherent anto-Stokes Raman scattering spectral interferometry: determination of the real and imaginary components of nonlinear susceptibility $\chi(3)$ for vibrational microscopy. Opt. Lett. **29**, 2923–2925 (2004)
12. C.L. Evans, E.O. Potma, M. Puorishaag, D. Côté, C.P. Lin, X.S. Xie, Chemical imaging of tissue in vivo with video-rate coherent anti-Stokes Raman scattering microscopy. Proc. Natl. Aca. Sci. **102**, 16807–16812 (2005)
13. Y. Fu, H. Wang, T.B. Huff, R. Shi, J.-X. Cheng, Coherent anti-Stokes Raman scattering imaging of myelin degradation reveals a calcium-dependent pathway in Lyso-PtdCho-induced demyelination. J. Neuroscience Research **85**, 2870–2881 (2007)
14. Y. Fu, T.B. Huff, H.-W. Wang, H. Wang, J.-X. Cheng, Ex vivo and in vivo imaging of myelin fibers in mouse brain by coherent anti-Stokes Raman scattering microscopy. Opt. Exp. **16**, 19396–19409 (2008)
15. K. Fujita, N.I. Smith, Label-free molecular imaging of living cells. Mol. Cells **26**, 530–535 (2008)
16. K. Hamada, K. Fujita, N. Smith, M. Kobayashi, Y. Inouye, S. Kawata, Raman microscopy for dynamic molecular imaging of living cells. J. Biomed. Opt. **13**, 044027 (2008)
17. M. Hashimoto, T. Araki, Three dimensional coherent and optical transfer functions of coherent anti-strokes Raman scattering microscopy. J. Opt. Soc. Am. A **18**, 771–776 (2001)
18. M. Hashimoto, T. Araki, S. Kawata, Molecular vibration imaging in the fingerprint region by use of coherent anti-Stokes Raman scattering microscopy with a collinear configuration. Opt. Lett. **25**, 1768 (2000)
19. A. Hartschuh, E.J. Sánchez, X.S. Xie, L. Novotny, High-resolution near-field Raman microscopy of single-walled carbon nanotubes. Phys. Rev. Lett. **90**, 095503 (2003)
20. N. Hayazawa, Y. Inouye, S. Kawata, Evanescent field excitation and measurement of dye fluorescence in a metallic probe near-field scanning optical microscope. J. Microsc. **194**, 472–476 (1999)
21. N. Hayazawa, Y. Inouye, Z. Sekkat, S. Kawata, Metallized tip amplification of near-field Raman scattering. Opt. Commun. **183**, 333 (2000)
22. N. Hayazawa, Y. Inouye, Z. Sekkat, S. Kawata, Near-field Raman scattering enhanced by a metallized tip. Chem. Phys. Lett. **335**, 369–374 (2001)
23. N. Hayazawa, T. Yano, H. Watanabe, Y. Inouye, S. Kawata, Detection of an individual single-wall carbon nanotube by tip-enhanced near-field Raman spectrocopy. Chem. Phys. Lett. **376**, 174–180 (2003)
24. N. Hayazawa, T. Ichimura, M. Hashimoto, Y. Inouye, S. Kawata, Amplification of coherent anti-Stokes Raman scattering by a metallic nano-structure for a high resolution vibration microscopy. J. Appl. Phys. **95**, 2676–2681 (2004)
25. N. Hayazawa, Y. Saito, S. Kawata, Detection and characterization of longitudinal field for tip-enhanced Raman spectroscopy. Appl. Phys. Lett. **85**, 6239–6241 (2004)
26. B. Hecht, H. Bielefeldt, Y. Inouye, D.W. Pohl, L. Novotny, Facts and artifacts in near-field optical microscopy. J. Appl. Phys. **81**, 2492–2498 (1997)
27. Y.-S. Huang, T. Kawashima, M. Yamamoto, H. Hamaguchi, Molecular-level pursuit of yeast mitosis by time- and space-resolved Raman spectroscopy. J. Raman Spectrosc. **34**, 1–3 (2003)

28. A. Ichihara, T. Tanaami, K. Isozaki, Y. Sugiyama, Y. Kosugi, K. Mikuriya, M. Abe, I. Uemura, High-speed confocal fluorescence microscopy using a nipkow scanner with microlenses for 3-D imaging of single fluorescent molecule in real time. Bioimages **42**, 57–62 (1996)
29. T. Ichimura, N. Hayazawa, M. Hashimoto, Y. Inouye, S. Kawata, Tip-enhanced coherent anti-Stokes Raman scattering for vibrational nanoimaging. Phys. Rev. Lett. **92**, 220801 (2004)
30. Y. Inouye, N. Hayazawa, K. Hayashi, Z. Sekkat, S. Kawata, Near-field scanning optical microscope using a metallized cantilever tip. Proc. SPIE **3791**, 40–48 (1999)
31. H. Kano, H. Hamaguchi, Supercontinuum dynamically visualizes a dividing single cell. Anal. Chem. **79**, 8967–8973 (2007)
32. T.W. Kee, M.T. Cicerone, Simple approach to one-laser, broadband coherent anti-Stokes Raman scattering microscopy. Opt. Lett. **29**, 2701–2703 (2004)
33. M. Kerker, D.-S. Wang, H. Chew, Surface enhanced Raman scattering (SERS) by molecules adsorbed at spherical particles: errata. Appl. Opt. **19**, 4159–4174 (1980)
34. M. Kobayashi, K. Fujita, T. Kaneko, T. Takamatsu, O. Nakamura, S. Kawata, Secondharmonic- generation microscope with a microlens array scanner. Opt. Lett. **27**, 1324–1326 (2002)
35. F. Lu, W. Zheng, Z. Huang, Phase-controlled polarization coherent anti-Stokes Raman scattering microscopy for high-sensitivity and high-contrast molecular imaging. Opt. Lett. **25**, 1907–1909 (2008)
36. P.D. Maker, R.W. Terhune, Study of optical effects due to an induced polarization third order in the electric field strength. Phys. Rev. **137**, A801–A818 (1965)
37. H.-J. Manen, Y.M. Kraan, D. Roos, C. Otto, Intracellular chemical imaging of hemecontaining exzymes involved in innate immunity using resonance Raman microscopy. J. Phys. Chem. B **108**, 18762–18771 (2004)
38. H.-J. Manen, Y.M. Kraan, C. Otto, Sinle-cell Raman and fluorescence microscopy reveal the association of lipid bodies with phagosomes in leukocytes. Proc. Natl. Acad. Sci. U. S. A. **102**, 10159–10164 (2005)
39. C. Matthäus, S. Boydston-White, M. Miljkovíc, M. Romeo, M. Diem, Raman and infrared microspectral imaging of mitotic cells. Appl. Spectrosc. **60**, 1–8 (2006)
40. C. Matthäus, T. Chernenko, J.A. Newmark, C.M. Warner, M. Diem, Label-free detection of mictochondorial distribution in cells by nonresonant Raman microspectroscopy. Biophys. J. **93**, 668–673 (2007)
41. D. Mehtani, N. Lee, R.D. Hartschuh, A. Kisliuk, M.D. Foster, A.P. Sokolov, J.F. Maguire, Nano-Raman spectroscopy with side-illumination optics. J. Raman Spectrosc. **36**, 1068 (2005)
42. T. Minamikawa, N. Tanimoto, M. Hashimoto, T. Araki, M. Kobayashi, K. Fujita, S. Kawata, Jitter reduction of two synchronized picosecond mode-locked lasers using balanced crosscorrelator with two-photon detectors. Appl. Phys. Lett. **89**, 191101 (2006)
43. T. Minamikawa, M. Hashimoto, K. Fujita, S. Kawata, T. Araki, Multi-focus excitation coherent anti-Stokes Raman scattering (CARS) microscopy and its applications for real-time imaging. Opt. Express **17**, 9526 (2009)
44. M. Müller, J.M. Schins, Imaging the thermodynamic state of lipid membranes with multiplex CARS microscopy. J. Phys. Chem. B **106**, 3715–3723 (2002)
45. X. Nan, E.A.M. Tonary, A. Stolow, X.S. Xie, J.P. Pezzacki, Intracellular imaging of HCV RNA and cellular lipids by using simultaneous two-photon fluorescence and coherent anti- Stokes Raman scattering microscopies. ChemBioChem **7**, 1895–1897 (2006)
46. M. Okuno, H. Kano, P. Leproux, V. Couderc, H. Hamaguchi, Ultrabroadband multiplex CARS microspectroscopy and imaging using a subnanosecond supercontinuum light source in the deep near infrared. Opt. Lett. **33**, 923–925 (2008)
47. G.J. Puppels, F.F.M. de Mul, C. Otto, J. Greve, M. Robert-Nicoud, D.J. Arndt-Jovin, T.M. Jovin, Studying single living cells and chromosomes by confocal Raman microspectroscopy. Nature **347**, 301–303 (1990)
48. G.J. Puppels, M. Grond, J. Greve, Direct imaging Raman microscope based on tunable wavelength excitation and narrow-band emission detection. Appl. Spectrosc. **47**, 1256–1267 (1993)

49. H.N. Paulsen, K.M. Hilligsoe, J. Thogersen, S.R. Keiding, J.J. Larsen, Coherent anti- Stokes Raman scattering microscopy with a photonics crystal fiber based light source: Opt. Lett. **28**, 1123–1125 (2003)
50. Y. Saito, T. Murakami, Y. Inouye, S. Kawata, Fabrication of silver probes for localized plasmon excitation in near-field Raman spectroscopy. Chem. Lett. **34**, 920–921 (2005)
51. Y. Saito, M. Kobayashi, D. Hiraga, K. Fujita, S. Kawano, N. Smith, Y. Inouye, S. Kawata, Z-polarization sensitive detection in micro Raman spectroscopy by radially polarized incident light. J. Raman Spectrosc. **39**, 1643–1648 (2008)
52. Y.R. Shen, *The Principles of Nonlinear Optics* (John Wiley and Sons, New York, 1984)
53. M.D. Levenson, *Introduction to Nonlinear Laser Spectroscopy* (Academic press, Orlando, 1988)
54. R.L. Shoemaker, P.H. Bartels, D.W. Hillman, J. Jonas, D. Kessler, R.V. Sshack, D. Vukiobratovich, An ultrafast laser scanner microscope for digital image analysis. IEEE Trans. Biomed. Eng. **29**, 82–91 (1982)
55. R.M. Stöckle, Y.D. Suh, V. Deckert, R. Zenobi, Nanoscale chemical analysis by Tipenhanced Raman Scattering. Chem. Phys. Lett. **318**, 131–136 (2000)
56. S. Tanaka, L.T. Cai, H. Tabata, T. Kawai, Formation of two-dimensional network structure of DNA molecules on Si substrate. Jpn. J. Appl. Phys. **40**, L407–L409 (2001)
57. R.Y. Tsien, B.J. Bacskai, Video-rate confocal microscopy, in *Handbook of Biological Confocal Microscopy*, ed. by J.B. Pawley (Plenum Press, New York, 1995), pp. 459–478
58. N. Uzunbajakava, C. Otto, Combined Raman and continuouswave-excited two-photon fluorescence cell imaging. Opt. Lett. **28**, 2073–2075 (2003)
59. A. Volkmer, J.-X. Cheng, X.S. Xie, Vibrational imaging with high sensitivity via Epidetected Coherent Anti-Stokes Raman Scattering (E-CARS) microscopy. Phys. Rev. Lett. **87**, 023901 (2001)
60. A. Volkmer, L.D. Book, X.S. Xie, Time-resolved coherent anti-Stokes Raman scattering microscopy: Imaging based on Raman free induction decay. Appl. Phys. Lett. **80**, 1505–1507 (2002)
61. T. Wilson, *Confocal Microscopy* (Academic Press, London, 1990), p. 58
62. H. Wang, Y. Fu, P. Zickmund, R. Shi, J.-X. Cheng, Coherent anti-Stokes Raman scattering imaging of axonal myelin in live spinal tissues. Biophysical Journal **89**, 692–591 (2005)
63. X.S. Xie, J. Yu, W.Y. Yang, Living cells as test tubes. Science **312**, 228–230 (2006)
64. A. Zumbusch, G.R. Holtom, X.S. Xie, Three-dimensional Vibrational Imaging by Coherent Anti-Stokes Raman Scattering. Phy. Rev. Lett. **82**, 4142–4145 (1999)

Chapter 12
Background-free Coherent Raman Imaging: The CARS and SRS Contrast Mechanisms

Hervé Rigneault and David Gachet

Abstract We present here the various methodologies that permit to achieve high speed Raman imaging through stimulated processes. Both coherent anti-Stokes Raman scattering (CARS) and stimulated Raman scattering (SRS) are considered. Because CARS imaging is affected by a non-resonant contribution, we concentrate here on methods that can remove this non-resonant background in CARS microscopy. We next present the principles of SRS imaging and its implementation in microscopy. We finally present a hyperspectral microscopy setup that is able to perform CARS and SRS together with other nonlinear optical contrasts [second harmonic generation (SHG), two-photon excited fluorescence (TPEF)] and show some results in the field of skin imaging.

12.1 Introduction

'Nonlinear magic', with this striking title multiphoton microscopy pioneers envisioned 10 years ago the revolution of nonlinear contrast mechanism in life sciences [1]. Indeed at that time implementation of coherent Raman techniques in microscopy was still in its early years since its first report in 1999 [2]. In this first paper coherent anti-Stokes Raman Scattering (CARS) was announced as being compatible with point scanning imaging modalities and could aspire to approach

H. Rigneault (✉)
Institut Fresnel, CNRS UMR 6133, Aix-Marseille Université,
Domaine Universitaire de St Jérôme,
13397 Marseille Cedex 20, France
e-mail: herve.rigneault@fresnel.fr

D. Gachet
Centre Interdisciplinaire des Nanosciences de Marseille (CINaM),
CNRS UPR 3118, Aix-Marseille Université, Campus de Luminy,
Case 913, 13288 Marseille Cedex 9, France
e-mail: gachet@cinam.univ-mrs.fr

A. Zoubir (ed.), *Raman Imaging*, Springer Series in Optical Sciences 168,
DOI: 10.1007/978-3-642-28252-2_12, © Springer-Verlag Berlin Heidelberg 2012

the dream of video rate 'Raman imaging'. Although very encouraging results were presented in this first i paper, 'Raman imaging' was not really reached in those early days, because of an inherent CARS background signal that spoils the measurements in the case of weak or lowly concentrated Raman scatters. This is the first goal of this chapter to present the advances that overcame the CARS non-resonant background limitation. We will review in detail the origin of this non-resonant background and present the various methods that have been implemented to overcome this difficulty.

The second part of this chapter deals with stimulated Raman scattering (SRS), a relatively new contrast mechanism implemented in microscopy [3]. SRS is concomitant to CARS, but presents the asset of being a priori background-free. This is a big advantage for life science imaging and SRS is, in this view, the stimulated Raman technique that can reach video rate 'Raman imaging' [4]. This is our goal to present here the principles of SRS in a comprehensive way, its practical implementation, and to show some recent results obtained in the field of skin imaging. CARS and SRS are now referred to as "coherent Raman scattering" (CRS) imaging. We will also present a hyperspectral microscopy setup that is able to perform CRS, second harmonic generation (SHG), and two-photon excitation fluorescence (TPEF) at the same time.

12.2 Non-resonant Background in CARS

Optical processes, such as elastic (Rayleigh) or inelastic (Brillouin or Raman) scattering, involve only one incident electromagnetic wave. In stark contrast, CARS is a higher order (nonlinear) scattering process that mixes multiple waves [5]. Wave-mixing effects include second and third harmonic generation (THG) to cite a few. In the case of CARS, three incoming waves, referred as pump, Stokes, and probe, interact in the medium of interest to give rise to a so-called blueshifted anti-Stokes wave. CARS is thus seen as a four-wave mixing process. Although this interaction can take many routes, two are noteworthy (Fig. 12.1).

In the first one (Fig. 12.1a), the pump and Stokes beams operate a frequency difference that can match a vibrational resonance of the medium. In the time domain, the interference of the pump and Stokes beams creates a beat that coherently drives the molecular vibration. The probe beam is then scattered by this driven vibration to give rise to the anti-Stokes beam. As it involves a vibrational resonance, this first route is referred as "resonant". In the second route (Fig. 12.2b), there is no resonant interaction between the vibration and the incoming beams. This "nonresonant" route (also referred as non-resonant four-wave mixing) takes its origin in the instantaneous electronic response in the medium of interest, similar to non-resonant harmonic generation for instance. For vibrational spectroscopy/microspectrometry, only the resonant route gives interesting information. Due to the coherent nature of CARS, the overall anti-Stokes field is the result of the interference between the anti-Stokes fields generated via the resonant and the non-resonant route [6] and it is hard to disentangle these two contributions.

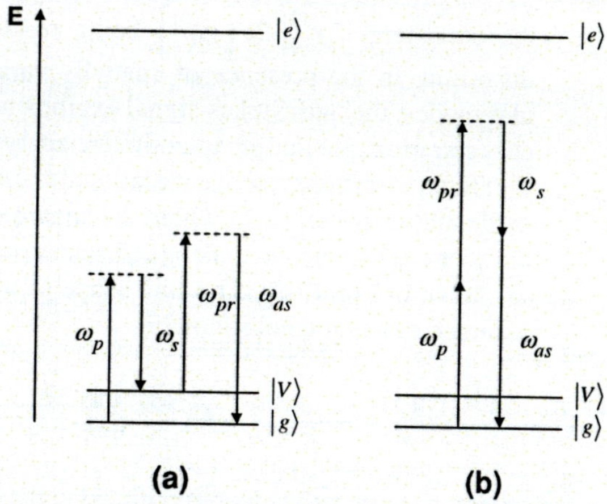

Fig. 12.1 Energetic diagram of the CARS process. **a** Resonant and **b** non-resonant CARS. $|V\rangle$, $|g\rangle$, and $|e\rangle$ stand for the molecular vibrational state, the molecular ground state, and the first molecular excited electronic level, respectively

12.2.1 Non-resonant Background Contribution in CARS Spectrum

On a formal way, the incoming pump, Stokes, and probe beams (with respective angular frequencies ω_p, ω_s and ω_{pr}) generate at any point $\vec{r}$ in the material a polarization oscillating at angular frequency $\omega_{as} = (\omega_p - \omega_s) + \omega_{pr}$ according to

$$\vec{P}^{(3)}(\vec{r}, \omega_{as}) = \chi^{(3)}(\vec{r}, -\omega_{as}; \omega_p, -\omega_s, \omega_{pr})\, \vec{E}_p(\vec{r}, \omega_p) : \vec{E}_s^*(\vec{r}, -\omega_s) : \vec{E}_{pr}(\vec{r}, \omega_{pr}).$$

Here, $\chi^{(3)}$ is the third-order nonlinear susceptibility of the material of interest. It is a fourth-order tensor that mathematically describes the resonant and non-resonant routes described before. It is thus the sum of a resonant and a non-resonant part following $\chi^{(3)} = \chi_R^{(3)} + \chi_{NR}^{(3)}$. The nonlinear polarization $\vec{P}^{(3)}(\vec{r}, \omega_{as})$ acts as a source for the anti-Stokes field $E_{as} = E_{as}^R + E_{as}^{NR}$, where E_{as}^R and E_{as}^{NR} are the anti-Stokes fields generated on a resonant and a non-resonant way, respectively. As such, these two fields follow the spectral behavior of $\chi_R^{(3)}$ and $\chi_{NR}^{(3)}$. Because $\chi_R^{(3)}$ describes a resonance, it exhibits a dispersive behavior as the molecular vibrations are driven out of phase compared to the excitation. It is thus a complex quantity. By contrast, $\chi_{NR}^{(3)}$ is associated with a flat spectral background that responds instantaneously to the excitation. It is thus a real quantity [6]. The spectral response of a vibrational resonance in CARS spectroscopy is distorted due to the spectral interference between $\chi_R^{(3)}$ and $\chi_{NR}^{(3)}$ [7], as seen in Fig. 12.2a. Indeed, the anti-Stokes signal is proportional to the quantity

$$\left|\chi^{(3)}\right|^2 = \left|\chi_R^{(3)} + \chi_{NR}^{(3)}\right|^2 = \left|\chi_R^{(3)}\right|^2 + \left|\chi_{NR}^{(3)}\right|^2 + 2\chi_{NR}^{(3)}\mathrm{Re}\left[\chi_R^{(3)}\right],$$

the cross-term $2\chi_{NR}^{(3)} Re\left[\chi_{R}^{(3)}\right]$ being responsible for the vibrational line shape distortion. In the presence of spatially variable concentration of resonant molecular modes, the anti-Stokes signal exhibits non-linear dependence on the molecular concentration, precluding quantitative analysis.

The spectral behavior of a vibrational resonance as detected in CARS spectroscopy is illustrated in Fig. 12.2a where the square modulus of $\chi^{(3)}$ together with its phase are plotted. It is better understood when plotting $\chi^{(3)}$ in the complex plane [8, 9] as shown in Fig. 12.2b. The line shape of a single vibrational resonance is usually assumed as Lorentzian following

$$\chi_{R}^{(3)} = \frac{a}{(\omega_p - \omega_s - \Omega_R) + i\Gamma},$$

where, a is a negative quantity related to the oscillator strength of the vibration, $\Omega_R$is the vibration frequency, and Γ is the half-width at half-maximum of the vibrational resonance [5]. When plotted in the complex plane, $\chi_{R}^{(3)}$ describes a circle (dashed line) as the excitation $\omega_p - \omega_s$ scans the resonance. The resonance ($\omega_p - \omega_s = \Omega_R$) is characterized by the quadrature between the excitation and the molecular vibration as $\chi_{R}^{(3)}$ becomes a purely imaginary quantity. Bringing the non-resonant part $\chi_{NR}^{(3)}$ into play, the resonance-related circle (continuous line) is shifted along the real axis with several important consequences. First, the maximum of anti-Stokes signal (referred as "peak", point 2) is now redshifted compared to the (Raman, point 3) vibrational resonance. In addition, a minimum of anti-Stokes signal (referred as "dip", point 5) is found immediately after the peak. They result from the constructive and destructive interference, respectively, between $\chi_{R}^{(3)}$ and $\chi_{NR}^{(3)}$. This "peak" and "dip" structure is reminiscent of a Fano profile, whereby a discrete level coherently interferes with a spectrally broad continuum [10].

12.2.2 *Non-resonant Background Contribution in CARS Image*

Targeting specific vibrational modes or identifying molecular species in CARS imaging is usually done with narrow and broad-band CARS setups. In both cases, the non-resonant background affects the collected vibrational information. It distorts both spatial and spectral contrasts, precluding quantitative analysis and limiting molecular identification capability. For instance, the non-resonant background from water adds a strong signal offset in CARS images of cells [11]. This is illustrated in Fig. 12.3 which presents a resonant (Fig. 12.3a) and a non-resonant (Fig. 12.3b) image of a giant unilamellar vesicle (GUV) immersed in water. The GUV is made of deuterated lipids and is fabricated by electro formation. It shows a distinct CD_2 related Raman band at 2105 cm^{-1}. At this resonance frequency (Fig. 12.3a) the GUV lipid bilayer appears bright whereas it remains visible off-resonance (Fig. 12.3b—2140 cm^{-1}) due to its non-resonant background contribution. In both images the background has a significant contribution due to the non-resonant background of water.

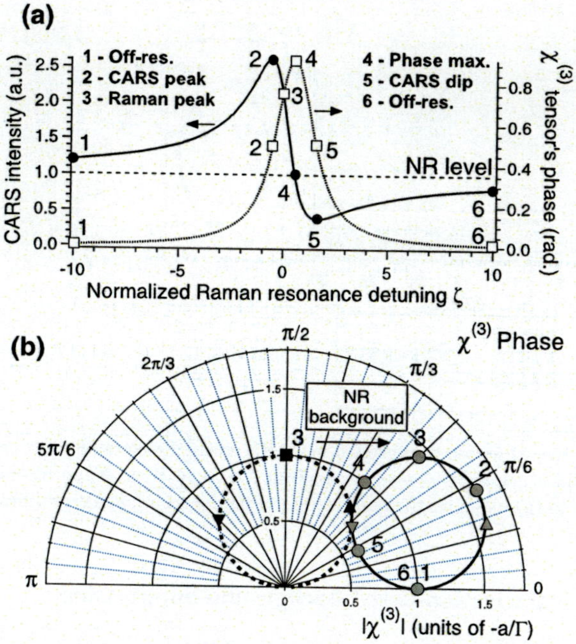

Fig. 12.2 Comparison between Raman and CARS resonance. **a** Intensity (*plain line*) and phase (*dashed line*) of the CARS signal close to a vibrational resonance, as a function of $\zeta = \omega_p - \omega_s - \Omega_R)/\Gamma$, ζ is the normalized detuning to the vibrational resonance. **b** Raman (*dashed line*) and CARS (*continuous line*) resonance depicted in the complex plane. *1*: far-before resonance; *2*: CARS signal peak; *3*: Raman resonance; *4*: maximum of phase; *5*: CARS signal dip; *6*: far-after resonance

The main task of background-free CARS techniques is to extract the pure vibrational response of the sample under survey (i.e., its Raman response) that is contained in the term $\mathrm{Im}\left[\chi_R^{(3)}\right]$ [12], or at least to reduce as much as possible the $\chi_{NR}^{(3)}$ contribution into the anti-Stokes signal. In doing so, specific vibrational contrast is reached and quantitative chemical analysis is possible. Many ideas implemented for background-free CARS microscopy take their very origin in CARS spectroscopy techniques developed in the 1970s.

12.2.3 Background-free CARS Methods

Background-free CARS methods take their origin in the physical differences existing between the resonant and the non-resonant part of the $\chi^{(3)}$ tensor. Important features such as the vibrational phase shift, the temporal coherence of the induced vibrational motion of molecules or the different polarization response in scattering come into play when trying to discriminate the resonant contribution from the non-resonant background.

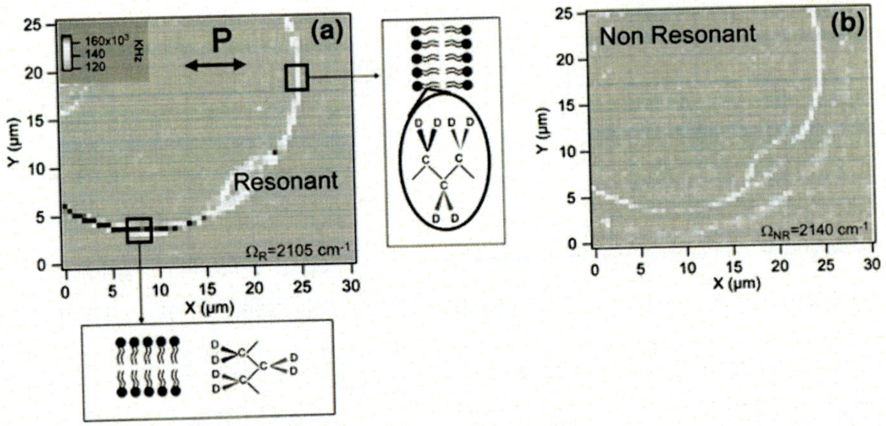

Fig. 12.3 Non-resonant background in CARS image. Deuterated giant unilamellar vesicle (GUV) immersed in water, **a** CD_2 resonant image at 2,105 cm^{-1} and **b** non-resonant image at 2,140 cm^{-1}. The *arrow* in (**a**) indicates the polarization of the incoming pump and Stokes fields

12.2.3.1 Polarization-based Background Suppression

Molecular vibrations are known to depolarize the scattered light in linear Raman processes. Each vibrational mode can be thus characterized by a specific "depolarization ratio". This property governs nonlinear Raman scattering as well. The resonant and the nonresonant part of the $\chi^{(3)}$ tensor scatter the anti-Stokes field in a different fashion. In particular, in a two-color (also referred as "degenerate") configuration (where the pump field acts as the probe beam $\omega_p = \omega_{pr}$), when the incident pump and Stokes fields impinge into the sample of interest with different linear polarizations (the difference in angle is usually set at 71.6° to optimize the effect, see Fig. 12.4a, P-CARS), the anti-Stokes fields scattered by resonant and non-resonant process exhibit linear polarizations differently oriented in space. Efficient suppression of the non-resonant background is thus insured by placing a polarizer perpendicular to the anti-Stokes field scattered in a non-resonant fashion. Such approach, named "Polarization-CARS", initially proposed in CARS spectroscopy [13–15], was successfully applied to CARS microscopy by Cheng et al. [16]. Despite the simplicity of this method, a non-negligible part of the resonant signal is lost after filtering by the polarizer. The proportion of "lost" signal is directly linked to the Raman depolarization ratio of the probed Raman line. It increases when the Raman depolarization ratio approaches one-third as the resonant and the non-resonant anti-Stokes fields tend to align.

12.2.3.2 Temporal Suppression of the Non-resonant Background

In the CARS process, the resonant and the non-resonant response have different physical origin. While the resonant response arises from the creation by the pump and

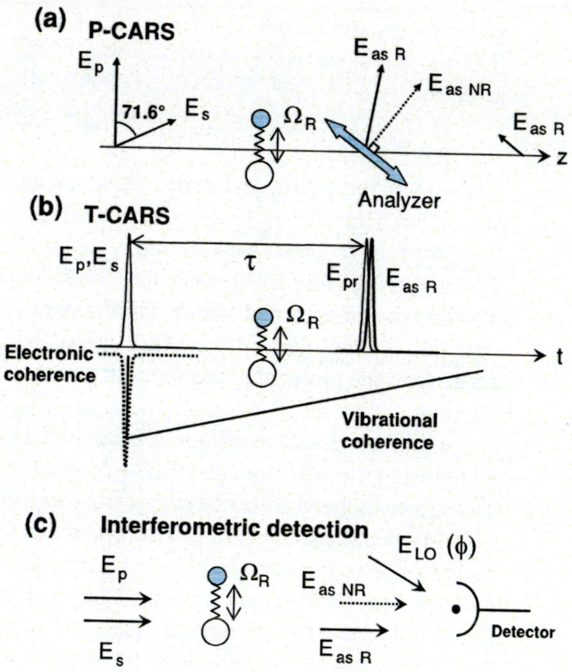

Fig. 12.4 Background-free methods for CARS imaging. **a** Polarization-CARS, **b** Temporal-CARS, **c** Interferometric detection of the CARS signal. See text for details

Stokes pulses of a vibrational coherence (that last typically a tens of picoseconds), the electronic non-resonant response is instantaneous and does not last after the pump and Stokes pulses. Control over the preferential generation of resonant or non-resonant signal is gained by introducing a time delay τ between the probe and the pump/Stokes pulse. Efficient suppression of the non-resonant background is thus achieved when τ is fixed larger than the pump/Stokes pulse duration and shorter than the vibrational coherence duration T_2 (Fig. 12.4b, T-CARS). This "Temporal-CARS" approach was first introduced in 1980 [17] in spectroscopy before being adapted to microscopy by Volkmer et al. [18]. The introduction of a two beam configuration [19] has recently simplified this approach.

12.2.3.3 Interferometry-based Methods

Any vibrational resonance induces a temporal delay of the resonant anti-Stokes field compared to the non-resonant one. This is accompanied by a phase shift φ between these two fields (transferred via the $\chi^{(3)}$ tensor). As already seen with the dashed circle in Fig. 12.2b, this phase shift equals 0, $\pi/2$ or π, depending if the excitation is spectrally far-before, at, or far-after the vibrational resonance, respectively. By means of interferometric measurement, it is possible to recover this phase shift and thus the resonant part of the CARS signal. For this, the anti-Stokes field E_{as}

generated by the sample is mixed with an external field E_{LO} (named local oscillator (LO)) at the same wavelength (see Fig. 12.4c). The interference detected signal is given by

$$S_{tot} = |E_{as}|^2 + |E_{LO}|^2 + 2\,|E_{LO}|\,|E_{as}|\cos\phi,$$

where ϕ is the phase difference between the anti-Stokes and the LO. Extraction of the last (interference) term by means of modulation brings three advantages. First, the non-resonant background is suppressed if ϕ is chosen adequately to make a destructive interference between the LO and the non-resonant anti-Stokes field. Then, the resonant signal, that can be significantly small in the case of dilute/biological samples, is amplified by the LO. Finally, the recovered signal has a linear dependence over the molecular species concentration, making quantitative CARS analysis possible.

First demonstrations of non-resonant background suppression by interferometry were brought in the domain of CARS spectroscopy [20, 21], following early work on Raman-induced Kerr effect spectroscopy (RIKES) [22].

Early demonstrations of interferometry in CARS imaging made use of LOs generated outside the sample [23–26], generally obtained by generating a pure non-resonant anti-Stokes field in an auxiliary sample. The anti-Stokes beam and the LO following two different optical paths raises the concern of mechanical stability in the interferometer arrangement. Simplified setups, where the LO is generated at the output of a non-collinear optical parametric amplifier (NOPA) [27] or an optical parametric oscillator (OPO) [28], improve interferometric detection as such schemes provide LOs with good intensity and phase stability. Other schemes, where the LO is generated in a photonic crystal fiber pumped by the pump and Stokes beam [29] or in situ [30], solve the stability issue. However, in situ LOs lack control on the amount of signal amplification. Very recently, based on a dual interferometry technique, recovery of the pure vibrational phase in CARS microscopy has been possible [31].

12.2.3.4 Phase-Matching-based Background-free Methods

As a nonlinear coherent process, the CARS signal builds up if the anti-Stokes waves generated at different locations in the sample are in phase (see Fig. 12.5a). Such phase-matching condition is easily fulfilled in the forward direction in a collinear configuration whereby the initially collinear pump, Stokes, and probe beams are tightly focused into the sample. The induced nonlinear polarization and the anti-Stokes wave thus travel in phase (see Fig. 12.5b, c). However, this property is not verified for the anti-Stokes waves generated into the backward ("Epi") direction as induced polarization and the anti-Stokes wave counter propagate. In this special case, Epi-CARS signal efficiently builds up only for objects with subwavelength extension along the optical axis [32, 33] (see Fig. 12.5b). Indeed, this is not the case of the bulk surrounding medium (often water in bioimaging) that builds up only in the forward direction. Epi detection is therefore free of the bulk non-resonant background. Strictly

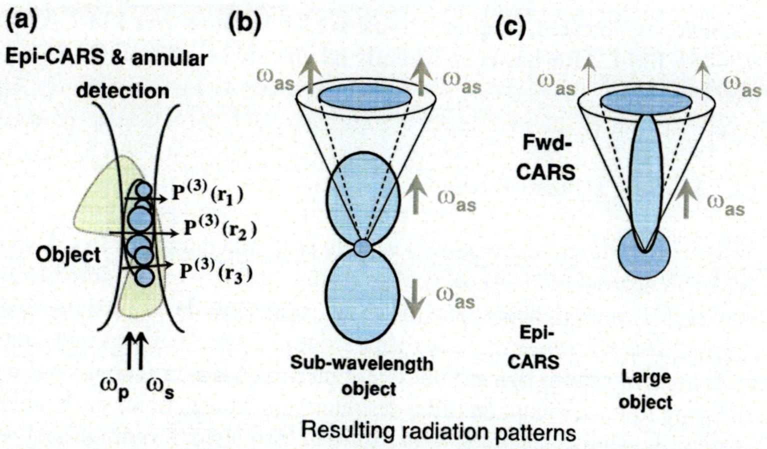

Fig. 12.5 Suppression of the bulk background by Epi-CARS and annular detection. **a** Coherent build up of the CARS signal in the excitation volume. **b** CARS far-field radiation pattern of a subwavelength object. A fraction of the CARS signal is detected in the epi direction or at large angles in the forward direction. An annular filter selects this large angular range. **c** CARS far-field radiation pattern of a large object. Neither Epi-CARS nor angularly large forward-CARS signal is emitted

speaking this Epi-detection scheme does not totally cancel out the non-resonant background as a portion of the intrinsic background of small objects is still detected. In addition, this detection mode is hindered by the partial back-reflection [34, 35] or the scattering [36] of forward-emitted signal.

Selective detection of subwavelength objects with non-resonant background attenuation is also possible with annular filtering of the forward-emitted CARS signal [37, 38]. This scheme takes advantage of the phasematching properties of the CARS process as well. Phase-matching condition makes the radiation pattern of large objects (compared to the emission wavelength) more directive than small ones' (for which the phase-matching condition is relaxed) [39, 40] (see Fig. 12.5b, c).

12.2.3.5 Frequency-Modulation Methods

Frequency-modulation (FM) spectroscopy has proved in the past as a powerful method to retrieve weak spectrally dependant (absorptive/dispersive) signals [41]. FM of the excitation leads to amplitude modulation of the signal. Locked-in detection of this amplitude modulation leads eventually to very sensitive measurements. Recent improvements in narrow-band light sources for CARS microscopy have led to dual pump sources that permit switching between two close pump wavelengths and subsequent amplitude modulation of the anti-Stokes signal. Due to its flat spectral dependence, purely non-resonant anti-Stokes signal is insensitive to such FM and is thus not detected. In contrast, the dispersive line-shape of CARS resonances is efficiently detected, leading to sensitive background-free FM-CARS microscopy [42–45]. FM-CARS has been recently adapted to broadband CARS microscopy as well [46–48].

12.2.3.6 Single-Pulse Methods

Using two or three laser beams is usually seen as a drawback in CARS microscopy. When broadband femtosecond pulses (ultra-short in the time domain) are used, they may contain pump, Stokes, and probe frequencies at the same time, enabling single-pulse CARS microscopy. In the time domain, these pulses excite impulsively the molecular vibrations that exhibit temporal period longer than the pulse duration. At the same time, resonant and non-resonant anti-Stokes fields are scattered from the same pulse. Due to the broadband nature of the pulses, the vibrational information is spectrally lost into the anti-Stokes field. A set of coherent control methods has been employed to recover the vibrational information. A first option is to selectively excite vibrational modes. This is accomplished by applying a sinusoidal spectral phase modulation to the pulse [49] or by creating two time-delayed replica of the original pulse [50, 51]. In the time domain, the pulse is split into a train of subpulses separated by a multiple of the temporal period of the targeted vibrational mode, efficiently driving this particular vibration. In addition to its selectivity, this method efficiently suppresses the non-resonant background. Narrowband probing of the exciting excited modes is the second method of choice. In this scheme, narrowband phase variations, by pure phase [52] or amplitude-phase gating [53], are introduced into the pulse. Characteristic interference patterns between the resonant anti-Stokes field and the strong non-resonant background thus arise in the scattered anti-Stokes spectrum, enabling vibrational identification. Subsequent refinements in this method have led to background-free CARS microscopy [54–56]. Combinations of single-pulse with interferometric approaches to increase vibrational sensitivity have been demonstrated as well [57–59].

12.2.3.7 Post-processing Methods

In the frame of multiplex CARS imaging, the vibrational information can be retrieved after data acquisition by post processing the CARS broadband spectra. Several approaches have been proposed so far, based on the maximum entropy method (MEM) [60, 61] or on Fourier-transform of spectra [62, 63].

12.2.3.8 Chemical Interface Detection

An original mode of contrast in CARS microscopy has recently emerged. Implemented with narrowband (picosecond) pulses, it highlights interfaces between media with different spectroscopic properties ("chemical" interfaces), i.e., a vibrationally

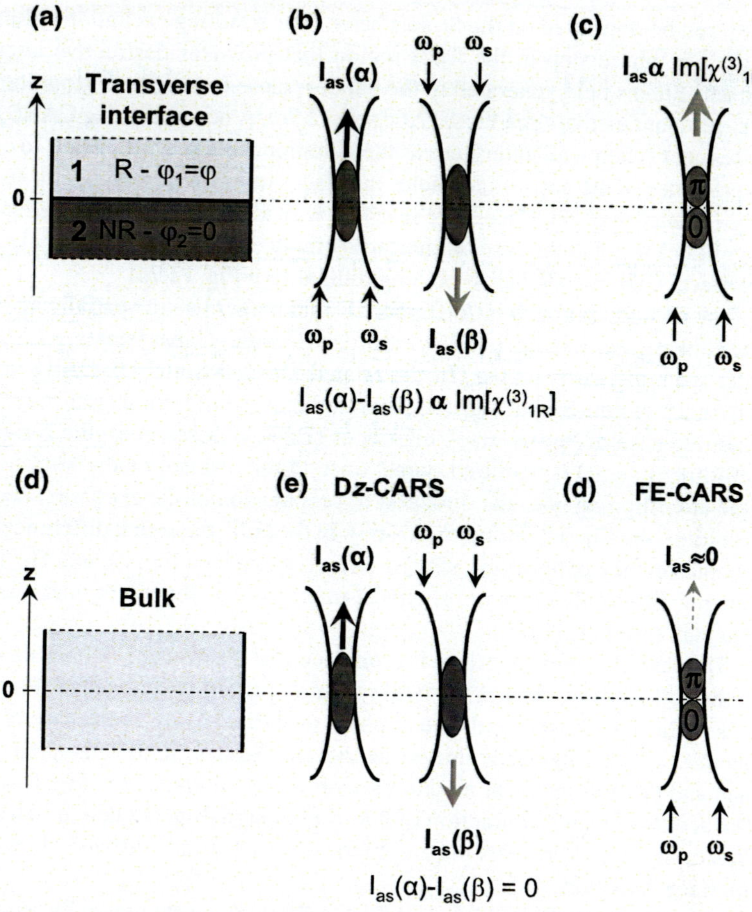

Fig. 12.6 Detection of transverse "chemical" interfaces. **a** Scheme of a transverse interface between a vibrationally resonant (*1*) and a pure non-resonant (*2*) medium. **b**, **e** Excitation and detection scheme with Dz-CARS: the excitation and detection are sequentially flipped compared to the interface. The two resulting CARS signals are subtracted to obtain background-free information. (**c**, **f**) Excitation and detection scheme with FE-CARS: a π phase step is axially introduced into the excitation volume. **d** Scheme of a bulk (vibrationally resonant or purely non-resonant) medium. For bulk media, the Dz-CARS and FE-CARS signals vanish

resonant (medium 1) and a pure non-resonant medium (medium 2). Two distinct modes of operation have been demonstrated so far. They rely on the fact that the anti-Stokes field emission is delayed in the resonant medium compared to the non-resonant one (i.e., as mentioned previously, there is a phase shift φ between these two fields).

The first and initial approach relies on "focus beam-engineering" (FE-CARS) whereby the Stokes laser beam is spatially phase shaped. High order Hermite-Gauss and Laguerre-Gauss modes [64, 65], as well as optical bottle beam (OBB) structure

[66], can be imprinted on the Stokes beam. The resulting excitation volume exhibits a sharp π phase step in the focal region that provokes destructive interference in the anti-Stokes field generated in the bulk (or close to interfaces separating two pure non-resonant media—see Figs. 12.6c and 12.7c). A non-vanishing CARS signal only arises near "chemical" interfaces as the vibrational phase shift φ leads to constructive interference in the anti-Stokes field. Interfaces oriented *perpendicular* to the optical axis ("transverse" interfaces) are revealed with a π-phase step oriented laterally (see Fig. 12.6c) while in the case of interfaces oriented *parallel* to the optical axis ("axial" interfaces), the π-step has to be longitudinal (see Fig. 12.7c).

The second approach is "differential" and makes use of spatially standard pump and Stokes laser beams, i.e., lowest order Hermite–Gauss beams, as found in most laser scanning microscopes. In the related Dz-CARS and Dk-CARS schemes, the difference of anti-Stokes signals generated in two different directions is operated.

In the case of "transverse" interfaces (Dz-CARS scheme), the forward-emitted anti-Stokes signal is acquired twice, once in the original configuration, the second time after flipping both the direction of propagation of the excitation beams and the detection (see Fig. 12.6b and e) [67–69]. In the bulk, the signal difference vanishes as the interference patterns are similar in both cases (see Fig. 12.6e). The interference patterns differ only close to "chemical interfaces" as the vibrational phase φ comes into play (see Fig. 12.6b).

The case of "axial" interface is reminiscent of the Young's double slit experiment. In the bulk, the anti-Stokes far-field radiation pattern is symmetric along the optical axis as the anti-Stokes fields are in phase (see Fig. 12.7e). In the case of "chemical" interfaces, it is angularly shifted as the anti-Stokes fields generated in the resonant and the non-resonant medium are no longer in phase (see Fig. 12.7b) [70, 71]. The acquisition and subtraction of anti-Stokes signals generated in two symmetrical directions $\mathbf{k}$ and $\mathbf{k}'$ reveals the amplitude of this angular shift and allows for "axial" interface detection.

The FE-CARS, Dz-CARS, and Dk-CARS schemes theoretically recover the pure Raman spectrum of the resonant medium that is part of the "chemical" interface as the anti-Stokes signals are proportional to $\mathrm{Im}\left[\chi^{(3)}_{1\mathrm{R}}\right]$, allowing background-free CARS microscopy. In addition, these signals are heterodyned by the non-resonant CARS signal coming from the non-resonant medium $\chi^{(3)}_{2\mathrm{NR}}$, boosting the image contrast. Finally, the linear dependence of the recovered signal over the number of scatterers gives access to quantitative analysis. These various properties have been recently demonstrated experimentally [67, 69, 70, 72, 73]. Latest development combine the detection of "transverse" and "axial" interfaces [74].

All these interface-sensitive schemes are convenient because they can be easily implemented in a commercial CARS microscope by inserting spatial light modulators (SLM) or slightly modifying the detection scheme. Nonetheless, it should be mentioned that they are sensitive to linear index mismatches. As such, they are relevant for imaging small organelles in cells for instance. As an example, Fig. 12.8 presents CARS images of living cells obtained with the conventional CARS scheme and the Dk-CARS scheme. In the case of conventional CARS detection the resonant image (Fig. 12.8b) is poorly contrasted because of the non-resonant background coming mainly from the water. The D*k*-CARS scheme brings a better contrast. Off-resonance (Fig. 12.8c) the contrast comes from the refraction index mismatches and has a pure refractive origin. In the case of the resonant image (Fig. 12.8d), the contrast is stronger as coming from the targeted CH_2 molecular bonds and is free of non-resonant background. The remaining refraction index contribution in Fig. 12.8d is weak as compared to the vibrational contribution [70].

12.2.4 Stimulated Raman Scattering

Although SRS has received much attention in its early years [75–79] since its discovery in 1962 [80], SRS microscopy is a relatively new technique that took benefit from the recent advances in the field of femtosecond-stimulated Raman spectroscopy (FSRS). Improvements in FSRS encompass sensitivity as well as spectral and temporal resolution [81]. FSRS increases the signal and rejects the fluorescent background as compared to spontaneous Raman. Quite recently, the first implementation of FSRS in microscopy by Ploetz and co-workers [3] has stimulated many works in the CARS community, bringing what is referred to as SRS microscopy to a mature level. Altogether, CARS and SRS are part of the CRS techniques and one refers to their implementation in microscopy as "CRS microscopy".

12.2.4.1 Principles

SRS is a nonlinear process similar to CARS where two degenerated beams, namely pump and Stokes, interact through a vibrational molecular level. Figure 12.9 presents SRS in the energy picture. In this process, the pump beam is depleted and is known as stimulated Raman loss (SRL), whereas the Stokes beam is amplified, the process is known as stimulated Raman gain (SRG).

In this wave mixing process the energy difference between the incoming pump and Stokes beams matches the molecular vibrational energy level; in this respect it is similar to CARS. Contrary to CARS, SRL and SRG processes are wave mixing routes that generate an induced polarization at the frequency of the pump and Stokes beam, respectively.

These induced polarizations can be written as

$$\vec{P}^{(3)\mathrm{SRL}}\left(\vec{r},-\omega_\mathrm{p}\right)=\chi^{(3)}\left(\vec{r},-\omega_\mathrm{p};\omega_\mathrm{p},-\omega_\mathrm{s},\omega_\mathrm{s}\right)\vec{E}_\mathrm{p}\left(\vec{r},\omega_\mathrm{p}\right):\vec{E}_\mathrm{s}^{*}\left(\vec{r},\omega_\mathrm{s}\right):\vec{E}_\mathrm{s}\left(\vec{r},\omega_\mathrm{s}\right)$$
$$\vec{P}^{(3)\mathrm{SRG}}\left(\vec{r},-\omega_\mathrm{s}\right)=\chi^{(3)}\left(\vec{r},-\omega_\mathrm{s};\omega_\mathrm{s},-\omega_\mathrm{p},\omega_\mathrm{p}\right)\vec{E}_\mathrm{s}\left(\vec{r},\omega_\mathrm{s}\right):\vec{E}_\mathrm{p}^{*}\left(\vec{r},\omega_\mathrm{p}\right):\vec{E}_\mathrm{p}\left(\vec{r},\omega_\mathrm{p}\right),$$

where, as previously, ω_p and ω_s are the angular frequencies of the incoming pump and Stokes beams, respectively. Because SRL and SRG mixing routes (Fig. 12.9) are resonant with the vibrational level, the fields $\vec{E}^{\mathrm{SRL}}\left(\vec{r},\omega_\mathrm{p}\right)$, and $\vec{E}^{\mathrm{SRG}}\left(\vec{r},\omega_\mathrm{s}\right)$ emitted

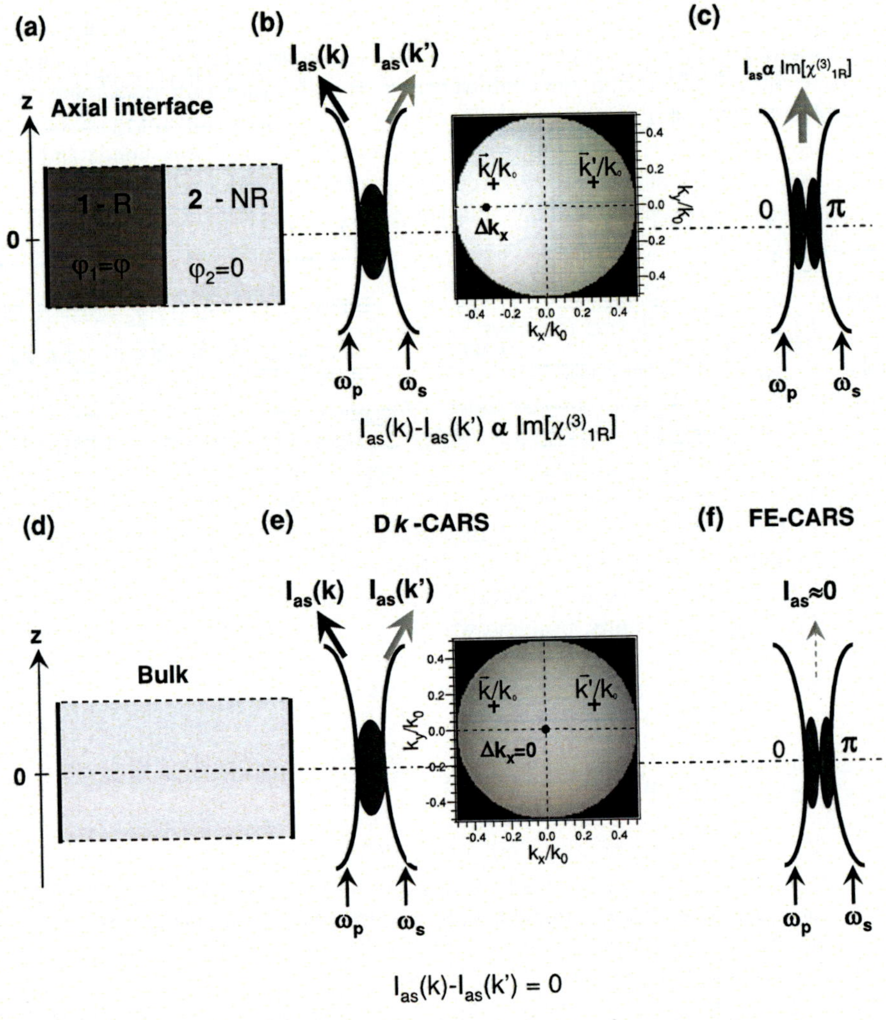

Fig. 12.7 Detection of axial "chemical" interfaces. **a** Scheme of an axial interface between a vibrationally resonant (*1*) and a pure non-resonant (*2*) medium. **b, e** Excitation and detection scheme with D*k*-CARS: CARS signals generated in two symmetrical directions **k** and **k′** are detected and then subtracted. Insets: CARS far-field radiation patterns. For a "chemical interface", it is angularly shifted whereas for a bulk medium, it is symmetrical compared to the optical axis. **c, f** Excitation and detection scheme with FE-CARS: a π-step is laterally introduced into the excitation volume. **d** Scheme of a bulk (vibrationally resonant or purely non-resonant) medium. For bulk media, the D*k*-CARS and FE-CARS signals vanish

by the induced polarizations experience a phase shift similar to the one that has been previously described and illustrated in Fig. 12.2. These induced fields $\vec{E}^{\mathrm{SRL}}(\vec{r}, \omega_{\mathrm{p}})$, and $\vec{E}^{\mathrm{SRG}}(\vec{r}, \omega_{\mathrm{s}})$ interfere with the incoming pump and Stokes fields, respectively.

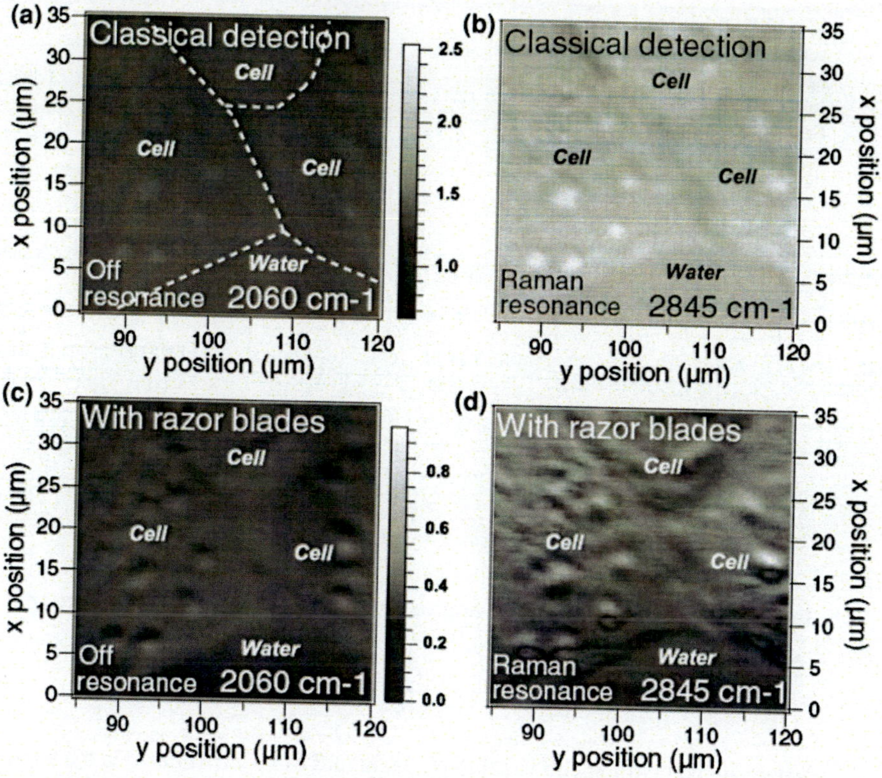

Fig. 12.8 Living COS7 cells as imaged off-resonance (**a–c**) and at the $2850\,\mathrm{cm}^{-1}$ CH_2 aliphatic bond resonance (**b–d**), classical CARS detection (**a–b**) and D*k*-CARS detection (**c–d**). The off-resonance D*k*-CARS image shows the refraction index sample variation whereas the D*k*-CARS resonant image exhibits the chemical specificity. Ref. [70]

Assuming a plane wave description of the incoming and the emitted fields, the total emitted intensities write

$$I_{\mathrm{p}}(\omega_{\mathrm{p}}) = \frac{1}{2}\varepsilon_0 n(\omega_{\mathrm{p}})c\left|E_{\mathrm{p}}(\omega_{\mathrm{p}}) + E^{SRL}(\omega_{\mathrm{p}})\right|^2$$
$$I_{\mathrm{s}}(\omega_{\mathrm{s}}) = \frac{1}{2}\varepsilon_0 n(\omega_{\mathrm{s}})c\left|E_{\mathrm{s}}(\omega_{\mathrm{s}}) + E^{SRG}(\omega_{\mathrm{s}})\right|^2,$$

where the incoming pump and Stokes beams act as LOs that homodyne with the weak generated SRL and SRG fields.

In the case of a homogeneous isotropic medium (with a 1/3 Raman depolarization ratio), the SRS susceptibility $\chi^{(3)}$ has a single non-zero value and the induced polarization can be written as

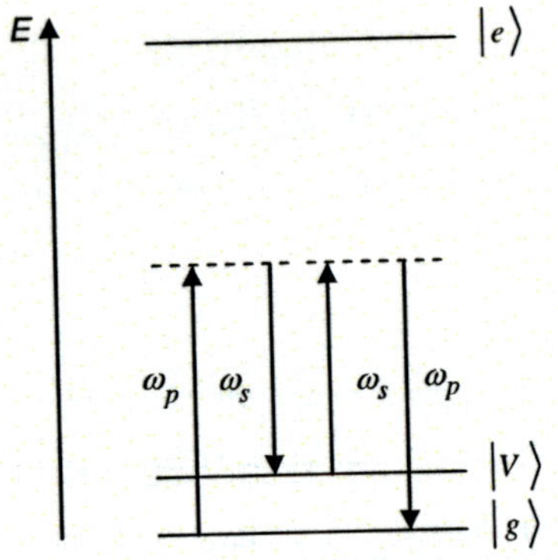

Fig. 12.9 Stimulated Raman scattering. The pump beam is depleted (stimulated Raman loss–SRL) and the Stokes beam is amplified (stimulated Raman gain–SRG)

$$P^{(3)\mathrm{SRL}}(\omega_\mathrm{p}) = \frac{3}{2}\varepsilon_0\chi^{(3)}(\omega_\mathrm{p})\,|E_\mathrm{s}|^2\,E_\mathrm{p}$$
$$P^{(3)\mathrm{SRG}}(\omega_\mathrm{s}) = \frac{3}{2}\varepsilon_0\chi^{(3)}(\omega_\mathrm{s})\,\left|E_\mathrm{p}\right|^2\,E_\mathrm{s}.$$

In the slowly varying envelope approximation the wave equation writes [82]

$$\frac{\partial}{\partial z}E(\omega) = \frac{i\omega^2\mu_0}{2k(\omega)}P^{(3)}(\omega)\exp(i\Delta kz),$$

where μ_0 is the magnetic vacuum constant and $k(\omega) = n(\omega)\omega/c$ is the wave vector. The wave vector mismatch between the pump and the Stokes field $\Delta k = k_\mathrm{p} - k_\mathrm{p} + k_\mathrm{s} - k_\mathrm{s} = 0$ vanishes identically. This is a singular feature of SRS that the relation between the phases of the induced polarization and the generated fields depends only on the argument of the complex third-order susceptibility $\chi^{(3)}$, and not on Δk.

After propagating over a distance L in the active medium, and in the limit of weak depletion, the pump and Stokes fields write [82]

$$E^{\mathrm{SRG}}(\omega_\mathrm{S}, L) = E^{\mathrm{SRG}}(\omega_\mathrm{S}, 0)\exp\left[\frac{i3\omega_\mathrm{s}\chi^{(3)}(\omega_\mathrm{s})}{2\varepsilon_0c^2n(\omega_\mathrm{s})n(\omega_\mathrm{p})}I_\mathrm{p}(\omega_\mathrm{p}, 0)\right]$$
$$E^{\mathrm{SRL}}(\omega_\mathrm{p}, L) = E^{\mathrm{SRL}}(\omega_\mathrm{p}, 0)\exp\left[\frac{i3\omega_\mathrm{p}\chi^{(3)}(\omega_\mathrm{p})}{2\varepsilon_0c^2n(\omega_\mathrm{p})n(\omega_\mathrm{s})}I_\mathrm{s}(\omega_\mathrm{s}, 0)\right].$$

Examination of the above-written equations shows that this is the imaginary part of $\chi^{(3)}$ that dictates if the fields are amplified or depleted that dictates the field amplification or depletion. Writing $\chi^{(3)}(\omega) = \mathrm{Re}\left[\chi^{(3)}\right] + i\mathrm{Im}\left[\chi^{(3)}\right]$ it can be shown from classical and quantum calculations [83] that $\mathrm{Im}\left[\chi^{(3)}(\omega_\mathrm{s})\right]$ is negative whereas $\mathrm{Im}\left[\chi^{(3)}(\omega_\mathrm{p})\right]$ is positive. This leads to the amplification of the Stokes beam and the depletion of the pump beam, justifying the SRG and SRL terminology. Very importantly the amplification (SRG) and depletion (SRL) depend explicitly on the

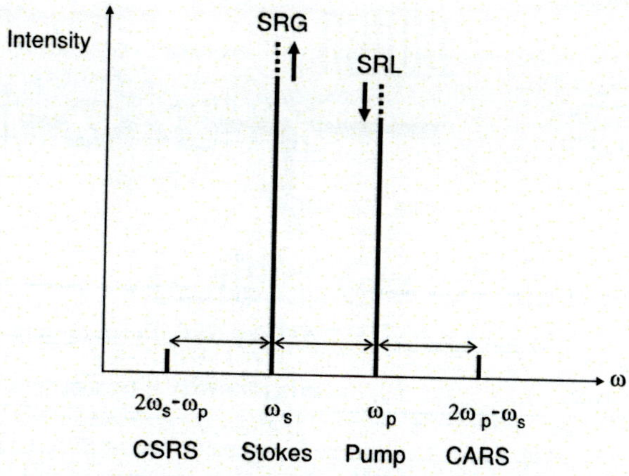

Fig. 12.10 SRS, CARS and CSRS in the spectral domain: the stokes beam is amplified (SRG), the pump beam is depleted (SRL) and two new frequencies appear (CARS and CSRS)

imaginary part of $\chi^{(3)}$ that is known to be equal to the pure Raman spectrum. Contrary to CARS, SRG and SRL are background-free and spectrally follow the molecular signature. Such a property has important consequences for application in bio-imaging.

The real part $\text{Re}\left[\chi^{(3)}\right]$ is known as 'cross-phase modulation'. This is a refraction index variation induced by the beam interaction and is the source of artifacts in SRS microscopy such as wavefront distortion, focusing, or defocusing. Figure 12.10 presents in the spectral domain the intensity variation of the incoming beams and the generated CARS and coherent Stokes Raman scattering (CSRS) signals.

12.2.4.2 Practical Implementation of SRS Microscopy

SRS microscopy has been first implemented in the femtosecond regime using an amplified laser system and a combination of broad- and narrow-band pulses [3]. In this case, the pulse peak pulse energy is of the order of a few μJ and the repetition rate 1 kHz. Because application in biology requires low pulse energy power (~nJ) and higher repetition rate (MHz), SRS microscopy has been further developed using picosecond pulse trains at 80 MHz [84, 85]. The retained scheme in these implementations detects the faint power loss in the pump beam (SRL) when the incoming Stokes beam is modulated. In practice the pump power loss is of the order of $10^{-5} - 10^{-6}$ and requires a fast Stokes intensity modulation frequency (~MHz) to get rid of low frequency electronic and laser noise. SRL is experimentally easier to detect than SRG when using near-IR beams because detectors are usually more efficient at lower wavelengths. Figure 12.10 presents the SRS scheme with intensity modulation. A faster modulation scheme using active sub harmonic synchronization has been reported to reach shot noise limited sensitivity in SRS [86].

Although picosecond pulses permit to achieve spectral selectivity of the order of a few cm^{-1}, SRS signal can be improved by one order of magnitude if femtosecond pulses are used [87] with a loss in spectral selectivity that is not prejudicial when

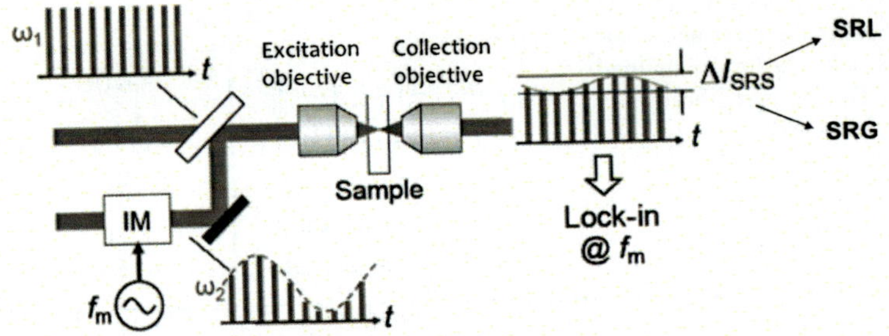

$$\Delta I_{SRS} \Rightarrow SRL \text{ if } \omega_1 = \omega_p \text{ and } \omega_2 = \omega_s$$

$$\Delta I_{SRS} \Rightarrow SRG \text{ if } \omega_1 = \omega_s \text{ and } \omega_2 = \omega_p$$

Fig. 12.11 Experimental implementation of SRS. One of the two incoming beams (*pump* and *Stokes*) is intensity modulated at high frequency ($f_m \sim 1$ MHz) while the weak modulation is detected on the other beam after crossing the sample thanks to a lock-in detection

imaging lipids in the broad spectral range of CH bond vibrations. Spectral selectivity can also be achieved using chirped femtosecond pulses as originally proposed in a CARS framework [88]; this approach is usually named 'spectral focusing'. Following this scheme, spectrally focused SRS has been recently demonstrated using negatively chirped femtosecond pulses coming from a femtosecond oscillator and a femtosecond soliton generated in a photonic crystal fiber [89].

12.2.4.3 Non-resonant Background in SRS and CARS Spectra

Figure 12.11 shows the CARS and SRS spectra of polystyrene beads around the C=C resonance at 1,000 cm^{-1}. Clearly the CARS spectrum is distorted due to the non-resonant background (see Fig. 12.2). Indeed, far away below or above the molecular resonance the CARS signal is nonzero and equals the non-resonant contribution. This is not the case of the SRS spectrum that nicely follows the Raman spectrum and shows no significant offset or distortion.

In practice, and in the case of weak vibrational signal, a non-resonant contribution can show up in SRS. This is mainly due to the cross-phase modulation between the pump and Stokes beams. In this Kerr-induced refraction effect, the interaction between the pump and Stokes beams affects the detected (pump or Stokes). For example, in SRL, the pump beam experiences a wavefront distortion that may change its divergence. If the sensitive area on the detector is not large enough, or if some apertures are placed on the pump beam path, this cross-phase-induced divergence can be misinterpreted as a loss in the pump signal and appears as a non-resonant SRS background. Other sources of background such as parasitic two-photon-induced fluorescence can affect the SRS spectra in a way that is similar to the baseline found in spontaneous Raman spectra.

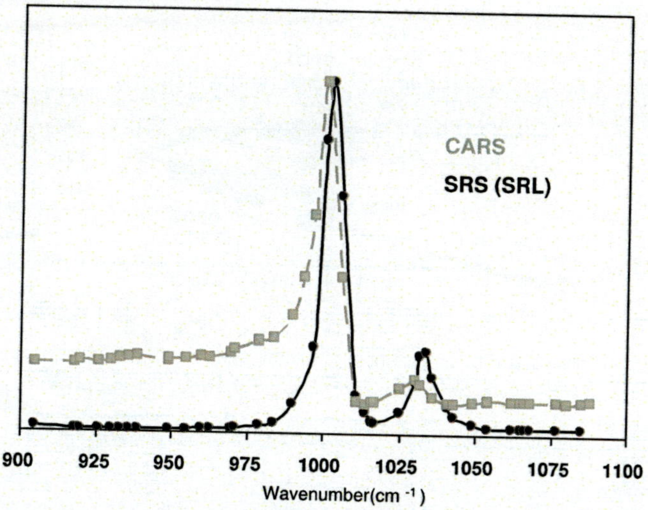

Fig. 12.12 CARS and SRS spectra of polystyrene beads showing how the non-resonant background present in CARS is removed in SRS

12.2.4.4 Hyperspectral CRS Imaging: An Example of Practical Implementation

To achieve broadband hyperspectral CRS imaging [90], we use the two signal beams that are generated within two optical parametric oscillators (OPO–Emerald, APE, Berlin, Germany) pumped by a picosecond mode-locked, frequency-doubled, Nd:YVO laser (HighQ laser) operating at 532 nm. Figure 12.12 presents the scheme of the experiment where OPO1 delivers the pump beam and is fixed at 735 nm while OPO2 delivers the tunable Stokes beam, with a wavelength that continuously varies between 765 and 990 nm, to address the 533–3500 cm^{-1} vibrational spectral range. The present pump and Stokes wavelength configuration has the advantage of setting the CARS anti-Stokes wavelength below 700 nm, a spectral range wherein conventional photomultiplier tubes are very sensitive. For the detection we use low-pass filters that enable the recording of the anti-Stokes line while scanning the tunable OPO2.

For SRS spectroscopy we monitor the loss at the pump wavelength (SRL at OPO1) with an analog photodiode (Thorlabs DET-100A) while the Stokes beam (OPO2) is modulated with an acousto-optical modulator (AOM: AA Optoelectronic: MT200-AO.2-800). The tiny pump depletion signal is retrieved with a fast lock-in amplifier (Signal Recovery 7280).

To perform wavelength scanning, we use a custom-made software that fully controls the OPO2 wavelength (for details see [90]). It takes typically 60 min to obtain a hyperspectral CARS image on the 900–3200 cm^{-1} range with 10 cm^{-1} resolution.

For CARS imaging and SRS spectroscopy, the following parameters are used:

Objective lens: Olympus LUCPFLN 40×; NA 0.6 (air).

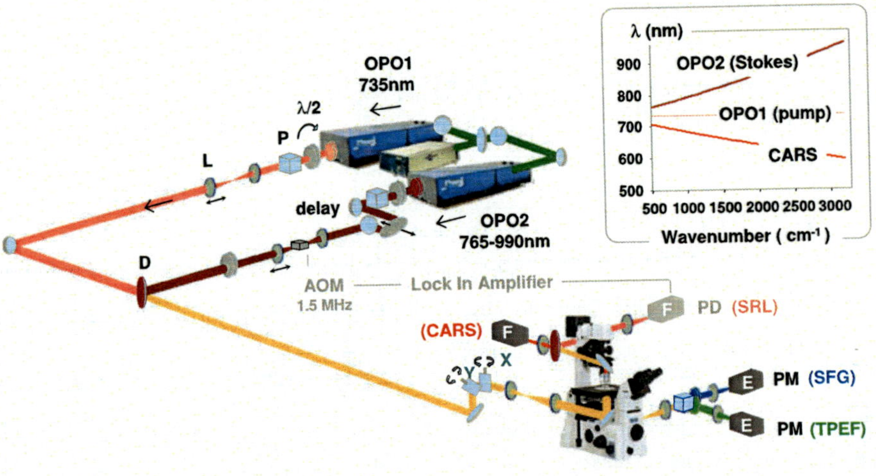

Fig. 12.13 CRS hyperspectral imaging setup with two OPOs, F and E stand for Forward and Epi (backward) detection, PM for photomultiplier tubes, P for polarizer, L for NIR achromatic doublets, D for dichroic filter, PD for photodiode, SRL for Stimulated Raman Loss, SFG for sum frequency generation and TPEF for two-photon excited fluorescence. Inset: OPO1 (*pump*), OPO2 (*tunable Stokes*) and anti-Stokes CARS wavelengths associated to Raman shifts

CARS: Dwell time: 50 μs/pixel; Number of pixel: 250 × 250 (∼3 s/image); 5 accumulations per spectral point; $P_\mathrm{p} = P_\mathrm{s} \sim 20\,\mathrm{mW}$.

SRS: AOM modulation frequency: 1.5 MHz; Integration time: 100 ms/pixel in the lipid spectral range; 10 s/pixel in the amide spectral range, $P_\mathrm{p} = P_\mathrm{s} \sim 20\,\mathrm{mW}$.

While performing CARS or SRS, the TPEF, and the sum frequency generation (SFG) between the pump and Stokes beams are detected in the Epi (backward) direction with photomultiplier tubes. For SRS spectroscopy, the pump beam modulation is detected in the forward direction with a photodiode.

Figure 12.13 shows an example of hyperspectral CRS imaging on a 10 μm thick cryosection of mouse skin tissue. Figure 12.13a shows the immunostained section while the CRS hyperspectral imaging is performed on an adjacent unstained cryosection. Figure 12.13b presents the CARS image at $2840\,\mathrm{cm}^{-1}$, a band that exhibits a strong lipid-related CRS signal. SRS image is also obtained simultaneously with the CARS image such as, for each pixel, CARS and SRS spectra are available. This is illustrated in Fig. 12.13c, d in the lipid and fingerprint range, respectively. Figure 12.13c concentrates on the lipid-rich skin border (stratum corneum) to show how SRS and CARS spectra differ, here again because of the non-resonant background contribution to the CARS signal. Figure 12.13d, obtained deeper in the tissue, exemplifies the ability of SRS to access relevant vibrational molecular signal in the spectrally congested region of the fingerprint. In this spectral range the Raman cross-section is weak and the CARS signal is completely overwhelmed by the non-resonant background precluding any vibrational bond identification.

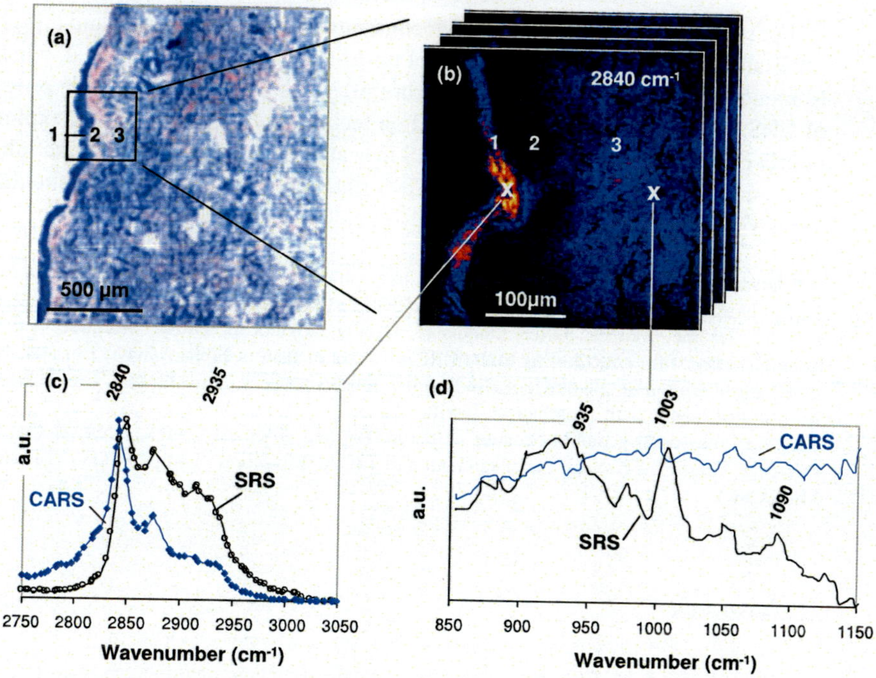

Fig. 12.14 Hyperspectral CRS image of mouse skin tissue. **a** Immunostained cryosection, **b** Hyperspectral CRS imaging showing here the CARS image at 2840 cm^{-1} (lipid), **c** CARS and SRS spectra on the lipid rich skin border (stratum corneum), and **d** in the fingerprint region deeper in the tissue

Nevertheless, and as stressed before, the SRS spectrum exhibits a base line that comes from cross-phase modulation and TPEF.

12.3 Conclusion

We have presented here some recent advances in the field of high speed, high spatial resolution Raman imaging as made possible by stimulated Raman techniques. Both CARS and SRS have been considered with the aim of reaching molecular information that is similar to the one obtained in spontaneous Raman, but fully compatible with scanning imaging modalities. CARS requires a special attention because it is inherently contaminated by a non-resonant electronic background. We have focused here our attention on CARS methodologies that permit the non-resonant background suppression and retrieve the pure Raman information. SRS implemented in microscopy is not subjected to this spurious non-resonant background and gives access to the awaited Raman information. We have described here its principle, its implementation, and gave some results on mouse skin imaging.

For sure, these stimulated Raman techniques are evolving rapidly and this chapter is certainly a snapshot that might be not up-to-date quite rapidly. Indeed, stimulated Raman principles have been known from 50 years and the spectacular progresses of CRS microscopy have mostly relied on technology. Nevertheless, unexplored or poorly explored areas exist, such as CRS in scattering media, CRS polarization, CRS photo-acoustics, CRS endoscopy, or CRS plasmonics. These gardens should keep the physicists at work for many years.

Acknowledgments The authors are grateful to Sophie Brustlein, Pascal Berto, Fabiana Munhoz, Nadia Djaker, Franck Billard, Nicolas Sandeau, Alicja Gąsecka, Sophie Brasselet, Patrick Ferrand, Jerôme Wenger, Julien Duboisset, Cyrille Billaudeau, Didier Marguet, Anne-Marie Schmitt-Verhulst for contributing to the CRS advances at Institut Fresnel (www.fresnel.fr/mosaic) and Centre d'Immunologie de Marseille-Luminy (http://www.ciml.univ-mrs.fr). They acknowledge fruitful discussions with Jonathan M. Levitt.
This work has been possible because of public funding from the CNRS, ANR, INCA, Region Provence Alpes Cote d'Azur, European Union FP7 (CARSExplorer), and EU-COST (MicroCARS, MicroCoR).

References

1. W.R. Zipfel, R.M. Williams, W.W. Webb, Nonlinear magic: multiphoton microscopy in the biosciences. Nat. Biotechnol. **21**, 1369–1377 (2003)
2. A. Zumbusch, G.R. Holtom, X.S. Xie, Vibrationnal microscopy using coherent anti-Stokes Raman scattering. Phys. Rev. Lett. **82**, 4014–4017 (1999)
3. E. Ploetz, S. Laimgruber, S. Berner, W. Zinth, P. Gilch, Femtosecond stimulated Raman microscopy. Appl. Phys. B **87**, 389–393 (2007)
4. B.G. Saar, C.W. Freudiger, J. Reichman, C.M. Stanley, G.R. Holtom, X.S. Xie, Video-rate molecular imaging in vivo with stimulated Raman scattering. Science **330**, 1368–1370 (2010)
5. Y.R. Shen, *The Principles of Non Linear Optics* (Wiley, New York, 1984)
6. H. Lotem, R.T. Lynch Jr, N. Bloembergen, Interference between Raman resonances in four-wave difference mixing. Phys. Rev. A **14**, 1748–1755 (1976)
7. M.D. Levenson, N. Bloembergen, Dispersion of the nonlinear optical susceptibility tensor in centrosymmetric media. Phys. Rev. A **14**, 4447–4463 (1974)
8. S.A.J. Druet, B. Attal, T.K. Gustafson, J.P. Taran, Electronic resonance enhancement of coherent anti-Stokes Raman scattering. Phys. Rev. A **18**, 1529–1557 (1978)
9. J.W. Fleming, C.S. Johnson Jr, A practical analysis for coherent anti-Stokes Raman scattering (CARS) spectra. J. Raman Spectr. **8**, 284–290 (1979)
10. U. Fano, Effects of configuration interaction on intensities and phase shifts. Phys. Rev. **124**, 1866–1878 (1961)
11. J.X. Cheng, Y.K. Jia, G. Zheng, X.S. Xie, Laser-scanning coherent anti-Stokes Raman scattering microscopy and applications to cell biology. Biophys. J. **83**, 502–509 (2002)
12. R.W. Hellwarth, Third-order optical suceptibilities of solids and liquids. Progr. Quant. Electr. **5**, 1–68 (1977)
13. J.J. Song, G.L. Easley, M.D. Levenson, Background suppression in coherent Raman spectroscopy. Appl. Phys. Lett. **29**, 567–569 (1976)
14. S.A. Akhmanov, A.F. Bunkin, S.G. Ivanov, N.I. Koroteev, D. Akimov, Polarization active Raman spectroscopy and coherent Raman ellipsometry. Sov. Phys. JETP **47**, 667–677 (1978)
15. J.L. Oudar, R.W. Smith, Y.R. Shen, Polarization—sensitive coherent anti-Stokes Raman spectroscopy. Appl. Phys. Lett. **34**, 758–760 (1979)

16. J.-X. Cheng, L.D. Book, X.S. Xie, Polarization coherent anti-Stokes Raman scattering microscopy. Opt. Lett. **26**, 1341–1343 (2001)
17. F.M. Kamga, M.G. Sceats, Pulse-sequenced coherent anti-Stokes Raman scattering spectroscopy: a method for suppression of the nonresonant background. Opt. Lett. **5**, 126–128 (1980)
18. A. Volkmer, L.D. Book, X.S. Xie, Time-resolved coherent anti-Stokes Raman scattering microscopy: imaging based on Raman free induction decay. Appl. Phys. Lett. **80**, 1505–1507 (2002)
19. Y.J. Lee, M.T. Cicerone, Vibrational dephasing time imaging by time-resolved broadband coherent anti-Stokes Raman scattering microscopy. Appl. Phys. Lett. **92**, 041108 (2008)
20. Y. Yacoby, R. Fitzgibbon, B. Lax, Coherent cancellation of background in four-wave mixing spectroscopy. J. Appl. Phys. **51**(6), 3072–3077 (1980)
21. G. Marowsky, G. Lüpke, CARS-background suppression by phase-controlled nonlinear interferometry. Appl. Phys. B **51**, 49–51 (1990)
22. G.L. Eesley, M.D. Levenson, W.M. Tolles, Optically heterodyned coherent Raman spectroscopy. J. Quant. Electr. **QE-14**(1), 45–49 (1978)
23. D.L. Marks, C. Vinegoni, J.S. Bredfeldt, S.A. Boppart, Interferometric differentiation between resonant coherent anti-Stokes Raman scattering and nonresonant four-wave-mixing processes. Appl. Phys. Lett. **85**, 5787–5789 (2004)
24. C. Vinegoni, J.S. Bredfeldt, D.L. Marks, S.A. Boppart, Nonlinear optical contrast enhancement for optical coherence tomography. Opt. Expr. **12**, 331–341 (2004)
25. C.L. Evans, E.O. Potma, X.S. Xie, Coherent anti-Stokes Raman scattering spectral interferometry: determination of the real and imaginary components of nonlinear susceptibility chi(3) for vibrational microscopy. Opt. Lett. **29**, 2923–2925 (2004)
26. E.O. Potma, C.L. Evans, X.S. Xie, Heterodyne coherent anti-Stokes Raman scattering (CARS) imaging. Opt. Lett. **31**, 241–243 (2006)
27. M. Greve, B. Bodermann, H.R.B. Telle, P. Riedle, E. High-contrast chemical imaging with gated heterodyne coherent anti-Stokes Raman scattering microscopy. Appl. Phys. B **81**, 875–879 (2005)
28. M. Jurna, J.P. Korterik, C. Otto, H.L. Offerhaus, Shot noise limited heterodyne detection of CARS signals. Opt. Expr. **15**(23), 15207–15213 (2007)
29. E.R. Andresen, S.R. Keiding, E.O. Potma, Picosecond anti-Stokes generation in a photonic-crystal fiber for interferometric CARS microscopy. Opt. Expr. **14**, 7246–7251 (2006)
30. F. Lu, W. Zheng, Z. Huang, Heterodyne polarization coherent anti-Stokes Raman scattering microscopy. Appl. Phys. Lett. **92**, 123901 (2008)
31. M. Jurna, J.P. Korterik, C. Otto, J.L. Herek, H.L. Offerhaus, Vibrational phase contrast microscopy by use of coherent anti-Stokes Raman scattering. Phys. Rev. Lett. **103**, 043905 (2009)
32. A. Volkmer, J.-X. Cheng, X.S. Xie, Vibrational imaging with high sensitivity via epidetected coherent anti-Stokes Raman scattering microscopy. Phys. Rev. Lett. **87**, 23901 (2001)
33. J.-X. Cheng, A. Volkmer, L.D. Book, X.S. Xie, An epi-detected coherent anti-Stokes Raman scattering (E-CARS) microscope with high spectral resolution and high sensitivity. J. Phys. Chem. B **105**, 1277–1280 (2001)
34. E.O. Potma, X.S. Xie, Detection of single lipid bilayers with coherent anti-Stokes Raman scattering (CARS) microscopy. J. Raman Spectr. **34**, 642–650 (2003)
35. N. Djaker, D. Gachet, N. Sandeau, P.F. Lenne, H. Rigneault, Refractive effects in coherent anti-Stokes Raman scattering microscopy. Appl. Opt. **45**, 7005–7011 (2006)
36. C.L. Evans, E.O. Potma, M. Puoris'haag, D. Cote, C.P. Lin, X.S. Xie, Chemical imaging of tissue in vivo with video-rate coherent anti-Stokes Raman scattering microscopy. Proc. Nat. Acad. Sci. **102**, 16807–16812 (2005)
37. J. Lin, F. Lu, H.F. Wang, W. Zheng, C.J.R. Sheppard, Z. Huang, J. Lin, Improved contrast radially polarized coherent anti-Stokes Raman scattering microscopy using annular aperture detection. Appl. Phys. Lett. **95**, 133703 (2009)

38. J. Lin, F. Lu, W. Zheng, Z. Huang, Annular aperture-detected coherent anti-Stokes Raman scattering microscopy for high contrast vibrational imaging. Appl. Phys. Lett. **97**, 083701 (2010)
39. J.-X. Cheng, A. Volkmer, X.S. Xie, Theoretical and experimental characterization of coherent anti-Stokes Raman scattering microscopy. J. Opt. Soc. Am. B **19**, 1363–1375 (2002)
40. D. Gachet, N. Sandeau, H. Rigneault, Far-field radiation pattern in coherent anti-Stokes Raman scattering (CARS) microscopy. Proc. SPIE **6093**, 62–68 (2005)
41. G.C. Bjorklund, Frequency-modulation spectroscopy: a new method for measuring weak absorptions and dispersions. Opt. Lett. **5**, 15–17 (1980)
42. F. Ganikhanov, S. Carrasco, X.S. Xie, M. Katz, W. Seitz, D. Kopf, Broadly tunable dual-wavelength light source for coherent anti-Stokes Raman scattering microscopy. Opt. Lett. **31**, 1292–1294 (2006)
43. F. Ganikhanov, C.L. Evans, B.G. Saar, X.S. Xie, High-sensitivity vibrational imaging with frequency modulation coherent anti-Stokes Raman scattering (FM CARS) microscopy. Opt. Lett. **31**, 1872–1874 (2006)
44. O. Burkacky, A. Zumbusch, C. Brackmann, A. Enejder, Dual-pump coherent anti-Stokes-Raman scattering microscopy. Opt. Lett. **31**, 3656–3658 (2006)
45. B.G. Saar, G.R. Holtom, C.W. Freudiger, C. Ackermann, W. Hill, X.S. Xie, Intracavity wavelength modulation of an optical parametric oscillator for coherent Raman microscopy. Opt. Expr. **17**, 12532–12539 (2009)
46. I. Rocha-Mendoza, W. Langbein, P. Watson, P. Borri, Differential coherent anti-Stokes Raman scattering microscopy with linearly chirped femtosecond laser pulses. Opt. Lett. **34**, 2258–2260 (2009)
47. W. Langbein, I. Rocha-Mendoza, P. Borri, Single source coherent anti-Stokes Raman microspectroscopy using spectral focusing. Appl. Phys. Lett. **95**, 081109 (2009)
48. B.-C. Chen, J. Sung, S.-H. Lim, Chemical imaging with frequency modulation coherent anti-Stokes Raman scattering microscopy at the vibrational fingerprint region. J. Phys. Chem. B **114**, 16871–16880 (2010)
49. N. Dudovich, D. Oron, Y. Silberberg, Single pulse coherently controlled nonlinear Raman spectroscopy and microscopy. Nature **418**, 512–514 (2002)
50. J.P. Ogilvie, E. Beaurepaire, A. Alexandrou, M. Joffre, Fourier-transform coherent anti-Stokes Raman scattering microscopy. Opt. Lett. **31**, 480–482 (2006)
51. K. Isobe, A. Suda, M. Tanaka, H. Hashimoto, F. Kannari, H. Kawano, H. Mizuno, A. Miyawaki, K. Midorikawa, Single-pulse coherent anti-Stokes Raman scattering microscopy employing an octave spanning pulse. Opt. Expr. **17**, 11259–11266 (2009)
52. D. Oron, N. Dudovich, Y. Silberberg, Single-pulse phase-contrast nonlinear Raman spectroscopy. Phys. Rev. Lett. **89**(27), 273001 (2002)
53. O. Katz, J.M. Levitt, E. Grinvald, Y. Silberberg, Single-beam coherent Raman spectroscopy and microscopy via spectral notch shaping. Opt. Expr. **18**, 22693–22701 (2010)
54. D. Oron, N. Dudovich, Y. Silberberg, Femtosecond phase-and-polarization control for background-free coherent anti-Stokes Raman spectroscopy. Phys. Rev. Lett. **90**(21), 213902 (2003)
55. S.-H. Lim, A.G. Caster, S.R. Leone, Single-pulse phase-control interferometric coherent anti-Stokes Raman scattering spectroscopy. Phys. Rev. A **72**, 041803 (2005)
56. S.H. Lim, A.G. Caster, O. Nicolet, S.R. Leone, Chemical imaging by single pulse interferometric coherent anti-Stokes Raman scattering microscopy. J. Phys. Chem. B **110**, 5196–5204 (2006)
57. M. Cui, M. Joffre, J. Skodack, J.P. Ogilvie, Interferometric Fourier transform coherent anti-Stokes Raman scattering. Opt. Expr. **14**, 8448–8458 (2006)
58. B. von Vacano, T. Buckup, M. Motzkus, Highly sensitive single-beam heterodyne anti-Stokes Raman scattering. Opt. Lett. **31**, 2495–2497 (2006)
59. C. Müller, T. Buckup, B. von Vacano, M. Motzkus, Heterodyne single-beam CARS microscopy. J. Raman Spectrosc. **40**, 809–816 (2009)

60. E.M. Vartiainen, H.A. Rinia, M. Müller, M. Bonn, Direct extraction of Raman line-shapes from congested CARS spectra. Opt. Expr. **14**, 3622–3630 (2006)
61. H.A. Rinia, M. Bonn, M. Müller, Quantitative Multiplex CARS Spectroscopy in Congested Spectral Regions. J. Phys. Chem. B **110**, 4472–4479 (2006)
62. S.-H. Lim, A.G. Caster, S.R. Leone, Fourier transform spectral interferometric coherent anti-Stokes Raman scattering (FTSI-CARS) spectroscopy. Opt. Lett. **32**, 1332–1334 (2007)
63. Y. Liu, Y.J. Lee, M.T. Cicerone, Broadband CARS spectral phase retrieval using a time-domain Kramers-Kronig transform. Opt. Lett. **34**, 1363–1365 (2009)
64. V.V. Krishnamachari, E.O. Potma, Focus-engineered coherent anti-Stokes Raman scattering microscopy: a numerical investigation. J. Opt. Soc. Am. A **24**, 1138–1147 (2007)
65. C. Liu, D.Y. Kim, Differential imaging in coherent anti-Stokes Raman scattering microscopy with Laguerre-Gaussian excitation beams. Opt. Expr. **15**, 10123–10134 (2007)
66. V.V. Krishnamachari, E.O. Potma, Imaging chemical interfaces perpendicular to the optical axis with focus-engineered coherent anti-Stokes Raman scattering microscopy. Chem. Phys. **341**, 81–88 (2007)
67. D. Gachet,, F. Billard, H. Rigneault, Focused field symmetries for background-free coherent anti-Stokes Raman spectroscopy. Phys. Rev. A **77**, 061802(R) (2008)
68. D. Gachet, F. Billard, H. Rigneault, Background-free coherent anti-Stokes Raman spectroscopy near transverse interfaces: a vectorial study. J. Opt. Soc. Am. B **25**, 1655–1666 (2008)
69. S. Brustlein, D. Gachet, F. Billard, H. Rigneault, Transverse chemical interface detection with coherent anti-Stokes Raman scattering microscopy. J. Biomed. Opt. **16**, 086006 (2011)
70. D. Gachet, S. Brustlein, H. Rigneault, Revisiting the Young's double slit experiment for background-free nonlinear Raman Spectroscopy and microscopy. Phys. Rev. Lett. **104**, 213905 (2010)
71. Gachet, D., Rigneault, H.: Detection of chemical interfaces in coherent anti-Stokes Raman scattering (CARS) microscopy: Dk-CARS part I, axial interfaces. J. Opt. Soc. Am. A **28**, (2011)
72. V.V. Krishnamachari, E.O. Potma, Detecting lateral interfaces with focus-engineered coherent anti-Stokes Raman scattering microscopy. J. Raman Spectr. **39**, 593–598 (2008)
73. V.V. Krishnamachari, E.O. Potma, Multi-dimensional differential imaging with FE-CARS microscopy. Vibr. Spectr. **50**, 043905 (2009)
74. D. Gachet, H. Rigneault, Detection of chemical interfaces in coherent anti-Stokes Raman scattering (CARS) microscopy: D-CARS part II, arbitrary interfaces. J. Opt. Soc. Am. A **28**, (2011)
75. G. Eckhardt, S.E. Schwarz, F.J. McClung, R.W. Hellwarth, E.J. Woodbury, D. Weiner, Stimulated Raman scattering from organic liquids. Phys. Rev. Lett. **9**, 455–457 (1962)
76. R.W. Hellwarth, Theory Of stimulated Raman scattering. Phys. Rev. **130**, 1850–1852 (1963)
77. H. Takuma, D.A. Jennings, Coherent Raman effect in off-axis Raman laser resonator. Appl. Phys. Lett. **4**, 185–186 (1964)
78. W.J. Jones, B.P. Stoicheff, Inverse Raman spectra—induced absorption at optical frequencies. Phys. Rev. Lett. **13**(22), 657–659 (1964)
79. N. Bloembergen, The stimulated Raman effect. Am. J. Phys. **35**, 989–1023 (1967)
80. E.J. Woodbury, W.K. Ng, Ruby laser operation in near IR. Proc. Inst. Radio Eng. **50**, 2367 (1962)
81. P. Kukura, D.W. McCamant, R.A. Mathies, Femtosecond stimulated Raman spectroscopy. Ann. Rev. Phys. Chem **58**, 461–488 (2007)
82. N. Butcher, C. Cotter, *The Elements of Nonlinear Optics* (Cambridge Studies in Modern Optics, Cambridge, 1991)
83. R.W. Boyd, *Nonlinear Optics* (Academic Press, Boston, 1992)
84. P. Nandakumar, A. Kovalev, A. Volkmer, Vibrational imaging based on stimulated Raman scattering microscopy. New J. Phys. **11**, 033026 (2009)
85. C.W. Freudiger, W. Min, B.G. Saar, S. Lu, G.R. Holtom, C. He, J.C. Tsai, J.X. Kang, X.S. Xie, Label-free biomedical imaging with high sensitivity by stimulated Raman scattering microscopy. Science **322**, 1857–1861 (2008)

86. Y. Ozeki, Y. Kitagawa, K. Sumimura, N. Nishizawa, W. Umemura, S. Kajiyama, K. Fukui, K. Itoh, Stimulated Raman scattering microscope with shot noise limited sensitivity using subharmonically synchronized laser pulses. Opt. Expr. **18**, 13708–13719 (2010)
87. D. Zhang, M.N. Sipchenko, J.X. Cheng, Highly sensitive vibrational imaging by femtosecond pulse stimulated Raman loss. J. Phys. Chem. Lett. **2**, 1248–1253 (2011)
88. T. Hellerer, A. Enejder, A. Zumbusch, Spectral focusing: high spectral resolution spectroscopy with broadbandwidth laser pulses. Appl. Phys. Lett. **85**, 25–27 (2004)
89. E.R. Andresen, P. Berto, H. Rigneault, Stimulated Raman scattering microscopy by spectral focusing and fiber-generated soliton as Stokes pulse. Opt. Lett. **36**, 2387–2389 (2011)
90. S. Brustlein, P. Ferrand, N. Walther, S. Brasselet, C. Billaudeau, D. Marguet, H. Rigneault, Optical parametric oscillator-based light source for coherent Raman scattering microscopy: practical overview. J. Biomed. Opt. **16**, 021106 (2011)

Index

A. Zoubir (ed.), *Raman Imaging*, Springer Series in Optical Sciences 168,
DOI: 10.1007/978-3-642-28252-2, © Springer-Verlag Berlin Heidelberg 2012

D

E

Q

R

S

W

X

Y

Z